McMINN'S COLOR ATLAS OF HUMAN ANATOMY

4TH EDITION

P H Abrahams
MB BS FRCS(Ed) FRCR
Consultant Clinical Anatomist
Fellow, Girton College, Cambridge, UK
Examiner to the Royal College of Surgeons
of Edinburgh
Family Practitioner, London, UK

R T Hutchings
Formerly Chief Medical Laboratory
Scientific Officer
Royal College of Surgeons of England
London, UK

S C Marks Jr
DDS PhD
Professor of Cell Biology, Radiology
and Surgery
University of Massachusetts Medical Center
Worcester, Massachusetts, USA

Mosby

London · Philadelphia · St Louis · Sydney · Tokyo

QM
25
.A27
1998

First edition published in 1977 by Wolfe Publishing

Second edition published in 1988

Third edition published in 1993 by Mosby-Wolfe, an imprint of Times Mirror International Publishers Ltd

This edition published in 1998 by Mosby, an imprint of Mosby International Limited

Copyright © 1998 Mosby International Limited

Printed in Spain by Grafos S.A. Arte sobre papel, Barcelona

ISBN 0 7234 2641 4 (hard cover)

ISBN 0 7234 2772 0 (soft cover)

For full details of all Mosby titles, please write to Mosby International Limited, Lynton House, 7-12 Tavistock Square, London WC1H 9LB, UK.

A CIP catalogue record for this book is available from the British Library.

Library of Congress Cataloging-in-Publication Data applied for.

Project Manager:	Dave Burin
Development Editor:	Simon Pritchard
Designer:	Ian Spick
Layout:	Tim Read
Cover design:	Rob Curran
	Ian Spick
Illustration:	Mike Saiz
	Paul Wilkinson
	Rob Dean
Production:	Siobhán Egan
Index:	Laurence Errington
Publisher:	Geoff Greenwood

Contents

Acknowledgements

FROM THE THIRD EDITION

For the preparation of new dissections for this edition we are most grateful to Bari Logan, Prosector to the University of Cambridge. For new artwork we are much indebted to Rosemary Watts, with additional material from Philip Ball. We also thank our models for surface anatomy, and Dr Umraz Khan and Dr Ravinder Ranger for all their valuable help in proofing the Atlas.

In addition to imaging material from Dr Oscar Craig, Dr Paul Grech, Dr Kim Fox and Dr Richard Underwood, new material has generously been provided by Dr Niall Moore, Professor Jamie Weir, Dr Philip Owen, Dr Phil Gishen and Dr Anna-Maria Belli, and endoscopic views of the knee and hip joints by Mr David J Dandy and Mr Richard R Villar, respectively. For the endoscopic views of the nose and larynx we are grateful to Mr J D Shaw and Mr J M Lancer,

and for the ophthalmoscopic view we thank Miss Erna Kritzinger.

We renew our thanks to all those who over many years contributed specimens to the Anatomy Museum of the Royal College of Surgeons of England, and especially to the late Dr D H Tompsett who also prepared the corrosion casts (full details of the methods used can be found in his book *Anatomical Techniques*, 2nd edition, 1970, Livingstone), and to the late Sir Edward Muir, who, as President of the College, allowed us to reproduce the illustrations of museum specimens; to Dr J L Cordingley, Professor T W Glenister, and Professor F R Johnson for the loan of osteological material; and to Mr V H Oswal for the coloured dissection of the ear.

FOR THE FOURTH EDITION

Special attention has been given to updating the imaging and adding new laparoscopic photographs; for these we are grateful to Professor A Darzi, of St Mary's Hospital, London, and Laparoscopic Tutor at the Royal College of Surgeons of England. The new radiological images were generously given by Professor J Weir, Dr P Dubbins, Dr A-M Belli, Dr M Hourihan, Dr G Lloyd, Dr N Moore and Dr J P Owen from the Second Edition of Weir and Abrahams' *Imaging Atlas of Human Anatomy* (Mosby, 1997).

We are most grateful for a lengthy correspondence from Professor Dr Wolfgang Zenker whose paper (*Zschr Anat Entwicklungsgesch* 118:335–368, 1955), the first to describe the functional anatomy of the deep (medial) head of the temporalis muscle (sphenomandibular portion), has settled a recent argument in the international and anatomical press about a 'new' muscle (see

page 31, D17). We are also most happy to acknowledge the careful proof reading and suggestions of numerous colleagues and former anatomy demonstrators: Theodore P Welch MB BS FRCS, S Roy Choudhury PhD FRCS(Ed), Jonathan D Spratt MA FRCS(Eng) FRCS(Glasg), Ravinder Singh-Ranger BSc FRCS, Deepak Singh-Ranger BSc MB BS, Mark J Shaffer MA(Elec.Eng.) MB BCh, Farres Haddad BSc FRCS FRCS(Ed), David Choi MA MB ChB FRCS, David Johnson MA BM BCh FRCS, Nick Matharu BSc MB ChB, Angela M Riddel BSc MB BS and Daron Smith MA BM BCh.

Finally, it has been a pleasure working closely with Lucy Hamilton, Simon Pritchard, Dave Burin, Paul Wilkinson and Geoff Greenwood at Mosby to ensure expeditious publication of this edition.

UNIVERSITY OF CAMBRIDGE DISSECTING TEAM

Bari Logan, University Prosector (BL), Lynette Nearn (LN), Carmen Bester (CB) and Martin Watson (MW)

We are most grateful for the diligent work and devotion to the art of dissection from the team above, who have produced all the new dissections in this edition. These new dissections are summarized here, listed by page number:

30A (LN), 31D (BL), 32A (LN), 33B (LN), 34A (LN), 34B (LN), 42A (BL), 42B (BL), 43D (BL), 43E (BL), 78A (BL), 79C (BL), 114B (BL), 132A (LN), 134B (MW), 142A (MW), 142B (MW), 152B (BL), 164A (BL), 176A (BL), 176B (BL), 188A (BL, LN), 189B (BL, LN), 190A (MW), 191B (LN), 193B (CB), 210B (BL), 217E (BL), 220A (BL), 227A (BL), 227B (CB), 232C (LN), 233E (CB), 280A (LN), 280B (LN), 284A (CB), 284B (CB), 309F (BL).

Preface

The guiding principle in preparing the fourth edition of *McMinn's Color Atlas of Human Anatomy* has been to emphasize clinical anatomy and to make this popular Atlas even more 'user-friendly'. This has been accomplished largely by adding several new features, reorganising some sections and selectively deleting peripheral figures and text.

To assist in this task the authors of the third edition, after the retirement of Professor McMinn and the untimely death of Professor Pegington, are pleased to welcome Professor Sandy Marks, Jr, to the team. Professor Marks brings to the new edition over three decades of teaching and research experience in anatomy, is a Founding Member and Past President of the American Association of Clinical Anatomists and is the North American Editor of *Clinical Anatomy*.

The general order of presentation has been preserved, but major new features include expansion of the dissections of the abdomen, pelvis and lymphatics, the addition of clinical notes, and the identification of core pages for several disciplines.

We have taken advantage of the combined wisdom of 15 eminent physiotherapists, teachers of advanced trauma/life-support procedures and clinical anatomists in the UK and North America to identify the most important 50–100 pages for each discipline. These are designated on the top of each page by the appropriate icon and the pages for each topic are summarized below. This will be useful for students both beginning and those reviewing their disciplines.

Over 20% of this edition is new with over 100 new plates, mainly dissection images and laparoscopic views. The new figures are taken from dissections performed at the University of Cambridge, UK, and were designed in part to facilitate introductions to regions. This has been enhanced further by illustrating bilateral structures from proximal to distal mainly on the same side of the body.

The clinical relevance of specific structures or procedures based on regional anatomy has been emphasized by 250 clinical notes distributed throughout the text. These notes cover the majority of clinical topics emphasized in the landmark publication by the American Association of Clinical Anatomists, 'A Clinical Anatomy Curriculum for the Medical Student of the 21st Century' (*Clinical Anatomy* 9:71–99, 1996), and underscore the increased clinical direction of this edition.

Finally, we thank our colleagues for their insights and suggestions which have appeared in this edition. We believe that even the best can be improved and hope that the users of this atlas will continue to communicate to us their ideas and suggestions for a future edition that is even more accurate and useful.

P H Abrahams
R T Hutchings
S C Marks Jr

Guide to icons

To help readers find key information quickly and easily, pages essential to students of basic anatomy, physical therapy and ATLS and ACLS courses are identified using green, purple and red icons at the head of the page (*see below*). The selection of pages for each of the three groups of readers has been identified by teachers and students, and serves as a guide to the material most essential to each area of study. A summary of the pages for each group appears below.

Medicine	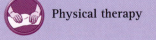 Physical therapy	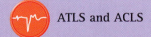 ATLS and ACLS
1, 4, 6, 7, 9, 11, 13, 14, 17, 18, 28–39, 41–43, 45, 48, 49, 51–53, 55, 56, 58, 59, 61, 62, 64, 66–72, 74, 76, 78, 81, 88, 89, 91, 93, 95, 99–101, 105–109, 111–113, 115–117, 122–124, 127, 128, 130–134, 138, 143, 145, 149–153, 157–161, 165, 167, 168, 170, 172, 173, 177, 179, 181, 185, 187–191, 196, 199, 201, 203, 205, 206, 208, 212, 217, 219, 220, 222, 226, 228–234, 236, 239– 242, 244–247, 249, 252, 253, 255, 256, 259, 260, 262–264, 267, 270–272, 276, 277, 279–283, 285, 286, 288, 290, 292, 293, 295, 298, 299, 303–306, 310, 311, 314–334	2, 3, 6, 7, 8, 10, 11, 19, 28, 29, 31–37, 41, 49, 51, 52, 58–60, 62, 65–72, 74–76, 78, 81, 82, 84–86, 88, 90, 92, 94, 96, 99, 100, 103, 105–112, 114–124, 126–134, 136–139, 147–150, 152, 153, 155, 157, 158, 161, 164, 166, 168, 170, 172, 173, 177, 183, 185–190, 226, 227, 229, 230, 231, 234, 239, 243, 248, 250, 252, 254, 257–259, 261, 262, 265–267, 269, 270, 272–274, 276–280, 282–286, 288, 290–295, 297, 298, 301–334	1, 3, 4, 5, 7, 9, 27, 30, 32–37, 41, 42, 48–52, 69–72, 76, 78, 80, 87, 88, 105–107, 114, 117, 122–125, 150, 152, 154, 156, 157, 164, 167, 177–179, 182, 187, 233, 266, 288, 300, 302, 304, 306, 315–334

Introduction

The body is made up of the head, trunk and limbs. The trunk consists of the neck, thorax (chest) and abdomen (belly). The lower part of the abdomen is the pelvis, but this word is also used to refer just to the bones of the pelvis. The lowest part of the pelvis (and lowest part of the trunk) is the perineum. The central axis of the trunk is the vertebral column (spine), and the upper part of it (cervical part) supports the head.

The main parts of the upper limb are the arm, forearm and hand. Note that in strict anatomical terms the word 'arm' means the upper arm, the part between the shoulder and elbow, although the word is commonly used to mean the whole of the upper limb.

The main parts of the lower limb are the thigh, leg and foot. Note that in strict anatomical terms the word 'leg' means the lower leg, the part between the knee and foot, although the word is commonly used to mean the whole of the lower limb.

For the description of the positions of structures in human anatomy, the body is assumed to be standing upright with the feet together and the head and eyes looking to the front, with the arms straight by the side and the palms of the hands facing forwards. this is the 'anatomical position' (see the illustration opposite), and structures are always described relative to one another using this as the 'standard' position, even when the body is, for example, lying on the back in bed or on a dissecting room table.

The 'median sagittal plane' is an imaginary vertical, longitudinal line through the middle of the body from front to back, dividing the body into right and left halves. The adjective 'medial' means nearer the median plane, and 'lateral' means farther from it. Thus, in the anatomical position, the little finger is on the medial side of the hand and the thumb is on the lateral side; the great toe is on the medial side of the foot and the little toe on the lateral side.

In the forearm where there are two bones, the radius on the lateral (thumb) side and the ulna on the medial side, the adjectives 'radial' and 'ulnar' can be used instead of lateral and medial. Similarly in the lower leg where there are two bones, the fibula on the lateral side and the tibia on the medial side, 'fibular' and 'tibial' are alternative adjectives.

'Anterior' and 'posterior' mean nearer the front or nearer the back of the body, respectively. Thus on the face the nose is anterior to the ears, and the ears are posterior to the nose. Sometimes 'ventral' is used instead of anterior, and 'dorsal' instead of posterior (terms from comparative anatomy that are appropriate for four-footed animals).

The hand and foot have special terms applied to them. The anterior or ventral surface of the hand is usually called the palm or palmar surface, and the posterior or dorsal surface is the dorsum. But in the foot the upper surface is the dorsal surface or dorsum, and the undersurface or sole is the plantar surface.

'Superior' and 'inferior' mean nearer the upper or lower end of the body, respectively; the nose is superior to the mouth and inferior to the forehead (even if the body is upside down; the upright 'anatomical position' is always the reference position).

'Superficial' means near the skin surface, and 'deep' means farther away from the surface.

'Proximal' and 'distal' mean nearer to and farther from the root of the structure: in the upper limb, the forearm is distal to the elbow and proximal to the hand.

The words 'sagittal' and 'coronal' describe certain planes of section, most often used in the head and brain. The 'sagittal plane' is any front-to-back plane that is parallel to the median plane, and the 'coronal plane' (sometimes called the frontal plane) is a vertical plane at right angles to the median plane.

The book is arranged in the general order 'head to toe'—head and neck (including the brain), followed by the vertebral column and spinal cord, thorax, upper limb, abdomen and pelvis, and lower limb. In each section the bones are considered first, followed by dissections and other illustrations.

Structures are labelled by overlying numbers which are identified in the key lists. Sometimes in crowded areas or for small structures, leader lines are necessary; an arrowhead at the end of a leader indicates that the item is just out of view beyond the tip of the arrow. Self-testing can be carried out by covering up the key.

For bones, the parts of each are first named, and then the pictures are repeated indicating the sites of attachments of muscles and ligaments. Although the details of individual skull bones are included, for most students knowledge of the skull as a whole is much more important.

Dissections and other items are introduced by a short commentary which draws attention to the most significant features, so helping to sort out 'the wood from the trees'. The commentary is supplemented by notes which again emphasize the more important features or help to explain difficult topics, but they are not intended to give a comprehensive description of everything seen. The book is designed to supplement existing texts, not to substitute for them.

The Systemic Review at the back of the book provides an overview of the musculature, vasculature and cranial nerves from a variety of perspectives. Diagrams and lists emphasize the continuity of systems between the regions of the body and provide part of the big picture necessary for understanding the relevance and relationships of the details illustrated elsewhere in this atlas.

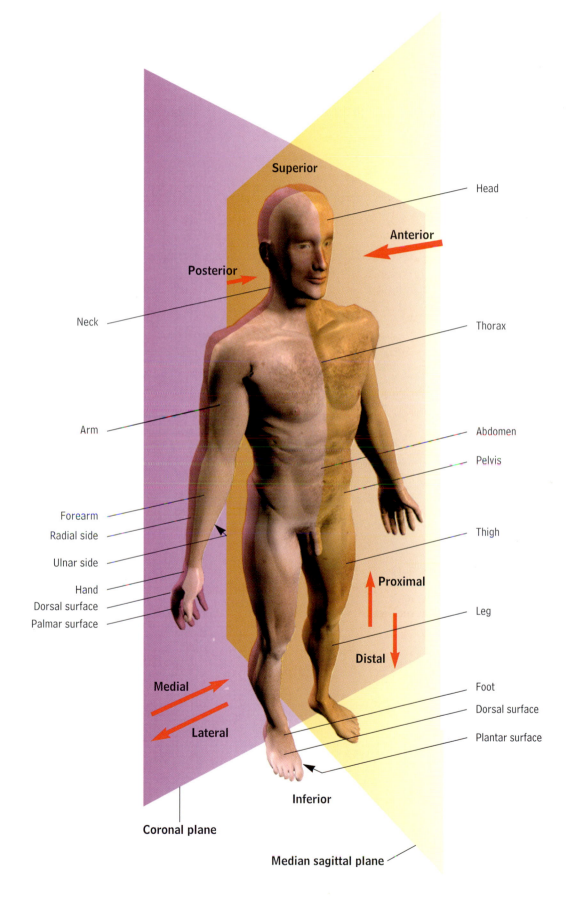

Superior

Head

Anterior

Posterior

Neck

Thorax

Arm

Abdomen

Pelvis

Forearm

Radial side

Ulnar side

Thigh

Proximal

Hand

Dorsal surface

Palmar surface

Leg

Distal

Medial

Foot

Dorsal surface

Lateral

Plantar surface

Inferior

Coronal plane

Median sagittal plane

Dedicated to the memory of Peter Wolfe

Chapter 1

Head, neck and brain

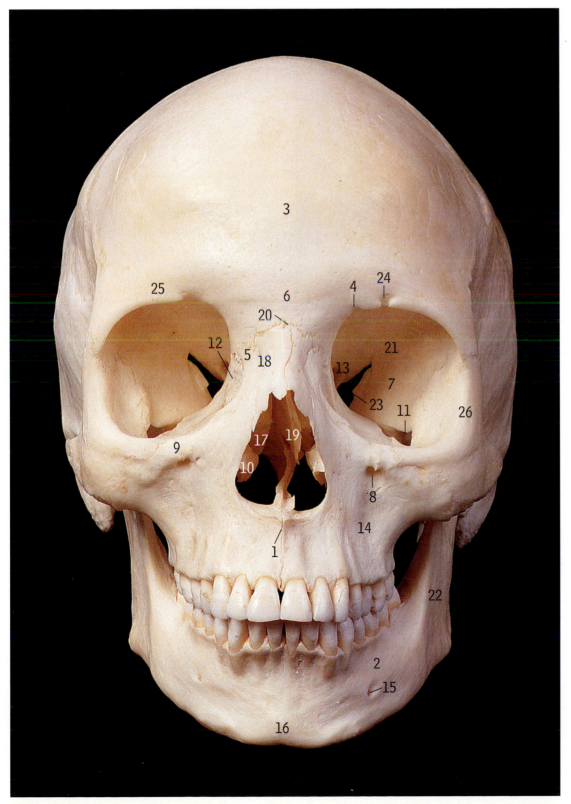

Skull, from the front

1 Anterior nasal spine
2 Body of mandible
3 Frontal bone
4 Frontal notch
5 Frontal process of maxilla
6 Glabella
7 Greater wing of sphenoid bone
8 Infra-orbital foramen
9 Infra-orbital margin
10 Inferior nasal concha
11 Inferior orbital fissure
12 Lacrimal bone
13 Lesser wing of sphenoid bone
14 Maxilla
15 Mental foramen
16 Mental protuberance
17 Middle nasal concha
18 Nasal bone
19 Nasal septum
20 Nasion
21 Orbit (orbital cavity)
22 Ramus of mandible
23 Superior orbital fissure
24 Supra-orbital foramen
25 Supra-orbital margin
26 Zygomatic bone

- The term 'skull' includes the mandible, and 'cranium' refers to the skull without the mandible, but these definitions are not always strictly observed.
- The calvaria is the vault of the skull (cranial vault or skull-cap) and is the upper part of the cranium that encloses the brain.
- The front part of the skull forms the facial skeleton.
- The supra-orbital, infra-orbital and mental foramina (24, 8 and 15) lie in approximately the same vertical plane.
- Details of individual skull bones are given on pages 18 to 27, of the bones of the orbit and nose on page 12, and of the teeth on page 13.

A Skull, from the front. Muscle attachments

1	Buccinator
2	Corrugator supercilii
3	Depressor anguli oris
4	Depressor labii inferioris
5	Levator anguli oris
6	Levator labii superioris
7	Levator labii superioris alaeque nasi
8	Masseter
9	Mentalis
10	Nasalis
11	Orbicularis oculi
12	Platysma
13	Procerus
14	Temporalis
15	Zygomaticus major
16	Zygomaticus minor

• The attachment of levator labii superioris (6) is above the infra-orbital foramen and that of levator anguli oris (5) is below it.
• The attachment of depressor labii inferioris (4) is in front of the mental foramen and that of depressor anguli oris (3) is below it.

B Skull radiograph. Occipitofrontal projection

 1 Basi-occiput
 2 Body of sphenoid
 3 Crista galli
 4 Ethmoidal air cells
 5 Floor of maxillary sinus (antrum)
 6 Foramen rotundum
 7 Frontal sinus
 8 Greater wing of sphenoid
 9 Internal acoustic meatus
10 Lambdoid suture
11 Lateral mass of atlas (first cervical vertebra)
12 Lesser wing of sphenoid
13 Mastoid process
14 Nasal septum
15 Odontoid process (dens) of axis (second cervical
     vertebra)
16 Petrous part of temporal bone
17 Sagittal suture
18 Sella turcica
19 Superior orbital fissure
20 Temporal surface of greater wing of sphenoid

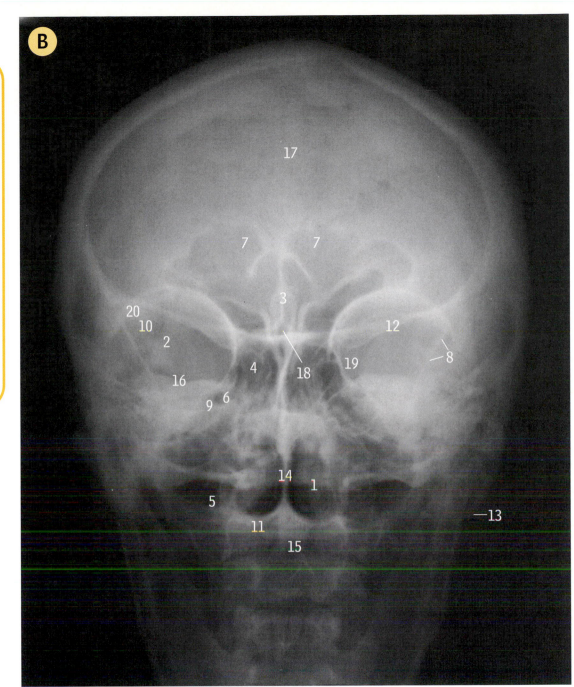

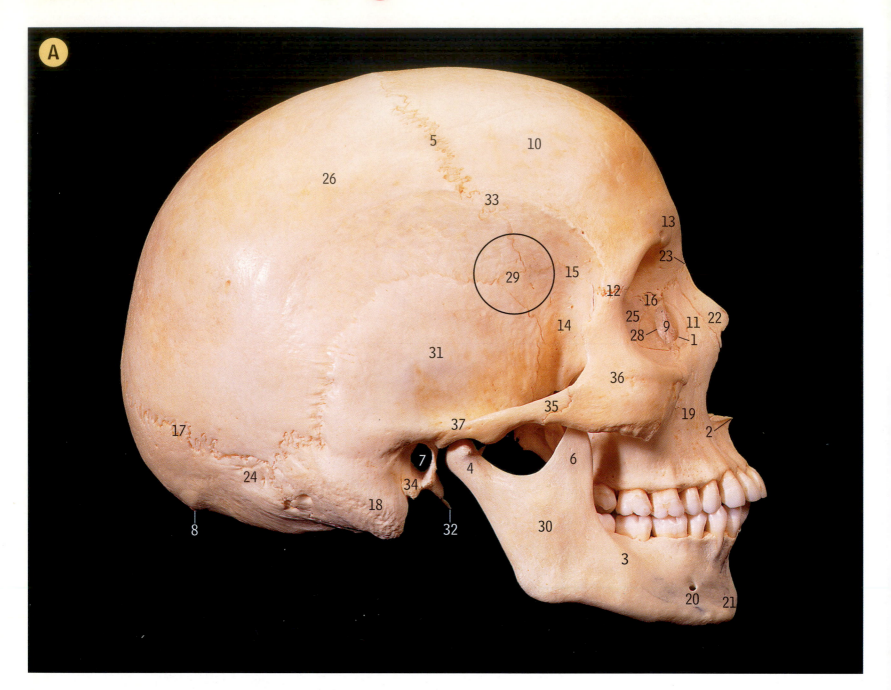

Skull, **A** from the right, **B** radiograph, lateral projection

1	Anterior lacrimal crest	20	Mental foramen
2	Anterior nasal spine	21	Mental protuberance
3	Body of mandible	22	Nasal bone
4	Condyle of mandible	23	Nasion
5	Coronal suture	24	Occipital bone
6	Coronoid process of mandible	25	Orbital part of ethmoid bone
7	External acoustic meatus of temporal bone	26	Parietal bone
8	External occipital protuberance (inion)	27	Pituitary fossa (sella turcica)
9	Fossa for lacrimal sac	28	Posterior lacrimal crest
10	Frontal bone	29	Pterion (encircled)
11	Frontal process of maxilla	30	Ramus of mandible
12	Frontozygomatic suture	31	Squamous part of temporal bone
13	Glabella	32	Styloid process of temporal bone
14	Greater wing of sphenoid bone	33	Superior temporal line
15	Inferior temporal line	34	Tympanic part of temporal bone
16	Lacrimal bone	35	Zygomatic arch
17	Lambdoid suture	36	Zygomatic bone
18	Mastoid process of temporal bone	37	Zygomatic process of temporal bone
19	Maxilla		

- Pterion (29) is not a single point but an area where the frontal (10), parietal (26), squamous part of the temporal (31) and greater wing of the sphenoid bone (14) adjoin one another. It is an important landmark for the anterior branch of the middle meningeal artery which underlies this area on the inside of the skull (page 17, 39).

Extradural haemorrhage in the skull is usually due to trauma at the pterion on the same or opposite side of the cranium, which causes tearing of the middle meningeal artery or of one of its divisions.

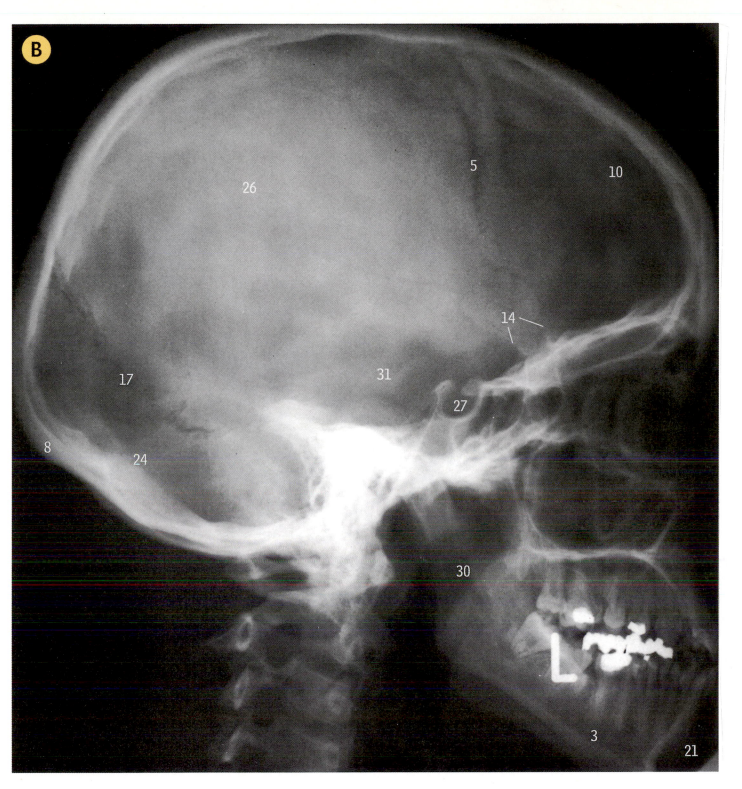

- The position of mandibular ramus (30) is superimposed by air shadow in the nasopharynx.

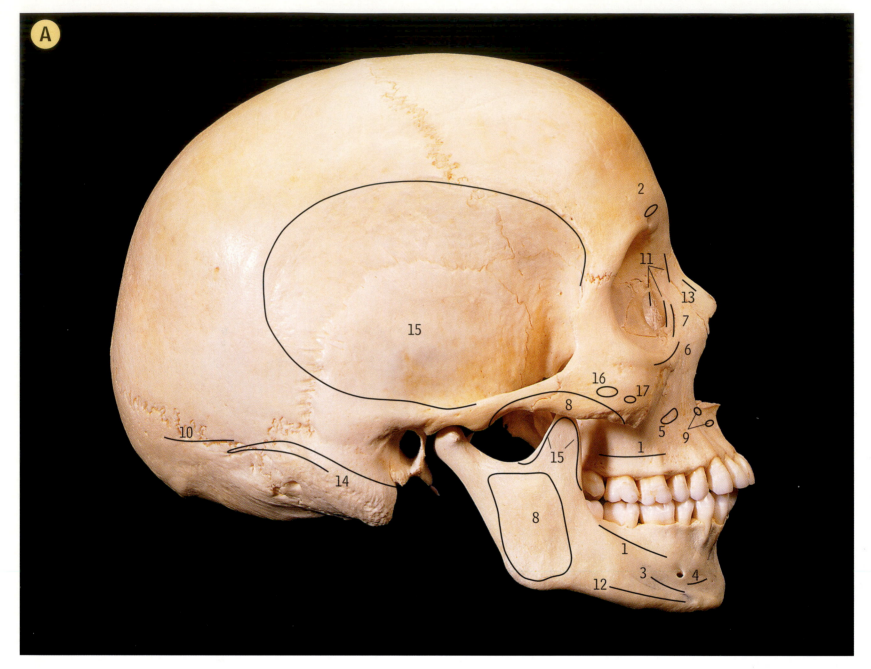

A Skull, from the right. Muscle attachments

1	Buccinator
2	Corrugator supercilii
3	Depressor anguli oris
4	Depressor labii inferioris
5	Levator anguli oris
6	Levator labii superioris
7	Levator labii superioris alaeque nasi
8	Masseter
9	Nasalis
10	Occipital part of occipitofrontalis
11	Orbicularis oculi
12	Platysma
13	Procerus
14	Sternocleidomastoid
15	Temporalis
16	Zygomaticus major
17	Zygomaticus minor

- The bony attachments of the buccinator muscle (1) are to the upper and lower jaws (maxilla and mandible) opposite the three molar teeth. (The teeth are identified on page 13, C.)
- The upper attachment of temporalis (upper 15) occupies the temporal fossa (the narrow space above the zygomatic arch at the side of the skull). The lower attachment of temporalis (lower 15) extends from the lowest part of the mandibular notch of the mandible, over the coronoid process and down the front of the ramus almost as far as the last molar tooth.
- Masseter (8) extends from the zygomatic arch to the lateral side of the ramus of the mandible.

B Skull, from behind

1 External occipital protuberance (inion)
2 Highest nuchal line
3 Inferior nuchal line
4 Lambda
5 Lambdoid suture
6 Occipital bone
7 Parietal bone
8 Parietal foramen
9 Sagittal suture
10 Superior nuchal line

Burr holes through the skull have been performed for many years (trephination), originally 'to let out the evil spirits', and are now used by neurosurgeons to release intracranial pressure.

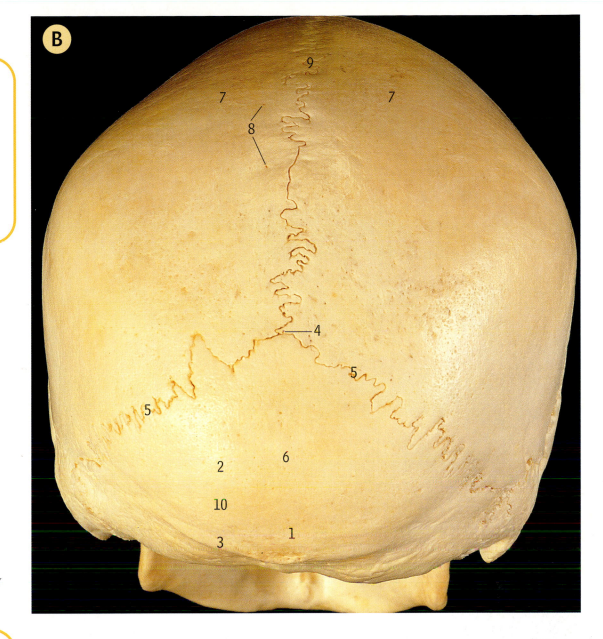

C Skull. Right infratemporal region, obliquely from below

1 Articular tubercle
2 External acoustic meatus
3 Horizontal plate of palatine bone
4 Inferior orbital fissure
5 Infratemporal crest
6 Infratemporal (posterior) surface of maxilla
7 Infratemporal surface of greater wing of sphenoid bone
8 Lateral pterygoid plate
9 Mandibular fossa
10 Mastoid notch
11 Mastoid process
12 Medial pterygoid plate
13 Occipital condyle
14 Occipital groove
15 Pterygoid hamulus
16 Pterygomaxillary fissure and pterygopalatine fossa
17 Pyramidal process of palatine bone
18 Spine of sphenoid bone
19 Styloid process and sheath
20 Third molar tooth
21 Tuberosity of maxilla
22 Vomer
23 Zygomatic arch

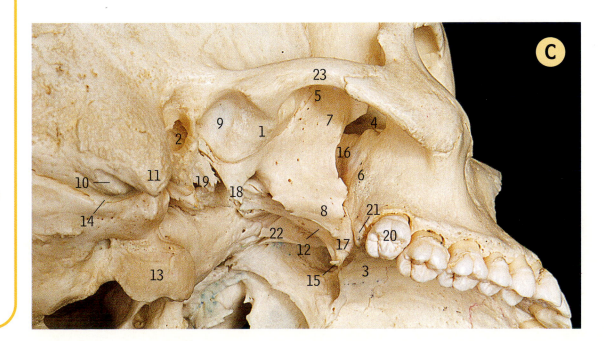

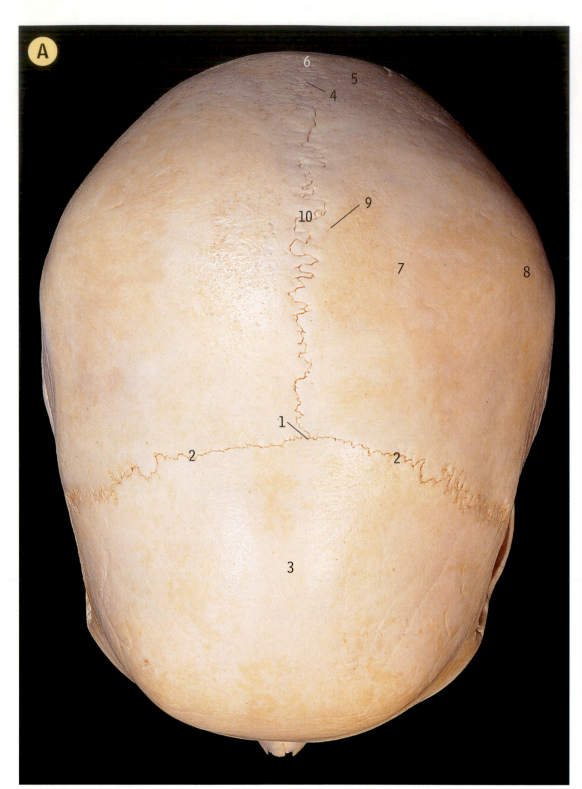

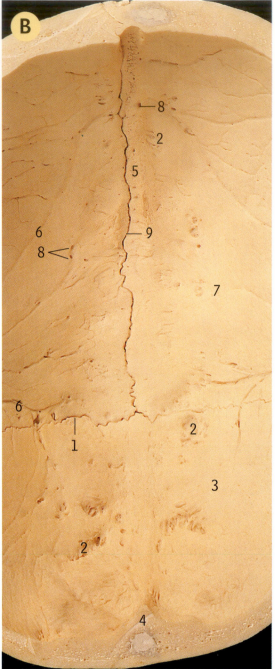

B Skull. Internal surface of the cranial vault, central part

A Skull, from above

1	Bregma
2	Coronal suture
3	Frontal bone
4	Lambda
5	Lambdoid suture
6	Occipital bone
7	Parietal bone
8	Parietal eminence
9	Parietal foramen
10	Sagittal suture

- In this skull the parietal eminences are prominent (A8).
- The point where the sagittal suture (A10) meets the coronal suture (A2) is the bregma (A1). At birth the unossified parts of the frontal and parietal bones in this region form the membranous anterior fontanelle (page 14, D1).
- The point where the sagittal suture (A10) meets the lambdoid suture (A5) is the lambda (A4). At birth the unossified parts of the parietal and occipital bones in this region form the membranous posterior fontanelle (page 14, C13).
- The label 3 in the centre of the frontal bone indicates the line of the frontal suture in the fetal skull (page 14, A5). The suture may persist in the adult skull and is sometimes known as the metopic suture.

1	Coronal suture
2	Depressions for arachnoid granulations
3	Frontal bone
4	Frontal crest
5	Groove for superior sagittal sinus
6	Grooves for middle meningeal vessels
7	Parietal bone
8	Parietal foramen
9	Sagittal suture

- The arachnoid granulations (page 58, B1), through which cerebrospinal fluid drains into the superior sagittal sinus, cause the irregular depressions (B2) on the parts of the frontal and parietal bones (B3 and 7) that overlie the sinus.

C Skull. External surface of the base

1 Apex of petrous part of temporal bone
2 Articular tubercle
3 Carotid canal
4 Condylar canal (posterior)
5 Edge of tegmen tympani
6 External acoustic meatus
7 External occipital crest
8 External occipital protuberance
9 Foramen lacerum
10 Foramen magnum
11 Foramen ovale
12 Foramen spinosum
13 Greater palatine foramen
14 Horizontal plate of palatine bone
15 Hypoglossal (anterior condylar) canal
16 Incisive fossa
17 Inferior nuchal line
18 Inferior orbital fissure
19 Infratemporal crest of greater wing of sphenoid bone
20 Jugular foramen
21 Lateral pterygoid plate
22 Lesser palatine foramina
23 Mandibular fossa
24 Mastoid foramen
25 Mastoid notch
26 Mastoid process
27 Medial pterygoid plate
28 Median palatine (intermaxillary) suture
29 Occipital condyle
30 Occipital groove
31 Palatine grooves and spines
32 Palatine process of maxilla
33 Palatinovaginal canal
34 Petrosquamous fissure
35 Petrotympanic fissure
36 Pharyngeal tubercle
37 Posterior border of vomer
38 Posterior nasal aperture (choana)
39 Posterior nasal spine
40 Pterygoid hamulus
41 Pyramidal process of palatine bone
42 Scaphoid fossa
43 Spine of sphenoid bone
44 Squamotympanic fissure
45 Squamous part of temporal bone
46 Styloid process
47 Stylomastoid foramen
48 Superior nuchal line
49 Transverse palatine (palatomaxillary) suture
50 Tuberosity of maxilla
51 Tympanic part of temporal bone
52 Vomerovaginal canal
53 Zygomatic arch

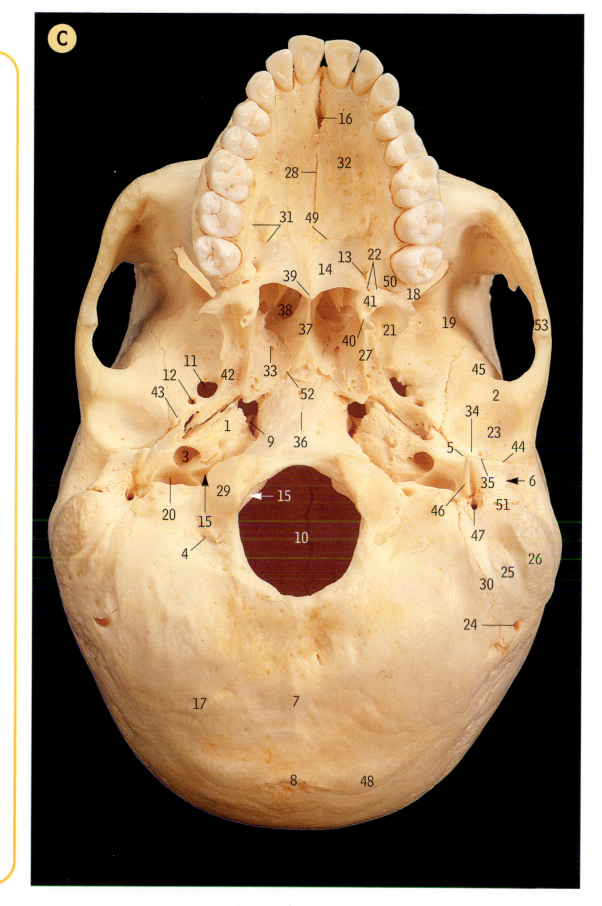

- The palatine process of the maxilla (32) and the horizontal plate of the palatine bone (14) form the hard palate (roof of the mouth and floor of the nose).
- The carotid canal (3), recognized by its round shape on the inferior surface of the petrous part of the temporal bone, does not pass straight upwards to open into the inside of the skull but takes a right-angled turn forwards and medially within the petrous temporal to open into the back of the foramen lacerum (9).

Intracranial spread of infections: scalp. Emissary veins traverse the skull, connecting scalp veins and intracranial sinuses. Foramina for the largest of these veins can be found in the occipital bone and near the mastoid process.

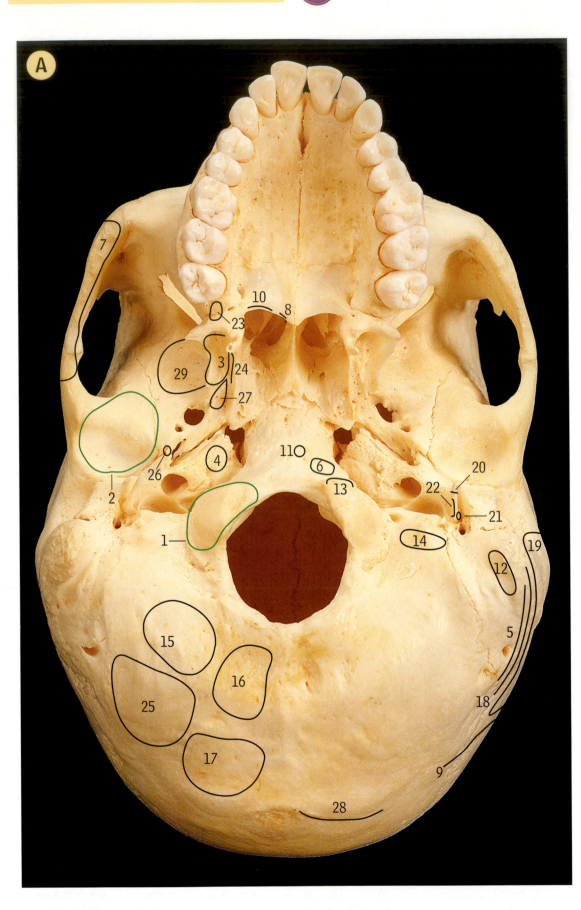

A Skull. External surface of the base. Muscle attachments

Green line = capsule attachments of atlanto-occipital and temporomandibular joints

 1 Capsule attachment of atlanto-occipital joint
 2 Capsule attachment of temporomandibular joint
 3 Deep head of medial pterygoid
 4 Levator veli palatini
 5 Longissimus capitis
 6 Longus capitis
 7 Masseter
 8 Musculus uvulae
 9 Occipital part of occipitofrontalis
10 Palatopharyngeus
11 Pharyngeal raphe
12 Posterior belly of digastric
13 Rectus capitis anterior
14 Rectus capitis lateralis
15 Rectus capitis posterior major
16 Rectus capitis posterior minor
17 Semispinalis capitis
18 Splenius capitis
19 Sternocleidomastoid
20 Styloglossus
21 Stylohyoid
22 Stylopharyngeus
23 Superficial head of medial pterygoid
24 Superior constrictor
25 Superior oblique
26 Tensor tympani
27 Tensor veli palatini
28 Trapezius
29 Upper head of lateral pterygoid

- The medial pterygoid plate has no pterygoid muscles attached to it. It passes straight backwards, giving origin at its lower end to part of the superior constrictor of the pharynx (24).
- The lateral pterygoid plate has both pterygoid muscles attached to it: medial and lateral muscles from the medial and lateral surfaces, respectively (3 and 28). The plate becomes twisted slightly laterally because of the constant pull of these muscles which pass backwards and laterally to their attachments to the mandible (page 19).

B Skull. Internal surface of the base (cranial fossae)

1 Anterior clinoid process
2 Arcuate eminence
3 Carotid groove
4 Clivus
5 Cribriform plate of ethmoid bone
6 Crista galli
7 Diploë
8 Dorsum sellae
9 Foramen caecum
10 Foramen lacerum
11 Foramen magnum
12 Foramen ovale
13 Foramen rotundum
14 Foramen spinosum
15 Frontal crest
16 Frontal sinus
17 Greater wing of sphenoid bone
18 Groove for anterior ethmoidal nerve and vessels
19 Groove for inferior petrosal sinus
20 Groove for sigmoid sinus
21 Groove for superior petrosal sinus
22 Groove for superior sagittal sinus
23 Groove for transverse sinus
24 Grooves for middle meningeal vessels
25 Hiatus and groove for greater petrosal nerve
26 Hiatus and groove for lesser petrosal nerve
27 Hypoglossal canal
28 Internal acoustic meatus
29 Internal occipital protuberance
30 Jugular foramen
31 Jugum of sphenoid bone
32 Lesser wing of sphenoid bone
33 Occipital bone
34 Optic canal
35 Orbital part of frontal bone
36 Parietal bone (postero-inferior angle only)
37 Petrous part of temporal bone
38 Pituitary fossa (sella turcica)
39 Posterior clinoid process
40 Prechiasmatic groove
41 Squamous part of temporal bone
42 Superior orbital fissure
43 Tegmen tympani
44 Trigeminal impression
45 Tuberculum sellae
46 Venous foramen

Anosmia is loss of smell commonly following forehead trauma, fracture of the cribriform plate of the ethmoid bone and resultant damage to the olfactory nerves.

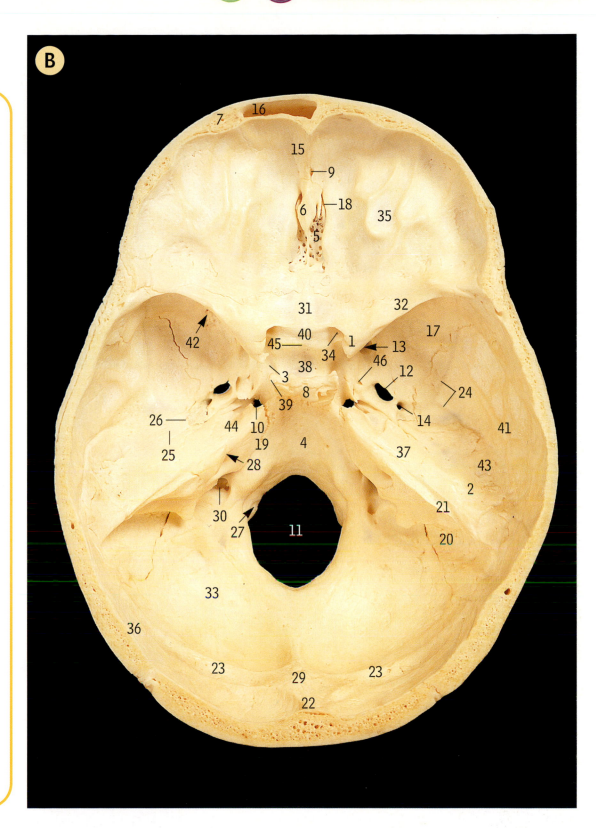

- The anterior cranial fossa is limited posteriorly on each side by the free margin of the lesser wing of the sphenoid (32) with its anterior clinoid process (1), and centrally by the anterior margin of the prechiasmatic groove (40).
- The middle cranial fossa is butterfly-shaped and consists of a central or median part and right and left lateral parts. The central part includes the pituitary fossa (38) on the upper surface of the body of the sphenoid, with the prechiasmatic groove (40) in front and the dorsum sellae (8) with its posterior clinoid processes (39) behind. Each lateral part extends from the posterior border of the lesser wing of the sphenoid (32) to the groove for the superior petrosal sinus (21) on the upper edge of the petrous part of the temporal bone.
- The posterior cranial fossa, whose most obvious feature is the foramen magnum (11), is behind the dorsum sellae (8) and the grooves for the superior petrosal sinuses (21).

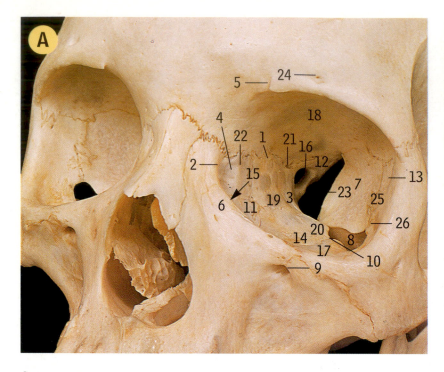

A Skull. Bones of the left orbit

B Lateral wall of the left nasal cavity
In this midline sagittal section of the skull, with the nasal septum removed, the superior and middle nasal conchae have been dissected away to reveal the air cells of the ethmoidal sinus, in particular the ethmoidal bulla (5)

1	Anterior ethmoidal foramen	
2	Anterior lacrimal crest	
3	Body of sphenoid bone, forming medial wall	
4	Fossa for lacrimal sac	
5	Frontal notch	
6	Frontal process of maxilla, forming medial wall	
7	Greater wing of sphenoid bone, forming lateral wall	
8	Inferior orbital fissure	
9	Infra-orbital foramen	
10	Infra-orbital groove	
11	Lacrimal bone, forming medial wall	
12	Lesser wing of sphenoid bone, forming roof	
13	Marginal tubercle	

14	Maxilla, forming floor
15	Nasolacrimal canal
16	Optic canal
17	Orbital border of zygomatic bone, forming floor
18	Orbital part of frontal bone, forming roof
19	Orbital plate of ethmoid bone, forming medial wall
20	Orbital process of palatine bone, forming floor
21	Posterior ethmoidal foramen
22	Posterior lacrimal crest
23	Superior orbital fissure
24	Supra-orbital foramen
25	Zygomatic bone forming lateral wall
26	Zygomatico-orbital foramen

1	Air cells of ethmoidal sinus
2	Clivus
3	Cribriform plate of ethmoid bone
4	Dorsum sellae
5	Ethmoidal bulla
6	Frontal sinus
7	Horizontal plate of palatine bone
8	Incisive canal
9	Inferior meatus
10	Inferior nasal concha
11	Lateral pterygoid plate
12	Left sphenoidal sinus

13	Medial pterygoid plate
14	Nasal bone
15	Nasal spine of frontal bone
16	Opening of maxillary sinus
17	Palatine process of maxilla
18	Perpendicular plate of palatine bone
19	Pituitary fossa (sella turcica)
20	Pterygoid hamulus
21	Right sphenoidal sinus
22	Semilunar hiatus
23	Sphenopalatine foramen
24	Uncinate process of ethmoid bone

- The fossa for the lacrimal sac (A4) is formed partly by the lacrimal groove of the frontal process of the maxilla (A6) and partly by the similar groove on the lacrimal bone (A11).

- The roof of the nasal cavity consists mainly of the cribriform plate of the ethmoid bone (B3) with the body of the sphenoid containing the sphenoidal sinuses (B21 and 12) behind, and the nasal bone (B14) and the nasal spine of the frontal bone (B15) at the front.
- The floor of the cavity consists of the palatine process of the maxilla (B17) and the horizontal plate of the palatine bone (B7).
- The medial wall is the nasal septum (page 17) which is formed mainly by two bones – the perpendicular plate of the ethmoid and the vomer – and the septal cartilage.
- The lateral wall consists of the medial surface of the maxilla with its large opening (B16), overlapped from above by parts of the ethmoid (B1, 5 and 24) and lacrimal bones, from behind by the perpendicular plate of the palatine (B18), and below by the inferior concha (B10).
- When covered by mucous membrane, the ethmoidal bulla (B5) and the uncinate process of the ethmoid (B24) form the upper and lower boundaries, respectively, of the semilunar hiatus (page 43, D15).

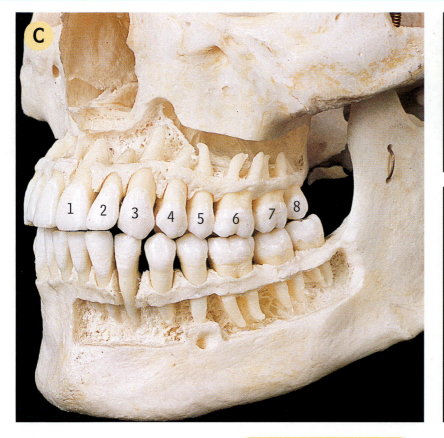

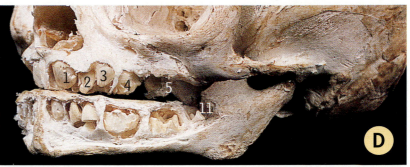

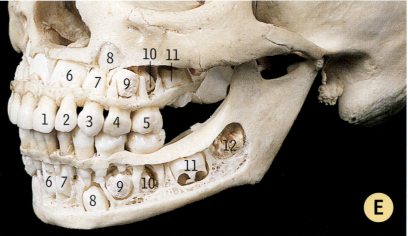

C Skull. Permanent teeth, from the left and in front

- The corresponding teeth of the upper and lower jaws have similar names. In clinical dentistry the teeth are often identified by the numbers 1 to 8 (as listed here) rather than by name.
- The third molar is sometimes called the wisdom tooth.

1 First (central) incisor
2 Second (lateral) incisor
3 Canine premolar
4 First premolar
5 Second premolar
6 First molar
7 Second molar
8 Third molar

Skull. Upper and lower jaws from the left and in front, **D** in the newborn with unerupted deciduous teeth, **E** in a four-year-old child with erupted deciduous teeth and unerupted permanent teeth

- The deciduous molars occupy the positions of the premolars of the permanent dentition.

1 First (central) incisor of deciduous dentition
2 Second (lateral) incisor of deciduous dentition
3 Canine of deciduous dentition
4 First molar of deciduous dentition
5 Second molar of deciduous dentition
6 First (central) incisor of permanent dentition
7 Second (lateral) incisor of permanent dentition
8 Canine of permanent dentition
9 First premolar of permanent dentition
10 Second premolar of permanent dentition
11 First molar of permanent dentition
12 Second molar of permanent dentition

F Skull. Edentulous mandible in old age, from the left

- With the loss of teeth the alveolar bone becomes absorbed, so that the mental foramen (3) and mandibular canal lie near the upper margin of the bone.
- The angle (1) between the ramus (4) and body (2) becomes more obtuse, resembling the infantile angle (as in D and E, above).

1 Angle
2 Body
3 Mental foramen
4 Ramus

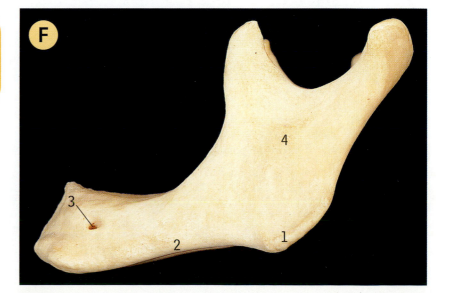

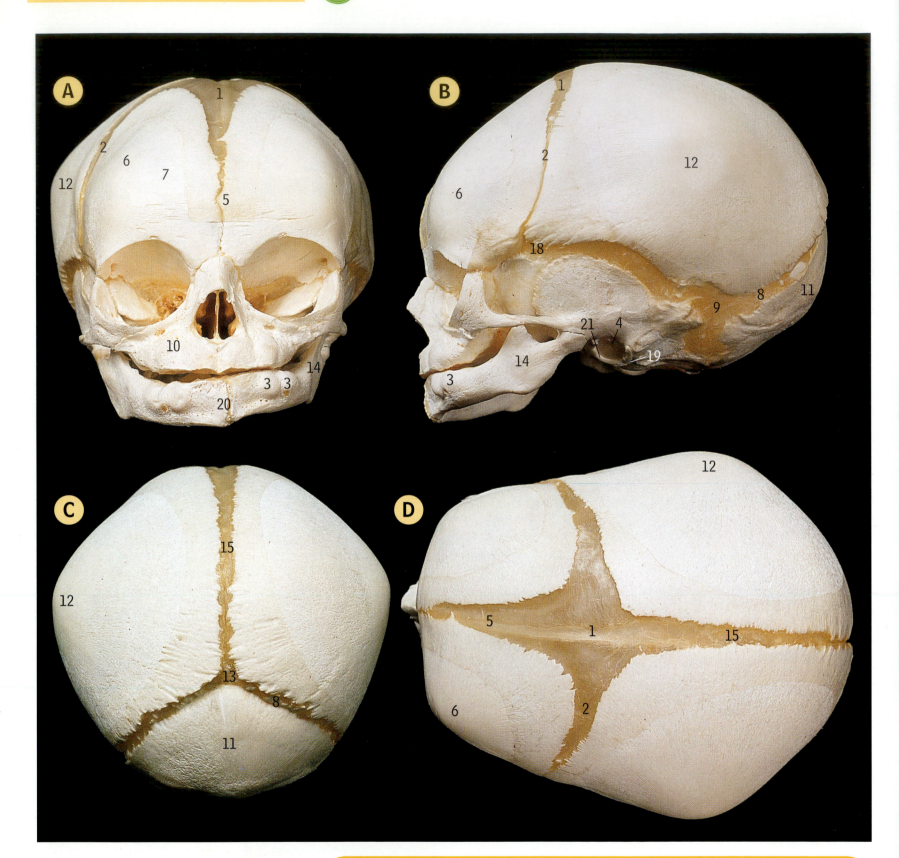

Skull of a full-term fetus, **A** from the front,
B from the left and slightly below, **C** from
behind, **D** from above. Fetal skull radiographs,
E frontal projection, **F** lateral projection

1 Anterior fontanelle	12 Parietal tuberosity
2 Coronal suture	13 Posterior fontanelle
3 Elevations over deciduous teeth in body of mandible	14 Ramus of mandible
4 External acoustic meatus	15 Sagittal suture
5 Frontal suture	16 Sella turcica
6 Frontal tuberosity	17 Semi-circular canals
7 Half of frontal bone	18 Sphenoidal fontanelle
8 Lambdoid suture	19 Stylomastoid foramen
9 Mastoid fontanelle	20 Symphysis menti
10 Maxilla	21 Tympanic ring
11 Occipital bone	

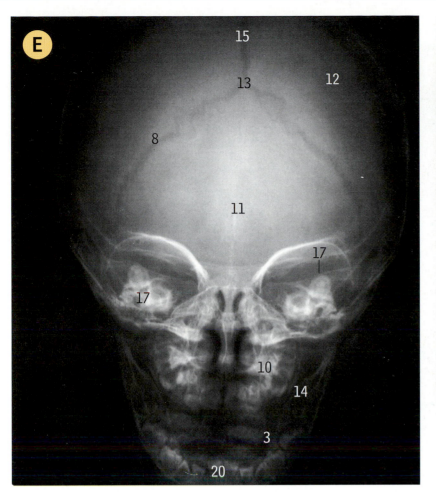

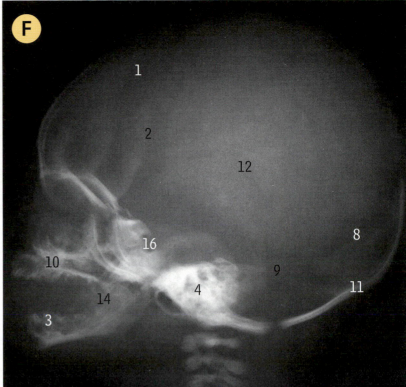

- The face at birth forms a relatively smaller proportion of the cranium than in the adult (about one eighth compared with one half) because of the small size of the nasal cavity and maxillary sinuses and the lack of erupted teeth.
- The posterior fontanelle (C13, E13) closes about two months after birth, the anterior fontanelle (A1, D1, F1) in the second year.
- Owing to the lack of the mastoid process (which does not develop until the second year) the stylomastoid foramen (B19) and the emerging facial nerve are relatively near the surface and unprotected.

Hydrocephalus. This condition of cranial enlargement is due to increased CSF pressure which produces dilatation of the cerebral ventricles. Typically, there is enlargement of the skull, prominence of the forehead and deterioration in mental capacity due to brain atrophy.

Scalp wounds. Cuts in the scalp tend to bleed profusely owing to its numerous anastomoses of branches of the internal and external carotid arteries. The fibrous arrangement of the connective tissue in the scalp tends to keep arteries open when cut. However, the rich blood supply also means that scalp wounds heal rapidly.

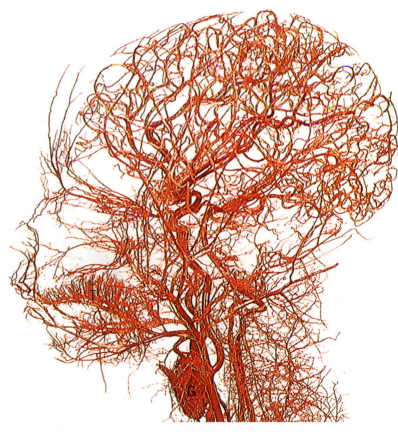

G Cast of the head and neck arteries of a full-term fetus, from the left
In this cast of fetal arteries, note in the front of the neck the dense arterial pattern indicating the thyroid gland (G), and above and in front of it the fine vessels outlining the tongue (T)

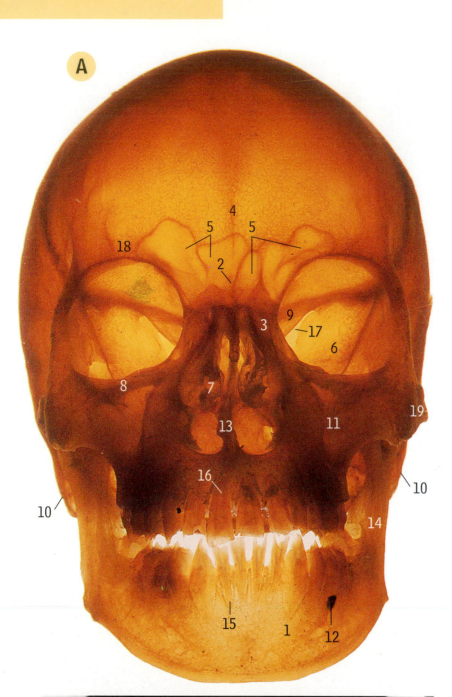

A

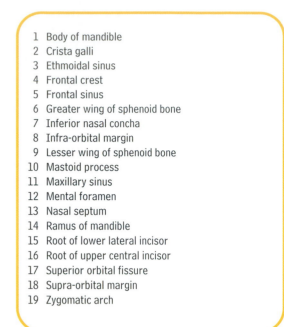

Skull, **A** cleared specimen from the front, illuminated from behind, **B** radiograph of facial bones, occipitofrontal view

 1 Body of mandible
 2 Crista galli
 3 Ethmoidal sinus
 4 Frontal crest
 5 Frontal sinus
 6 Greater wing of sphenoid bone
 7 Inferior nasal concha
 8 Infra-orbital margin
 9 Lesser wing of sphenoid bone
10 Mastoid process
11 Maxillary sinus
12 Mental foramen
13 Nasal septum
14 Ramus of mandible
15 Root of lower lateral incisor
16 Root of upper central incisor
17 Superior orbital fissure
18 Supra-orbital margin
19 Zygomatic arch

• Compare with the skull on page 1.

'Blow-out' fractures of orbit are usually produced by direct trauma to the eye. The eye itself is rarely ruptured but the thin orbital floor is often fractured and the eye and its surrounding fat are pushed into the roof of the maxillary sinus. This may be a cause of diplopia.

Mastoiditis, infection of the mastoid air cells, is rarely seen now since the common use of antibiotics. However, once the mastoid air cells are infected it is difficult to eradicate the condition and there is always the risk of spread internally towards the sigmoid sinus.

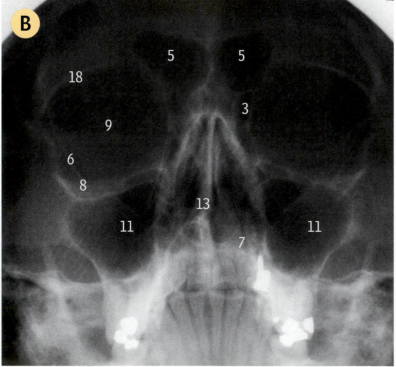

B

Left half of the skull. Sagittal section
The inside of the left half of the skull is seen from the right, with the bony part of the nasal septum (36 and 45) preserved

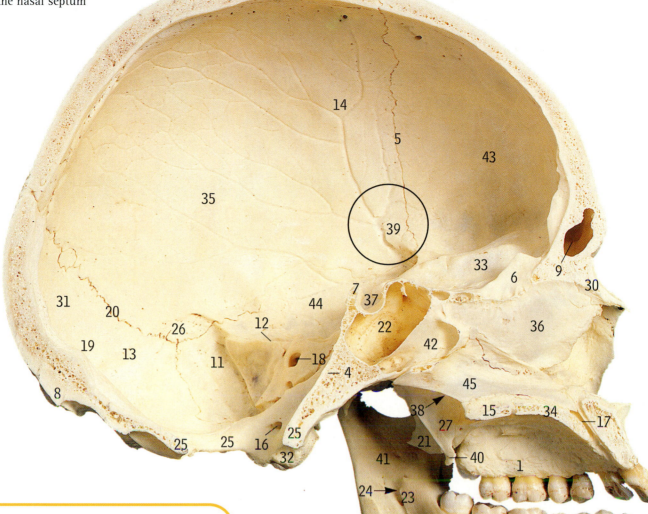

1	Alveolar process of maxilla	24	Mandibular foramen
2	Angle of mandible	25	Margin of foramen magnum
3	Body of mandible	26	Mastoid (posterior inferior) angle of
4	Clivus		parietal bone
5	Coronal suture	27	Medial pterygoid plate
6	Crista galli of ethmoid bone	28	Mental protuberance
7	Dorsum sellae	29	Mylohyoid line
8	External occipital protuberance	30	Nasal bone
9	Frontal sinus	31	Occipital bone
10	Groove for mylohyoid nerve	32	Occipital condyle
11	Groove for sigmoid sinus	33	Orbital part of frontal bone
12	Groove for superior petrosal sinus	34	Palatine process of maxilla
13	Groove for transverse sinus	35	Parietal bone
14	Grooves for middle meningeal	36	Perpendicular plate of ethmoid bone
	vessels (anterior division)	37	Pituitary fossa (sella turcica)
15	Horizontal plate of palatine bone	38	Posterior nasal aperture (choana)
16	Hypoglossal canal	39	Pterion (encircled)
17	Incisive canal	40	Pterygoid hamulus of medial
18	Internal acoustic meatus in petrous		pterygoid plate
	part of temporal bone	41	Ramus of mandible
19	Internal occipital protuberance	42	Right sphenoidal sinus
20	Lambdoid suture	43	Squamous part of frontal bone
21	Lateral pterygoid plate	44	Squamous part of temporal bone
22	Left sphenoidal sinus	45	Vomer
23	Lingula		

- The bony part of the nasal septum consists of the vomer (45) and the perpendicular plate of the ethmoid bone (36). The anterior part of the septum consists of the septal cartilage (page 42, A6).
- In this skull the sphenoidal sinuses (42 and 22) are large, and the right one (42) has extended to the left of the midline. The pituitary fossa (37) projects down into the left sinus (22).
- The grooves for the middle meningeal vessels (14) pass upwards and backwards. The circle (39) marks the region of the pterion, and corresponds to the position shown on the outside of the skull on page 4, A29.

Pituitary tumour. Presenting with endocrine derangements, a tumour in the pituitary fossa (sella turcica) may affect the optic chiasma, causing bitemporal hemianopia. Surgical access to the sella via the sphenoidal sinus does not leave a visible facial scar. An advanced condition may be seen on a lateral skull X-ray as enlargement and erosion of the fossa.

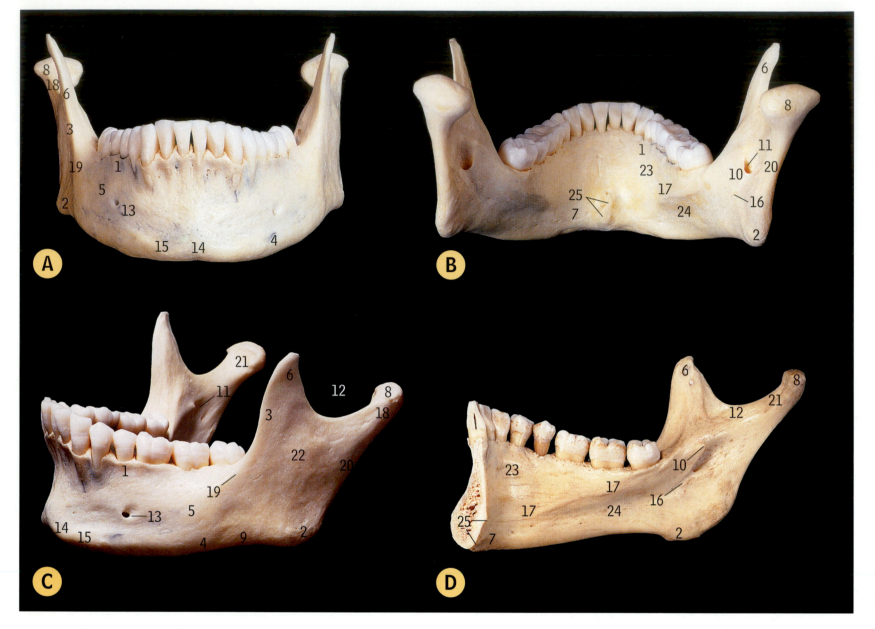

Mandible, **A** from the front, **B** from behind and above, **C** from the left and front, **D** internal view from the left

1	Alveolar part	14	Mental protuberance
2	Angle	15	Mental tubercle
3	Anterior border of ramus	16	Mylohyoid groove
4	Base	17	Mylohyoid line
5	Body	18	Neck
6	Coronoid process	19	Oblique line
7	Digastric fossa	20	Posterior border of ramus
8	Head	21	Pterygoid fovea
9	Inferior border of ramus	22	Ramus
10	Lingula	23	Sublingual fossa
11	Mandibular foramen	24	Submandibular fossa
12	Mandibular notch	25	Superior and inferior mental spines (genial tubercles)
13	Mental foramen		

- The head (8) and the neck (18, including the pterygoid fovea, 21) constitute the condyle.
- The alveolar part (1) contains the sockets for the roots of the teeth.
- The base (4) is the inferior border of the body (5), and becomes continuous with the inferior border (9) of the ramus (22).

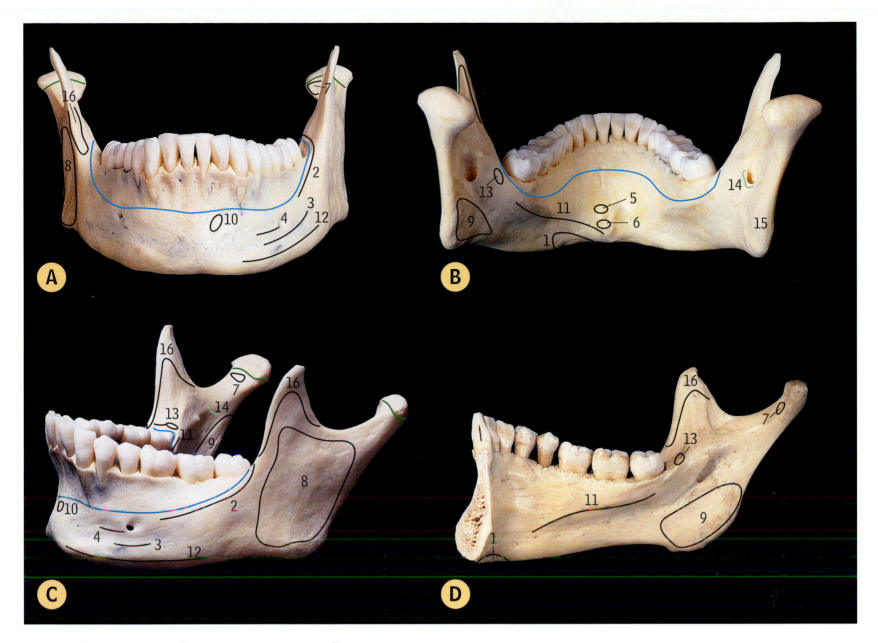

Mandible, **A** from the front, **B** from behind and above, **C** from the left and front, **D** internal view from the left. Muscle attachments
Green line = capsular attachment of temporomandibular joint;
blue line = limit of attachment of the oral mucous membrane; pale green line = ligament attachment

1   Anterior belly of digastric
2   Buccinator
3   Depressor anguli oris
4   Depressor labii inferioris
5   Genioglossus
6   Geniohyoid
7   Lateral pterygoid
8   Masseter
9   Medial pterygoid
10   Mentalis
11   Mylohyoid
12   Platysma
13   Pterygomandibular raphe and superior constrictor
14   Sphenomandibular ligament
15   Stylomandibular ligament
16   Temporalis

- The lateral pterygoid (A7) is attached to the pterygoid fovea on the neck of the mandible (and also to the capsule of the temporomandibular joint and the articular disc – see page 31, A8 and B3).
- The medial pterygoid (B9, C9) is attached to the medial surface of the angle of the mandible, below the groove for the mylohyoid nerve.
- Masseter (C8) is attached to the lateral surface of the ramus.
- Temporalis (C16) is attached over the coronoid process, extending back as far as the deepest part of the mandibular notch and downwards over the front of the ramus almost as far as the last molar tooth.
- Buccinator (C2) is attached opposite the three molar teeth, at the back reaching the pterygomandibular raphe (C13).
- Genioglossus (B5) is attached to the upper mental spine and geniohyoid (B6) to the lower.
- Mylohyoid (11) is attached to the mylohyoid line.
- The attachment of the lateral temporomandibular ligament to the lateral aspect of the neck of the condyle is not shown.

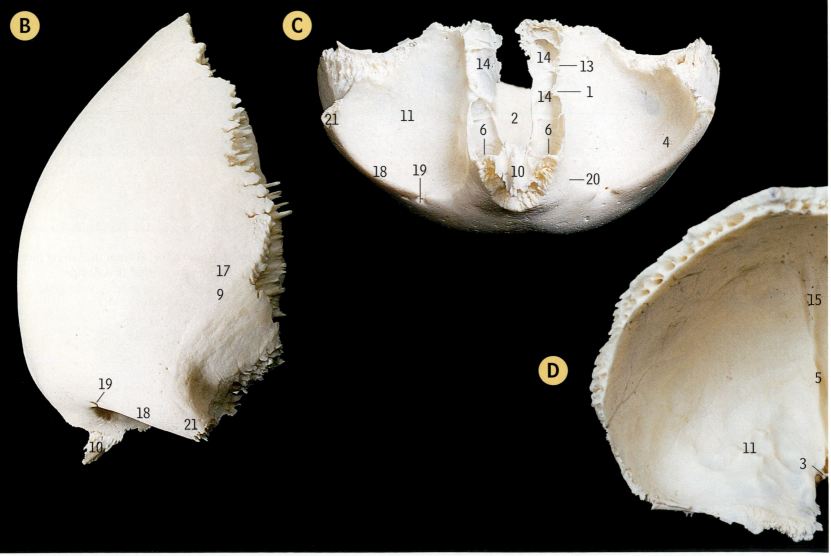

Frontal bone. **A** external surface from the front, **B** external surface from the left, **C** from below, **D** internal surface from above and behind (right half removed; ethmoidal notch is inferior)

 1 Anterior ethmoidal canal (position of groove)
 2 Ethmoidal notch
 3 Foramen caecum
 4 Fossa for lacrimal gland
 5 Frontal crest
 6 Frontal sinus
 7 Frontal tuberosity
 8 Glabella
 9 Inferior temporal line
10 Nasal spine
11 Orbital part
12 Position of frontal notch or foramen
13 Posterior ethmoidal canal (position of groove)
14 Roof of ethmoidal air cells
15 Sagittal crest
16 Superciliary arch
17 Superior temporal line
18 Supra-orbital margin
19 Supra-orbital notch or foramen
20 Trochlear fovea (or tubercle)
21 Zygomatic process

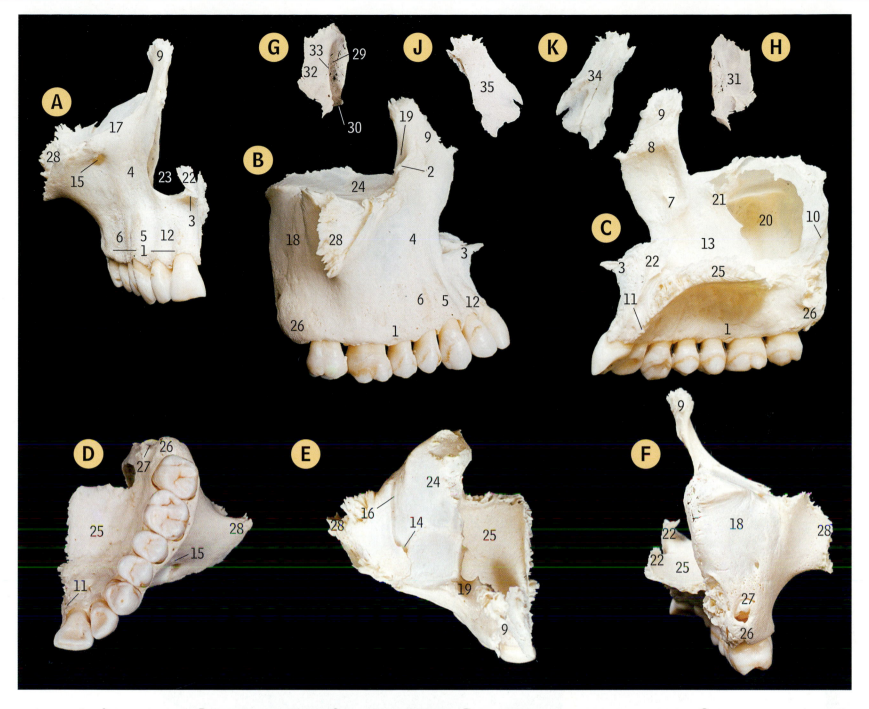

Right maxilla, **A** from the front, **B** from the lateral side, **C** from the medial side, **D** from below, **E** from above, **F** from behind

Right lacrimal bone, **G** from the lateral (orbital) side, **H** from the medial (nasal) side

1 Alveolar process	15 Infra-orbital foramen
2 Anterior lacrimal crest	16 Infra-orbital groove
3 Anterior nasal spine	17 Infra-orbital margin
4 Anterior surface	18 Infratemporal surface
5 Canine eminence	19 Lacrimal groove
6 Canine fossa	20 Maxillary hiatus and sinus
7 Conchal crest	21 Middle meatus
8 Ethmoidal crest	22 Nasal crest
9 Frontal process	23 Nasal notch
10 Greater palatine canal (position of groove)	24 Orbital surface
11 Incisive canal	25 Palatine process
12 Incisive fossa	26 Tuberosity
13 Inferior meatus	27 Unerupted third molar tooth
14 Infra-orbital canal	28 Zygomatic process

29 Lacrimal groove
30 Lacrimal hamulus
31 Nasal surface
32 Orbital surface
33 Posterior lacrimal crest

Right nasal bone, **J** from the lateral side, **K** from the medial side

34 Internal surface and groove for anterior ethmoidal nerve
35 Lateral surface

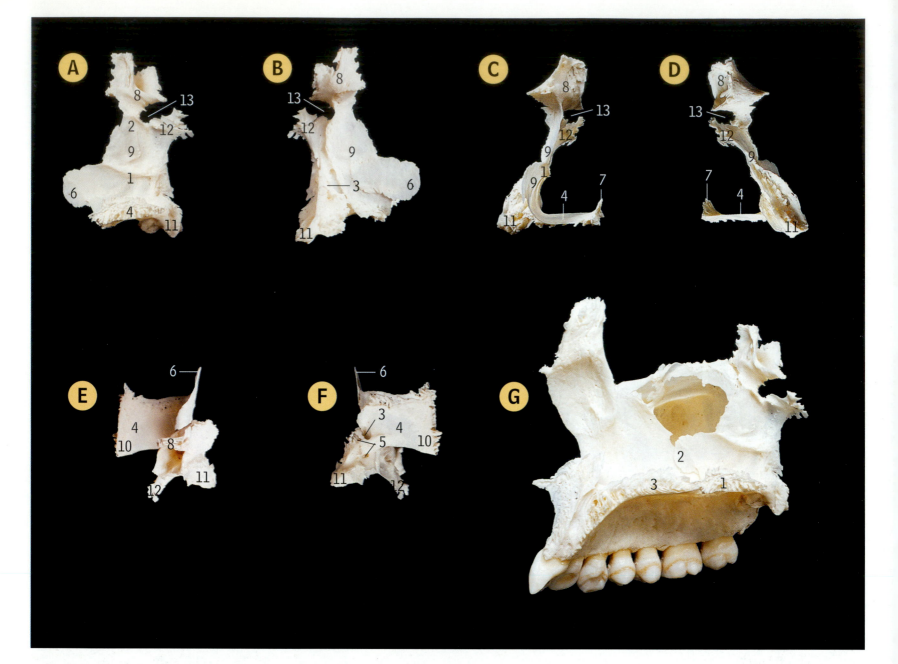

Right palatine bone, **A** from the medial side,
B from the lateral side, **C** from the front,
D from behind, **E** from above, **F** from below

G Articulation of the right maxilla and the
palatine bone, from the medial side

1	Conchal crest
2	Ethmoidal crest
3	Greater palatine groove
4	Horizontal plate
5	Lesser palatine canals
6	Maxillary process
7	Nasal crest
8	Orbital process
9	Perpendicular plate
10	Posterior nasal spine
11	Pyramidal process
12	Sphenoidal process
13	Sphenopalatine notch

1	Horizontal plate of palatine
2	Maxillary process of palatine
3	Palatine process of maxilla

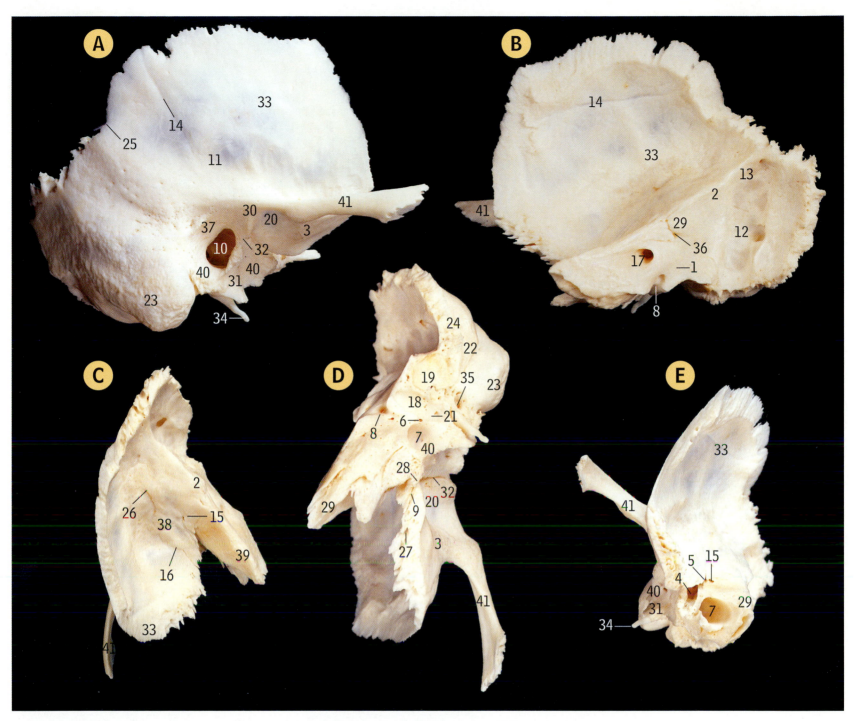

Right temporal bone, **A** external aspect, **B** internal aspect, **C** from above, **D** from below, **E** from the front

1 Aqueduct of vestibule	15 Hiatus and groove for greater petrosal nerve	29 Petrous part
2 Arcuate eminence	16 Hiatus and groove for lesser petrosal nerve	30 Postglenoid tubercle
3 Articular tubercle	17 Internal acoustic meatus	31 Sheath of styloid process
4 Auditory tube	18 Jugular fossa	32 Squamotympanic fissure
5 Canal for tensor tympani	19 Jugular surface	33 Squamous part
6 Canaliculus for tympanic branch of	20 Mandibular fossa	34 Styloid process
glossopharyngeal nerve	21 Mastoid canaliculus for auricular branch	35 Stylomastoid foramen
7 Carotid canal	of vagus nerve	36 Subarcuate fossa
8 Cochlear canaliculus	22 Mastoid notch	37 Suprameatal triangle
9 Edge of tegmen tympani	23 Mastoid process	38 Tegmen tympani
10 External acoustic meatus	24 Occipital groove	39 Trigeminal impression on apex of petrous part
11 Groove for middle temporal artery	25 Parietal notch	40 Tympanic part
12 Groove for sigmoid sinus	26 Petrosquamous fissure (from above)	41 Zygomatic process
13 Groove for superior petrosal sinus	27 Petrosquamous fissure (from below)	
14 Grooves for branches of middle meningeal vessels	28 Petrotympanic fissure	

Right parietal bone, **A** external surface, **B** internal surface

1 Frontal (anterior) border
2 Frontal (anterosuperior) angle
3 Furrows for frontal branch of middle meningeal vessels (anterior division)
4 Furrows for parietal branch of middle meningeal vessels (posterior division)
5 Groove for sigmoid sinus at mastoid angle
6 Inferior temporal line
7 Mastoid (postero-inferior) angle
8 Occipital (posterior) border
9 Occipital (posterosuperior) angle
10 Parietal foramen
11 Parietal tuberosity
12 Sagittal (superior) border
13 Sphenoidal (antero-inferior) angle
14 Squamosal (inferior) border
15 Superior temporal line

Right zygomatic bone, **C** lateral surface, **D** from the medial side, **E** from behind

1 Frontal process
2 Marginal tubercle
3 Maxillary border
4 Orbital border
5 Orbital surface
6 Temporal border
7 Temporal process
8 Temporal surface
9 Zygomatico-orbital foramen
10 Zygomaticofacial foramen
11 Zygomaticotemporal foramen

- The zygomatic process of the temporal bone (page 23, 41) and the temporal process of the zygomatic bone (C7, D7) form the zygomatic arch (page 4, A35).

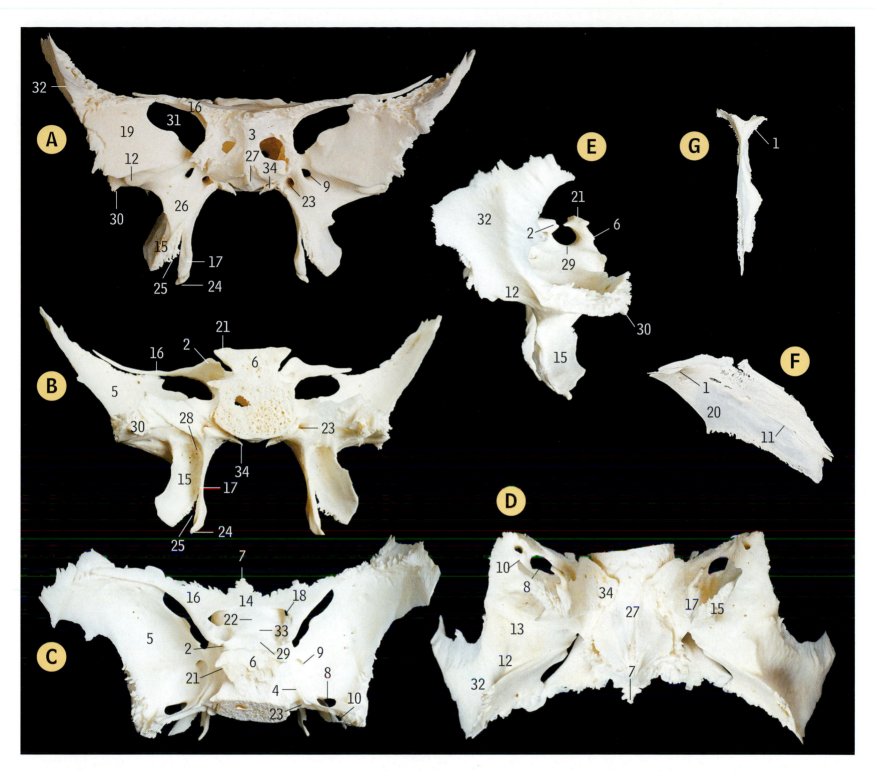

Sphenoid bone, **A** from the front, **B** from behind, **C** from above and behind, **D** from below, **E** from the left. Vomer, **F** from the right, **G** from behind

1 Ala	13 Infratemporal surface of greater wing	24 Pterygoid hamulus
2 Anterior clinoid process	14 Jugum	25 Pterygoid notch
3 Body with openings of sphenoidal sinuses	15 Lateral pterygoid plate	26 Pterygoid process
4 Carotid groove	16 Lesser wing	27 Rostrum
5 Cerebral surface of greater wing	17 Medial pterygoid plate	28 Scaphoid fossa
6 Dorsum sellae	18 Optic canal	29 Sella turcica (pituitary fossa)
7 Ethmoidal spine	19 Orbital surface of greater wing	30 Spine
8 Foramen ovale	20 Posterior border	31 Superior orbital fissure
9 Foramen rotundum	21 Posterior clinoid process	32 Temporal surface of greater wing
10 Foramen spinosum	22 Prechiasmatic groove	33 Tuberculum sellae
11 Groove for nasopalatine nerve and vessels	23 Pterygoid canal	34 Vaginal process
12 Infratemporal crest of greater wing		

• In C, the anterior and middle clinoid processes are almost joined together (C2)

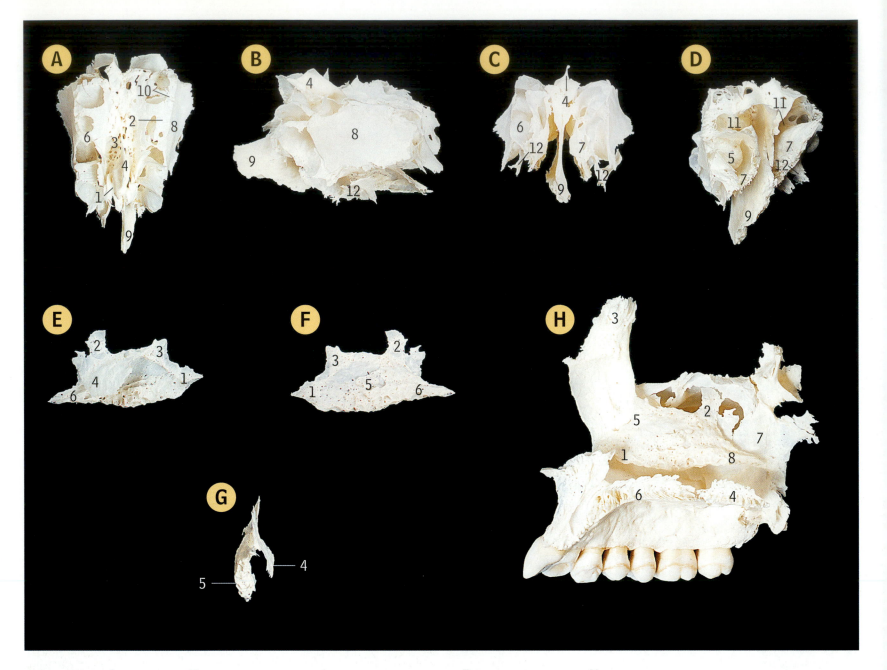

Ethmoid bone, **A** from above, **B** from the left, **C** from the front, **D** from the left, below and behind

Right inferior nasal concha, **E** from the lateral side, **F** from the medial side, **G** from behind

H Articulation of right maxilla, palatine bone and inferior nasal concha, from the medial side

1 Ala of crista galli
2 Anterior ethmoidal groove
3 Cribriform plate
4 Crista galli
5 Ethmoidal bulla
6 Ethmoidal labyrinth (containing ethmoidal air cells)
7 Middle nasal concha
8 Orbital plate
9 Perpendicular plate
10 Posterior ethmoidal groove
11 Superior nasal concha (meatus)
12 Uncinate process

1 Anterior end
2 Ethmoidal process
3 Lacrimal process
4 Maxillary process
5 Medial surface
6 Posterior end

1 Anterior end of inferior nasal concha
2 Ethmoidal process of inferior nasal concha
3 Frontal process of maxilla
4 Horizontal plate of palatine
5 Lacrimal process of inferior nasal concha
6 Palatine process of maxilla
7 Perpendicular plate of palatine
8 Posterior end of inferior nasal concha

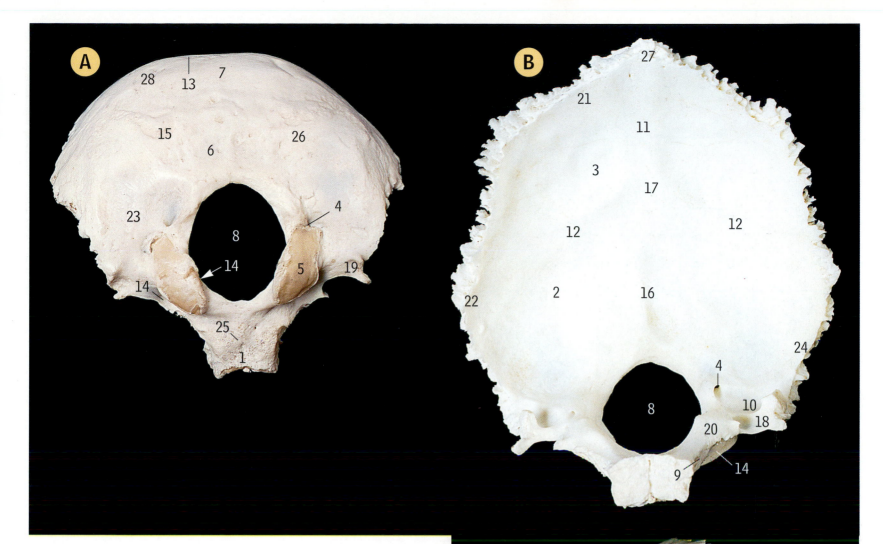

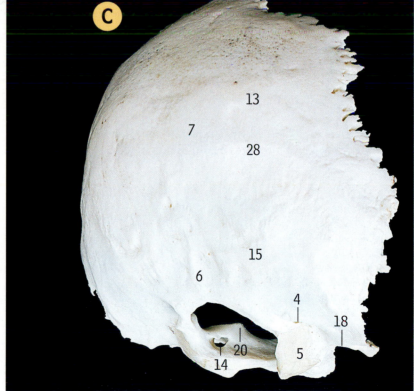

Occipital bone, **A** external surface from below, **B** internal surface, **C** external surface from the right and below.

1	Basilar part	15	Inferior nuchal line
2	Cerebellar fossa	16	Internal occipital crest
3	Cerebral fossa	17	Internal occipital protuberance
4	Condylar fossa (and condylar canal in B and C)	18	Jugular notch
5	Condyle	19	Jugular process
6	External occipital crest	20	Jugular tubercle
7	External occipital protuberance	21	Lambdoid margin
8	Foramen magnum	22	Lateral angle
9	Groove for inferior petrosal sinus	23	Lateral part
10	Groove for sigmoid sinus	24	Mastoid margin
11	Groove for superior sagittal sinus	25	Pharyngeal tubercle
12	Groove for transverse sinus	26	Squamous part
13	Highest nuchal line	27	Superior angle
14	Hypoglossal canal	28	Superior nuchal line

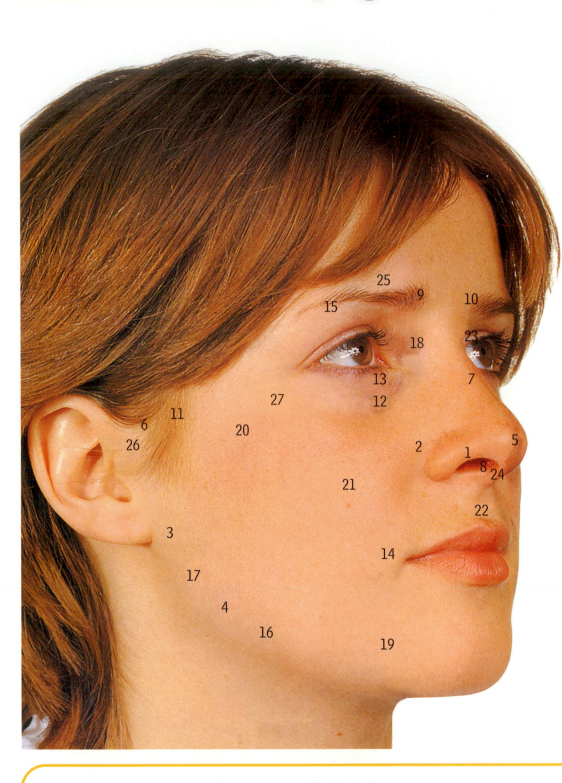

Face. Surface markings on the front and right side

- The pulsation of the superficial temporal artery (6) is palpable in front of the tragus of the ear (26).
- The parotid duct (20 and 21) lies under the middle third of a line drawn from the tragus of the ear (26) to the midpoint of the philtrum (22).
- The pulsation of the facial artery (4) is palpable where the vessel crosses the lower border of the mandible at the anterior margin of the masseter muscle, about 2.5 cm (1 in) in front of the angle of the mandible (3).

> Ophthalmic herpes zoster, also known as shingles, is a cutaneous viral eruption that maps out the distribution of the ophthalmic division of the trigeminal nerve (scalp, forehead, upper eyelid, nose and possibly as far down as the philtrum). Involvement of the corneal membrane is an indication for immediate antiviral treatment.

> The philtrum is the small flat area bounded by two vertical ridges below the nose where the two maxillary processes meet the frontonasal process in the developing face. It is along the ridges of the philtrum that the common congenital abnormality of harelip occurs.

1 Ala of external nose	10 Glabella of external nose	20 Parotid duct emerging from gland
2 Alar groove of external nose	11 Head of mandible	21 Parotid duct turning medially at anterior border of masseter
3 Angle of mandible	12 Infra-orbital foramen, nerve and vessels	22 Philtrum
4 Anterior border of masseter and facial vessels	13 Infra-orbital margin	23 Root of external nose
5 Apex of external nose	14 Lateral angle of mouth	24 Septum of external nose
6 Auriculotemporal nerve and superficial temporal vessels	15 Lateral part of supra-orbital margin	25 Supra-orbital notch (or foramen), nerve and vessels
7 Dorsum of external nose	16 Lower border of body of mandible	26 Tragus
8 External aperture of external nose	17 Lower border of ramus of mandible	27 Zygomatic arch
9 Frontal notch and supratrochlear nerve and vessels	18 Medial palpebral ligament anterior to lacrimal sac	
	19 Mental foramen, nerve and vessels	

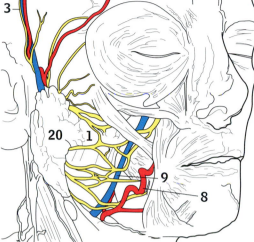

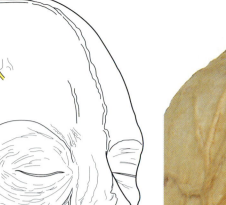

1 Accessory parotid gland overlying parotid duct
2 Anterior branch of superficial temporal artery
3 Auriculotemporal nerve and superficial temporal
   vessels
4 Body of mandible
5 Buccinator and buccal branches of facial nerve
6 Depressor anguli oris
7 Depressor labii inferioris
8 Facial artery
9 Facial vein
10 Frontalis part of occipitofrontalis
11 Great auricular nerve
12 Levator anguli oris
13 Levator labii superioris
14 Levator labii superioris alaeque nasi
15 Marginal mandibular branch of facial nerve
16 Masseter
17 Nasalis
18 Orbicularis oculi
19 Orbicularis oris
20 Parotid gland
21 Procerus
22 Sternocleidomastoid
23 Supra-orbital nerve
24 Supratrochlear nerve
25 Temporal branch of facial nerve
26 Temporalis underlying temporal fascia
27 Zygomatic branch of facial nerve
28 Zygomaticus major
29 Zygomaticus minor

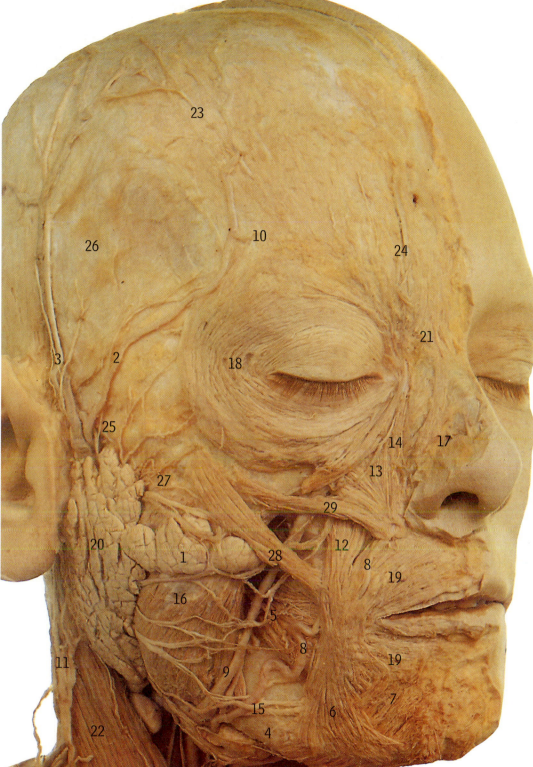

Face. Superficial dissection from the front and the right

Intracranial spread of infections: face. Ophthalmic veins connect the cavernous sinus to facial veins. These valveless connections mean that superficial facial infections can easily become serious intracranial ones.

Bell's palsy is a facial nerve palsy of unknown aetiology first described by Sir Charles Bell. The site of this lower motor neurone lesion may be diagnosed precisely by careful evaluation as to whether the stapedius, petrosal nerves and chorda tympani are involved.

Surgical flaps of the scalp. Plastic surgeons have devised numerous flaps based upon different vascular pedicles within the scalp. All depend on the rich blood supply of the scalp, which contains numerous anastomoses between internal and external carotid branches. Approximately 10–12 named arteries supply the scalp periphery, so that any one single artery will often be enough to give nourishment for recovery from scalp injuries and flap construction.

Right lower face and upper neck,
A parotid and upper cervical regions,
B submandibular region

Mumps is an acute infection of the parotid and submandibular salivary glands. It is extremely painful and both opening the mouth and chewing may be restricted.

Parotid papilla. Just lateral to the crown of the maxillary second molar, the parotid duct enters the oral cavity at the tip of a small papilla.

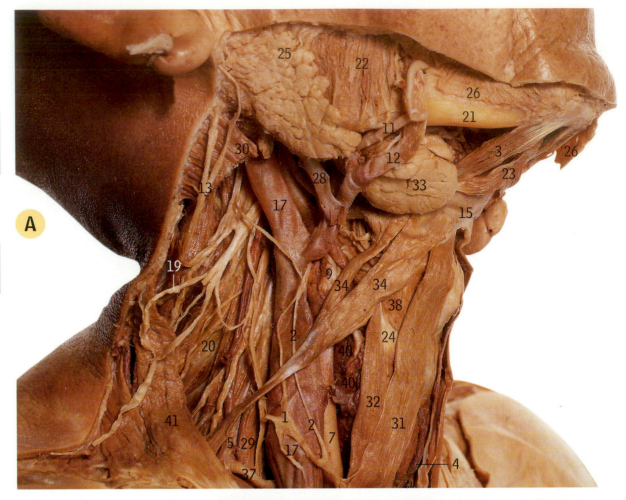

1	Ansa cervicalis, inferior root	22	Masseter
2	Ansa cervicalis, superior root	23	Mylohyoid
3	Anterior belly of digastric	24	Oblique line of the thyroid cartilage
4	Anterior jugular vein	25	Parotid gland and facial nerve
5	Brachial plexus (roots)		branches at anterior border
6	Buccinator	26	Platysma
7	Common carotid artery	27	Posterior belly of digastric
8	Depressor anguli oris	28	Retromandibular vein
9	External carotid artery	29	Scalenus anterior
10	External jugular vein	30	Sternocleidomastoid
11	Facial artery	31	Sternohyoid
12	Facial vein	32	Sternothyroid
13	Great auricular nerve	33	Submandibular gland
14	Greater horn of hyoid bone	34	Superior belly of omohyoid (bifid)
15	Hyoid bone	35	Superior laryngeal artery
16	Hypoglossal nerve	36	Superior thyroid artery
17	Internal jugular vein	37	Suprascapular artery
18	Internal laryngeal nerve	38	Thyrohyoid
19	Lesser occipital nerve	39	Thyrohyoid membrane
20	Levator scapulae	40	Thyroid gland
21	Mandible	41	Trapezius

Parotidectomy (surgical removal of the parotid gland) requires tedious dissection to minimize damage to branches of the facial nerve. The use of a facial nerve stimulator (a special pair of forceps conducting a low voltage) helps the surgeon avoid these branches. The deep portion of the gland abuts the pharyngeal wall.

Parotid tumours may involve the retromandibular vein or the superficial temporal artery which lie within its substance, but the most common effect is involvement of the facial nerve as its numerous branches pass from deep to superficial through this gland. A facial paralysis associated with a swelling in the parotid gland is a condition to be taken seriously. The pain of a tumour here may often be referred to the temporomandibular joint via the auriculotemporal nerve which carries the parotid's secretomotor fibres.

Right infratemporal fossa, **A** with the pterygoid muscles intact, **B** after removal of the lateral pterygoid, **C** after removal of the lateral and medial pterygoids

1 Accessory meningeal artery	22 Lingual nerve
2 Accessory nerve	23 Lower head of lateral pterygoid
3 Articular disc of temporomandibular joint	24 Mandibular nerve
4 Ascending pharyngeal artery	25 Masseter
	26 Maxillary artery
5 Auriculotemporal nerve	27 Medial pterygoid
6 Buccal nerve	28 Middle meningeal artery
7 Buccinator	29 Molar glands
8 Capsule of temporomandibular joint	30 Nerve to medial pterygoid
9 Chorda tympani	31 Nerve to mylohyoid
10 Deep temporal artery	32 Parotid duct
11 Deep temporal nerve	33 Posterior superior alveolar artery
12 External acoustic meatus	34 Roots of auriculotemporal nerve
13 External carotid artery	35 Styloglossus
14 Facial artery	36 Stylohyoid
15 Facial vein	37 Styloid process
16 Inferior alveolar artery	38 Stylopharyngeus and glossopharyngeal nerve
17 Inferior alveolar nerve	39 Superficial temporal artery
18 Infratemporal surface of maxilla	40 Superior constrictor of pharynx
19 Internal jugular vein	41 Temporal fascia
20 Lateral pterygoid plate	42 Temporalis
	43 Tensor veli palatini
21 Levator veli palatini	44 Upper head of lateral pterygoid

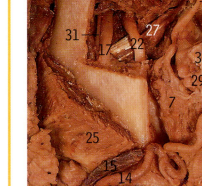

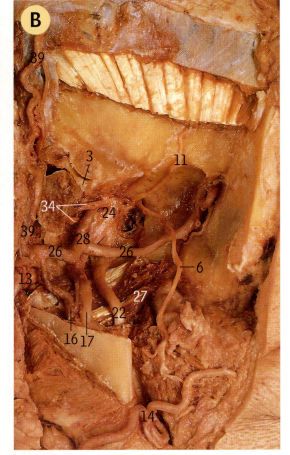

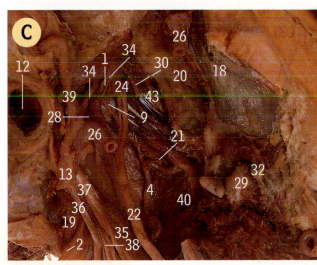

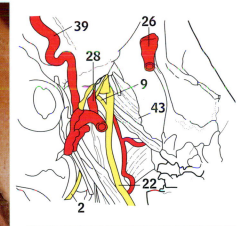

D Coronal section to show deep head of temporalis

1 Buccinator	13 Palate
2 Greater wing of sphenoid	14 Sphenoidal sinus
3 Lateral pterygoid	15 Temporal bone
4 Lateral rectus	16 Temporal lobe, brain
5 Lesser wing of sphenoid	17 Temporalis, deep head (sphenomandibularis – Zenker 1955)
6 Mandible	18 Temporalis, insertion
7 Masseter	19 Temporalis, superficial head
8 Maxilla	20 Tongue
9 Maxillary air (paranasal) sinus	21 Vestibule of oral cavity
10 Maxillary artery, muscular branches	22 Zygoma
11 Nasal septum	
12 Optic nerve	

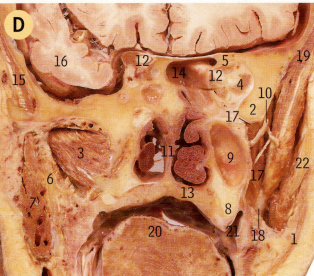

Temporomandibular joint (TMJ) reduction. Following dislocation of the TMJ, the mandibular condyle rests anterior to the articular tubercle of the temporal bone and the resultant spasm in the masticatory muscles is painful. Reduction of the condyle into the mandibular fossa requires inferior and posterior movements, accomplished by firmly grasping the mandibular body, and can be aided by local anaesthesia or muscle relaxants.

Inferior alveolar nerve block is a procedure that anaesthetizes this nerve as it enters the mandibular foramen on the medial side of the ramus of the mandible. Using an intra-oral approach the anaesthetic is deposited near the lingula, producing anaesthesia of the ipsilateral mandibular dentition, lip and cheek.

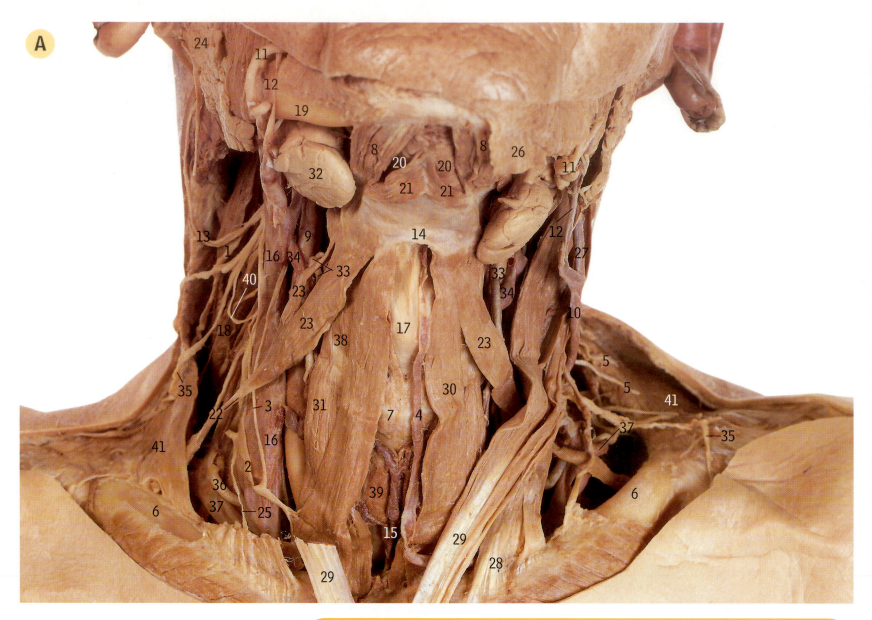

Front of the neck. A Superficial dissection

- Midline landmarks in the neck include the body of the hyoid bone (14), the laryngeal prominence (Adam's apple, 17) and the arch of the cricoid cartilage (7).
- The site for feeling the carotid pulse (pulsation of the common carotid artery) is in the lower front part of the neck, between the anterior border of sternocleidomastoid and the side of the larynx.

Tracheostomy is an opening (stoma) into the trachea often used for patients who cannot breathe on their own. Anatomically, it involves splitting the strap muscles, passing through the deep cervical fascia and removing or clamping the isthmus of the thyroid gland. A low tracheostomy below the isthmus may endanger an unusually high left brachiocephalic vein (especially in children) or the rare thyroidea ima artery. A safer emergency procedure is to pass a needle or tube through the cricothyroid membrane, just above the cricoid cartilage.

1	Accessory nerve	22	Omohyoid, intermediate tendon
2	Ansa cervicalis, lower root	23	Omohyoid, superior belly
3	Ansa cervicalis, upper root	24	Parotid gland
4	Anterior jugular vein	25	Phrenic nerve
5	Cervical nerves to trapezius	26	Platysma
6	Clavicle	27	Retromandibular vein
7	Cricoid cartilage	28	Sternocleidomastoid, clavicular head
8	Digastric, anterior belly	29	Sternocleidomastoid, sternal head
9	External carotid artery	30	Sternohyoid
10	External jugular vein	31	Sternothyroid
11	Facial artery	32	Submandibular gland
12	Facial vein	33	Superior thyroid artery
13	Great auricular nerve	34	Superior thyroid vein
14	Hyoid bone, body	35	Supraclavicular nerve
15	Inferior thyroid vein	36	Suprascapular artery
16	Internal jugular vein	37	Suprascapular vein
17	Laryngeal prominence	38	Thyrohyoid
18	Levator scapulae	39	Thyroid gland, isthmus
19	Mandible	40	Transverse cervical nerve
20	Mylohyoid	41	Trapezius
21	Mylohyoid, anomalous fibres		

Front of the neck. **B** Deeper dissection

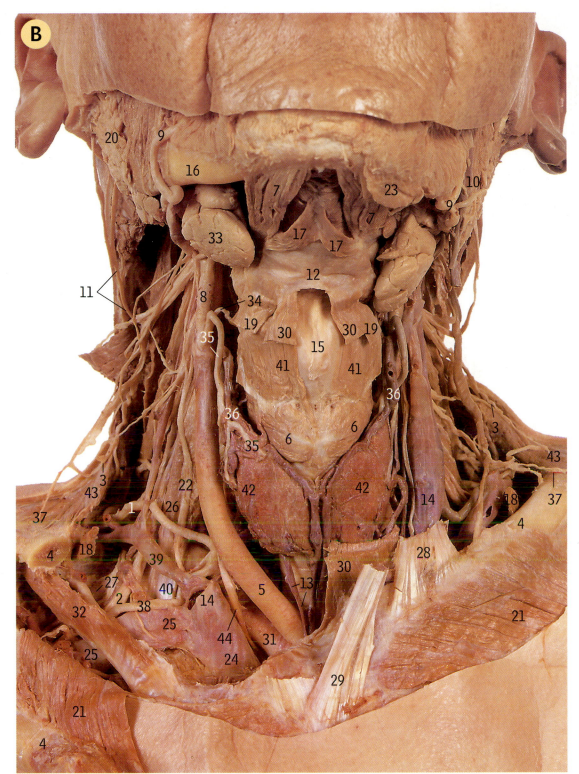

1 Accessory nerve
2 Brachial plexus (roots)
3 Cervical nerves to trapezius
4 Clavicle
5 Common carotid artery
6 Cricothyroid
7 Digastric, anterior belly
8 External carotid artery
9 Facial artery
10 Facial vein
11 Great auricular nerve
12 Hyoid bone, body
13 Inferior thyroid vein
14 Internal jugular vein
15 Laryngeal prominence
16 Mandible
17 Mylohyoid, anomalous fibres
18 Omohyoid, inferior belly
19 Omohyoid, superior belly
20 Parotid gland
21 Pectoralis major
22 Phrenic nerve
23 Platysma
24 Right brachiocephalic vein
25 Right subclavian vein
26 Scalenus anterior
27 Scalenus medius
28 Sternocleidomastoid, clavicular head
29 Sternocleidomastoid, sternal head
30 Sternohyoid
31 Subclavian artery
32 Subclavius
33 Submandibular gland
34 Superior laryngeal artery
35 Superior thyroid artery
36 Superior thyroid vein
37 Supraclavicular nerve
38 Suprascapular artery
39 Suprascapular vein
40 Tendon of scalenus anterior
41 Thyrohyoid
42 Thyroid gland, lateral lobe
43 Trapezius
44 Vagus nerve

● On the right hand side the clavicle (4) has been cut and retracted forwards to reveal the underlying subclavius (32).

Accessory nerve paralysis. Lesions of the cranial branch of this nerve are rare but associated with vagus nerve problems and may be related to bulbar palsy. This is because all accessory nerve motor fibres are transferred to the vagus as it exits the skull. The spinal portion is more commonly damaged in the mid-portion of the posterior triangle of the neck from stab wounds, or from spread of malignancy along the accessory lymph node chain. This injury can cause paralysis of the sternocleidomastoid and trapezius muscles, with weakness in shoulder raising and a permanent droop on that side and also difficulty in turning the head towards the opposite side

A goitre is a non-specific enlargement of the thyroid gland, making it easily palpable and often visible. There are numerous causes but world-wide the most common is a dietary deficiency of iodine. A very large goitre may cause compression of lower cervical and superior mediastinal structures including the trachea, making breathing difficult.

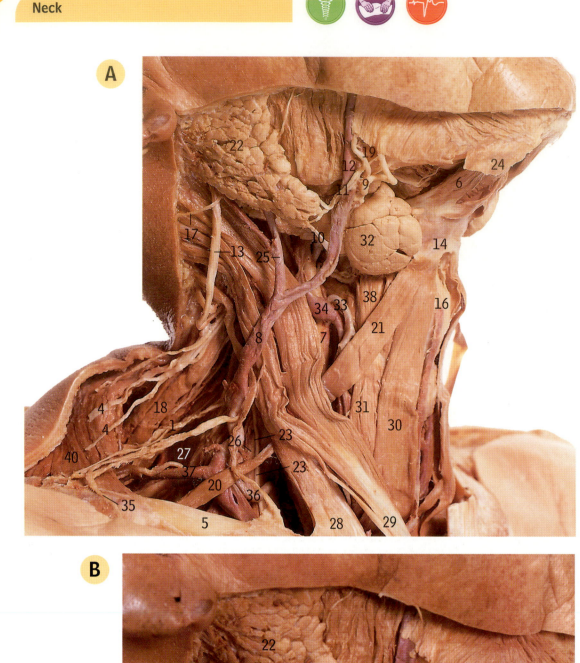

Right side of the neck, A superficial, B deep

1   Accessory nerve
2   Ansa cervicalis, upper root
3   Brachial plexus (roots)
4   Cervical nerves to trapezius
5   Clavicle
6   Digastric, anterior belly
7   External carotid artery
8   External jugular vein
9   Facial artery
10 Facial nerve, cervical branch
11 Facial nerve, marginal mandibular branch
12 Facial vein
13 Great auricular nerve
14 Hyoid bone
15 Internal jugular vein
16 Laryngeal prominence
17 Lesser occipital nerve
18 Levator scapulae
19 Mandible
20 Omohyoid, inferior belly
21 Omohyoid, superior belly
22 Parotid gland
23 Phrenic nerve
24 Platysma
25 Retromandibular vein
26 Scalenus anterior
27 Scalenus medius
28 Sternocleidomastoid, clavicular head
29 Sternocleidomastoid, sternal head
30 Sternohyoid
31 Sternothyroid
32 Submandibular gland
33 Superior thyroid artery
34 Superior thyroid vein
35 Supraclavicular nerve
36 Suprascapular artery
37 Suprascapular vein
38 Thyrohyoid
39 Thyroid gland, lateral lobe
40 Trapezius

- The spinal part of the accessory nerve is in official anatomical nomenclature the ramus externus of the truncus nervi accessorii. The cells of origin are in the anterior horn of the upper five or six cervical segments of the spinal cord, and the fibres supply sternocleidomastoid and trapezius. The cranial part of the accessory nerve, the ramus internus of the truncus nervi accessorii, is derived from the nucleus ambiguus in the medulla oblongata and joins the vagus nerve to supply muscles of the soft palate and larynx.
- The motor nerve supply of trapezius (40) is usually the accessory nerve (1), with the branches from the cervical plexus to the muscle (4) being afferent only, but in some cases the cervical branches do appear to be motor.

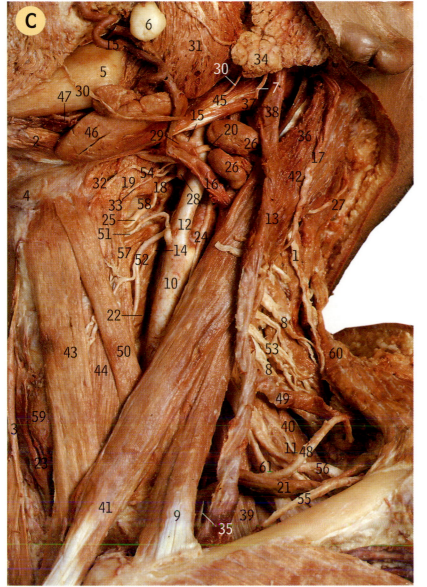

C **Left side of the neck, from the left and front**
Platysma and the deep cervical fascia have been removed

- In 20 per cent of faces, as in this specimen, the marginal mandibular branch of the facial nerve (30) arches downwards off the face for part of its course and overlies the submandibular gland (46).

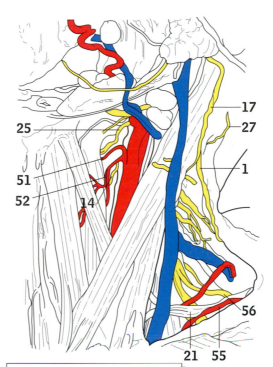

Carotid artery bruits. Extra sounds heard through the stethoscope just lateral to the larynx are often the result of turbulent arterial blood flow due to stenosis of the carotid artery. Early detection of narrowing of this essential artery is often accomplished by colour Doppler ultrasound.

Cervical lymph node enlargement. Any infection in the head and neck can cause lymph node enlargement, the most common site being the jugulodigastric nodes located just below the angle of the jaw.

1 Accessory nerve	23 Inferior thyroid vein	44 Sternothyroid
2 Anterior belly of digastric	24 Internal carotid artery and superior root of ansa cervicalis	45 Stylohyoid
3 Anterior jugular vein		46 Submandibular gland
4 Body of hyoid bone	25 Internal laryngeal nerve	47 Submental artery and vein
5 Body of mandible	26 Jugulodigastric lymph nodes	48 Superficial cervical artery
6 Buccal fat pad	27 Lesser occipital nerve	49 Superficial cervical vein
7 Cervical branch of facial nerve	28 Lingual artery	50 Superior belly of omohyoid
8 Cervical nerves to trapezius	29 Lingual vein	51 Superior laryngeal artery
9 Clavicular head of sternocleidomastoid	30 Marginal mandibular branch of facial nerve	52 Superior thyroid artery
10 Common carotid artery	31 Masseter	53 Supraclavicular nerve (cut upper edge)
11 Dorsal scapular nerve	32 Mylohyoid	54 Suprahyoid artery
12 External carotid artery	33 Nerve to thyrohyoid	55 Suprascapular artery
13 External jugular vein	34 Parotid gland	56 Suprascapular nerve
14 External laryngeal nerve	35 Phrenic nerve (on scalenus anterior)	57 Thyrohyoid
15 Facial artery	36 Posterior auricular vein	58 Thyrohyoid membrane
16 Facial vein	37 Posterior belly of digastric	59 Thyroid gland
17 Great auricular nerve	38 Posterior branch of retromandibular vein	60 Trapezius
18 Greater horn of hyoid bone (underlying 25)	39 Scalenus anterior	61 Upper trunk of brachial plexus
19 Hyoglossus	40 Scalenus medius	
20 Hypoglossal nerve	41 Sternal head of sternocleidomastoid	
21 Inferior belly of omohyoid	42 Sternocleidomastoid	
22 Inferior constrictor of pharynx	43 Sternohyoid	

A Neck. Surface markings of the front and right side

1 Accessory nerve emerging from sternocleidomastoid
2 Accessory nerve passing under anterior border of trapezius
3 Angle of mandible
4 Anterior border of masseter and facial artery
5 Anterior jugular vein
6 Arch of cricoid cartilage
7 Body of hyoid bone
8 Clavicle
9 Clavicular head of sternocleidomastoid
10 Deltoid
11 External jugular vein
12 Hypoglossal nerve
13 Inferior belly of omohyoid
14 Infraclavicular fossa and cephalic vein
15 Internal laryngeal nerve
16 Isthmus of thyroid gland
17 Jugular notch and trachea
18 Laryngeal prominence (Adam's apple)
19 Lowest part of parotid gland
20 Mastoid process
21 Pectoralis major
22 Site for palpation of common carotid artery
23 Sternal head of sternocleidomastoid
24 Sternoclavicular joint and union of internal jugular and subclavian veins to form brachiocephalic vein
25 Sternocleidomastoid
26 Submandibular gland
27 Tip of greater horn of hyoid bone
28 Tip of transverse process of atlas
29 Upper trunk of brachial plexus
30 Vocal fold

- The pulsation of the common carotid artery (A22, B8) can be felt by backward pressure in the angle between the lower anterior border of sternocleidomastoid and the side of the larynx and trachea.
- The cricoid cartilage (A6) is about 5 cm (2 in) above the jugular notch of the manubrium of the sternum (A17).
- The lower end of the internal jugular vein lies behind the interval between the sternal (A23) and clavicular (A9) heads of sternocleidomastoid (when viewed from the front), just above the point where it joins the subclavian vein to form the brachiocephalic vein (A24).
- The uppermost part of the brachial plexus (A29) can be felt as a cord-like structure in the lower part of the posterior triangle.

Torticollis is an abnormal sustained contraction of the neck musculature causing the head to be pulled to one side. It is commonly seen after a difficult forceps delivery, when the infant may develop a cervical swelling – usually a haematoma of the trapezius or sternocleidomastoid.

Internal jugular vein catheterization is often performed by anaesthetists on unconscious patients, and has two common routes: the first is directly through the sternocleidomastoid muscle halfway down the neck; the second is through the gap between the two heads of the sternocleidomastoid muscle, deep to which lies the termination of the internal jugular vein. The needle is passed at 45° to the skin in the direction of the ipsilateral nipple. It provides emergency intravenous access for example during cardiac arrest.

Subclavian vein catheterization takes advantage of the vascular relations on the superior aspect of the first rib to place a central venous line, normally by an infraclavicular route. The tip of the needle should be pointed as anteriorly as possible towards the jugular notch to avoid injury to posterior structures (apex of the lung, subclavian artery and the brachial plexus). A supraclavicular approach puts the needle into the origin of the brachiocephalic vein.

B Right side of the neck. Deep dissection

- The lingual nerve (27) lies superficial to hyoglossus (17) and at this level is a flattened band rather than a typical round nerve, with the deep part of the submandibular gland (10) below it. The nerve crosses underneath the submandibular duct (51), lying first lateral to the duct and then medial to it.
- The thyrohyoid membrane (60) is pierced by the internal laryngeal nerve (23) and the superior laryngeal artery (55).
- Apart from supplying muscles of the tongue, the hypoglossal nerve (19) gives branches to geniohyoid (14) and thyrohyoid (59) and forms the upper root of the ansa cervicalis (62). These three branches consist of the fibres from the first cervical nerve that have joined the hypoglossal nerve higher in the neck; they are not derived from the hypoglossal nucleus. The C1 fibres in the upper root of the ansa contribute to the supply of sternohyoid (45) and omohyoid (21, 54).

Sialolithiasis. Formation of salivary calculi or stones depends on the composition of the fluid being secreted by each salivary gland. Most commonly found in the submandibular gland and rarely in the parotid, these small calcium stones may completely block the duct system and cause a painful swelling of the gland. Surgical removal is usually via the floor of the mouth.

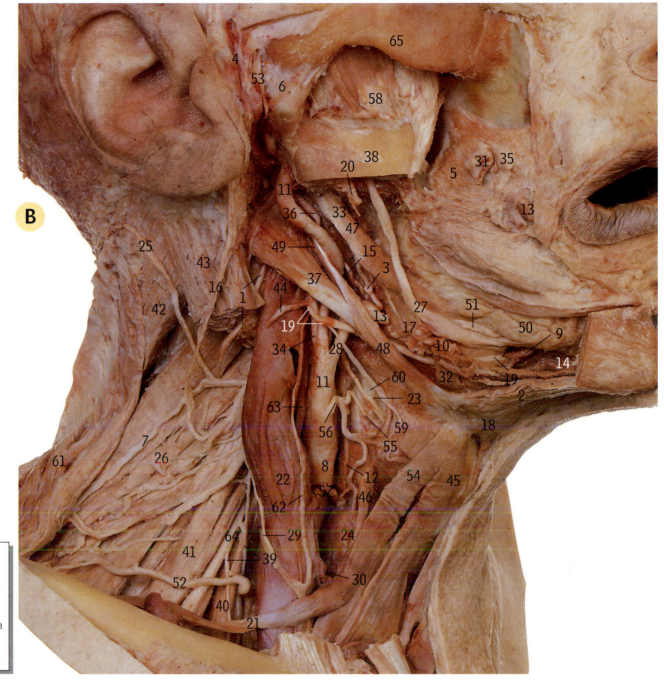

1	Accessory nerve	18	Hyoid bone
2	Anterior belly of digastric and nerve	19	Hypoglossal nerve
3	Ascending palatine artery	20	Inferior alveolar nerve
4	Auriculotemporal nerve	21	Inferior belly of omohyoid
5	Buccinator	22	Internal jugular vein
6	Capsule of temporomandibular joint	23	Internal laryngeal nerve
7	Cervical nerves to trapezius	24	Lateral lobe of thyroid gland
8	Common carotid artery	25	Lesser occipital nerve
9	Deep lingual artery	26	Levator scapulae
10	Deep part of submandibular gland	27	Lingual nerve
11	External carotid artery	28	Linguofacial trunk
12	External laryngeal nerve	29	Lower root of ansa cervicalis
13	Facial artery	30	Middle thyroid vein
14	Geniohyoid	31	Molar glands
15	Glossopharyngeal nerve	32	Mylohyoid and nerve
16	Great auricular nerve	33	Nerve to mylohyoid
17	Hyoglossus	34	Occipital artery

35	Parotid duct	51	Submandibular duct
36	Posterior auricular artery	52	Superficial cervical artery
37	Posterior belly of digastric	53	Superficial temporal artery
38	Ramus of mandible	54	Superior belly of omohyoid
39	Roots of phrenic nerve	55	Superior laryngeal artery
40	Scalenus anterior	56	Superior thyroid artery
41	Scalenus medius	57	Superior thyroid vein
42	Splenius capitis	58	Temporalis
43	Sternocleidomastoid	59	Thyrohyoid and nerve
44	Sternocleidomastoid branch of occipital artery	60	Thyrohyoid membrane
45	Sternohyoid	61	Trapezius
46	Sternothyroid	62	Upper root of ansa cervicalis
47	Styloglossus	63	Vagus nerve
48	Stylohyoid	64	Ventral ramus of fifth cervical nerve
49	Stylohyoid ligament	65	Zygomatic arch
50	Sublingual gland		

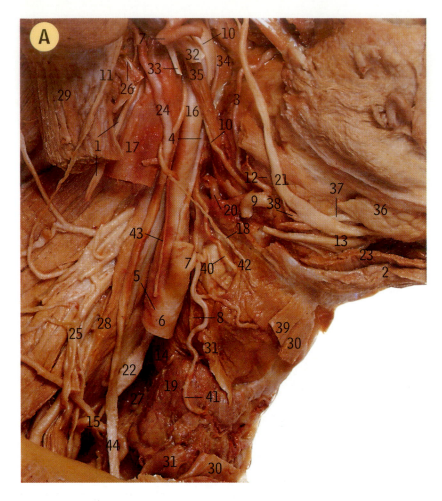

A Right side of the neck. Deep dissection

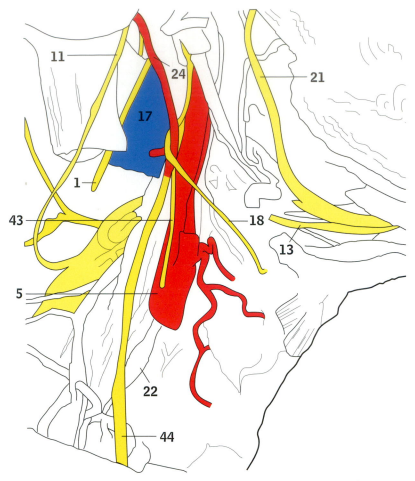

1 Accessory nerve	24 Occipital artery
2 Anterior belly of digastric	25 Phrenic nerve
3 Ascending palatine artery	26 Posterior belly of digastric
4 Ascending pharyngeal artery	27 Recurrent laryngeal nerve
5 Carotid sinus	28 Scalenus anterior
6 Common carotid artery	29 Sternocleidomastoid
7 External carotid artery	30 Sternohyoid
8 External laryngeal nerve	31 Sternothyroid
9 Facial artery	32 Styloglossus
10 Glossopharyngeal nerve	33 Stylohyoid (cut end displaced
11 Great auricular nerve	medially)
12 Hyoglossus	34 Stylohyoid ligament
13 Hypoglossal nerve	35 Stylopharyngeus
14 Inferior constrictor	36 Sublingual gland
15 Inferior thyroid artery	37 Submandibular duct
16 Internal carotid artery	38 Submandibular ganglion
17 Internal jugular vein	39 Superior belly of omohyoid
18 Internal laryngeal nerve	40 Superior laryngeal artery
19 Lateral lobe of thyroid gland	41 Superior thyroid artery
20 Lingual artery	42 Thyrohyoid and nerve
21 Lingual nerve	43 Upper root of ansa cervicalis
22 Middle cervical sympathetic ganglion	44 Vagus nerve
23 Mylohyoid	

Carotid endarterectomy is the removal of atherosclerotic plaque from the narrowed lumen of common and internal cartoid arteries.

- The hypoglossal nerve (13) passes downwards, curling around the occipital artery (24) and lying superficial to the external carotid (7) and lingual (20) arteries.
- The glossopharyngeal nerve (10) passes downwards and forwards, curling round the lateral side of stylopharyngeus (35).
- The removal of parts of the sternohyoid (30), omohyoid (39) and sternothyroid (31) displays the lateral lobe of the thyroid gland (19). Note the inferior thyroid artery (15) behind the lower part of the lobe, with the recurrent laryngeal nerve (27) passing deep to this looping vessel to enter the pharynx beneath the inferior constrictor (14). The nerve may lie behind or in front of the artery or pass between branches of it.

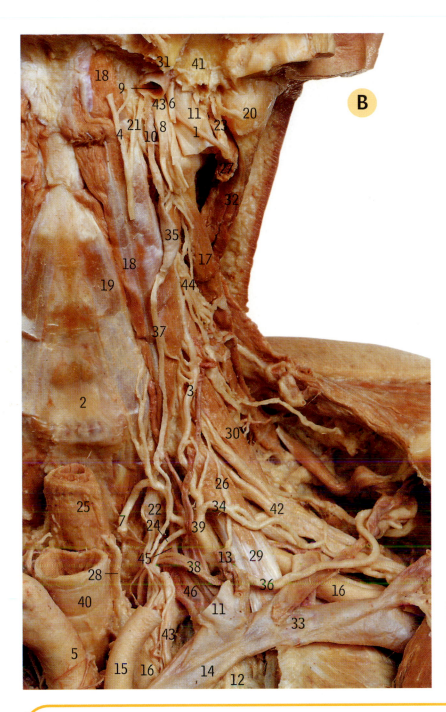

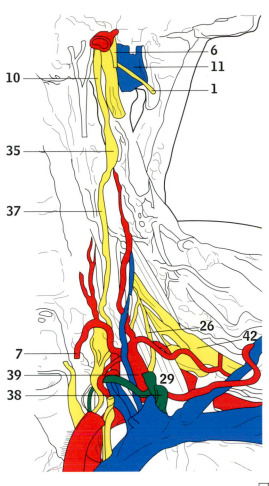

B Left prevertebral region

Cervical sympathectomy is performed through either the neck or the axilla to sever the sympathetic nerves to the upper limb. A common reason is to prevent gangrene of the fingertips in Raynaud's disease.

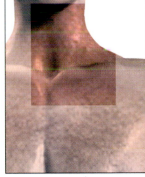

1 Accessory nerve	17 Levator scapulae	32 Sternocleidomastoid
2 Anterior longitudinal ligament	18 Longus capitis	33 Subclavian vein
3 Ascending cervical vessels	19 Longus colli	34 Superficial cervical artery
4 Ascending pharyngeal artery	20 Mastoid process	35 Superior cervical ganglion
5 Brachiocephalic trunk	21 Meningeal branch of ascending pharyngeal artery	36 Suprascapular artery
6 Glossopharyngeal nerve	22 Middle cervical ganglion	37 Sympathetic trunk
7 Inferior thyroid artery	23 Occipital artery	38 Thoracic duct
8 Inferior vagal ganglion	24 Oesophageal branch (large) of the inferior thyroid artery	39 Thyrocervical trunk
9 Internal carotid artery		40 Trachea
10 Internal carotid nerve (sympathetic)	25 Oesophagus	41 Tympanic part of temporal bone
11 Internal jugular vein	26 Phrenic nerve	42 Upper trunk of brachial plexus
12 Internal thoracic artery	27 Posterior belly of digastric	43 Vagus nerve
13 Jugular lymphatic trunk	28 Recurrent laryngeal nerve	44 Ventral ramus of third cervical nerve
14 Left brachiocephalic vein	29 Scalenus anterior	45 Vertebral artery
15 Left common carotid artery	30 Scalenus medius	46 Vertebral vein
16 Left subclavian artery	31 Spine of sphenoid bone	

Right trigeminal nerve branches,
from the left

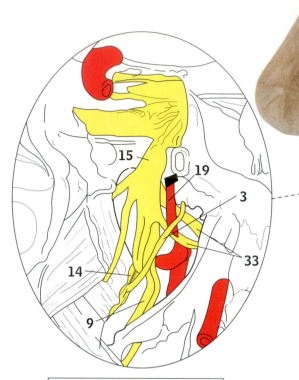

Hypoglossal nerve paralysis. Because all intrinsic and most extrinsic muscles of the tongue are supplied by the twelfth cranial nerve, its paralysis causes atrophy of the ipsilateral half of the tongue. On protrusion the tip of the tongue deviates towards the paralysed, injured side, owing to the unapposed action of the genioglossus muscle of the other side.

1 Abducent nerve
2 Body of hyoid bone
3 Chorda tympani
4 External carotid artery
5 Genioglossus
6 Geniohyoid
7 Hyoglossus
8 Hypoglossal nerve
9 Inferior alveolar nerve
10 Inferior nasal concha
11 Internal carotid artery
12 Jugular bulb
13 Lingual artery
14 Lingual nerve
15 Mandibular branch of trigeminal nerve
16 Marker in auditory tube
17 Maxillary branch of trigeminal nerve
18 Medial pterygoid
19 Middle meningeal artery
20 Middle nasal concha
21 Motor root of trigeminal nerve
22 Mylohyoid
23 Nasal septum (cartilaginous part)

24 Nerve to medial pterygoid
25 Nerve to mylohyoid
26 Oculomotor nerve
27 Ophthalmic branch of trigeminal nerve
28 Optic nerve
29 Parotid gland
30 Petrous part of temporal bone
31 Pons
32 Posterior belly of digastric
33 Roots of auriculotemporal nerve
34 Sphenomandibular ligament and maxillary artery
35 Stylohyoid ligament
36 Sublingual gland
37 Submandibular duct
38 Submandibular ganglion
39 Submandibular gland
40 Superior nasal concha
41 Supreme nasal concha
42 Tensor veli palatini
43 Trigeminal ganglion
44 Trigeminal nerve
45 Trochlear nerve

- The occasional supreme (highest) nasal concha (41) is present in this specimen.
- The chorda tympani (3) leaves the skull through the petrotympanic fissure and joins the posterior aspect of the lingual nerve (14) about 2 cm below the skull.
- The right cavernous sinus has been opened up from the medial side, so revealing from this aspect the nerves that lie in the sinus – the ophthalmic (27), maxillary (17), oculomotor (26), trochlear (45) and abducent (1).
- The lower end of the lingual nerve (14, in the mouth) is seen hooking under the submandibular duct (37), lying first lateral to the duct and then medial to it.
- The mandibular nerve (15) is labelled just after it has passed through the foramen ovale and where it divides into its various branches. Note the inferior alveolar nerve (9) entering the mandibular foramen after giving off the nerve to mylohyoid (25), with the lingual nerve (14) anterior to it and being joined by the chorda tympani (3).
- The middle meningeal artery (19) runs upwards between the two roots of the auriculotemporal nerve (33) to reach the foramen spinosum.

Sagittal section of the head, **A** right half, from the left, **B** MR (magnetic resonance) image.

1 Anterior arch of atlas
2 Anterior cerebral artery
3 Arachnoid granulations
4 Cerebellomedullary cistern (cisterna magna)
5 Cerebellum
6 Choana (posterior nasal aperture)
7 Corpus callosum
8 Dens of axis
9 Epiglottis
10 Falx cerebri
11 Fourth ventricle
12 Great cerebral vein
13 Hard palate
14 Hyoid bone
15 Inlet of larynx
16 Intervertebral disc between axis and third cervical vertebra
17 Laryngeal part of pharynx
18 Left ethmoidal air cells
19 Left frontal sinus
20 Mandible
21 Margin of foramen magnum
22 Medial surface of right cerebral hemisphere
23 Medulla oblongata
24 Midbrain
25 Nasal septum (bony part)
26 Nasopharynx
27 Opening of auditory tube
28 Optic chiasma
29 Oral part of pharynx (oropharynx)
30 Pharyngeal (nasopharyngeal) tonsil
31 Pituitary gland
32 Pons
33 Posterior arch of atlas
34 Soft palate
35 Sphenoidal sinus
36 Spinal cord
37 Straight sinus
38 Superior sagittal sinus
39 Tentorium cerebelli
40 Thyroid cartilage
41 Tongue
42 Vallecula

• The falx cerebri (10) separates the two cerebral hemispheres. The tentorium cerebelli (39) separates the posterior parts of the cerebral hemispheres from the cerebellum (5).

Adenoid enlargement. The adenoids, properly known as the nasopharyngeal tonsils, lie on the posterior nasopharyngeal wall near the openings of the auditory tubes. If repeatedly swollen they may cause prolonged otitis media and need to be removed to allow drainage of the middle ear by the auditory tube.

Middle ear pressure equalization. The auditory tube is normally closed and opens to atmospheric pressure momentarily during swallowing, allowing equilibration of pressures between the middle ear and the atmosphere. Changes in external pressures (in an aeroplane or when diving) require pressure equilibration to avoid the severe pain of a stretched tympanic membrane.

Nasogastric intubation involves passing a small plastic tube via the nose and nasopharynx into the stomach and may be used to obtain gastric secretions for analysis or to remove an accumulation of gastric secretions when the bowel is obstructed.

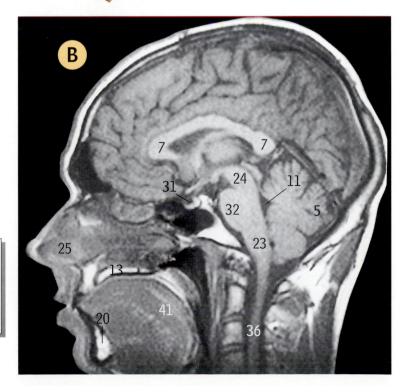

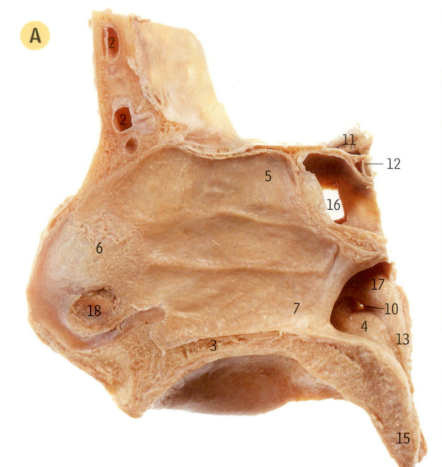

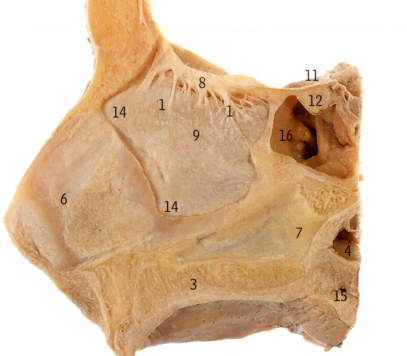

Nose, in sagittal section, from the left, **A, B** with the nasal septum intact, **D, E** with the nasal septum removed

1	Dural covering, olfactory fibres
2	Frontal sinus
3	Hard palate
4	Nasal part of pharynx (nasopharynx)
5	Nasal septum: perpendicular plate of ethmoid
6	Nasal septum: septal cartilage
7	Nasal septum: vomer
8	Olfactory bulb
9	Olfactory epithelium
10	Opening of auditory tube
11	Optic nerve
12	Pituitary gland
13	Salpingopharyngeal fold
14	Septal bone, cut edge
15	Soft palate
16	Sphenoidal sinus
17	Tubal elevation
18	Vestibule

Palatoglossal folds cover the palatoglossus muscles, travel between the palate and the tongue anterior to the palatine tonsils and mark the division between the mouth and oropharynx.

Epistaxis. Nose bleeds are most commonly found on the anteromedial septum (Little's area), a site of rich anastomoses (Kiesselbach) from external and internal carotid branches (facial, palatine and ophthalmic).

Palatine tonsils
The pits on the medial surfaces of these operation specimens from a child aged 14 years are the openings of the tonsillar crypts. The arrows indicate the intratonsillar clefts (the remains of the embryonic second pharyngeal pouch)

- The palatine tonsils (commonly called 'the tonsils') are masses of lymphoid tissue that are frequently enlarged in childhood but become much reduced in size in later life. Together with the lymphoid tissue in the posterior part of the tongue (lingual tonsil) and in the posterior wall of the nasopharynx (pharyngeal tonsil) and the tubal tonsil they form a protective 'ring' of lymphoid tissue (Waldeyer's ring) at the upper end of the respiratory and alimentary tracts.

Tonsillitis. This common condition results from infection of the palatine tonsils (which lie between the palatoglossal and palatopharyngeal folds). The inflamed cherry-like tonsil may be packed with pus, particularly in the crypts of this lymphoid mass, or even develop into an abscess (quinsy). Pain from this condition is often referred to the ear via the glossopharyngeal nerve which lies in the tonsillar bed (the ninth cranial nerve also supplies sensation to the middle ear). Swallowing is also made difficult. Haemorrhage after surgical removal may be severe from the tonsillar branch of the facial artery or from the external palatine vein.

D Lateral wall of the right nasal cavity

1 Anterior arch of atlas
2 Clivus
3 Cut edge of nasal concha
4 Dens of axis
5 Ethmoidal bulla
6 Ethmoidal infundibulum
7 Inferior meatus
8 Inferior nasal concha
9 Middle meatus
10 Opening of anterior ethmoidal air cells
11 Opening of auditory tube
12 Opening of maxillary sinus
13 Opening of nasolacrimal duct
14 Pituitary gland
15 Semilunar hiatus
16 Sphenoethmoidal recess
17 Sphenoidal sinus
18 Superior meatus
19 Superior nasal concha
20 Vestibule

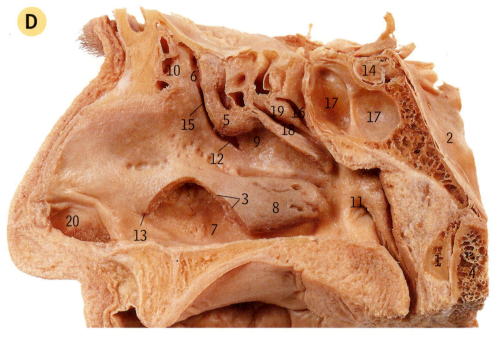

E Right nasal cavity and pterygopalatine ganglion, from the left

1 Abducent nerve
2 Clivus
3 Cribriform plate of ethmoid
4 Ethmoidal air cell (anterior)
5 Frontal sinus
6 Greater palatine nerve
7 Incisive foramen
8 Inferior nasal concha, cut edge of mucoperiosteum
9 Lesser palatine nerves
10 Middle nasal concha, cut
11 Nerve of pterygoid canal
12 Olfactory nerve fibres
13 Opening of auditory tube
14 Optic nerve
15 Pharyngeal branch to ganglion
16 Premaxilla
17 Pterygopalatine ganglion
18 Trigeminal nerve
19 Vertical plate of ethmoid
20 Vestibule

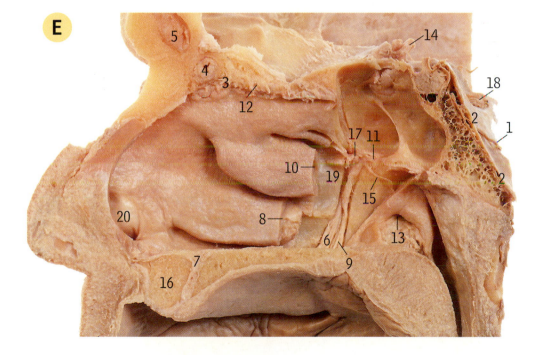

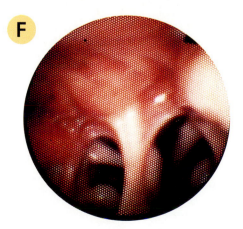

F Endoscopic view of the posterior nasal apertures (choanae) of the nose
The nasal septum is seen in the midline, with the conchae projecting from the lateral wall into the nasal cavity on each side.

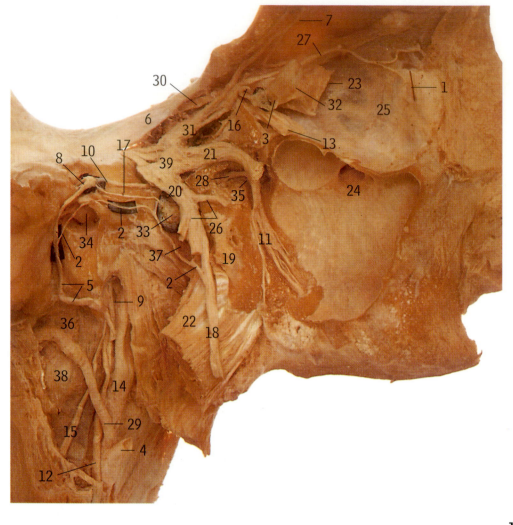

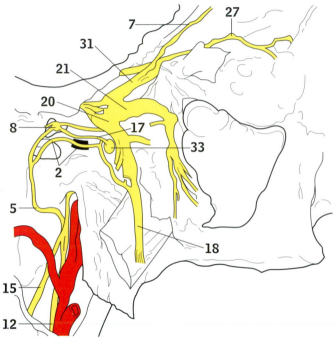

Right trigeminal, facial and petrosal nerves, with associated ganglia
Viewed from the right, much of the right side of the skull has been removed leaving the medial sides of the right orbit (25) and the maxillary sinus (24). Posterior to the sinus are seen the three branches of the trigeminal nerve: ophthalmic (31), maxillary (21) and mandibular (20)

1	Bristle in lacrimal canaliculus	21	Maxillary nerve
2	Chorda tympani	22	Medial pterygoid
3	Ciliary ganglion	23	Medial rectus
4	External carotid artery	24	Medial wall of maxillary sinus and ostium
5	Facial nerve		
6	Free margin of tentorium cerebelli	25	Medial wall of orbit
7	Frontal nerve	26	Muscular branches of mandibular nerve
8	Genicular ganglion of facial nerve		
9	Glossopharyngeal nerve	27	Nasociliary nerve
10	Greater petrosal nerve	28	Nerve of pterygoid canal
11	Greater and lesser palatine nerves	29	Occipital artery
12	Hypoglossal nerve	30	Oculomotor nerve
13	Inferior rectus	31	Ophthalmic nerve
14	Internal carotid artery	32	Optic nerve
15	Internal jugular vein and accessory nerve	33	Otic ganglion
		34	Position of tympanic membrane
16	Lacrimal nerve	35	Pterygopalatine ganglion
17	Lesser petrosal nerve	36	Rectus capitis lateralis
18	Lingual nerve	37	Tensor veli palatini
19	Lower head of lateral pterygoid and lateral pterygoid plate	38	Transverse process of atlas
		39	Trigeminal ganglion
20	Mandibular nerve		

- The greater petrosal nerve (10) is a branch of the geniculate ganglion of the facial nerve (8) and can be remembered as the nerve of tear secretion (though it also supplies nasal glands). It carries preganglionic fibres from the superior salivary nucleus in the pons, and runs in the groove on the floor of the middle cranial fossa (page 11, B25) to enter the foramen lacerum and become the nerve of the pterygoid canal (28) which joins the pterygopalatine ganglion (35). Postganglionic fibres leave the ganglion to join the maxillary nerve and enter the orbit by the zygomatic branch which communicates with the lacrimal nerve, supplying the gland (page 54, D9).

- The lesser petrosal nerve (17), although having a communication with the facial nerve, is a branch of the glossopharyngeal nerve, being derived from the tympanic branch which supplies the mucous membrane of the middle ear by the tympanic plexus (page 56, B19). Its fibres are derived from the inferior salivary nucleus in the pons, and after leaving the middle ear and running in its groove on the floor of the middle cranial fossa (17, and page 11, B26), the nerve reaches the otic ganglion (33) via the foramen ovale. From the ganglion secretomotor fibres join the mandibular nerve (20) to be distributed to the parotid gland by filaments from the auriculotemporal nerve.

- The chorda tympani (2) arises from the facial nerve before the latter leaves the stylomastoid foramen (5, upper leader line). It crosses the upper part of the tympanic membrane (34) underneath its mucosal covering and runs through the temporal bone, emerging from the petrotympanic fissure (page 9, C35) to join the lingual nerve (18). It carries preganglionic fibres to the submandibular ganglion (page 38, A39) for the submandibular and sublingual salivary glands, and also taste fibres for the anterior two-thirds of the tongue.

- The otic ganglion (33), which normally adheres to the deep surface of the mandibular nerve (20), has been teased off from the nerve and a black marker has been placed behind it.

Pharynx, from behind

The back of the pharynx has been exposed by removing the vertebral column and prevertebral muscles. On the left only the uppermost parts of the main vessels and nerves have been preserved; the glossopharyngeal nerve (8) is seen winding around stylopharyngeus (27). On the right the internal carotid artery (12) has been displaced slightly laterally to show the pharyngeal branches of the glossopharyngeal and vagus nerves (21 and 22) which form the pharyngeal nerve plexus on the surface of the middle constrictor (17). On the left the accessory nerve (1) passes behind the internal jugular vein (13) – its usual relationship; on the right it runs in front of the vein, a common variation. Both pairs of parathyroid glands (30 and 10) are seen behind the lateral lobes of the thyroid gland (15). Compare the dissection with the MR image on page 46, A

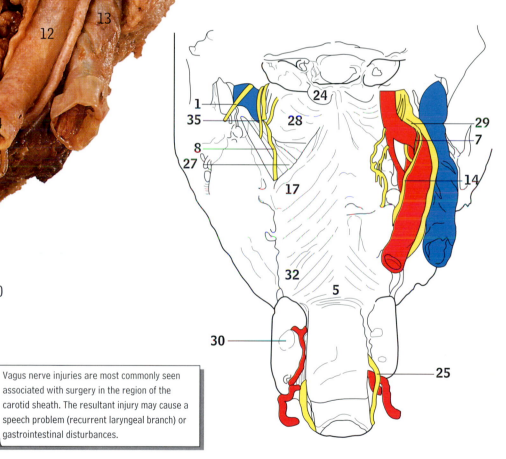

Vagus nerve injuries are most commonly seen associated with surgery in the region of the carotid sheath. The resultant injury may cause a speech problem (recurrent laryngeal branch) or gastrointestinal disturbances.

1 Accessory nerve	13 Internal jugular vein	24 Pharyngobasilar fascia
2 Ascending pharyngeal artery	14 Internal laryngeal nerve	25 Recurrent laryngeal nerve
3 Branches of glossopharyngeal and vagus nerves to carotid body and carotid sinus	15 Lateral lobe of thyroid gland	26 Spinal cord
	16 Margin of foramen magnum	27 Stylopharyngeus
4 Common carotid artery	17 Middle constrictor	28 Superior constrictor
5 Cricopharyngeal part of inferior constrictor	18 Occipital condyle	29 Superior laryngeal nerve
6 External carotid artery	19 Oesophagus	30 Superior parathyroid gland
7 External laryngeal nerve	20 Part of buccopharyngeal fascia and pharyngeal venous plexus	31 Sympathetic trunk
8 Glossopharyngeal nerve		32 Thyropharyngeal part of inferior constrictor
9 Hypoglossal nerve	21 Pharyngeal plexus: pharyngeal branch of glossopharyngeal nerve	33 Tip of greater horn of hyoid bone
10 Inferior parathyroid gland		34 Trachea
11 Inferior thyroid artery	22 Pharyngeal plexus: pharyngeal branch of vagus nerve	35 Vagus (inferior ganglion)
12 Internal carotid artery	23 Pharyngeal raphe	

Pharynx, A coronal MR image

1 Common carotid artery
2 Cricopharyngeal part of inferior constrictor
3 Inferior thyroid artery
4 Internal carotid artery
5 Middle constrictor
6 Oesophagus
7 Thyropharyngeal part of inferior constrictor

- The pharynx extends from the base of the skull to the level of the sixth cervical vertebra, and consists of nasal, oral and pharyngeal parts whose internal features (detailed below) are best seen in a sagittal section.
- The nasal part (nasopharynx), as far down as the lower border of the soft palate, contains the openings of the auditory tubes, the pharyngeal tonsil and pharyngeal recesses, and opens anteriorly into the nasal cavity.
- The oral part (oropharynx), between the soft palate and the upper border of the epiglottis, contains the (palatine) tonsils and the palatopharyngeal arch and opens anteriorly into the mouth. The palatoglossal arch is the boundary between the mouth and the oropharynx.
- The laryngeal part (laryngopharynx) below the upper border of the epiglottis, contains the piriform recess on either side of the larynx, which bulges backwards into the pharynx with the laryngeal inlet below and behind the epiglottis.
- The lower end of the pharynx becomes continuous with the oesophagus, at the same level (opposite the sixth cervical vertebra) as the larynx continues as the trachea.

> Gag reflex. Stimulation of the posterior third of the tongue or posterior oropharynx sends afferent stimuli via the glossopharyngeal (ninth cranial) nerve. The efferent pathway involves the vagus (tenth cranial) and accessory (eleventh cranial) nerves, causing elevation of the soft palate and contraction of the pharyngeal muscles. Putting a spatula against the back of the mouth will normally elicit this reflex and thus test three different cranial nerves.

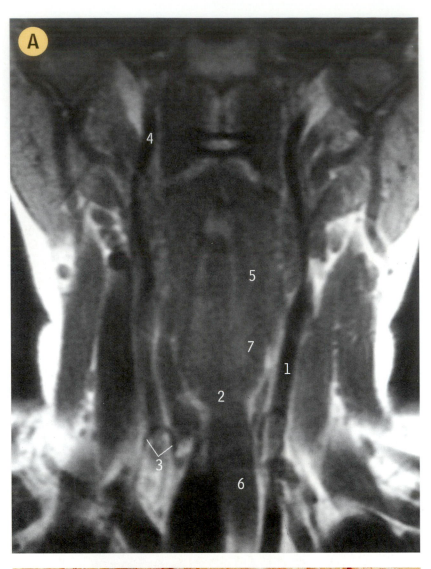

B The posterior surface of the pharynx

1 Accessory nerve	18 Pharyngeal veins
2 Ascending pharyngeal artery	19 Pharyngobasilar fascia
3 Attachment of pharyngeal raphe to pharyngeal tubercle of base of skull	20 Posterior meningeal artery
4 Carotid sinus	21 Stylopharyngeus
5 Common carotid artery	22 Superior cervical sympathetic ganglion
6 Cricopharyngeal part of inferior constrictor	23 Superior constrictor
7 External laryngeal nerve	24 Superior laryngeal branch of vagus nerve
8 Glossopharyngeal nerve	25 Superior thyroid artery
9 Hypoglossal nerve	26 Sympathetic trunk
10 Inferior ganglion of vagus nerve	27 Thyropharyngeal part of inferior constrictor
11 Internal carotid artery	28 Tip of greater horn of hyoid bone
12 Internal jugular vein	29 Upper border of inferior constrictor
13 Internal laryngeal nerve	30 Upper border of middle constrictor
14 Lateral lobe of thyroid gland	31 Upper border of superior constrictor
15 Middle constrictor	32 Vagal branch to carotid body
16 Pharyngeal branch of glossopharyngeal nerve	33 Vagus nerve
17 Pharyngeal branch of vagus nerve	

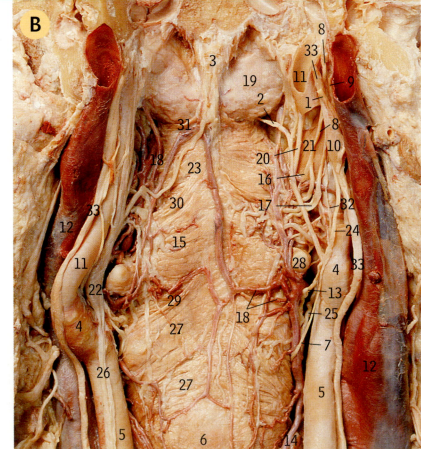

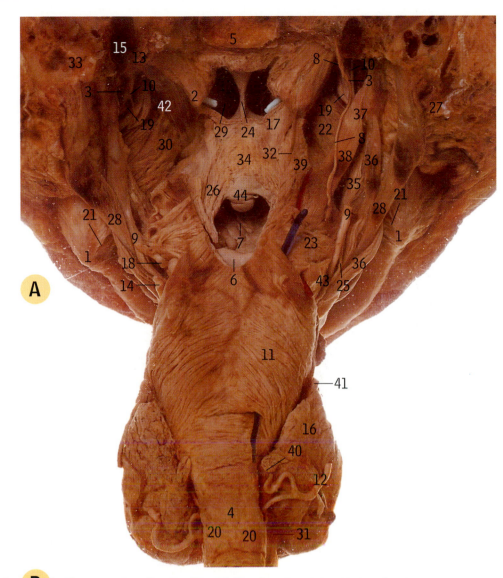

A Pharynx, from behind
By removing the upper part of the posterior pharyngeal wall, the posterior nasal apertures (29) can be seen above the back of the soft palate (34), while below it are the back of the tongue (7) and the tip of the epiglottis (6), seen through the arch formed by the two palatopharyngeus muscles (26) and the central uvula (44)

1	Angle of mandible	23	Middle constrictor (overlying red marker)
2	Cartilaginous part of auditory tube (marker in opening)	24	Nasal septum (vomer)
3	Chorda tympani	25	Nerve to thyrohyoid
4	Circular muscle of oesophagus	26	Palatopharyngeus
		27	Parotid gland
5	Clivus	28	Posterior belly of digastric
6	Epiglottis	29	Posterior nasal aperture (choana)
7	Foramen caecum in dorsum of tongue	30	Pterygoid hamulus
8	Glossopharyngeal nerve	31	Recurrent laryngeal nerve
9	Hypoglossal nerve	32	Salpingopharyngeus
10	Inferior alveolar nerve	33	Sigmoid sinus
11	Inferior constrictor (overlying blue marker)	34	Soft palate
		35	Styloglossus
12	Inferior thyroid artery	36	Stylohyoid
13	Internal carotid artery	37	Styloid process
14	Internal laryngeal nerve	38	Stylopharyngeus
15	Jugular bulb	39	Superior constrictor (cut edge)
16	Lateral lobe of thyroid gland	40	Superior parathyroid gland
17	Levator veli palatini	41	Superior thyroid artery
18	Lingual artery	42	Tensor veli palatini
19	Lingual nerve	43	Tip of greater horn of hyoid bone
20	Longitudinal muscle of oesophagus		
		44	Uvula
21	Masseter		
22	Medial pterygoid		

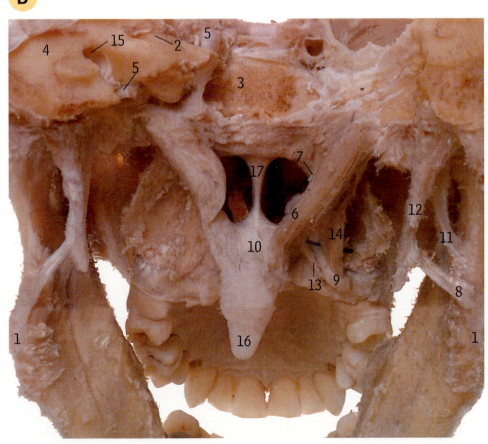

B Soft palate, from behind
Part of the base of the skull together with the pharynx and most other soft parts have been removed, leaving the central part of the soft palate (10)

1	Angle of mandible	9	Pterygoid hamulus
2	Apex of petrous part of temporal bone	10	Soft palate
		11	Sphenomandibular ligament
3	Clivus	12	Styloid process
4	Groove for sigmoid sinus	13	Tendon of tensor veli palatini
5	Internal carotid artery	14	Tensor veli palatini
6	Levator veli palatini	15	Tympanic membrane
7	Marker in auditory tube	16	Uvula
8	Part of stylomandibular ligament	17	Vomer (nasal septum)

- All the muscles of the pharynx and soft palate are supplied by the cranial part of the accessory nerve through the branches of the vagus that join the pharyngeal plexus, except for the stylopharyngeus which is supplied by the glosspharyngeal nerve and the tensor veli palatini by a branch from the nerve to the medial pterygoid muscle (mandibular nerve).
- Levator veli palatini (6), formerly called levator palati, is a short round muscle; tensor veli palatini (14), formerly called tensor palati, is a flat triangular muscle ending in a tendon (13) which hooks round the pterygoid hamulus (9) and then expands, to become with its fellow of the opposite side, the palatine aponeurosis

Hyoid bone, A from above and in front, **B** with muscle attachments

1 Body
2 Genioglossus
3 Geniohyoid
4 Greater horn
5 Hyoglossus
6 Lesser horn
7 Middle constrictor
8 Mylohyoid
9 Omohyoid
10 Sternohyoid
11 Stylohyoid
12 Stylohyoid ligament
13 Thyrohyoid

F Arytenoid cartilages, from behind

1 Apex
2 Articular surface for cricoid cartilage
3 Muscular process
4 Vocal process

Cricoid cartilage and muscle attachments, G from behind and below, **H** from the right

1 Arch
2 Articular surface for arytenoid cartilage
3 Articular surface for inferior horn of thyroid cartilage
4 Cricothyroid
5 Inferior constrictor
6 Lamina
7 Posterior crico-arytenoid
8 Tendon of oesophagus

C Cartilage of the epiglottis, from the front. **D** Thyroid cartilage from the front, **E** from the right with attachments

1 Cricothyroid
2 Inferior constrictor
3 Inferior horn
4 Inferior tubercle
5 Lamina
6 Laryngeal prominence (Adam's apple)
7 Sternothyroid
8 Superior horn
9 Superior tubercle
10 Thyrohyoid
11 Thyroid notch

K Tongue and the inlet of the larynx, from above

1 Corniculate cartilage in aryepiglottic fold
2 Cuneiform cartilage in aryepiglottic fold
3 Epiglottis
4 Foramen caecum
5 Fungiform papilla
6 Lateral glosso-epiglottic fold
7 Median glosso-epiglottic fold
8 Pharyngeal part of dorsum of tongue
9 Posterior wall of pharynx
10 Sulcus terminalis
11 Vallate papilla
12 Vallecula
13 Vestibular fold
14 Vocal fold

- The V-shaped sulcus terminalis (10), behind the row of vallate papillae (11), is not well marked in this tongue.

J Larynx, from behind
The left lamina of the thyroid cartilage has been reflected forwards and a glass rod (seen below the label 8 on the epiglottis) holds the pharynx open. Black markers underlie filaments from the recurrent and internal laryngeal nerves (3 and 1)

1 Anastomosis between internal and recurrent laryngeal nerves
2 Branch of internal laryngeal nerve
3 Branches of recurrent laryngeal nerve
4 Circular fibres of oesophagus
5 Corniculate cartilage in aryepiglottic fold
6 Cricothyroid muscle (reflected forwards with lamina of thyroid cartilage)
7 Cuneiform cartilage in aryepiglottic fold
8 Epiglottis
9 Posterior pharyngeal wall
10 Tendon of oesophagus
11 Transverse arytenoid muscle

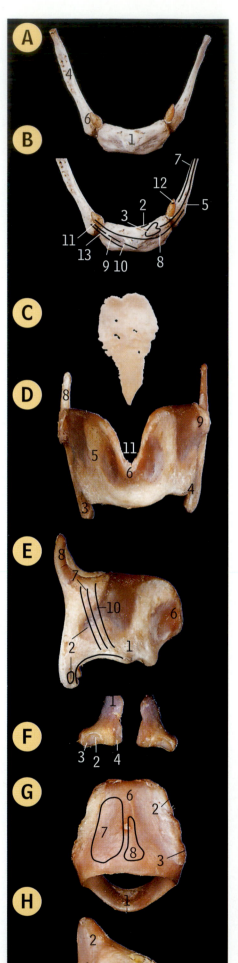

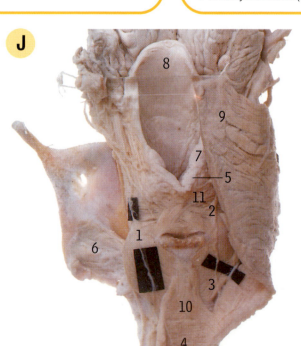

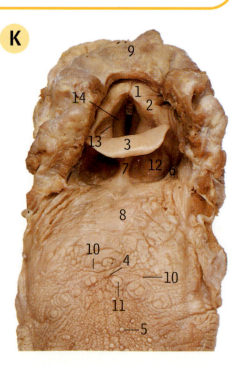

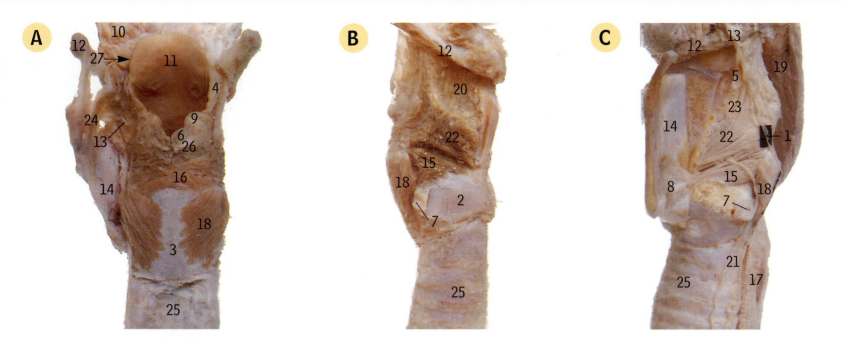

Intrinsic muscles of the larynx, A from behind, **B** from the right,
C from the left
In B the right lamina of the thyroid cartilage has been removed, and in C
part of the thyroid lamina has been turned forwards

1 Anastomosis of internal and recurrent laryngeal nerves	13 Internal laryngeal nerve
2 Arch of cricoid cartilage	14 Lamina of thyroid cartilage
3 Area on lamina of cricoid cartilage for tendon of oesophagus	15 Lateral crico-arytenoid muscle
	16 Oblique arytenoid muscle
4 Aryepiglottic fold	17 Oesophagus
5 Aryepiglottic muscle	18 Posterior crico-arytenoid muscle
6 Corniculate cartilage	19 Posterior wall of pharynx
7 Cricothyroid joint	20 Quadrangular membrane
8 Cricothyroid muscle (reflected from cricoid attachment)	21 Recurrent laryngeal nerve
	22 Thyro-arytenoid muscle
9 Cuneiform cartilage	23 Thyro-epiglottic muscle
10 Dorsum of tongue	24 Thyrohyoid membrane
11 Epiglottis	25 Trachea
12 Greater horn of hyoid bone	26 Transverse arytenoid muscle
	27 Vallecula

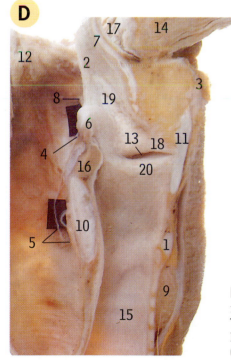

Recurrent laryngeal nerve damage.
A complication of thyroid surgery, this condition
causes paralysis of the vocal cords. When the
paralysis is bilateral the voice is almost absent
as the two vocal folds cannot be adducted. A
unilateral recurrent laryngeal nerve injury may
not be detected in normal speech.

D Larynx, in sagittal section,
from the right.
The vocal fold (vocal cord, 20)
lies below the vestibular fold
(false vocal cord, 18)

1 Arch of cricoid cartilage	9 Isthmus of thyroid gland
2 Aryepiglottic fold and inlet of larynx	10 Lamina of cricoid cartilage
3 Body of hyoid bone	11 Lamina of thyroid cartilage
4 Branches of internal laryngeal nerve anastomosing with recurrent laryngeal nerve	12 Pharyngeal wall
	13 Sinus of larynx
	14 Tongue
5 Branches of recurrent laryngeal nerve	15 Trachea
	16 Transverse arytenoid muscle
6 Corniculate cartilage and apex of arytenoid cartilage	17 Vallecula
	18 Vestibular fold
7 Epiglottis	19 Vestibule of larynx
8 Internal laryngeal nerve entering piriform recess	20 Vocal fold

- The space between the vestibular and vocal folds is the sinus of the larynx (D13), and this is continuous with the saccule, a small pouch that extends upwards for a few millimetres between the vestibular fold and the inner surface of the thyro-arytenoid muscle (B22).
- The fissure between the two vestibular folds (D18) is the rima of the vestibule. The fissure between the vocal folds is the rima of the glottis.
- The vestibular folds are sometimes called the false vocal cords.
- The intrinsic muscles of the larynx are supplied by the recurrent laryngeal nerve (C21), except the cricothyroid (page 33, B6) which is supplied by the external laryngeal nerve (page 35, C14).

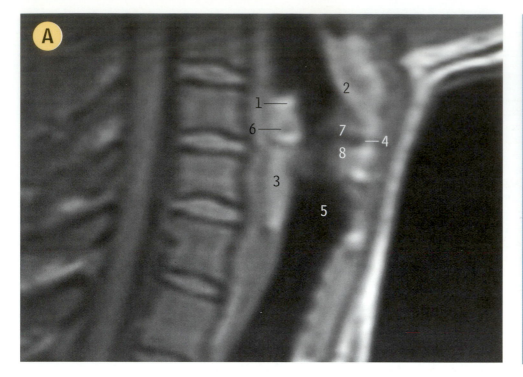

A MR image of the neck

> 1 Corniculate cartilage and apex of arytenoid cartilage
> 2 Epiglottis
> 3 Lamina of cricoid cartilage
> 4 Sinus of larynx
> 5 Trachea
> 6 Transverse arytenoid muscle
> 7 Vestibular fold
> 8 Vocal fold

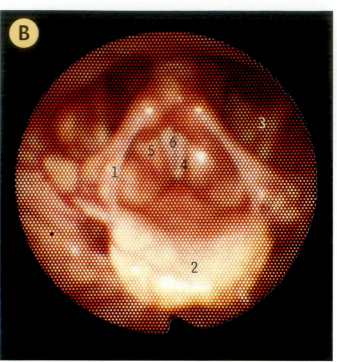

B Endoscopic view of the laryngeal inlet and vocal folds

> 1 Aryepiglottic fold
> 2 Epiglottis
> 3 Piriform recess of pharynx
> 4 Rima of glottis
> 5 Vestibular fold
> 6 Vocal fold

C Ligaments and membranes of the right side of the larynx, from the left
Most of the left side of the larynx has been removed but the whole of the cricoid cartilage remains intact

> 1 Apex of arytenoid cartilage
> 2 Arch of cricoid cartilage
> 3 Cricovocal membrane
> 4 Epiglottis
> 5 Hyo-epiglottic ligament
> 6 Hyoid bone
> 7 Lamina of cricoid cartilage
> 8 Lamina of thyroid cartilage
> 9 Quadrangular membrane
> 10 Thyro-epiglottic ligament
> 11 Vocal process of arytenoid cartilage

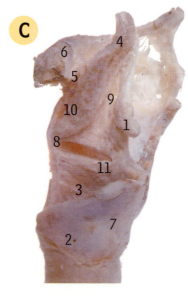

- The mucous membrane of the larynx above the level of the vocal folds is supplied by the internal laryngeal nerve, and below the vocal folds by the recurrent laryngeal nerve (D4 and 5 on page 49).
- The recurrent laryngeal nerve (C21 on page 49) enters the larynx by passing beneath the lower border of the inferior constrictor of the pharynx, and here it lies immediately behind the cricothyroid joint (C7 on page 49).
- The anterior part of the vocal fold (D20 on page 49; B6 above) is formed by the upper margin of the cricovocal membrane (C3), and the posterior part by the vocal process of the arytenoid cartilage (C11).
- The vestibular fold (false vocal cord, D18 on page 49; B5 above) is formed by the lower margin of the quadrangular membrane (C9), whose upper margin forms the aryepiglottic fold (A4 and D2 on page 49; B2 above).
- The central (anterior) part of the cricothyroid membrane is usually known as the conus elasticus but sometimes this term is used for the cricovocal membrane.

A Cerebral dura mater, outer surface

The right half of the cranial vault has been removed to show branches of the middle meningeal vessels on the outer surface of the dura, i.e. in the extradural space. These vessels do not supply the brain

Cavernous sinus thrombosis. Owing to the great number of structures passing through or in the lateral wall of the cavernous sinus, a blockage or infection in this region has serious consequences which may include damage to the third, fourth, fifth and sixth cranial nerves. Infections around the face and forehead may travel into the cavernous sinus.

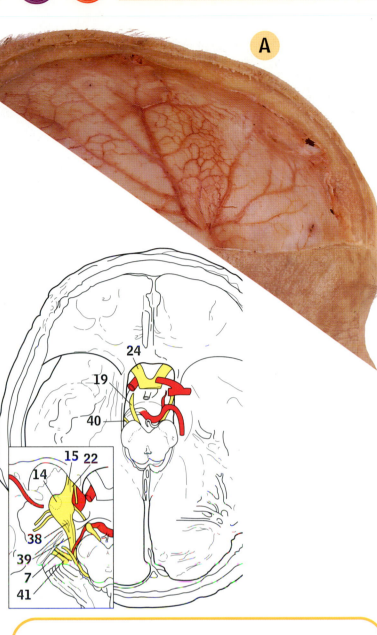

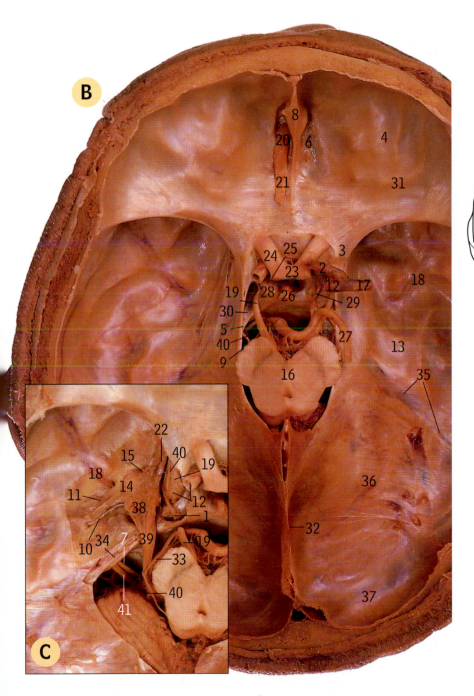

Cranial fossae, **B** with dura mater intact, **C** with some dura removed

1	Abducent nerve	22	Ophthalmic nerve
2	Anterior cerebral artery	23	Optic chiasma
3	Anterior clinoid process	24	Optic nerve
4	Anterior cranial fossa	25	Optic tract
5	Attached margin of tentorium cerebelli	26	Pituitary stalk
6	Cribriform plate of ethmoid bone	27	Posterior cerebral artery
7	Facial nerve	28	Posterior clinoid process
8	Falx cerebri attached to crista galli	29	Posterior communicating artery
9	Free margin of tentorium cerebelli	30	Roof of cavernous sinus
10	Hiatus for greater petrosal nerve	31	Sphenoparietal sinus (at posterior border of lesser wing of sphenoid bone)
11	Hiatus for lesser petrosal nerve	32	Straight sinus (at junction of falx cerebri and tentorium cerebelli)
12	Internal carotid artery	33	Superior cerebellar artery
13	Lateral part of middle cranial fossa	34	Superior petrosal sinus
14	Mandibular nerve	35	Superior petrosal sinus (at attached margin of tentorium cerebelli)
15	Maxillary nerve	36	Tentorium cerebelli
16	Midbrain (superior colliculus level)	37	Transverse sinus (at attached margin of tentorium cerebelli)
17	Middle cerebral artery	38	Trigeminal ganglion
18	Middle meningeal vessels	39	Trigeminal nerve
19	Oculomotor nerve	40	Trochlear nerve
20	Olfactory bulb	41	Vestibulocochlear nerve
21	Olfactory tract		

A Cerebral dura mater and cranial nerves

In this oblique view from the left and behind, the brain has been removed and a window has been cut in the posterior part of the falx cerebri (7) to show the upper surface of the tentorium cerebelli (29)

1 Abducent nerve
2 Arachnoid granulations
3 Attached margin of tentorium cerebelli
4 Choana (posterior nasal aperture)
5 Clivus
6 Dens of axis
7 Falx cerebri
8 Free margin of tentorium cerebelli
9 Glossopharyngeal, vagus and accessory nerves
10 Inferior sagittal sinus
11 Internal carotid artery
12 Margin of foramen magnum
13 Medulla oblongata
14 Motor root of facial nerve
15 Nasal septum
16 Oculomotor nerve
17 Olfactory tract
18 Optic nerve
19 Pituitary gland
20 Posterior arch of atlas
21 Rootlets of hypoglossal nerve
22 Sensory root (nervus intermedius) of facial nerve
23 Sphenoidal sinus
24 Sphenoparietal sinus
25 Spinal cord
26 Spinal part of accessory nerve
27 Straight sinus
28 Superior sagittal sinus
29 Tentorium cerebelli
30 Transverse sinus
31 Trigeminal nerve
32 Trochlear nerve
33 Vestibulocochlear nerve

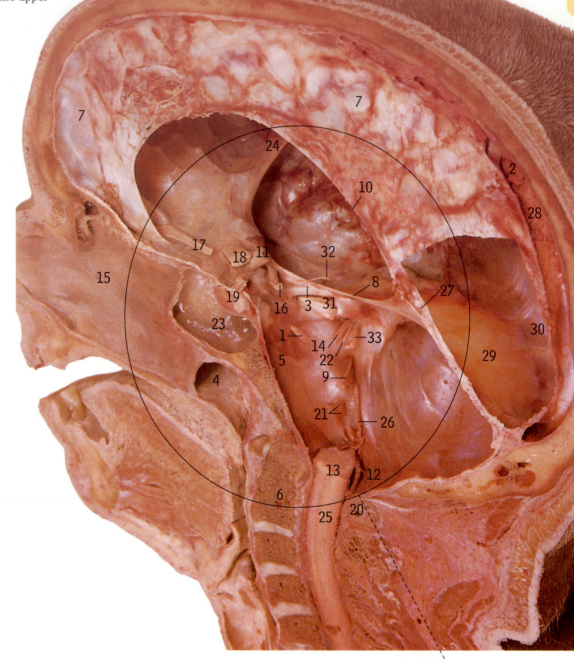

Cerebrospinal fluid rhinorrhoea. CSF dripping from the nose is most probably due to a traumatic tear of the olfactory nerve fibres as they pass through the cribriform plate of the ethmoid bone. A fracture of the ethmoid is most common in traffic accidents. To test whether the 'runny nose' is a cold or CSF, a simple dipstick will reveal a high glucose content in CSF. A late complication may be anosmia.

Subdural haemorrhages, being of venous origin, develop slowly following trauma but can have consequences as serious as those of extradural (arterial) haemorrhages.

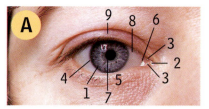

A Right eye. Surface features
With the eyelids in the normal open position, the lower margin of the upper lid (9) overlaps approximately the upper half of the iris (1); the margin of the lower lid (5) is level with the lower margin of the iris (1)

1 Iris behind cornea
2 Lacrimal caruncle
3 Lacrimal papilla
4 Limbus (corneoscleral junction)
5 Lower lid
6 Plica semilunaris
7 Pupil behind cornea
8 Sclera
9 Upper lid

- The cornea is the transparent anterior part of the outer coat of the eyeball and is continuous with the sclera (8) at the limbus (4).
- The pupil (7) is the central aperture of the iris (1), the circular pigmented diaphragm that lies in front of the lens.
- Each lacrimal papilla (3) contains the lacrimal punctum, the minute opening of the lacrimal canaliculus (page 55, C8) which runs medially to open into the lacrimal sac, lying deep to the medial palpebral ligament (page 55, C10) and continuing downwards as the nasolacrimal duct (page 55, C12) within the nasolacrimal canal.

Corneal reflex is closure of the eyelid after stimulation of the thin anterior transparent membrane of the eye and is controlled by two cranial nerves: the sensory component is via the ophthalmic division of the trigeminal nerve and the motor component (closing of the eye) by the facial.

Right extra-ocular muscles, B from above, C from the right
The upper and lateral walls of the orbit have been removed, together with all fat, vessels and nerves, leaving only the muscles

1 Anterior clinoid process
2 Ethmoidal air cells
3 Eyeball
4 Inferior oblique
5 Inferior rectus
6 Lateral rectus
7 Levator palpebrae superioris
8 Optic canal
9 Optic nerve
10 Pituitary fossa (sella turcica)
11 Posterior clinoid process
12 Superior oblique
13 Superior rectus
14 Tendinous ring
15 Tendon of superior oblique
16 Trochlea

B

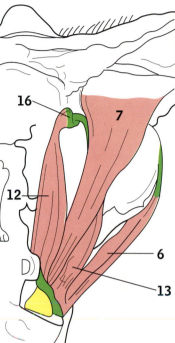

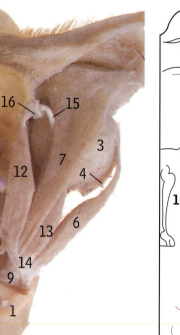

C

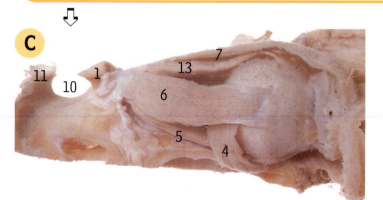

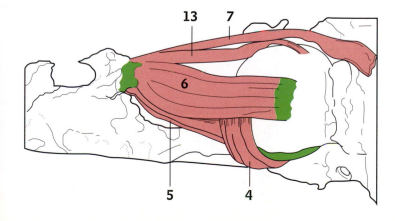

Trochlear nerve paralysis. This rare condition affects only the superior oblique muscle of the eye and commonly presents as a squint or with the patient complaining of diplopia on looking downwards or at the end of the nose.

Abducent nerve paralysis. The abducent nerve innervates only the lateral rectus muscle of the eye, and damage results in an inability to move the eye laterally in the horizontal plane. This symptom may be an indication of increased intracranial pressure, owing to the long intracranial course of the sixth nerve.

Oculomotor nerve paralysis. If complete, this condition affects most eye muscles and especially the levator palpebrae superioris and the sphincter pupillae. Consequently, the upper eyelid droops (ptosis), there is a fully dilated non-reactive pupil, and the eyeball tends to be looking downwards and outwards owing to the unopposed action of the lateral rectus and superior oblique muscles.

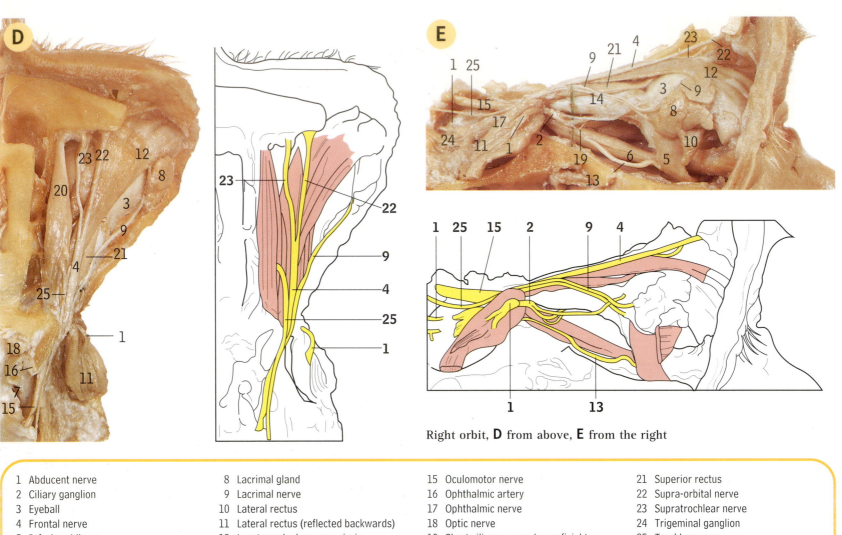

Right orbit, **D** from above, **E** from the right

1 Abducent nerve	8 Lacrimal gland	15 Oculomotor nerve	21 Superior rectus
2 Ciliary ganglion	9 Lacrimal nerve	16 Ophthalmic artery	22 Supra-orbital nerve
3 Eyeball	10 Lateral rectus	17 Ophthalmic nerve	23 Supratrochlear nerve
4 Frontal nerve	11 Lateral rectus (reflected backwards)	18 Optic nerve	24 Trigeminal ganglion
5 Inferior oblique	12 Levator palpebrae superioris	19 Short ciliary nerves (superficial to marker)	25 Trochlear nerve
6 Inferior rectus	13 Nerve to inferior oblique	20 Superior oblique	
7 Internal carotid artery	14 Nerve to medial rectus		

F Orbits, from above

Both orbits have been exposed from above, and most of levator palpebrae superioris (13) and the superior rectus (22) have been removed. On the right, as is usual, the ophthalmic artery (17) and

nasociliary nerve (16) cross above the optic nerve (19) from lateral to medial; on the left the artery has crossed below the nerve, which is uncommon. The supra-orbital artery (23) is unusually small on the left and is absent on the right

1 Anterior cerebral artery
2 Anterior communicating artery
3 Anterior ethmoidal artery and nerve
4 Cribriform plate of ethmoid bone
5 Eyeball
6 Frontal nerve
7 Infratrochlear nerve and ophthalmic artery
8 Internal carotid artery
9 Lacrimal artery
10 Lacrimal gland
11 Lacrimal nerve
12 Lateral rectus
13 Levator palpebrae superioris
14 Medial rectus
15 Middle cerebral artery
16 Nasociliary nerve
17 Ophthalmic artery
18 Optic chiasma
19 Optic nerve (with overlying short ciliary nerves in left orbit)
20 Posterior ciliary artery
21 Superior oblique
22 Superior rectus
23 Supra-orbital artery
24 Supra-orbital nerve
25 Supratrochlear nerve
26 Trochlear nerve

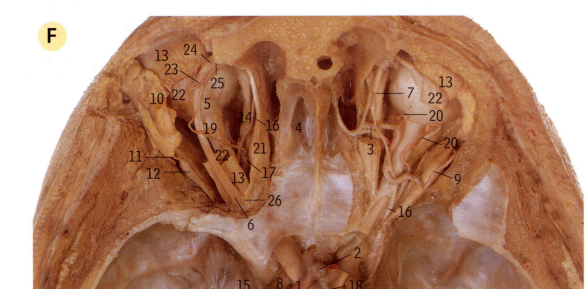

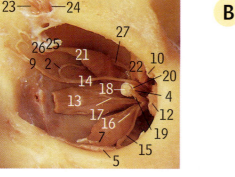

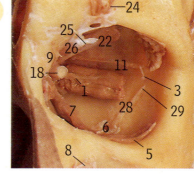

A The left orbit, from the front and left, **B** The left orbit, from the front and right

The views in A and B show muscles and nerves in relation to the orbital walls after removal of the eye. In A note the extension of the subarachnoid space (20) and the dural sheath (4) round the optic nerve (18).

1 Abducent nerve	10 Lacrimal gland	20 Subarachnoid space
2 Anterior ethmoidal nerve	11 Lacrimal nerve	21 Superior oblique
3 Communication between 11 and 28	12 Lateral rectus	22 Superior rectus
	13 Medial rectus	23 Supra-orbital artery
4 Dural sheath of optic nerve	14 Nasociliary nerve	24 Supra-orbital nerve
5 Inferior oblique	15 Nerve to inferior oblique	25 Tendon of superior oblique
6 Inferior orbital fissure	16 Nerve to inferior rectus	26 Trochlea
7 Inferior rectus	17 Nerve to medial rectus	27 Trochlear nerve
8 Infra-orbital nerve	18 Optic nerve	28 Zygomatic nerve
9 Infratrochlear nerve	19 Short ciliary nerves	29 Zygomatico-orbital foramen

> Central retinal artery occlusion is caused by a small thrombus or embolus within this branch of the ophthalmic artery. Blindness ensues unless there is immediate treatment.

> Pupillary reflex is constriction of the pupil on exposure of the retina to bright light. The sensory pathway is via the second cranial nerve; the motor pathway is from the parasympathetic fibres of the third cranial nerve that originate in the Edinger–Westphal nucleus. Other stimuli may also alter the size of the pupil; for example, emotional stimulation, fear and excitement cause dilatation of the pupil via the sympathetic fibres.

C The nasolacrimal duct

In C, the facial muscles and part of the skull have been dissected away to display the nasolacrimal duct (12) opening into the meatus of the nose (13)

D Macrodacryocystogram

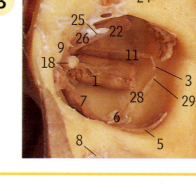

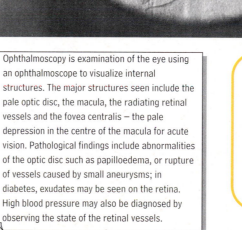

1 Aponeurosis of levator palpebrae superioris	11 Muscle fibres of levator palpebrae superioris
2 Cut edge of orbital septum and periosteum	12 Nasolacrimal duct
3 Dorsal nasal artery	13 Opening of nasolacrimal duct (anterior wall removed) in inferior meatus of nose
4 Inferior oblique	
5 Infra-orbital nerve	14 Orbital fat pad
6 Lacrimal gland	15 Supra-orbital artery
7 Lacrimal sac (upper extremity)	16 Supra-orbital nerve
8 Lower lacrimal canaliculus	17 Tendon of superior oblique
9 Lower lacrimal papilla and punctum	18 Trochlea
10 Medial palpebral ligament	

> Ophthalmoscopy is examination of the eye using an ophthalmoscope to visualize internal structures. The major structures seen include the pale optic disc, the macula, the radiating retinal vessels and the fovea centralis – the pale depression in the centre of the macula for acute vision. Pathological findings include abnormalities of the optic disc such as papilloedema, or rupture of vessels caused by small aneurysms; in diabetes, exudates may be seen on the retina. High blood pressure may also be diagnosed by observing the state of the retinal vessels.

1 Common canaliculus
2 Hard palate
3 Inferior canaliculus
4 Lacrimal catheters
5 Lacrimal sac
6 Nasolacrimal duct
7 Site of lacrimal punctum
8 Superior canaliculus

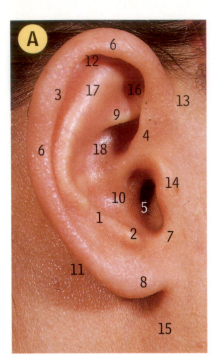

A Right external ear

1	Antihelix
2	Antitragus
3	Auricular tubercle
4	Crus of helix
5	External acoustic meatus
6	Helix
7	Intertragic notch
8	Lobule
9	Lower crus of antihelix
10	Lower part of concha
11	Mastoid process
12	Scaphoid fossa
13	Superficial temporal vessels and auriculotemporal nerve
14	Tragus
15	Transverse process of atlas
16	Triangular fossa
17	Upper crus of antihelix
18	Upper part of concha

Otalgia (referred pain). Pain from the ear itself is a common complaint but referred otalgia can be a diagnostic nightmare. Any structure that has the same nerve supply as the pinna or middle ear may have its pain referred to the ear. These include numerous branches of C2, C3 of the cervical plexus, and the fifth, seventh, ninth and tenth cranial nerves. Conditions that can cause this complaint include myocardial infarction, oesophagitis, tonsilitis, arthritis of the cervical spine, malocclusion, dental caries, sinusitis and carcinoma of the larynx or pharynx. A painful ear with a normal ear examination is therefore an anatomical challenge.

B Right temporal bone and ear

The bone has been bisected and opened out like a book, with some removal of the upper part of the petrous part. The section has opened up the tympanic (middle ear) cavity. On the left side of the figure the lateral wall of the middle ear, which includes the tympanic membrane (26), is seen from the medial side, while on the right the main features of the medial wall are in view.

Hyperacusis is a very acute sense of hearing (lowered threshold), most commonly caused by damage to the stapedius muscle (seventh cranial nerve) or the tensor tympani muscle (fifth cranial nerve). Occasionally this is a symptom associated with Bell's palsy.

1	Aditus to mastoid antrum	15	Mastoid antrum
2	Anterior semicircular canal	16	Mastoid process
3	Bony part of auditory tube	17	Part of carotid canal (red)
4	Canal for facial nerve (yellow)	18	Part of jugular bulb (blue)
5	Carotid canal (red)	19	Promontory with overlying tympanic plexus
6	Epitympanic recess	20	Stapes in oval window and stapedius muscle
7	Groove for greater petrosal nerve (yellow)	21	Styloid process
8	Groove for middle meningeal vessels	22	Stylomastoid foramen
9	Incus	23	Tegmen tympani
10	Jugular bulb (blue)	24	Tensor tympani muscle in its canal
11	Lateral semicircular canal	25	Tympanic branch of glossopharyngeal nerve entering its canaliculus
12	Lesser petrosal nerve	26	Tympanic membrane
13	Malleus		
14	Mastoid air cells		

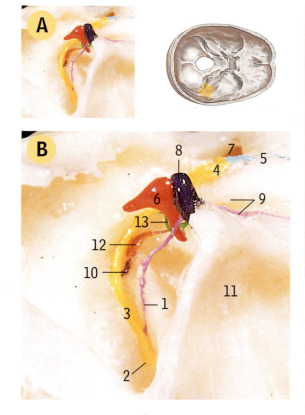

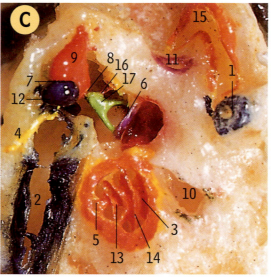

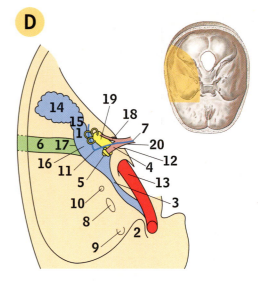

Right temporal bone. A Middle ear and the facial nerve and branches, B enlarged view of A
This dissection is seen from the right and above, looking forwards and medially. Bone has been removed to show the upper parts of the malleus (8) and incus (6), which normally project up into the epitympanic recess. The upper part of the facial canal (2) has been opened to show the facial nerve (3) giving off the chorda tympani (1) and the nerve to stapedius (10). The geniculate ganglion of the facial nerve (4) is seen giving off the greater petrosal nerve (5)

1 Chorda tympani
2 Facial canal leading to stylomastoid foramen
3 Facial nerve
4 Geniculate ganglion of facial nerve
5 Greater petrosal nerve
6 Incus
7 Internal acoustic meatus
8 Malleus
9 Margin of auditory tube
10 Nerve to stapedius
11 Paraffin wax (for support) overlying tympanic membrane
12 Stapedius
13 Stapes

• The stapedius (12) tendon emerges from a small conical projection on the posterior wall of the tympanic cavity, the pyramid (here dissected away).

C Right temporal bone. Middle ear and inner ear, enlarged
This dissection is viewed from above, looking slightly backwards and laterally. Within the cavity of the middle ear are the three auditory ossicles—malleus (12), incus (9) and stapes (17). The tympanic membrane and external acoustic meatus are not seen but lie below the label 7. The cochlea has been opened up to show its internal bony structure (3, 5, 13 and 14)

1 Anterior semicircular canal
2 Auditory tube
3 Bony canal of cochlea
4 Chorda tympani
5 Cupola of cochlea
6 Footplate of stapes in oval window of vestibule
7 Incudomallealar joint
8 Incudostapedial joint
9 Incus
10 Internal acoustic meatus
11 Lateral semicircular canal
12 Malleus
13 Modiolus of cochlea
14 Osseous spiral lamina of cochlea
15 Posterior semicircular canal
16 Stapedius muscle
17 Stapes

• The spiral organ (the end organ of hearing) lies on the basilar membrane, which stretches between the free edge of the osseous spiral lamina (14) and the side of the bony cochlear canal.
• The modiolus (13) is the central axis of the cochlea, and the cupola (5) is its apex.

Right ear, from above. D diagram of parts
The schematic diagram of part of the right side of the base of the skull (D) indicates the position of the parts of the ear within the temporal bone. (The auditory ossicles have been omitted from the middle ear cavity, 16.) The external acoustic meatus (6) is at a right angle to the side of the skull, and the internal acoustic meatus (12) is level with it on the inner side of the temporal bone. The line (from front to back) of the auditory tube (3), middle ear cavity (16), mastoid antrum (1 and 15) and mastoid air cells (14) lies at about 60° to the line of the external meatus. The cochlear part of the inner ear (5) is in front of the vestibular part (19). The facial nerve (7) runs immediately above the vestibulocochlear nerve (20) and takes a right-angled turn backwards at the geniculate ganglion (11) to pass below the lateral semicircular canal in the medial wall of the middle ear and then turns downwards in the medial wall of the aditus to the antrum (1) to reach the stylomastoid foramen

1 Aditus to mastoid antrum
2 Anterior clinoid process
3 Auditory tube
4 Cochlear nerve
5 Cochlear part of inner ear
6 External acoustic meatus
7 Facial nerve
8 Foramen ovale
9 Foramen rotundum
10 Foramen spinosum
11 Geniculate ganglion of facial nerve
12 Internal acoustic meatus
13 Internal carotid artery emerging from foramen lacerum
14 Mastoid air cells
15 Mastoid antrum
16 Middle ear
17 Tympanic membrane
18 Vestibular nerve
19 Vestibular part of inner ear
20 Vestibulocochlear nerve

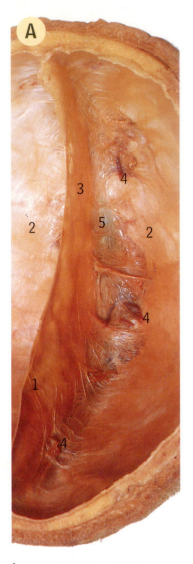

A Cranial vault and falx, from below
Looking up into the cranial vault from below, the falx cerebri (3) is seen to be continuous with the dura over the vault (2), and has been cut off at the back (1) from the tentorium cerebelli.

1 Cut edge of falx cerebri
2 Dura mater over cranial vault
3 Falx cerebri
4 Superior cerebral veins
5 Superior sagittal sinus

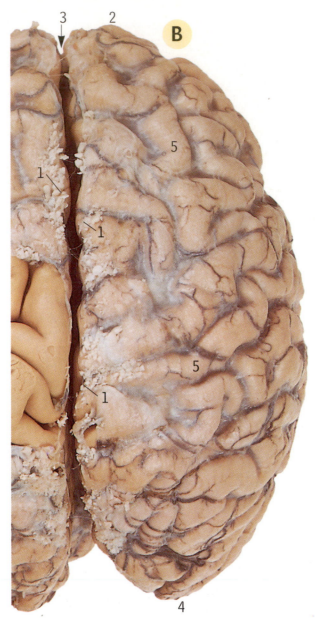

B Brain, from above
The right cerebral hemisphere is seen with the overlying arachnoid mater and arachnoid granulations (1) adjacent to the longitudinal fissure (3). Over the small part of the left hemisphere shown, a window has been cut in the arachnoid

1 Arachnoid granulations
2 Frontal pole
3 Longitudinal fissure
4 Occipital pole
5 Superolateral surface

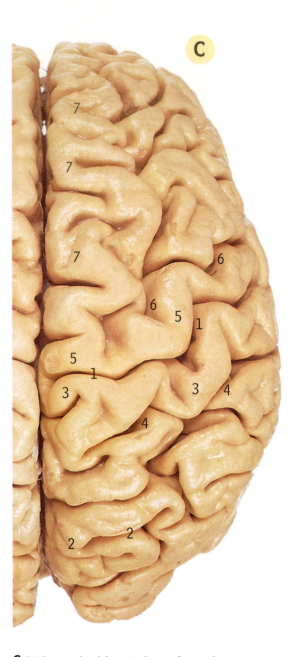

C Right cerebral hemisphere, from above
Removal of the arachnoid and the underlying vessels displays the gyri and sulci. Only a small number are named here; the most important are the central sulcus (1) and the precentral and postcentral gyri (5 and 3)

1 Central sulcus
2 Parieto-occipital sulcus
3 Postcentral gyrus
4 Postcentral sulcus
5 Precentral gyrus
6 Precentral sulcus
7 Superior frontal gyrus

Subarachnoid haemorrhage. Because the subarachnoid space contains the CSF, this condition may be diagnosed at lumbar puncture. It can be due to rupture of a congenital ('berry') aneurysm, often in the region of the 'Circle of Willis'.

A

B

B Right cerebral hemisphere, from the right
The arachnoid mater has been removed, leaving some of the larger branches of the middle cerebral artery (unlabelled) after they have emerged from the lateral sulcus (7). Only the main gyri and sulci are named here: the most important are the precentral and postcentral gyri (16 and 13) and the central and lateral sulci (3 and 7)

A Brain, from the right
As in B (page 58), the arachnoid mater has been left intact and vessels are seen beneath it; the larger ones are veins (as at 7)

1	Frontal pole
2	Inferior cerebral veins
3	Medulla oblongata and vertebral artery
4	Occipital pole
5	Pons and basilar artery
6	Right cerebellar hemisphere
7	Superficial middle cerebral vein overlying lateral sulcus
8	Superior cerebral veins
9	Superolateral surface of right cerebral hemispheres
10	Temporal pole

1	Anterior ramus of lateral sulcus	8	Lunate sulcus	16	Precentral gyrus
2	Ascending ramus of lateral sulcus	9	Middle frontal gyrus	17	Precentral sulcus
3	Central sulcus	10	Middle temporal gyrus	18	Superior frontal gyrus
4	Inferior frontal gyrus	11	Parieto-occipital sulcus	19	Superior temporal gyrus
5	Inferior temporal gyrus	12	Pars triangularis	20	Superior temporal sulcus
6	Inferior temporal sulcus	13	Postcentral gyrus	21	Supramarginal gyrus
7	Lateral sulcus (posterior ramus)	14	Postcentral sulcus		
		15	Pre-occipital notch		

- The brain consists of the forebrain (cerebrum, comprising the two cerebral hemispheres), the midbrain, and the hindbrain (comprising the pons, medulla oblongata and cerebellum).
- The midbrain, pons and medulla oblongata constitute the brainstem.
- The central sulcus (C1 (page 58) and B3 (above)) marks the boundary between the frontal and parietal lobes.

- An arbitrary line from the pre-occipital notch (B15) to the parieto-occipital sulcus (B11) marks the boundary between the parietal and occipital lobes, and the part of the hemisphere in front of this line and below the lateral sulcus (strictly, the posterior ramus of the lateral sulcus, B7) forms the temporal lobe.
- The precentral and postcentral gyri (B16 and 13) contain the classically described 'motor' and

'sensory' areas of the cortex.
- The motor speech areas (usually in the left cerebral hemisphere) are in the region of the ascending and anterior rami of the lateral sulcus and the pars triangularis (B2, 1 and 12).
- The auditory areas of the cortex probably comprise parts of the superior temporal gyrus (B19), especially the upper surface of it within the lateral sulcus (B7).

A

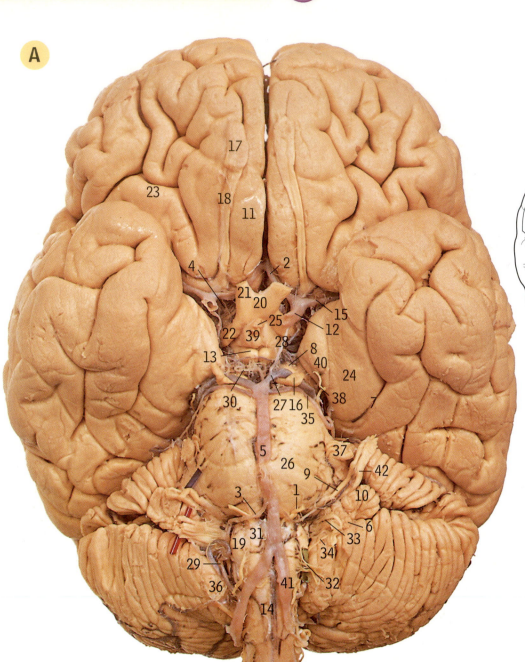

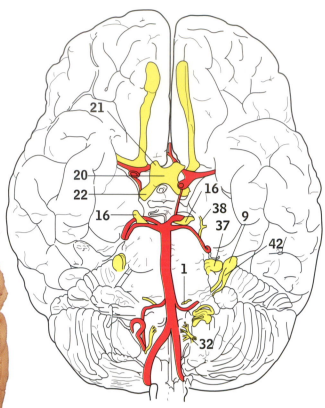

A Brain, from below

1	Abducent nerve	16	Oculomotor nerve
2	Anterior cerebral artery	17	Olfactory bulb
3	Anterior inferior cerebellar artery	18	Olfactory tract
4	Anterior perforated substance	19	Olive of medulla oblongata
5	Basilar artery	20	Optic chiasma
6	Choroid plexus from lateral recess of fourth ventricle	21	Optic nerve
7	Collateral sulcus	22	Optic tract
8	Crus of cerebral peduncle	23	Orbital sulcus
9	Facial nerve	24	Parahippocampal gyrus
10	Flocculus of cerebellum	25	Pituitary stalk (infundibulum)
11	Gyrus rectus	26	Pons
12	Internal carotid artery	27	Posterior cerebral artery
13	Mamillary body	28	Posterior communicating artery
14	Medulla oblongata	29	Posterior inferior cerebellar artery
15	Middle cerebral artery	30	Posterior perforated substance

31	Pyramid of medulla oblongata
32	Rootlets of hypoglossal nerve (superficial to marker)
33	Roots of glossopharyngeal, vagus and accessory nerves
34	Spinal part of accessory nerve
35	Superior cerebellar artery
36	Tonsil of cerebellum
37	Trigeminal nerve
38	Trochlear nerve
39	Tuber cinereum and median eminence
40	Uncus
41	Vertebral artery
42	Vestibulocochlear nerve

- A blue marker has been placed behind the right flocculus and the overlying facial and vestibulocochlear nerves (labelled 10, 9 and 42 on the left side).

- A red marker has been placed behind the roots of the right glossopharyngeal, vagus and accessory nerves (labelled 33 on the left side).

B Brain, from below

This is the view of the under-surface of the brain as typically seen when first removed from the skull, without any dissection. Arachnoid mater, torn in places and with blood vessels beneath it, remains on the outer surface.

1 Abducent nerve
2 Anterior perforated substance
3 Arachnoid mater overlying mamillary bodies
4 Basilar artery
5 Cerebellar hemisphere
6 Crus of cerebral peduncle (midbrain)
7 Facial nerve
8 Frontal pole
9 Gyrus rectus
10 Inferior surface of frontal lobe
11 Inferior surface of temporal nerve
12 Internal carotid artery
13 Longitudinal fissure
14 Medulla oblongata
15 Oculomotor nerve
16 Olfactory bulb
17 Olfactory tract
18 Optic chiasma
19 Optic nerve
20 Pituitary stalk (infundibulum)
21 Pons
22 Posterior communicating artery
23 Spinal part of accessory nerve
24 Temporal pole
25 Trigeminal nerve
26 Uncus
27 Vertebral artery
28 Vestibulocochlear nerve

C Optic tract and geniculate bodies, from below

The brainstem has been mostly removed, leaving only the upper part of the midbrain. The most medial parts of each cerebral hemisphere have also been dissected away. To find the geniculate bodies (4 and 6), which are on the under-surface of the posterior part (pulvinar, 13) of the thalamus, identify the optic chiasma (8) and then follow the optic tract (10) backwards round the side of the midbrain (3)

1 Anterior perforated substance
2 Aqueduct of midbrain
3 Crus of midbrain
4 Lateral geniculate body
5 Mamillary body
6 Medial geniculate body
7 Olfactory tract
8 Optic chiasma
9 Optic nerve
10 Optic tract
11 Pituitary stalk (infundibulum)
12 Posterior perforated substance
13 Pulvinar of thalamus
14 Splenium of corpus callosum
15 Substantia nigra of midbrain
16 Tectum of midbrain
17 Tegmentum of midbrain
18 Tuber cinereum

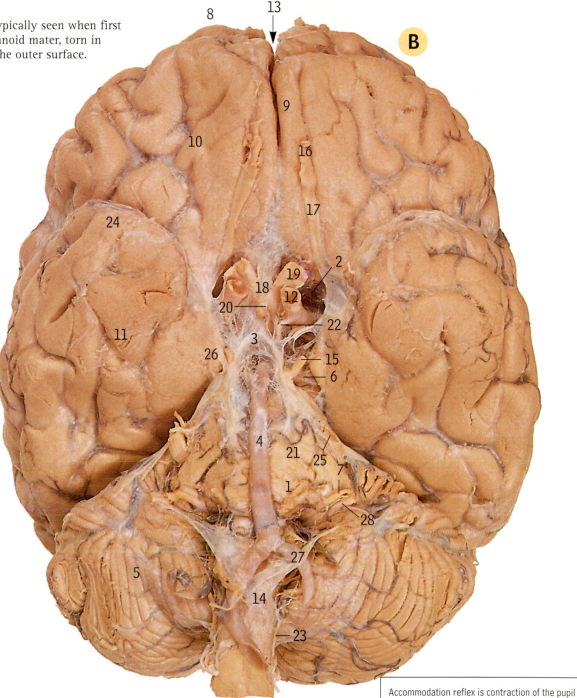

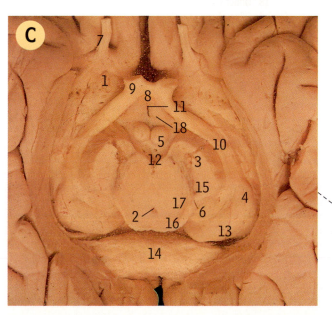

Accommodation reflex is contraction of the pupil when trying to focus on a near object and is controlled by the parasympathetic nerve fibres carried in the third cranial nerve from the Edinger–Westphal nucleus of the midbrain (synapse in ciliary ganglion) which act on the sphincter pupillae muscle to cause reduction in pupil diameter and on the ciliary muscle to cause relaxation of the suspensory ligament, allowing the lens to adopt a more spherical shape for near focusing.

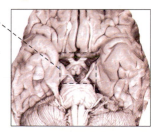

A Right half of the brain, in a midline sagittal section, from the left

In this typical half-section of the brain, the medial surface of the right cerebral hemisphere is seen, together with the sectioned brainstem (midbrain, 4, 20, 44, 47; pons, 36; and medulla oblongata, 29). The septum pellucidum, which is a midline structure and whose cut edge (12) is seen below the body of the corpus callosum (6), has been removed to show the interior of the body of the lateral ventricle (7). The third ventricle has the thalamus (48) and hypothalamus (19) in its lateral wall, while in its floor from front to back are the optic chiasma (32), the base of the pituitary stalk (21), the median eminence (49), the mamillary bodies (27), and the posterior perforated substance (40)

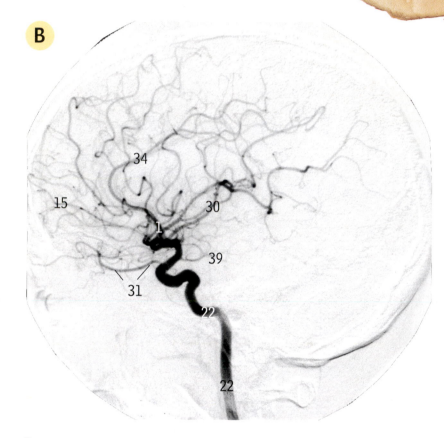

B Digitally subtracted arterial phase of carotid arteriogram, lateral projection

1 Anterior cerebral artery	25 Lamina terminalis
2 Anterior column of fornix	26 Lingual gyrus
3 Anterior commissure	27 Mamillary body
4 Aqueduct of midbrain	28 Median aperture of fourth ventricle
5 Basilar artery	29 Medulla oblongata
6 Body of corpus callosum	30 Middle cerebral artery
7 Body of lateral ventricle	31 Ophthalmic artery
8 Calcarine sulcus	32 Optic chiasma
9 Central sulcus	33 Parieto-occipital sulcus
10 Cerebellum	34 Pericallosal artery
11 Cingulate gyrus	35 Pineal body
12 Cut edge of septum pellucidum	36 Pons
13 Fornix	37 Postcentral gyrus
14 Fourth ventricle	38 Posterior commissure
15 Frontopolar artery	39 Posterior communicating artery
16 Genu of corpus callosum	40 Posterior perforated substance
17 Great cerebral vein	41 Precentral gyrus
18 Hypothalamic sulcus	42 Rostrum of corpus callosum
19 Hypothalamus	43 Splenium of corpus callosum
20 Inferior colliculus of midbrain	44 Superior colliculus of midbrain
21 Infundibular recess (base of pituitary stalk)	45 Supra-optic recess
	46 Suprapineal recess
22 Internal carotid artery	47 Tegmentum of midbrain
23 Interthalamic connexion	48 Thalamus
24 Interventricular foramen and choroid plexus	49 Tuber cinereum and median eminence

- The third ventricle is the cavity which has in its lateral wall the thalamus (A48) and hypothalamus (A19).
- The fourth ventricle (A14) is largely between the pons (A36) and cerebellum (A10), although its lower end is behind the upper part of the medulla oblongata (A29) (see page 65, D).
- The aqueduct of the midbrain (A4) connects the third and fourth ventricles; cerebrospinal fluid

normally flows through it from the third to the fourth ventricle.
- The interventricular foramen (A24) connects the third to the lateral ventricle, and is bounded in front by the anterior column of the fornix (A2) and behind by the thalamus (A48).
- The median eminence (A49) in the floor of the third ventricle is of great importance as the site of the neurosecretory cells whose products are released

into the portal system of pituitary blood vessels (hypophysial portal system) and which control the secretion of anterior pituitary hormones.
- Posterior pituitary hormones are manufactured by cells of the supra-optic and paraventricular nuclei in the lateral wall of the hypothalamus (A19). The axons of these cells pass down the whole length of the pituitary stalk into the posterior part of the gland; the hormones are stored within the axons.

C Medial surface of the right cerebral hemisphere

The brainstem has been removed through the midbrain (9) so that the lower part of the hemisphere can be seen; in A, on page 62, the brainstem hides this part

1 Anterior column of fornix
2 Anterior horn of lateral ventricle
3 Calcarine sulcus
4 Collateral sulcus
5 Corpus callosum
6 Hypothalamus in lateral wall of third ventricle
7 Interventricular foramen
8 Lingual gyrus
9 Midbrain
10 Parahippocampal gyrus
11 Parieto-occipital sulcus
12 Pineal body
13 Splenium of corpus callosum
14 Thalamus in lateral wall of third ventricle
15 Uncus

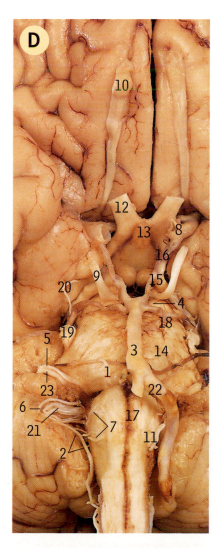

D Cranial nerves

In this ventral view of the central part of the brain, the right vertebral artery (on the left of the picture) has been removed almost at the junction with its fellow (22). The filaments of the first nerve (olfactory) are not seen entering the olfactory bulb (10) as they are torn off when removing the brain. The roots forming the glossopharyngeal, vagus and accessory nerves (6, 21 and 2) cannot be clearly identified from one another, but the spinal part of the accessory nerve (2) is seen running up beside the medulla to join the cranial part

1 Abducent nerve	9 Oculomotor nerve	17 Pyramid of medulla oblongata
2 Accessory nerve, spinal root	10 Olfactory bulb	18 Superior cerebellar artery
3 Basilar artery	11 Olive of medulla oblongata	19 Trigeminal nerve
4 Crus of cerebral peduncle	12 Optic nerve	20 Trochlear nerve
5 Facial nerve	13 Pituitary stalk	21 Vagus nerve
6 Glossopharyngeal nerve	14 Pons	22 Vertebral artery
7 Hypoglossal nerve	15 Posterior cerebral artery	23 Vestibulocochlear nerve
8 Internal carotid artery	16 Posterior communicating artery	

- The oculomotor nerve (D9) emerges on the medial side of the crus of the cerebral peduncle (D4), and the trochlear nerve (D20) winds round the lateral side of the peduncle. Both nerves pass between the posterior cerebral and superior cerebellar arteries (D15 and 18).
- The trochlear nerve (D20) is the only cranial nerve to emerge from the dorsal surface of the brainstem.
- The trigeminal nerve (D19) emerges from the lateral side of the pons (D14).
- The abducent nerve (D1) emerges between the pons and the pyramid (D14 and 17).

- The facial and vestibulocochlear nerves (D5 and 23) emerge from the lateral pontomedullary angle.
- The glossopharyngeal and vagus nerves (D6, 21, 2) and the cranial root of the accessory nerve emerge from the medulla oblongata lateral to the olive (D11).
- The hypoglossal nerve (D7) emerges as two series of rootlets from the medulla oblongata between the pyramid (D17) and the olive (D11).
- The spinal part of the accessory nerve emerges from the lateral surface of the upper five or six cervical segments of the spinal cord, dorsal to the denticulate ligament (page 65, E27).

1 Abducent nerve
2 Anterior cerebral
3 Anterior choroidal
4 Anterior communicating
5 Anterior inferior cerebellar
6 Anterior spinal
7 Basilar with pontine branches
8 Filaments of glossopharyngeal, vagus and accessory nerves
9 Internal carotid
10 Medulla oblongata
11 Middle cerebral
12 Oculomotor nerve
13 Olfactory tract
14 Olive
15 Optic nerve
16 Pons
17 Posterior cerebral
18 Posterior communicating
19 Posterior inferior cerebellar
20 Pyramid
21 Rootlets of first cervical nerve
22 Spinal cord
23 Spinal part of accessory nerve
24 Superior cerebellar
25 Trigeminal nerve
26 Unusually large branch of 5 overlying facial and vestibulocochlear nerves
27 Vertebral

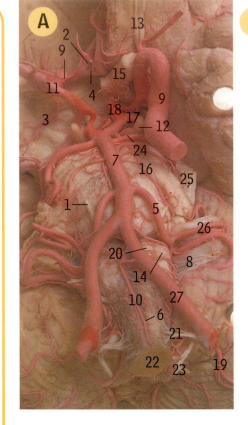

A Injected arteries of the base of the brain
Part of the right cerebral hemisphere (on the left of the picture) has been removed to show the right middle cerebral artery (11)

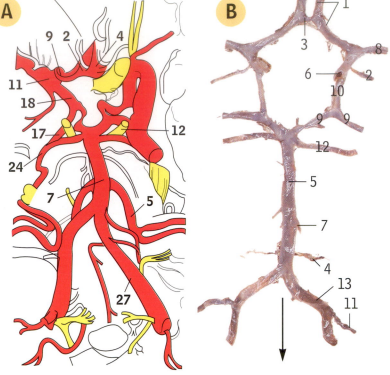

B Arterial circle and basilar artery
The anastomising vessels have been removed from the base of the brain and spread out in their relative positions

C Brainstem and cerebellum in sagittal section, from the left
The left half of the cerebellum has been removed by sagittal section in the midline and by transecting the left cerebellar peduncles (8, 12 and 25).

- The central part of the cerebellum constitutes the vermis (nodule, uvula and pyramid – C13, 30 and 21), which is continuous laterally with the hemispheres.
- The folia of the cerebellar cortex are considerably narrower than the gyri of the cerebral cortex.
- The largest of the subcortical nuclei of the cerebellar hemisphere is the dentate nucleus, whose axons constitute the main efferent pathway from the cerebellum and leave in the superior peduncle (C25).

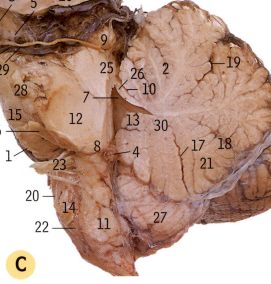

1 Anterior cerebral
2 Anterior choroidal
3 Anterior communicating
4 Anterior inferior cerebellar
5 Basilar
6 Internal carotid
7 Labyrinthine
8 Middle cerebral
9 Posterior cerebral
10 Posterior communicating
11 Posterior inferior cerebellar
12 Superior cerebellar
13 Vertebral

- The internal carotid artery (B6) gives off the anterior cerebral (B1) which passes forwards and medially to join its fellow by the anterior communicating artery (B3). The middle cerebral (B8) passes laterally from the carotid and the posterior communicating (B10) passes backwards to join the posterior cerebral (B9) which is the terminal branch of the basilar artery (B5).
- The basilar artery (B5) is formed by the union of the two vertebrals (B13).

1 Abducent nerve
2 Anterior lobe
3 Basal cerebral vein
4 Choroid plexus in lateral recess
5 Crus of cerebral peduncle
6 Facial and vestibulocochlear nerves
7 Fourth ventricle
8 Inferior cerebellar peduncle
9 Inferior colliculus
10 Lingula
11 Medulla oblongata
12 Middle cerebellar peduncle
13 Nodule of vermis
14 Olive
15 Pons
16 Posterior cerebral artery
17 Postpyramidal fissure
18 Prepyramidal fissure
19 Primary fissure
20 Pyramid of medulla oblongata
21 Pyramid of vermis
22 Rootlets of hypoglossal nerve
23 Roots of glossopharyngeal, vagus and accessory nerves
24 Superior cerebellar artery
25 Superior cerebellar peduncle
26 Superior medullary velum
27 Tonsil
28 Trigeminal nerve
29 Trochlear nerve
30 Uvula of vermis

D Brainstem and floor of the fourth ventricle

In this view of the dorsal surface of the brainstem, it has been cut off from the rest of the brain at the top of the midbrain, just above the superior colliculi (15). The cerebellum has been removed by transecting the superior (14), middle (12) and inferior (6) cerebellar peduncles

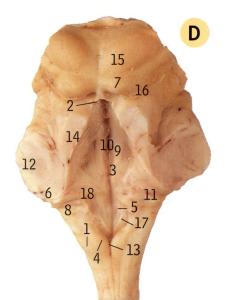

1 Cuneate tubercle	7 Inferior colliculus	14 Superior cerebellar peduncle
2 Cut edge of superior medullary velum	8 Lateral recess	15 Superior colliculus
	9 Medial eminence	16 Trochlear nerve
3 Facial colliculus	10 Median sulcus	17 Vagal triangle
4 Gracile tubercle	11 Medullary striae	18 Vestibular area
5 Hypoglossal triangle	12 Middle cerebellar peduncle	
6 Inferior cerebellar peduncle	13 Obex	

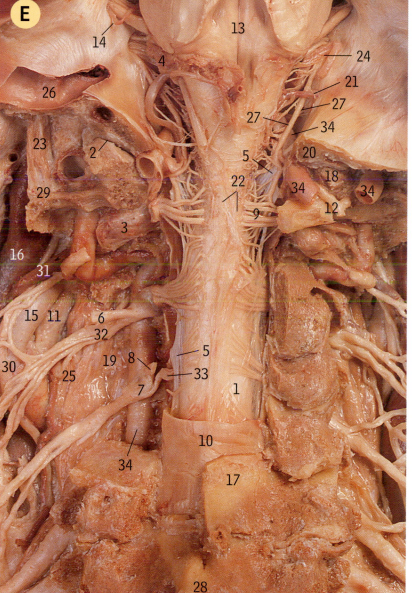

E Brainstem and upper part of the spinal cord, from behind

The posterior parts of the skull and upper vertebrae have been removed to show the continuity of the brainstem with the spinal cord, from which dorsal nerve rootlets are seen to emerge (as at 9). The spinal part of the accessory nerve (27) runs up through the foramen magnum (20) to join the cranial part in the jugular foramen (24). Ventral nerve rootlets (as at 33), ventral to the denticulate ligament (5), unite to form a ventral nerve root which joins with a dorsal nerve root (8, whose formative rootlets dorsal to the ligament have been cut off from the cord in order to make the ventral roots visible) to form a spinal nerve immediately beyond the dorsal root ganglion (7). The nerve immediately divides into ventral and dorsal rami (as at 32 and 6)

1 Arachnoid mater	17 Lamina of sixth cervical vertebra
2 Atlanto-occipital joint	18 Lateral mass of atlas
3 Capsule of lateral atlanto-axial joint	19 Longus capitus
4 Choroid plexus emerging from lateral recess of fourth ventricle	20 Margin of foramen magnum
5 Denticulate ligament	21 Posterior inferior cerebellar artery
6 Dorsal ramus of third cervical nerve	22 Posterior spinal arteries
7 Dorsal root ganglion of fourth cervical nerve	23 Rectus capitis lateralis
8 Dorsal root of fourth cervical nerve	24 Roots of glossopharyngeal, vagus and cranial part of accessory nerves and jugular foramen
9 Dorsal rootlets of second cervical nerve	25 Scalenus anterior
10 Dura mater	26 Sigmoid sinus
11 External carotid artery	27 Spinal part of accessory nerve
12 First cervical nerve and posterior arch of atlas	28 Spinous process of seventh cervical vertebra
13 Floor of the fourth vehicle	29 Transverse process of atlas
14 Internal acoustic meatus with facial and vestibulocochlear nerves and labyrinthe artery	30 Vagus nerve
	31 Vein from vertebral venous plexuses
15 Internal carotid artery	32 Ventral ramus of third cervical nerve
16 Internal jugular vein	33 Ventral rootlets of fourth cervical nerve
	34 Vertebral artery

- The lower part of the diamond-shaped floor of the fourth ventricle containing the hypoglossal and vagal triangles (D5 and 17) is part of the medulla oblongata; the rest of the floor is part of the pons.
- The gracile and cuneate tubercles (D4 and 1) are caused by the underlying gracile and cuneate nuclei, where the fibres of the gracile and cuneate tracts (posterior white columns) end by synapsing with the cells of the nuclei. The fibres from these cells form the medial lemniscus which runs through the brainstem to the thalamus.

- The facial colliculus (D3), at the lower end of the medial eminence (D9) in the floor of the fourth ventricle, is caused by fibres of the facial nerve overlying the abducent nerve nucleus; it is not produced by the facial nerve nucleus, which lies at a deeper level in the pons.
- After emerging from the foramen in the transverse process of the atlas the vertebral artery (E34) winds backwards round the lateral mass of the atlas (E18) on its posterior arch before turning upwards to enter the skull.

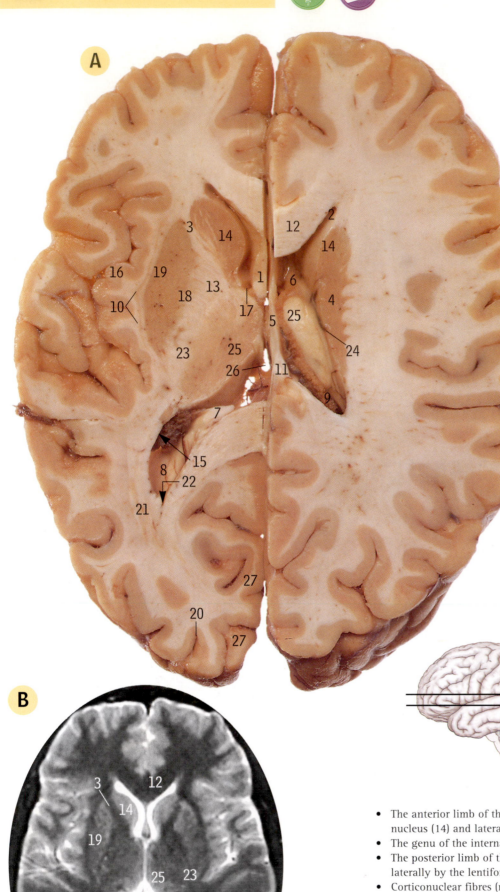

A

Cerebral hemispheres, A sectioned horizontally, B axial MR image

Viewed from above, the left cerebral hemisphere has been sectioned on a level with the interventricular foramen (17), and that on the right about 1.5 cm higher. The most important feature seen in the left hemisphere is the internal capsule (3, 13 and 23), situated between the caudate (14) and lentiform (18 and 19) nuclei and the thalamus (25). On the right side a large part of the corpus callosum (11) has been removed, so opening up the lateral ventricle (6) from above and showing the caudate nucleus (14 and 4) arching backwards over the thalamus (25), with the thalamostriate vein (24) and choroid plexus (9) in the shallow groove between them

1   Anterior column of fornix
2   Anterior horn of lateral ventricle
3   Anterior limb of internal capsule
4   Body of caudate nucleus
5   Body of fornix
6   Body of lateral ventricle
7   Bulb
8   Calcar avis
9   Choroid plexus
10 Claustrum
11 Corpus callosum
12 Forceps minor (corpus callosum)
13 Genu of internal capsule
14 Head of caudate nucleus
15 Inferior horn of lateral ventricle
16 Insula
17 Interventricular foramen
18 Lentiform nucleus: globus pallidus
19 Lentiform nucleus: putamen
20 Lunate sulcus
21 Optic radiation
22 Posterior horn of lateral ventricle
23 Posterior limb of internal capsule
24 Thalamostriate vein
25 Thalamus
26 Third ventricle
27 Visual area of cortex

- The anterior limb of the internal capsule (3) is bounded medially by the head of the caudate nucleus (14) and laterally by the lentiform nucleus (putamen and globus pallidus, 18 and 19).
- The genu of the internal capsule (13) lies at the most medial edge of the globus pallidus (18).
- The posterior limb of the internal capsule (23) is bounded medially by the thalamus (25) and laterally by the lentiform nucleus (18 and19).
- Corticonuclear fibres (motor fibres from the cerebral cortex to the motor nuclei of cranial nerves) pass through the genu of the internal capsule (13).
- Corticospinal fibres (motor fibres from the cerebral cortex to anterior horn cells of the spinal cord) pass through the anterior two-thirds of the posterior limb of the internal capsule (23).
- The genu and the posterior limb of the internal capsule, supplied by the striate branches of the anterior and middle cerebral arteries, are of the greatest clinical importance as they are the common sites for cerebral haemorrhage or thrombosis ('stroke').
- The choroid plexus of the third ventricle passes through the interventricular foramen into the body of the lateral ventricle and then into the inferior horn; there is no choroid plexus in the anterior or posterior horns.
- The optic radiation is alternatively known as the geniculocalcarine tract, and passes from the lateral geniculate body to the calcarine area of the cortex.

B

Brain, A coronal section, from the front, B coronal MR image
This coronal section is not quite vertical but passes slightly
backwards, through the third ventricle (25) and bodies of the lateral
ventricles (3) from a level about 0.5 cm behind the interventricular
foramina, and down through the pons (17) and the pyramid of the
medulla (19). It has been cut in this way to show the path of the
important corticospinal (motor) fibres passing down through the
internal capsule (11) and pons (17) to form the pyramid of the
medulla (19). Compare with features in the MR image

1 Body of caudate nucleus
2 Body of fornix
3 Body of lateral ventricle
4 Choroid plexus of inferior horn of lateral ventricle
5 Choroid plexus of lateral ventricle
6 Choroid plexus of third ventricle
7 Choroidal fissure
8 Corpus callosum
9 Hippocampus
10 Insula
11 Internal capsule
12 Interpeduncular cistern
13 Lentiform nucleus: globus pallidus
14 Lentiform nucleus: putamen
15 Olive of medulla oblongata
16 Optic tract
17 Pons
18 Posterior cerebral artery
19 Pyramid of medulla oblongata
20 Septum pellucidum
21 Substantia nigra
22 Tail of caudate nucleus
23 Thalamostriate vein
24 Thalamus
25 Third ventricle

**C Sectioned cerebral hemispheres and the
brainstem, from above and behind**
The cerebral hemispheres have been sectioned
horizontally just above the level of the
interventricular foramina, and the posterior parts
of the hemispheres have been removed, together
with the whole of the cerebellum, to show the tela
choroidea (12) of the posterior part of the roof of
the third ventricle and the underlying internal
cerebral veins (10)

1 Anterior horn of lateral ventricle
2 Anterior limb of internal capsule
3 Choroid plexus and junction of inferior and
   posterior horn of lateral ventricle
4 Floor of fourth vehicle
5 Forceps minor
6 Genu of internal capsule
7 Head of caudate nucleus
8 Inferior colliculus
9 Insula
10 Internal cerebral vein
11 Posterior limb of internal capsule
12 Tela choroidea of roof of third ventricle
13 Thalamus
14 Third ventricle
15 Trochlear nerve

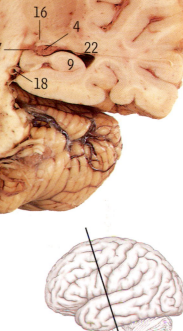

A Inferior horn of right lateral ventricle

Brain substance above the front part of the lateral sulcus has been removed, displaying the middle cerebral artery (9) running laterally over the upper surface of the front of the temporal lobe (14). Part of the temporal lobe has been opened up from above to show the hippocampus (11 and 8) in the floor of the inferior horn

1	Anterior cerebral artery	9	Middle cerebral artery
2	Anterior choroidal artery	10	Optic nerve
3	Choroid plexus	11	Pes hippocampi
4	Collateral eminence	12	Posterior horn
5	Collateral trigone	13	Tapetum
6	Fimbria	14	Temporal pole of
7	Fornix		temporal lobe
8	Hippocampus	15	Thalamus

B Dissection of the right cerebral hemisphere, from above

Much of the cerebral substance has been dissected away to show the caudate nucleus (3), thalamus (13) and lentiform nucleus (9). The intervening gap (8) is occupied by the internal capsule. The optic radiation (10) has also been dissected out; it runs backwards lateral to the posterior horn of the lateral ventricle. Compare this three-dimensional view of these structures with the brain sections on page 67

1	Bulb	9	Lentiform nucleus
2	Calcar avis	10	Optic radiation
3	Caudate nucleus	11	Posterior horn of lateral
4	Collateral trigone		ventricle
5	Forceps major	12	Splenium of corpus
6	Forceps minor		callosum
7	Fornix	13	Thalamus
8	Internal capsule		

C Cast of the cerebral ventricles, from the left

In this side view the left lateral ventricle largely overlaps the right one

1	Anterior horn of lateral ventricle
2	Aqueduct of midbrain
3	Body of lateral ventricle
4	Fourth ventricle
5	Inferior horn of lateral ventricle
6	Infundibular recess of third ventricle
7	Interventricular foramen
8	Lateral recess
9	Posterior horn of lateral ventricle
10	Supra-optic recess of third ventricle
11	Suprapineal recess of third ventricle
12	Third ventricle (with gap for interthalamic connexion)

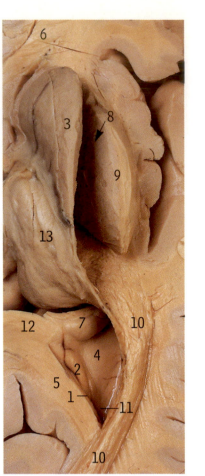

- The third ventricle (C12) communicates at its upper front end with each lateral ventricle through the interventricular foramen (C7).
- The main part of the lateral ventricle is the body (C3). The part in front of the interventricular foramen (C7) is the anterior horn (C1) which extends into the frontal lobe of the brain. At its posterior end the body divides into the posterior horn (C9) which extends backwards into the occipital lobe, and the inferior horn (C5) which passes downwards and forwards into the temporal lobe.
- The lower posterior part of the third ventricle (C12) communicates with the fourth ventricle (C4) through the aqueduct of the midbrain (C2).
- The floor of the inferior horn consists of the hippocampus (A11 and 8) medially and the collateral eminence (A4) laterally. At its junction with the posterior horn (A12 and B11) the eminence broadens into the collateral trigone (A5, B4).
- The collateral eminence (A4) is produced by the inward projection of the collateral sulcus (page 63, C4).
- In the medial wall of the posterior horn, the bulb (B1) is produced by fibres of the corpus callosum, and the calcar avis (B2) by the inward projection of the calcarine sulcus (page 63, C3).

Chapter 2

Vertebral column and spinal cord

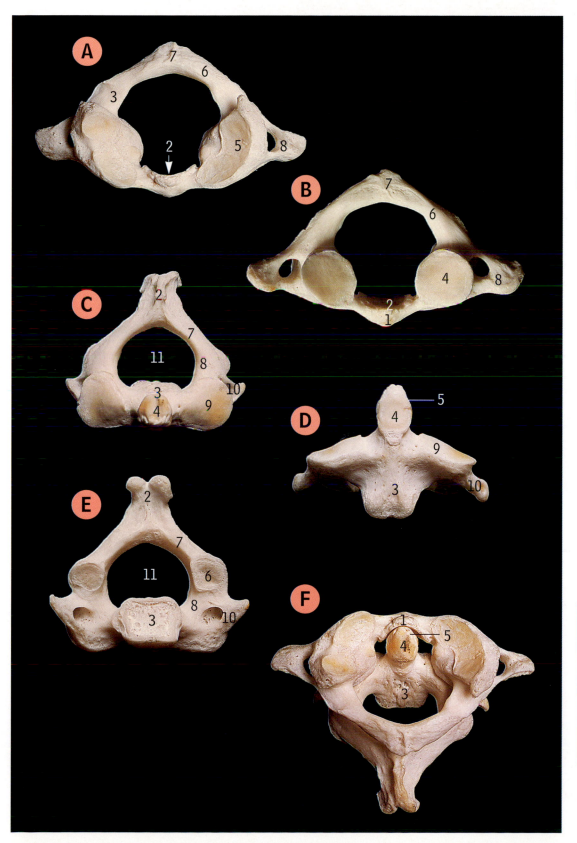

Atlas (first cervical vertebra), **A** from above, **B** from below

1. Anterior arch and tubercle
2. Facet for dens of axis
3. Groove for vertebral artery
4. Lateral mass with inferior articular facet
5. Lateral mass with superior articular facet
6. Posterior arch
7. Posterior tubercle
8. Transverse process and foramen

- The superior articular facets (5) are concave and kidney-shaped.
- The inferior articular facets (4) are round and almost flat.
- The anterior arch (1) is straighter and shorter than the posterior arch (6) and contains on its posterior surface the facet for the dens of the axis (2).
- The atlas is the only vertebra that has no body.

Axis (second cervical vertebra), **C** from above, **D** from the front, **E** from below, **F** articulated with the atlas, from above and behind

1. Anterior arch of atlas
2. Bifid spinous process
3. Body
4. Dens
5. Impression for alar ligament
6. Inferior articular process
7. Lamina
8. Pedicle
9. Superior articular surface
10. Transverse process and foramen
11. Vertebral foramen

- The axis is unique in having the dens (4) which projects upwards from the body, representing the body of the atlas.

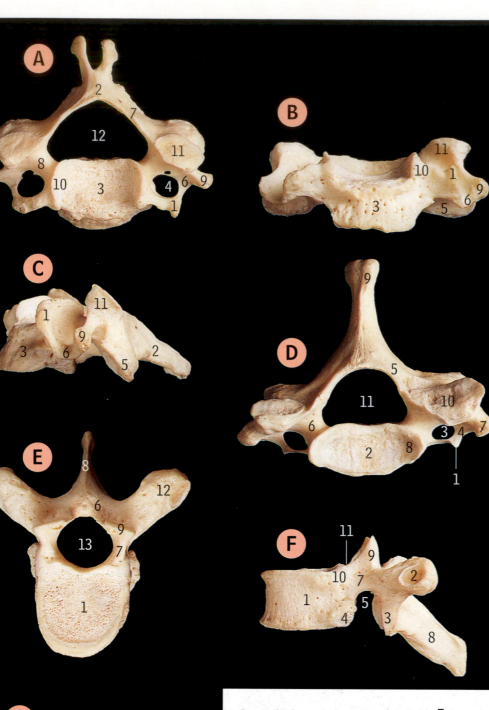

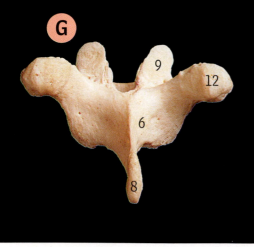

1 Anterior tubercle of transverse process
2 Bifid spinous process
3 Body
4 Foramen of transverse process
5 Inferior articular process
6 Intertubercular lamella of transverse process
7 Lamina
8 Pedicle
9 Posterior tubercle of transverse process
10 Posterolateral lip (uncus)
11 Superior articular process
12 Vertebral foramen

D Seventh cervical vertebra (vertebra prominens), from above

1 Anterior tubercle of transverse process
2 Body
3 Foramen of transverse process
4 Intertubercular lamella of transverse process
5 Lamina
6 Pedicle
7 Posterior tubercle of transverse process
8 Posterolateral lip (uncus)
9 Spinous process with tubercle
10 Superior articular process
11 Vertebral foramen

- All cervical vertebrae (first to seventh) have a foramen in each transverse process (as A4).
- Typical cervical vertebrae (third to sixth) have superior articular processes that face backwards and upwards (A11, C11), posterolateral lips on the upper surface of the body (A10), a triangular vertebral foramen (A12) and a bifid spinous process (A2).
- The anterior tubercle of the transverse process of the sixth cervical vertebra is large and known as the carotid tubercle.
- The seventh cervical vertebra (vertebra prominens) has a spinous process that ends in a single tubercle (D9).
- The rib element of a cervical vertebra is represented by the anterior root of the transverse process, the anterior tubercle, the intertubercular lamella (with its groove for the ventral ramus of a spinal nerve) and the anterior part of the posterior tubercle (as at D1, 4 and 7).

Seventh thoracic vertebra (typical), **E** from above, **F** from the left, **G** from behind

1 Body	7 Pedicle
2 Costal facet of transverse process	8 Spinous process
3 Inferior articular process	9 Superior articular process
4 Inferior costal facet	10 Superior costal facet
5 Inferior vertebral notch	11 Superior vertebral notch
6 Lamina	12 Transverse process
	13 Vertebral foramen

- Typical thoracic vertebrae (second to ninth) are characterized by costal facets on the bodies (F10, 4), costal facets on the transverse processes (F2), a round vertebral foramen (E13), a spinous process that points downwards as well as backwards (F8, G8) and superior articular processes that are vertical, flat and face backwards and laterally (E9, F9, G9).

First thoracic vertebra, A from above, B from the front and the left

1 Body
2 Inferior articular process
3 Inferior costal facet
4 Lamina
5 Pedicle
6 Posterolateral lip (uncus)
7 Spinous process
8 Superior articular process
9 Superior costal facet
10 Transverse process with costal facet
11 Vertebral foramen

Tenth thoracic vertebra, C, and eleventh thoracic vertebra, D, from the left

1 Body
2 Costal facet
3 Inferior articular process
4 Inferior vertebral notch
5 Pedicle
6 Spinous process
7 Superior articular process
8 Transverse process

Twelfth thoracic vertebra, E from the left, F from above, G from behind

1 Body
2 Costal facet
3 Inferior articular process
4 Inferior tubercle
5 Lateral tubercle
6 Pedicle
7 Spinous process
8 Superior articular process
9 Superior tubercle

- The atypical thoracic vertebrae are the first, tenth, eleventh and twelfth.
- The first thoracic vertebra has a posterolateral lip (A6, B6) on each side of the upper surface of the body and a triangular vertebral foramen (features like typical cervical vertebrae), and complete (round) superior costal facets (B9) on the sides of the body.
- The tenth, eleventh and twelfth thoracic vertebrae are characterized by a single complete costal facet on each side of the body that in successive vertebrae comes to lie increasingly far from the upper surface of the body and encroaches increasingly onto the pedicle (C2, D2 and E2). There is also no articular facet on the transverse process.

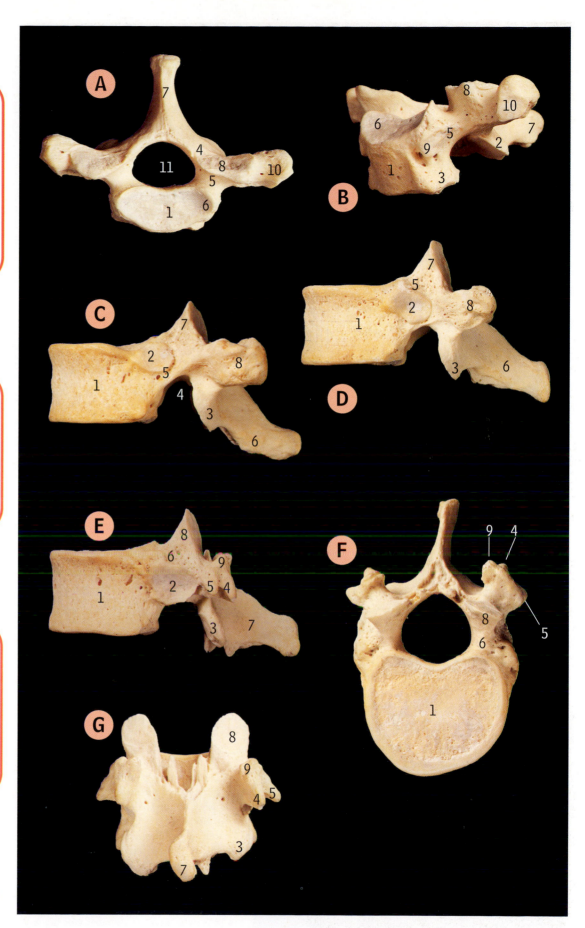

Spondylolisthesis is the anterior displacement or slipping of the body of one vertebra on another, usually in the lower lumbar region.

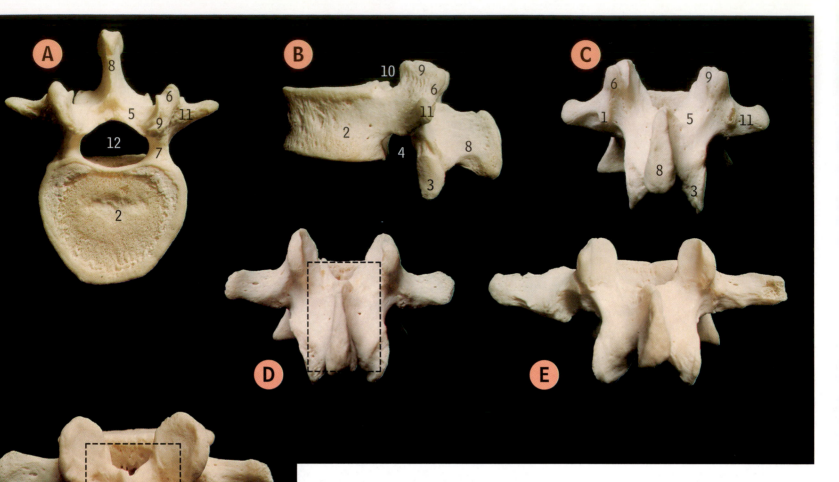

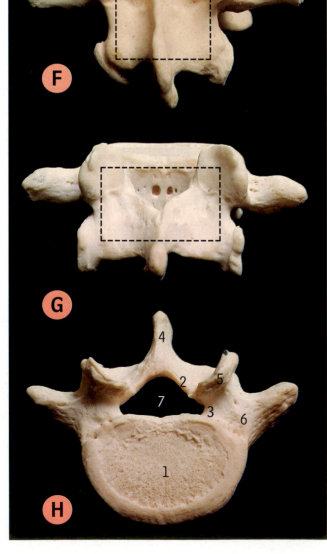

First lumbar vertebra, **A** from above, **B** from the left, **C** from behind

1 Accessory process
2 Body
3 Inferior articular process
4 Inferior vertebral notch
5 Lamina
6 Mamillary process
7 Pedicle
8 Spinous process
9 Superior articular process
10 Superior vertebral notch
11 Transverse process
12 Vertebral foramen

D Second lumbar vertebra, posterior view
E third lumbar vertebra, posterior view
F fourth lumbar vertebra, posterior view
G fifth lumbar vertebra, posterior view

- Viewed from behind, the four articular processes of the first and second lumbar vertebrae make a pattern (indicated by the interrupted line) of a vertical rectangle; those of the third or fourth vertebra make a square, and those of the fifth lumbar vertebra make a horizontal rectangle.

Laminectomy is removal of vertebral laminae to treat disc protrusion but recently has been superseded by discectomy and microdiscectomy. A laminectomy may still be indicated for spinal stenosis.

- Lumbar vertebrae are characterized by the large size of the bodies, the absence of costal facets on the bodies and the transverse processes, a triangular vertebral foramen (A12), a spinous process that points backwards and is quadrangular or hatchet-shaped (B8) and superior articular processes that are vertical, curved, face backwards and medially (A9) and possess a mamillary process at their posterior rim (A6).
- The rib element of a lumbar vertebra is represented by the transverse process (A11).
- The level at which facet joint orientation changes between the thoracic and lumbar regions is variable.

H Fifth lumbar vertebra, from above

1 Body
2 Lamina
3 Pedicle
4 Spinous process
5 Superior articular process
6 Transverse process fusing with pedicle and body
7 Vertebral foramen

- The fifth lumbar vertebra is unique in that the transverse process (H6) unites directly with the side of the body (H1) as well as with the pedicle (H3).

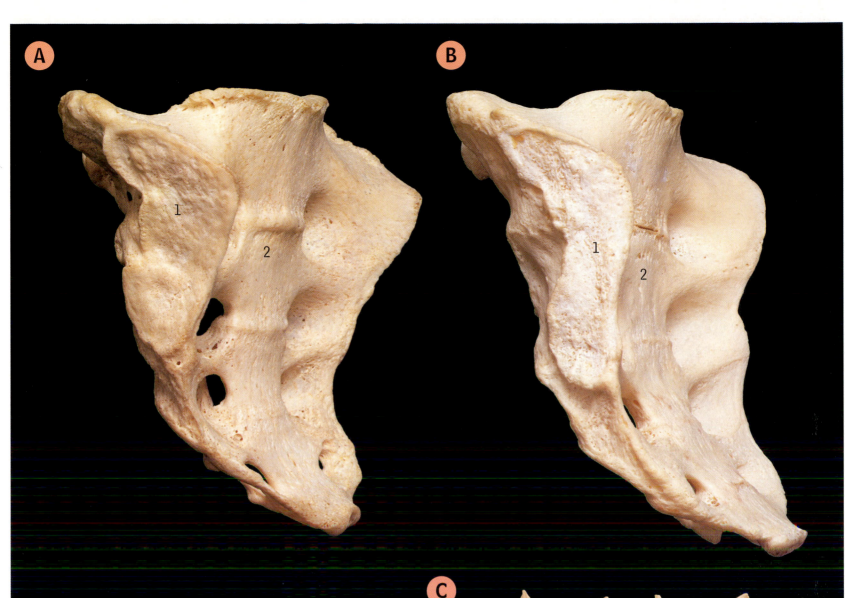

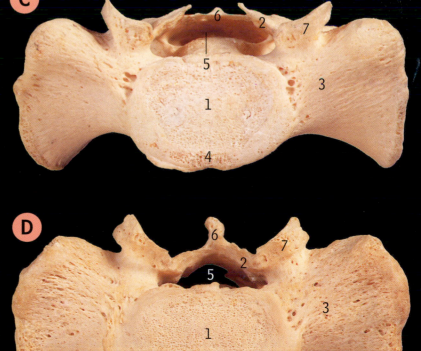

Sacrum, from the front and the right, A in the female, B in the male

| 1 | Auricular surface |
| 2 | Pelvic surface |

- In the female the pelvic surface is relatively straight over the first three sacral vertebrae and becomes more curved below. In the male the pelvic surface is more uniformly curved.
- The capsule of the sacro-iliac joint is attached to the margin of the auricular (articular) surface (A1, B1).

Base of the sacrum, upper surface, C in the female, D in the male

1	Body of first sacral vertebra	5	Sacral canal
2	Lamina	6	Spinous tubercle of median sacral crest
3	Lateral part (ala)	
4	Promontory	7	Superior articular process

- In the male the body of the first sacral vertebra (judged by its transverse diameter) forms a greater part of the base of the sacrum than in the female (compare D1 with C1).
- In C there is some degree of spina bifida (non-fusion of the laminae, 2, in the vertebral arch of the first sacral vertebra). Compare with the complete arch in D.

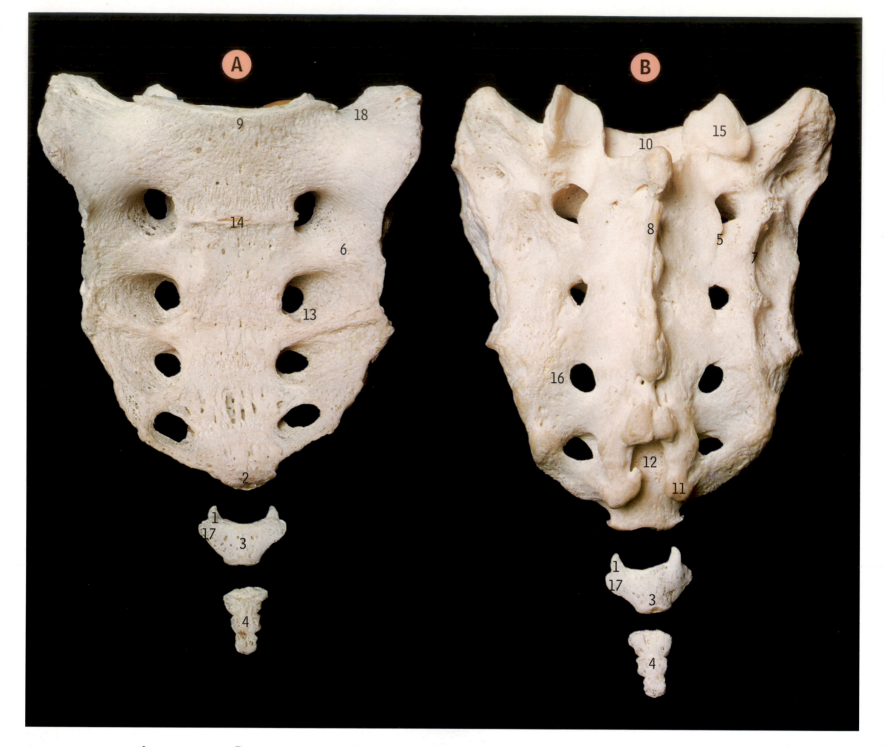

Sacrum and coccyx, **A** pelvic surface, **B** dorsal surface

1	Coccygeal cornu	11	Sacral cornu
2	Facet for coccyx	12	Sacral hiatus
3	First coccygeal vertebra	13	Second pelvic sacral foramen
4	Fused second to fourth vertebrae	14	Site of fusion of first and second
5	Intermediate sacral crest		sacral vertebrae
6	Lateral part	15	Superior articular process
7	Lateral sacral crest	16	Third dorsal sacral foramen
8	Median sacral crest	17	Transverse process
9	Promontory	18	Upper surface of lateral part (ala)
10	Sacral canal		

Coccydynia is pain in the region of the coccyx commonly found in women after childbirth and is caused by stretching and possible fracture of coccygeal segments. It may also occur following a sudden fall directly onto the buttocks.

- The sacrum is formed by the fusion of the five sacral vertebrae. The median sacral crest (B8) represents the fused spinous processes, the intermediate crest (B5) the fused articular processes, and the lateral crest (B7) the fused transverse processes.
- The sacral hiatus (B12) is the lower opening of the sacral canal (B10).
- The coccyx is usually formed by the fusion of four rudimentary vertebrae but the number varies from three to five. In this specimen the first piece of the coccyx (3) is not fused with the remainder (4).

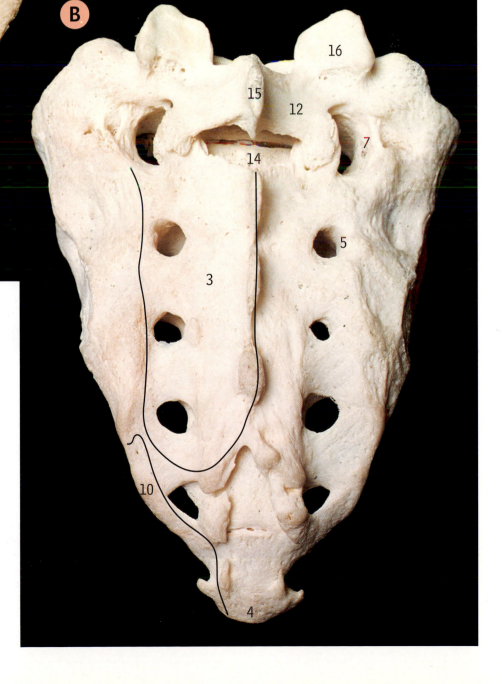

- In sacralization of the fifth lumbar vertebra, that vertebra (A1) is (usually incompletely) fused with the sacrum. In the more rare condition of lumbarization of the first sacral vertebra (not illustrated) the first piece of the sacrum is incompletely fused with the remainder.
- In this specimen, as well as fusion of the fifth lumbar vertebra with the top of the sacrum, the body of the first coccygeal vertebra (4) is fused with the apex of the sacrum.

Caudal anaesthesia is a special type of epidural (extradural) anaesthesia, the needle entering the epidural space via the sacral hiatus in the natal cleft for anaesthesia of the lower lumbar and sacral roots.

Sacrum with sacralization of the fifth lumbar vertebra, **A** pelvic surface, **B** dorsal surface, and sacral muscle attachments

1 Body of fifth lumbar vertebra
2 Coccygeus
3 Erector spinae
4 First coccygeal vertebra fused to apex of sacrum
5 First dorsal sacral foramen
6 First pelvic sacral foramen
7 Foramen for dorsal ramus of fifth lumbar nerve
8 Foramen for ventral ramus of fifth lumbar nerve
9 Fusion of transverse process and lateral part of sacrum
10 Gluteus maximus
11 Iliacus
12 Lamina
13 Piriformis
14 Sacral canal
15 Spinous process of fifth lumbar vertebra
16 Superior articular process of fifth lumbar vertebra

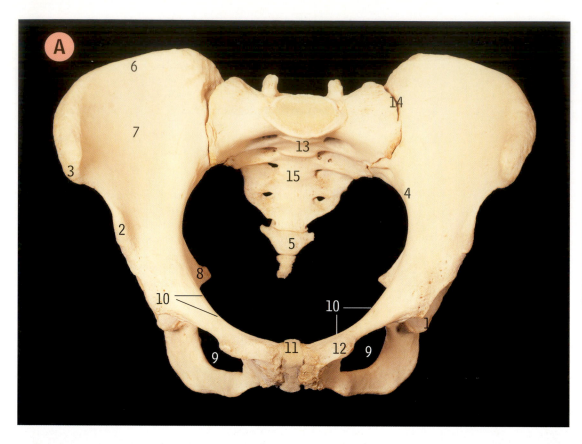

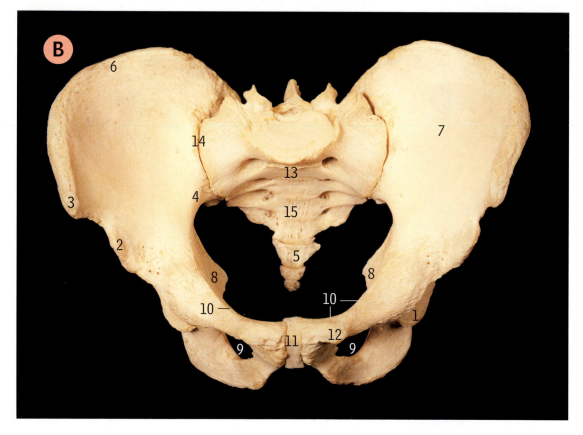

Bony pelvis, from in front and above, **A** female, **B** male

1 Acetabulum
2 Anterior inferior iliac spine
3 Anterior superior iliac spine
4 Arcuate line
5 Coccyx
6 Iliac crest
7 Iliac fossa
8 Ischial spine
9 Obturator foramen
10 Pectineal line
11 Pubic symphysis
12 Pubic tubercle
13 Sacral promontory
14 Sacro-iliac joint
15 Sacrum

- The pelvic inlet (brim) is bounded by the sacral promontory, arcuate and pectineal lines, the crest of the pubic bones and anteriorly the pubic symphysis.
- The female brim is more circular, the male more heart-shaped.
- The female sacrum is wider, shorter and less curved.
- The female ischial spines are further apart.

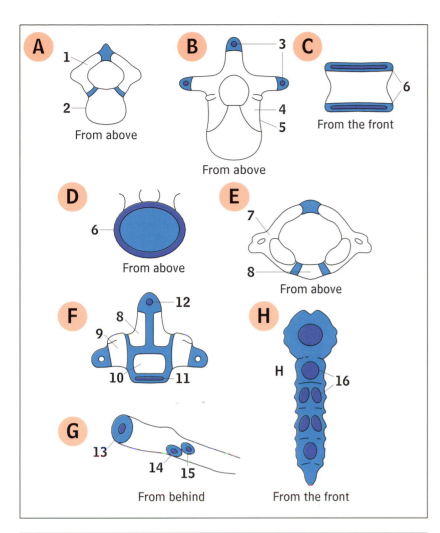

A
1
2
From above

B
3
4
5
From above

C
6
From the front

D
6
From above

E
7
8
From above

F
12
8
9
10   11
From behind

G
13
14   15

H
H
16
From the front

Ossification of vertebrae, ribs and sternum, **A** typical vertebra in a six-month fetus, **B** at four years, **C** and **D** during puberty, **E** atlas at four years, **F** axis, primary and secondary centres, **G**, typical rib, secondary centres, **H** sternum at birth, with primary centres

A typical vertebra, which is first cartilaginous, ossifies in early fetal life from three primary centres – one for most of the body (the centrum, A2) and one for each half of the neural arch (A1). The part of the adult body to which the pedicle is attached (B4) is part of the centre for the arch; the site in the developing vertebra where they meet is the neurocentral junction (B5). The two halves of the arch and the neurocentral junctions unite at variable times between birth and six years. Ossification spreads into the transverse processes and spine which grow out from the arch, but secondary centres (B3) appear at their tips during puberty and become fused at about twenty-five years. (Lumbar vertebrae have similar additional secondary centres for the mamillary processes.) There are also ring-like epiphyses on the periphery of the upper and lower surfaces of the vertebral bodies (C6 and D6).

The atlas has a primary centre (E7) for each lateral mass and the adjacent half of the posterior arch, and one for the anterior arch (E8). Fusion is complete by about eight years.

The axis has five primary centres – one for most of the body (F10), one for each lateral mass (F9), and one for each half of the dens and adjacent part of the body (F8). They should all fuse by about three years. There are secondary centres for the tip of the dens (F12, appearing by about two years and fusing at twelve) and the lower surface of the body (F11, appearing during puberty and fusing at about twenty-five years).

The seventh cervical vertebra, in addition to the typical vertebral centres, has additional centres for the costal elements, appearing during the first year and fusing at about five years.

The sacrum, representing five fused sacral vertebrae, has many ossification centres, corresponding to the centrum, neural arch halves and costal elements of each vertebra, as well as ring epiphyses for the vertebral bodies and for the auricular surfaces. Most have fused by about twenty years, but some not until middle age or later.

A typical rib has a primary centre for the body with secondary centres for the head (G13) and the articular and non-articular parts of the tubercle (G14 and 15), appearing during puberty and uniting at about twenty years.

The sternum has a variable number of primary centres (H16), one or two in the manubrium and in each of the four pieces of the body. Fusion occurs between puberty and twenty-five years. 'Bullet holes' in the sternum (sternal foramina) may occur when fusion is incomplete.

- The nucleus pulposus of an intervertebral disc as well as the apical ligament of the dens represent the remains of the notochord.
- The annulus fibrosus of an intervertebral disc is derived from the mesenchyme between adjacent vertebral bodies.

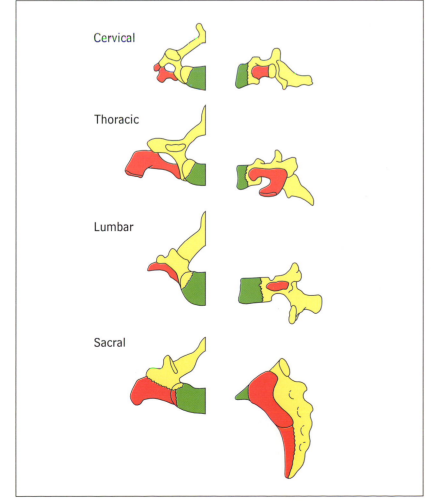

Cervical

Thoracic

Lumbar

Sacral

Developmental origins of the vertebrae
Red: costal elements
Green: centrum
Yellow: neural arch
Parts of the cervical, lumbar and sacral vertebrae represent the ribs that articulate with thoracic vertebrae. These costal elements are indicated here in red.
Cervical: anterior and posterior tubercles and the intertubercular lamella.
Thoracic: the true rib articulates with the vertebra.
Lumbar: the anterior part of the transverse process.
Sacral: the lateral part, including the auricular surface.

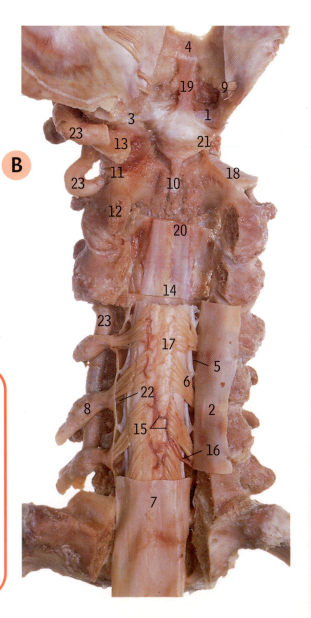

A Vertebral column, cervical region, from the front
The vertebral artery (14) is seen within foramina of cervical transverse processes

1   Anterior longitudinal ligament
2   Anterior tubercle of transverse process
3   Axis
4   Body of fifth cervical vertebra
5   Cut edge of the pleura
6   Intertubercular lamella of transverse process
7   Intervertebral disc
8   Joint of head of first rib
9   Lateral mass of atlas
10 Posterior tubercle of transverse process
11 Scalenus anterior muscle
12 Transverse process of atlas
13 Ventral ramus of third cervical nerve
14 Vertebral artery

B Vertebral column, cervical region, from behind
Much of the skull, the vertebral arches, brainstem and the upper part of the spinal cord have been removed to show the cruciform, transverse and alar ligaments (19, 10, 21 and 1). Lower down, the arachnoid and dura mater (2) have been reflected to show dorsal and ventral nerve roots (as at 6 and 22)

1   Alar ligament
2   Arachnoid and dura mater (reflected)
3   Atlanto-occipital joint
4   Basilar part of occipital bone and position of attachment of tectorial membrane
5   Denticulate ligament
6   Dorsal rootlets of spinal nerve
7   Dura mater
8   Dural sheath over dorsal root ganglion
9   Hypoglossal nerve and canal
10 Inferior longitudinal band of cruciform ligament
11 Lateral atlanto-axial joint
12 Pedicle of axis
13 Posterior arch of atlas
14 Posterior longitudinal ligament
15 Posterior spinal arteries
16 Radicular artery
17 Spinal cord
18 Superior articular surface of axis
19 Superior longitudinal band of cruciform ligament
20 Tectorial membrane
21 Transverse ligament of atlas (transverse part of cruciform ligament)
22 Ventral rootlets of spinal nerve
23 Vertebral artery

C Vertebral column, cervical and upper thoracic regions, from the right
Ventral and dorsal rami of spinal nerves (as at 16 and 4) are seen emerging from intervertebral foramina (as at 7)

1 Anterior tubercle of transverse process of fifth cervical vertebra
2 Body of first thoracic vertebra
3 Body of seventh cervical vertebra
4 Dorsal ramus of first cervical nerve
5 First cervical nerve
6 First rib
7 Intervertebral foramen
8 Lateral atlanto-axial joint
9 Lateral mass of atlas
10 Posterior arch of atlas
11 Seventh cervical nerve
12 Spinous process of second cervical vertebra
13 Spinous process of seventh cervical vertebra
14 Transverse process of atlas
15 Tubercle of first rib
16 Ventral ramus of fifth cervical nerve
17 Vertebral artery
18 Zygapophyseal joint

- The first and second vertebral nerves pass respectively above and below the posterior arch of the atlas.
- On its upward course from the subclavian artery the vertebral artery enters the foramen of the transverse process of the sixth cervical vertebra.
- For the joints of the ribs with thoracic vertebrae see page 183.

D Vertebral column, cervical region, from the left
Soft tissue has been removed to show the boundaries of intervertebral foramina (as at 5). Compare with the cleared specimens of thoracic vertebrae on page 82, A

1 Anterior tubercle of transverse process of fifth cervical vertebra
2 Body of third cervical vertebra
3 Intertubercular lamella of transverse process of fifth cervical vertebra
4 Intervertebral disc
5 Intervertebral foramen
6 Pedicle
7 Posterior tubercle of transverse process of fifth cervical vertebra
8 Zygapophyseal joint

- Each intervertebral foramen (as at D5) is bounded in front by a vertebral body and intervertebral disc (D2 and 4), above and below by pedicles (D6), and behind by a zygapophyseal joint (D8).
- In the thoracic and lumbar regions there are the same number of pairs of spinal nerves as there are vertebrae (twelve thoracic and five lumbar), and spinal nerves are numbered from the vertebra beneath whose pedicles they emerge. In the cervical region there are seven cervical vertebrae and eight cervical nerves. The first nerve emerges between the occipital bone of the skull and the atlas, and the eighth below the pedicle of the seventh cervical vertebra.
- The zygapophyseal joints between the articular processes of adjacent vertebrae are commonly called facet joints (between the articular facets of those processes).

E Vertebral column and spinal cord, lower cervical and upper thoracic regions, from behind
The vertebral arches and most of the dura mater and arachnoid have been removed, to show dorsal nerve rootlets (5) emerging from the spinal cord (9) to unite as a dorsal nerve root and enter the dural sheath (as at 7). Ventral nerve roots do the same from the ventral aspect of the cord but are not seen in this view as they are obscured by the dorsal roots

1 Angulation of nerve roots entering dural sheath
2 Dorsal ramus of fifth thoracic nerve
3 Dorsal root ganglion of eighth cervical nerve
4 Dorsal root ganglion of second thoracic nerve
5 Dorsal rootlets of eighth cervical nerve
6 Dura mater
7 Dural sheath of second thoracic nerve
8 Pedicle of first thoracic vertebra
9 Spinal cord and posterior spinal vessels
10 Ventral ramus of fifth thoracic nerve

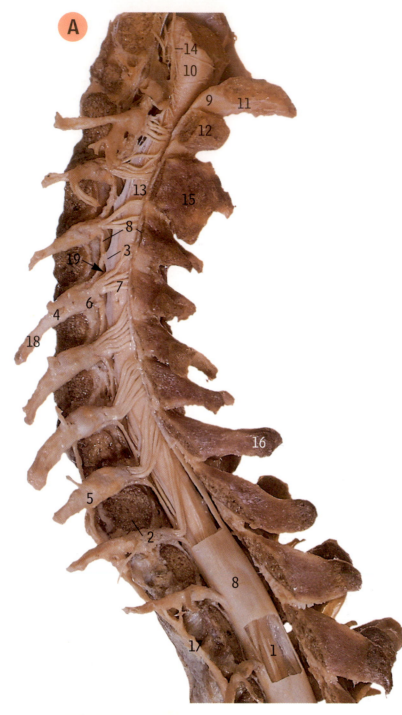

A Vertebral column and spinal cord, cervical and upper thoracic regions, from the left
Parts of the vertebral arches and meninges have been removed to show the denticulate ligament (3). Dorsal nerve rootlets lie behind it (as at 7) and ventral nerve rootlets in front of it (as at 19 but largely hidden in this view)

1	Arachnoid mater	11	Occipital bone
2	Body of first thoracic vertebra	12	Posterior arch of atlas
3	Denticulate ligament	13	Spinal cord
4	Dorsal ramus of fifth cervical nerve	14	Spinal part of accessory nerve
5	Dorsal root ganglion of eighth cervical nerve	15	Spinous process of axis (abnormally large)
6	Dorsal root ganglion of fifth cervical nerve	16	Spinous process of seventh cervical vertebra
7	Dorsal rootlets of fifth cervical nerve	17	Sympathetic trunk
8	Dura mater	18	Ventral ramus of fifth cervical nerve
9	Foramen magnum	19	Ventral rootlets of fifth cervical nerve
10	Medulla oblongata		

- The spinal cord is properly called the spinal medulla (not to be confused with medulla oblongata, the lowest part of the brainstem, which continues as the spinal medulla).
- Each spinal nerve is formed by the union of ventral and dorsal nerve roots.
- Each nerve root is formed by the union of several rootlets (as at A7).
- The union of ventral and dorsal nerve roots to form a spinal nerve occurs immediately distal to the ganglion on the dorsal root (as at A6), within the intervertebral foramen, and the nerve at once divides into a ventral and a dorsal ramus (formerly called ventral and dorsal primary rami) (as at A18 and 4). The spinal nerve proper is thus only a millimetre or two in length, but is often so short that the rami appear to be branches of the ganglion itself.
- The lowest cervical and upper thoracic nerve roots become acutely angled in order to enter their dural sheaths.

> Lordosis. The cervical and lumbar regions of the normal spine have a natural lordosis (a concavity anteriorly).

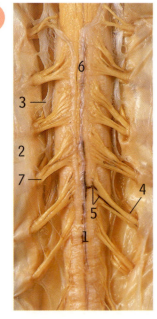

B Cervical region of the spinal cord, from the front
For this ventral view of the upper part of the spinal cord (6), the dura and arachnoid mater have been incised longitudinally and turned aside (2) to show the ventral nerve rootlets and roots (as at 7) passing laterally in front of the denticulate ligament (3) to enter meningeal nerve sheaths with dorsal roots (as at 4) and form a spinal nerve. On some roots, branches of radicular vessels (as at 5) are seen anastomosing with anterior spinal vessels (1)

1	Anterior spinal vessels
2	Arachnoid and dura mater
3	Denticulate ligament
4	Dorsal root of sixth cervical nerve
5	Radicular vessels
6	Spinal cord
7	Ventral root of fifth cervical nerve entering dural sheath

- The denticulate ligament (B3) is composed of pia mater. The ventral and dorsal nerve roots pass respectively ventral and dorsal to the ligament, which extends laterally from the side of the cord and is attached by its spiky denticulations (as at B3) to the arachnoid and dura mater in the intervals between dural nerve sheaths. The highest denticulation is above the first cervical nerve and the lowest below the twelfth thoracic nerve.
- For the continuity of the spinal cord with the brainstem see page 65, B.
- The spinal cord usually ends at the level of the first lumbar vertebra.
- The subarachnoid space ends at the level of the second sacral vertebra.
- The conus medullaris (C2) is the lower, pointed end of the spinal cord.
- The cauda equina (C1) consists of the dorsal and ventral roots of the lumbar, sacral and coccygeal nerves. Note that it is nerve roots which form the cauda, not the spinal nerves themselves; these are not formed until ventral and dorsal roots unite at the level of an intervertebral foramen, immediately distal to the dorsal root ganglion (as at C3).

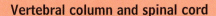

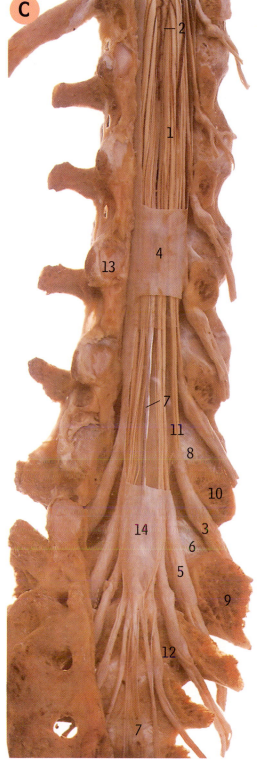

C

Vertebral column, C lumbar and sacral regions, from behind, D lumbar radiculogram
Parts of the vertebral arches and meninges have been removed, to show the cauda equina (1) and nerve roots entering their meningeal sheaths (as at 11), outlined as linear bands by contrast medium in the radiculogram

1 Cauda equina
2 Conus medullaris of spinal cord
3 Dorsal root ganglion of fifth lumbar nerve
4 Dura mater
5 Dural sheath of first sacral nerve roots
6 Fifth lumbar (lumbosacral) intervertebral disc
7 Filum terminale
8 Fourth lumbar intervertebral disc
9 Lateral part of sacrum
10 Pedicle of fifth lumbar vertebra
11 Roots of fifth lumbar nerve
12 Second sacral vertebra
13 Superior articular process of third lumbar vertebra
14 Thecal sac

• If the fifth lumbar intervertebral disc protrudes backwards (the commonest 'slipped disc') it may irritate the roots of the first sacral nerve (C5). This is the general rule for any part of the vertebral column—a protruded disc may irritate the roots of the nerve numbered one below the disc. Note for example that the fifth lumbar nerve roots (C11) within their dural sheath pass laterally immediately below the pedicle of the fifth lumbar vertebra (C10) and so do not come to lie immediately behind the fifth lumbar disc (C6); it is the first sacral roots (C5) which lie in this position. The fifth nerve roots (C11) lie behind the fourth disc (C8).

E Vertebral column and spinal cord, lower thoracic and upper lumbar regions
The specimen is seen from the left with parts of the vertebral arches and meninges removed, to show (at the front) part of the sympathetic trunk (13) on the vertebral bodies and (at the back) the spinous ligaments (7 and 11)

1 Body of first lumbar vertebra
2 Cauda equina
3 Dorsal root ganglion of tenth thoracic nerve
4 Dura mater
5 First lumbar intervertebral disc
6 Greater splanchnic nerve
7 Interspinous ligament
8 Rami communicantes
9 Spinal cord
10 Spinous process of tenth thoracic vertebra
11 Supraspinous ligament
12 Sympathetic ganglion
13 Sympathetic trunk

Spinal anaesthesia uses the same technique as lumbar puncture to insert a needle into the CSF and then introduce anaesthetic agents into the spinal canal. Occasionally, in cases of spinal tumours, an anaesthetic/analgesic may be directly administered via a syringe-driven pump system to control pain in the lower body.

Epidural anaesthesia is regional anaesthesia produced by pharmacological interruption of nerve transmission after placing an anaesthetic agent just outside the dura mater. It is usually carried out in the lower thoracic or lumbar regions by introduction of a needle in exactly the same fashion as a lumbar puncture, except that the needle tip is in the epidural (extradural) space. A small cannula is inserted and the anaesthetic agent dripped in at that level. Tipping of the patient will enable the anaesthetic to reach different spinal levels by gravity.

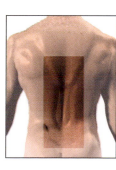

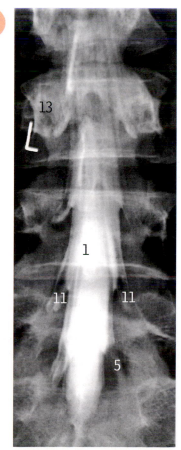

D

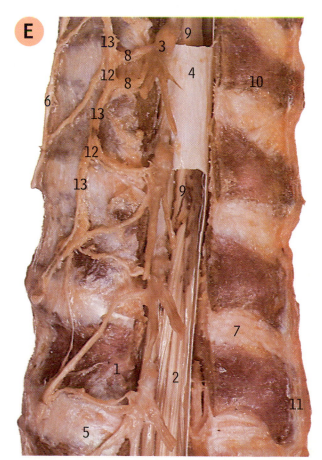

E

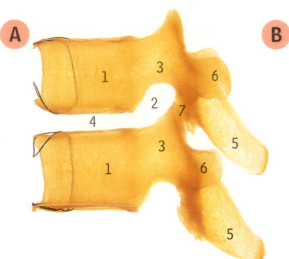

B Vertebral column, lumbar region. Posterior longitudinal ligament
The vertebral arches of the three upper lumbar vertebrae have been cut away through their pedicles (as at 4) and the meninges have been removed to show the posterior longitudinal ligament. Part of the internal vertebral venous plexus has been preserved

1 Internal vertebral venous plexus
2 Intervertebral disc
3 Intervertebral foramen
4 Pedicle of first lumbar vertebra
5 Posterior longitudinal ligament

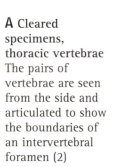

A Cleared specimens, thoracic vertebrae
The pairs of vertebrae are seen from the side and articulated to show the boundaries of an intervertebral foramen (2)

1 Body
2 Intervertebral foramen
3 Pedicle
4 Space for intervertebral disc
5 Spinous process
6 Transverse process
7 Zygapophyseal joint

Vertebral venous plexus (Batson's valveless plexus) is a multichannel set of veins that interconnects both inside and outside the spinal canal from the pelvis to the skull and serves as a transport system for metastases from the breast, prostate, ovary and uterus to the vertebral bodies and the cranial cavity.

• The posterior longitudinal ligament (B5) is broad where it is firmly attached to the intervertebral discs (B2), but narrow and less firmly attached to the vertebral bodies, leaving vascular foramina patent and allowing the basivertebral veins which emerge from them to enter the internal vertebral venous plexus (B1).
• The anterior longitudinal ligament (D1) is uniformly broad and firmly attached to discs and vertebral bodies.

• The intervertebral foramen (A2) is bounded in front by the lower part of the vertebral body (A1) and the intervertebral disc (A4), above and below by the pedicles (A3), and behind by the zygapophyseal joint (A7)

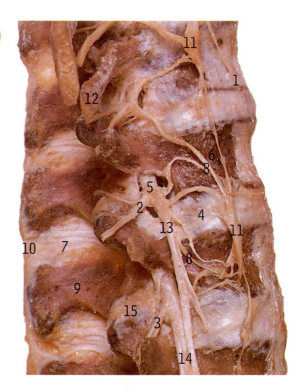

C Vertebral column, lower lumbar region, from the front
At the top the anterior longitudinal ligament (1) has a marker behind it, and part of it lower down has been reflected off an intervertebral disc (4) and vertebral bodies (2 and 3)

Compression of a spinal nerve causes pain, anaesthesia and/or paralysis in its field of distribution. Common causes include a prolapsed intervertebral disc or, in the older patient, the impingement of an osteophyte making the intervertebral foramen too tight to allow free movement of the existing spinal nerve.

1 Anterior longitudinal ligament
2 Body of fifth lumbar vertebra
3 Body of fourth lumbar vertebra
4 Fourth lumbar intervertebral disc
5 Lateral part of sacrum
6 Ventral ramus of fifth lumbar nerve

D Vertebral column, upper lumbar region, from the right
This side view shows lumbar nerves emerging from intervertebral foramina (as at 5)

1 Anterior longitudinal ligament
2 Dorsal ramus of first lumbar nerve
3 Dorsal ramus of second lumbar nerve
4 First lumbar intervertebral disc
5 First lumbar nerve emerging from intervertebral foramen
6 First lumbar vertebra
7 Interspinous ligament
8 Rami communicantes
9 Spinous process of second lumbar vertebra
10 Supraspinous ligament
11 Sympathetic trunk ganglion
12 Twelfth rib
13 Ventral ramus of first lumbar nerve
14 Ventral ramus of second lumbar nerve
15 Zygapophyseal joint

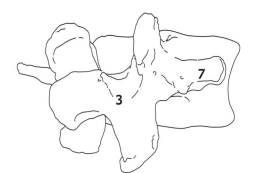

A Vertebral column, lumbar region, from the right and behind

This posterolateral view of the right side of some lumbar vertebrae shows ligamenta flava (as at 4), which pass between the laminae of adjacent vertebrae (as at 2 and 3)

1 Interspinous ligament
2 Lamina of second lumbar vertebra
3 Lamina of third lumbar vertebra
4 Ligamentum flavum
5 Spinous process of second lumbar vertebra
6 Supraspinous ligament
7 Transverse process of third lumbar vertebra
8 Zygapophyseal joint

Discectomy is a 'minimally invasive surgery' technique for disc disease in which, instead of removing the bony lamina, the operation is performed using a microscope to remove just that part of a protruding disc that is pressing against a nerve root and causing limb pain.

Lumbar puncture (spinal tap) is a procedure used to obtain cerebrospinal fluid (CSF) for diagnosis or to access the CSF for drug delivery. Entering the midline of the back in the space between the L4 and L5 vertebrae, the needle passes through the skin, subcutaneous tissue, supraspinous ligament and interspinous ligament between the ligamenta flava, and into the epidural space. It then passes through the tough dura mater into the subarachnoid space.

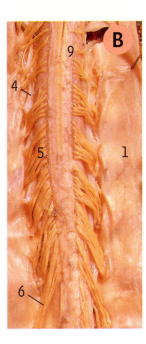

Spinal cord and cauda equina, B dorsal surface of upper end, C dorsal surface of lower end with cauda

The dura and arachnoid mater have been incised longitudinally and turned outwards (B1 and C1) to show the nerve roots entering their dural sheaths (as at B6 and C7). Below the level of the conus medullaris (the lower end of the spinal cord, C3) the nerve roots constitute the cauda equina (C2). Compare B with the ventral surface of the cervical part of the cord (page 80, B)

1 Annulus fibrosus
2 Nucleus pulposus
3 Plate of hyaline cartilage

- The nucleus pulposus of an intervertebral disc represents the remains of the notochord.
- The annulus fibrosus of an intervertebral disc is derived from the mesenchyme between adjacent vertebral bodies.

1 Arachnoid overlying dura mater
2 Cauda equina
3 Conus medullaris
4 Denticulate ligament
5 Dorsal rootlets of fifth cervical nerve
6 Eighth cervical nerve roots entering dural sheath
7 Fifth lumbar nerve roots entering dural sheath
8 Filum terminale
9 Spinal cord

- The filum terminale (C8), which consists of connective tissue, not neural elements, extends from the tip of the conus medullaris (C3) through the subarachnoid space to the level of the second sacral vertebra where it fuses with the dura mater and continues downwards to become attached to the first piece of the coccyx.

D Intervertebral disc

This disc, on the upper surface of the body of a lumbar vertebra, has been cut horizontally to show the central nucleus pulposus (2) and the concentric fibrocartilaginous laminae of the surrounding annulus fibrosus (1). At the back the annulus has been shaved off to reveal part of the plate of hyaline cartilage (3) on the surface of the vertebra.

Herniated disc, known also as a 'slipped disc', is due to protrusion of the nucleus pulposus through the annulus fibrosus, most commonly seen in the lower lumbar region.

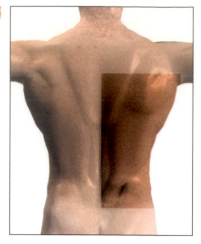

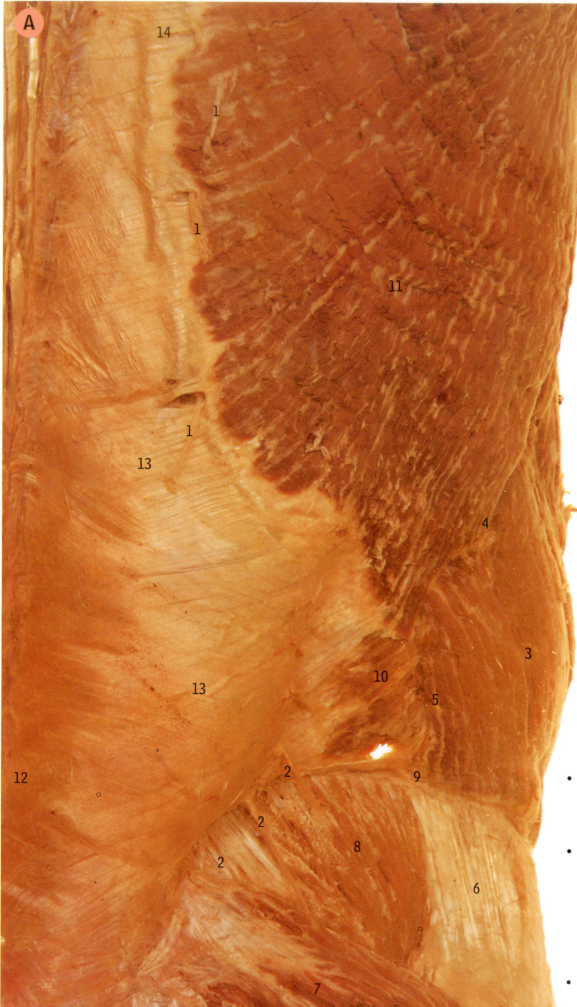

A Muscles of the vertebral column. Right erector spinae and thoracolumbar fascia
The thoracolumbar fascia (14 and 13) covers erector spinae and laterally gives origin to latissimus dorsi (11, a muscle of the upper limb) and internal oblique (10, a muscle of the anterolateral abdominal wall)

1 Branches of dorsal rami of thoracic nerves
2 Cutaneous branches of dorsal rami of first three lumbar nerves
3 External oblique
4 Free lateral border of latissimus dorsi
5 Free posterior border of external oblique
6 Gluteal fascia
7 Gluteus maximus
8 Gluteus medius
9 Iliac crest
10 Internal oblique
11 Latissimus dorsi
12 Level of fourth lumbar spinous process
13 Posterior layer of lumbar part of thoracolumbar fascia overlying erector spinae
14 Thoracic part of thoracolumbar fascia overlying erector spinae

- The thoracolumbar fascia consists of a thoracic part, single-layered, covering the thoracic part of the erector spinae (14), and a lumbar part (where there are no ribs) which is commonly called simply the lumbar fascia and which consists of three layers.
- The posterior layer (13) is continuous with the thoracic part (14). The middle and anterior layers are usually studied with the posterior abdominal wall; the quadratus lumborum muscle lies between them, and psoas major is in front of the anterior layer. The lumbar part of erector spinae is between the posterior and middle layers. The three layers come together approximately at the lateral border of erector spinae
- For other parts of erector spinae see opposite and page 86.

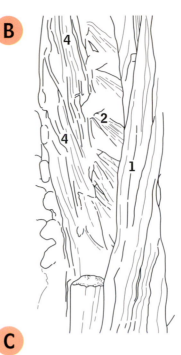

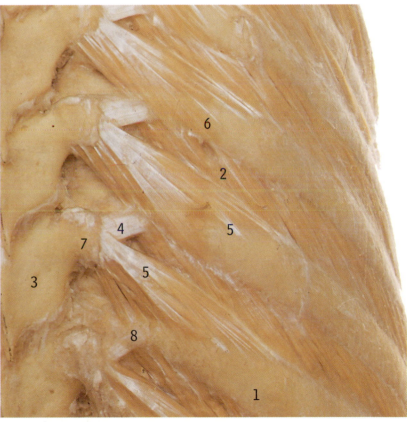

B Muscles of the vertebral column, right thoracolumbar region
Some of the iliocostalis (1), longissimus (3) and spinalis (4) parts of erector spinae are shown, together with levator costae muscles (2)

- In the upper lumbar region, erector spinae divides into three muscle masses: iliocostalis (1) laterally, an intermediate longissimus (3, mostly removed in this specimen), and spinalis (4) medially.
- The levator costae muscles (2) are classified as muscles of the thorax, not of the vertebral column. They are revealed here because much of the longissimus part of the erector spinae (3) has been removed.

1	Iliocostalis
2	Levator costae
3	Lower part of longissimus
4	Parts of spinalis
5	Spine of eighth thoracic vertebra

C Muscles of the thorax. Levator costae muscles, right side, from behind
The left erector spinae and latissimus dorsi have been removed from the left side of the vertebral column and adjacent ribs to show the levator costae muscles (as at 5) and the medial ends of the external intercostals (as at 2)

1	Angle of ninth rib
2	External intercostal
3	Lamina of eighth thoracic vertebra
4	Lateral costotransverse ligament
5	Levator costae
6	Seventh rib
7	Transverse process of eighth thoracic vertebra
8	Tubercle of ninth rib

Muscles of the back

Muscles of the back and thorax, A left erector spinae and serratus posterior inferior, from behind, **B** surface anatomy

In the view (A) of the lower left thorax and lumbar region from behind, latissimus dorsi has been removed to display serratus posterior inferior (9), part of whose aponeurotic origin from vertebral spines has also been removed to uncover part of erector spinae (7, 10 and 2) (which belongs to the vertebral column group of muscles—pages 84 and 85). The surface view in B, with the arms abducted, shows well-developed erector spinae muscles

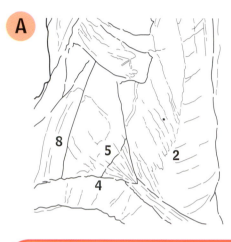

1 Dorsal rami of lower thoracic and upper lumbar nerves
2 Erector spinae
3 External intercostal
4 Iliac crest
5 Internal oblique
6 Latissimus dorsi
7 Longissimus part of erector spinae
8 Posterior (free) border of external oblique
9 Serratus posterior inferior
10 Spinalis part of erector spinae
11 Tenth rib

• The medial part of serratus posterior *inferior* (9) (arising from the last two thoracic and upper two lumbar spinous processes and the supraspinous ligament, and blending with the underlying lumbar part of the thoracolumbar fascia) has been removed, so displaying the medial and intermediate parts of erector spinae (7 and 10) which belongs to the muscles of the vertebral column (page 84). The lateral (iliocostalis) part of erector spinae is under cover of the lateral part of the serratus muscle, which becomes attached to the lower four ribs lateral to their angles.

• The serratus posterior *superior* muscle (not illustrated) passes to the second to fifth ribs lateral to their angles, under cover of the rhomboid muscles (page 110), having arisen from the lower part of the ligamentum nuchae and the spinous processes of the seventh cervical and upper two or three thoracic vertebrae and the supraspinous ligament.

• On each side there is one serratus *anterior* muscle (belonging to the group connecting the upper limb to the trunk) and two serratus *posterior* muscles (belonging to the muscles of the thorax).

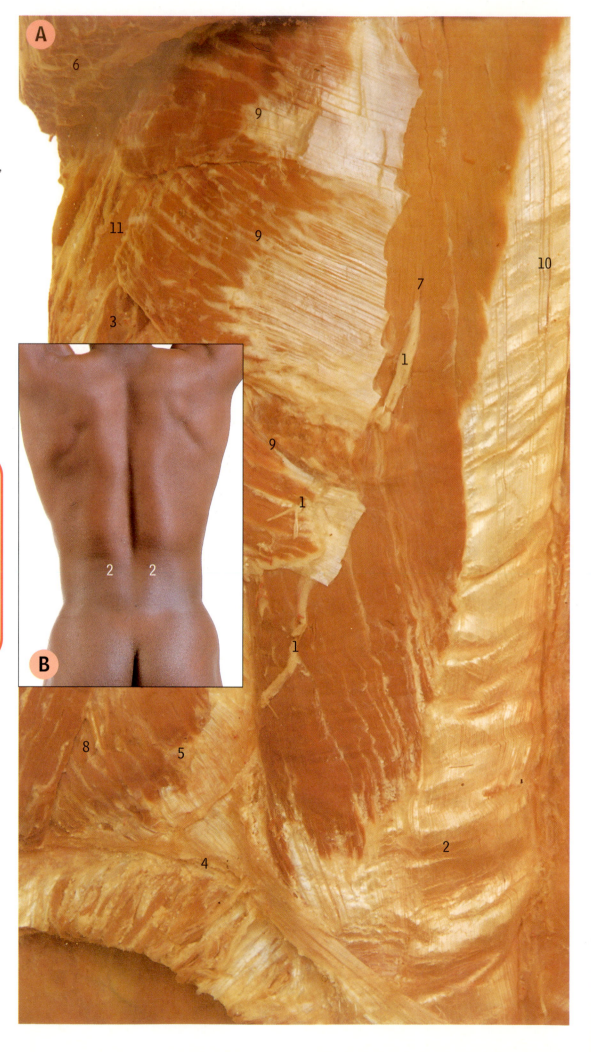

A Radiograph of upper cervical vertebrae
This is a standard radiographic view of the axis and its dens (4). The correct angle must be chosen with the mouth open to avoid overlying shadows of the teeth and jaws. The surfaces of the lateral atlanto-axial joints (5 and 7) do not appear congruent because the hyaline cartilage which covers the bony surfaces is not radio-opaque (this applies to any synovial joint). The outlines of the arches of the atlas are seen faintly between the sides of the shadow of the dens (4) and the lateral masses of the atlas (5)

1 Bifid spinous process of axis
2 Body of axis
3 Body of third cervical vertebra
4 Dens of axis
5 Inferior articular surface of lateral mass of atlas
6 Lateral atlanto-axial joint
7 Superior articular surface of axis

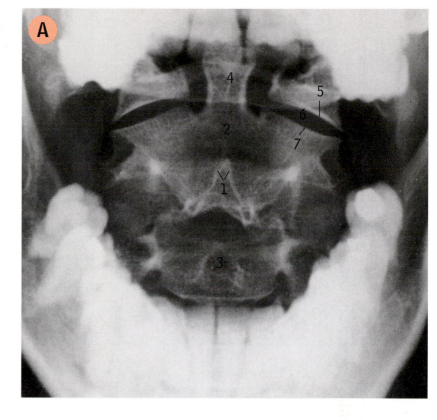

B Radiograph of lower cervical and upper thoracic vertebrae, from the front
Note the tracheal shadow produced by the translucency of its contained air

1 Body of first thoracic vertebra
2 Body of sixth cervical vertebra
3 Head of first rib
4 Head of second rib
5 Margin of tracheal shadow
6 Neck of first rib
7 Neck of second rib
8 Shaft of first rib
9 Transverse process of first thoracic vertebra
10 Tubercle of first rib
11 Tubercle of second rib

Cervical spinal immobilization. Following actual or suspected injuries to the cervical spine it is extremely important to keep the spine immobile and under slight traction to prevent spinal cord compression and, therefore, paraplegia or quadriplegia while transporting the patient to hospital.

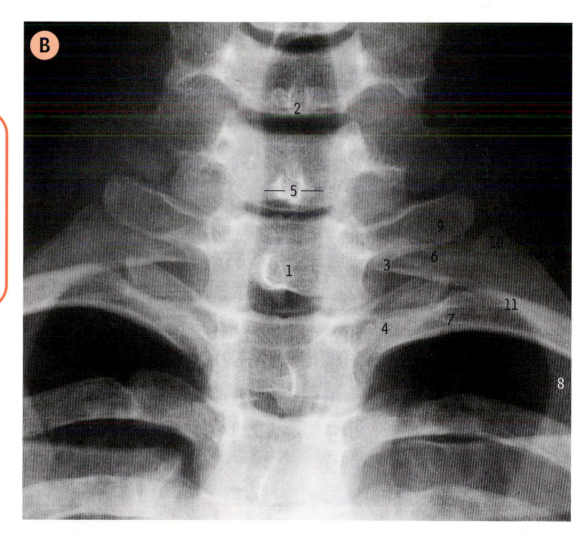

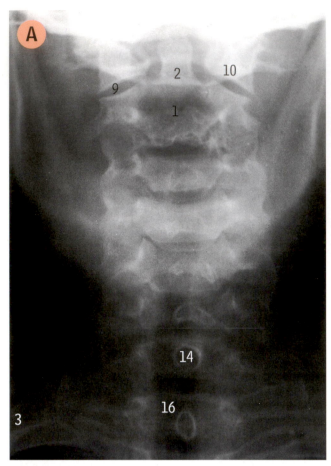

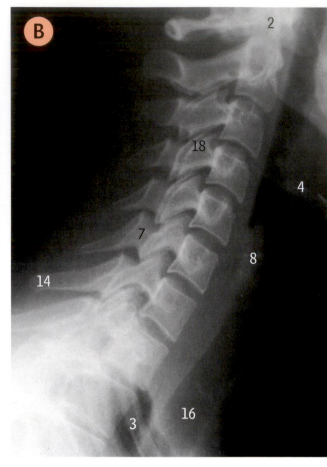

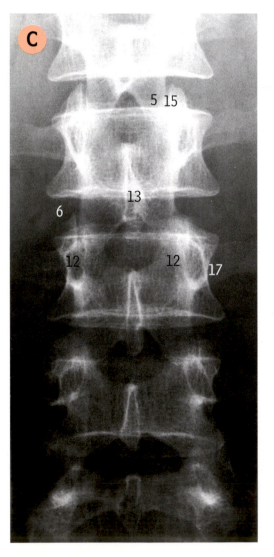

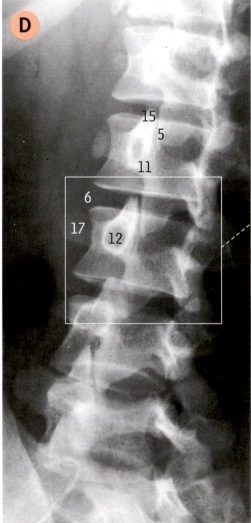

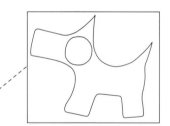

Vertebral radiographs, **A** cervical spine, anteroposterior projection, **B** cervical spine, lateral projection, **C** lumbar spine, anteroposterior projection, **D** lumbar spine, oblique projection

1 Body of axis
2 Dens of axis
3 First rib
4 Hyoid bone
5 Inferior articular process of first lumbar vertebra
6 Intervertebral disc space L2/3 level
7 Lamina of sixth cervical vertebra
8 Larynx
9 Lateral atlanto-axial joint
10 Lateral mass of atlas
11 Pars interarticularis of second lumbar vertebra
12 Pedicle of third lumbar vertebra
13 Spinous process of second lumbar vertebra
14 Spinous process of seventh cervical vertebra
15 Superior articular process of second lumbar vertebra
16 Trachea
17 Transverse process of third lumbar vertebra
18 Zygapophyseal joint

• The Scottie dog is seen on the oblique projection lumbar spine. The nose (17) is the transverse process, the ear (15) is the superior articular process, the eye (12) is the pedicle and the neck (11) is the pars interarticularis which may be incomplete in spondylolisthesis.

Vertebral fractures may occur anywhere along the vertebral column. A common lumbar site, due to a congenital malformation in the pars interarticularis (spondylolisthesis), is best diagnosed by an oblique lumbar spine radiograph, which reveals a little Scottie dog profile with the fracture across the dog's collar.

Chapter 3

Upper limb

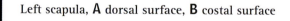

Left scapula, **A** dorsal surface, **B** costal surface

1	Acromial angle	9	Neck (and spinoglenoid notch on dorsal surface)
2	Acromion		
3	Coracoid process	10	Spine
4	Inferior angle	11	Subscapular fossa
5	Infraspinous fossa	12	Superior angle
6	Lateral border	13	Superior border
7	Margin of glenoid cavity	14	Suprascapular notch
8	Medial border	15	Supraspinous fossa

- The spine (A10) of the scapula projects from its dorsal surface with the acromion (A2) at the lateral end of the spine.
- The glenoid cavity (A7) is at the upper lateral angle, for articulation with the head of the humerus.

- The shoulder joint is the glenohumeral joint, and is the articulation between the glenoid cavity of the scapula and the head of the humerus.
- The suprascapular notch is bridged by the superior transverse scapular ligament (15).
- The conoid (1) and trapezoid (20) ligaments together form the coracoclavicular ligament, which attaches the coracoid process of the scapula to the under-surface of the lateral end of the clavicle (C2 and C6, opposite).
- The coracohumeral ligament (page 92, A3) reinforces the upper part of the capsule of the shoulder joint.
- The coraco-acromial ligament (2) passes between the coracoid process and the acromion, forming with these bony processes an arch above the shoulder joint.

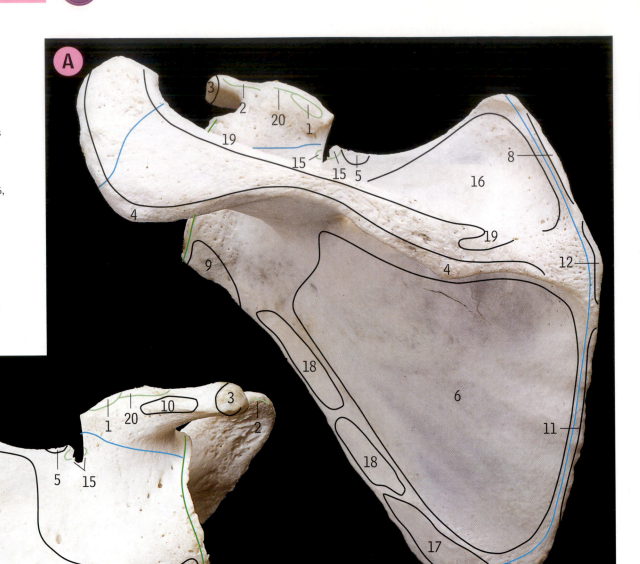

Left scapula, A dorsal surface, B costal surface. Attachments
Blue lines = epiphysial lines; green lines = capsular attachments of shoulder joint; pale green lines = ligament attachments

1 Conoid ligament of coraco-clavicular ligament	12 Rhomboid minor
2 Coraco-acromial ligament	13 Serratus anterior
3 Coracobrachialis and short head of biceps	14 Subscapularis
4 Deltoid	15 Superior transverse scapular ligament
5 Inferior belly of omohyoid	16 Supraspinatus
6 Infraspinatus	17 Teres major
7 Latissimus dorsi	18 Teres minor and intervening groove for circumflex scapular artery
8 Levator scapulae	19 Trapezius
9 Long head of triceps	20 Trapezoid ligament of coraco-clavicular ligament
10 Pectoralis minor	
11 Rhomboid major	

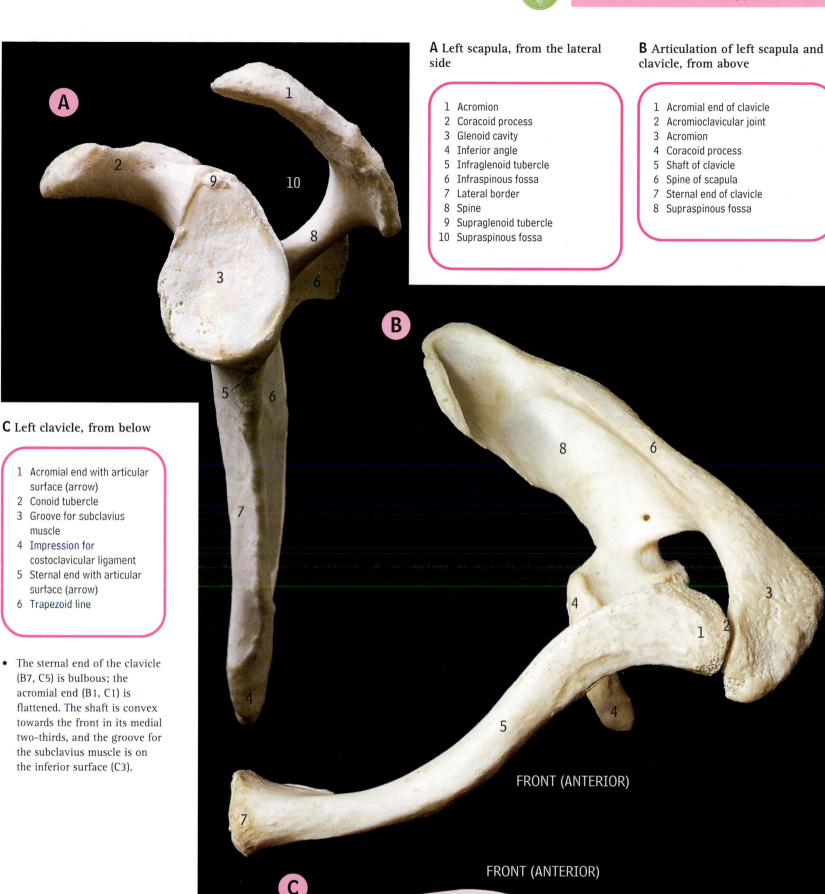

A Left scapula, from the lateral side

1 Acromion
2 Coracoid process
3 Glenoid cavity
4 Inferior angle
5 Infraglenoid tubercle
6 Infraspinous fossa
7 Lateral border
8 Spine
9 Supraglenoid tubercle
10 Supraspinous fossa

B Articulation of left scapula and clavicle, from above

1 Acromial end of clavicle
2 Acromioclavicular joint
3 Acromion
4 Coracoid process
5 Shaft of clavicle
6 Spine of scapula
7 Sternal end of clavicle
8 Supraspinous fossa

C Left clavicle, from below

1 Acromial end with articular surface (arrow)
2 Conoid tubercle
3 Groove for subclavius muscle
4 Impression for costoclavicular ligament
5 Sternal end with articular surface (arrow)
6 Trapezoid line

• The sternal end of the clavicle (B7, C5) is bulbous; the acromial end (B1, C1) is flattened. The shaft is convex towards the front in its medial two-thirds, and the groove for the subclavius muscle is on the inferior surface (C3).

FRONT (ANTERIOR)

FRONT (ANTERIOR)

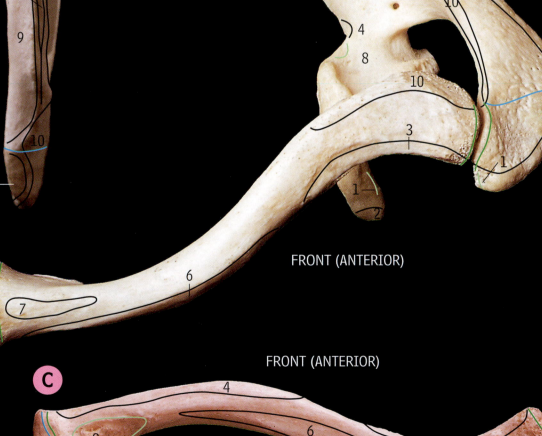

A Left scapula, from the lateral side. Attachments
Blue lines = epiphysial lines; green lines = capsular attachments of shoulder joint; pale green lines = ligament attachments

1 Coraco-acromial ligament	7 Long head of triceps
2 Coracobrachialis and short head of biceps	8 Serratus anterior
3 Coracohumeral ligament	9 Subscapularis
4 Deltoid	10 Teres major
5 Infraspinatus	11 Teres minor (with intervening groove for circumflex scapular artery)
6 Long head of biceps	

B Articulation of left scapula and clavicle, from above
Blue lines = epiphysial lines; green lines = capsular attachments of sternoclavicular and acromioclavicular joints; pale green lines = ligament attachments

1 Coraco-acromial ligament
2 Coracobrachialis and short head of biceps
3 Deltoid
4 Inferior belly of omohyoid
5 Levator scapulae
6 Pectoralis major
7 Sternocleidomastoid
8 Superior transverse scapular ligament
9 Supraspinatus
10 Trapezius

C Left clavicle, from below. Attachments
Blue lines = epiphysial lines; green lines = capsular attachments of sternoclavicular and acromioclavicular joints; pale green lines = ligament attachments

1 Conoid ligament
2 Costoclavicular ligament
3 Deltoid
4 Pectoralis major
5 Sternohyoid
6 Subclavius and clavipectoral fascia
7 Trapezius
8 Trapezoid ligament

FRONT (ANTERIOR)

FRONT (ANTERIOR)

Right humerus, upper end, **A** from the front, **B** from behind, **C** from the medial side, **D** from the lateral side, **E** from above

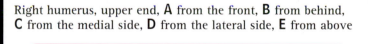

1	Anatomical neck	7	Lateral lip of intertubular groove
2	Deltoid tuberosity		
3	Greater tubercle	8	Lesser tubercle
4	Groove for radial nerve	9	Medial lip of intertubercular groove
5	Head		
6	Intertubercular groove	10	Surgical neck

- The intertubercular (bicipital) groove (A6) is on the front of the upper end and is occupied by the tendon of the long head of biceps. (For attachments see next page.)

Dislocation of the humerus. Dislocation of the humeral head at the shoulder joint is a relatively common injury and is in an inferior (no muscle support) or anterior (rotator cuff tear) direction, risking injury to the axillary nerve. An anterior dislocation is usually associated with a capsular or labral injury.

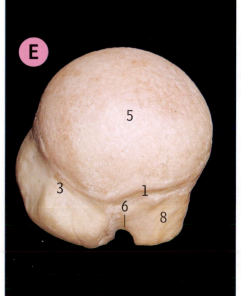

Right humerus, upper end, **A** from the front, **B** from behind, **C** from the medial side, **D** from the lateral side, **E** from above.
Attachments
Blue lines = epiphysial lines; green lines = capsular attachment of shoulder joint

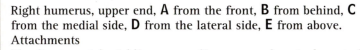

1 Brachialis	7 Medial head of triceps
2 Coracobrachialis	8 Pectoralis major
3 Deltoid	9 Subscapularis
4 Infraspinatus	10 Supraspinatus
5 Lateral head of triceps	11 Teres major
6 Latissimus dorsi	12 Teres minor

- Deltoid is attached to the V-shaped deltoid tuberosity (A3 and D3) on the *lateral* surface of the middle of the shaft.
- Coracobrachialis is attached to the *medial* surface of the middle of the shaft (C2) (opposite the deltoid tuberosity).
- Note the relative positions of the epiphysial and capsular lines: the epiphysis is partly intracapsular and partly extracapsular at the upper end of the humerus.

Right humerus, lower end, **A** from the front, **B** from behind, **C** from below, **D** from the medial side, **E** from the lateral side

1 Anterior surface
2 Capitulum
3 Coronoid fossa
4 Lateral edge of capitulum
5 Lateral epicondyle
6 Lateral supracondylar ridge
7 Medial epicondyle
8 Medial supracondylar ridge
9 Medial surface of trochlea
10 Olecranon fossa
11 Posterior surface
12 Radial fossa
13 Trochlea

- The medial epicondyle (7) is more prominent than the lateral (5).
- The medial part of the trochlea (13) is more prominent than the lateral part.
- The olecranon fossa (10) on the posterior surface is deeper than the radial and coronoid fossae on the anterior surface (12 and 3).

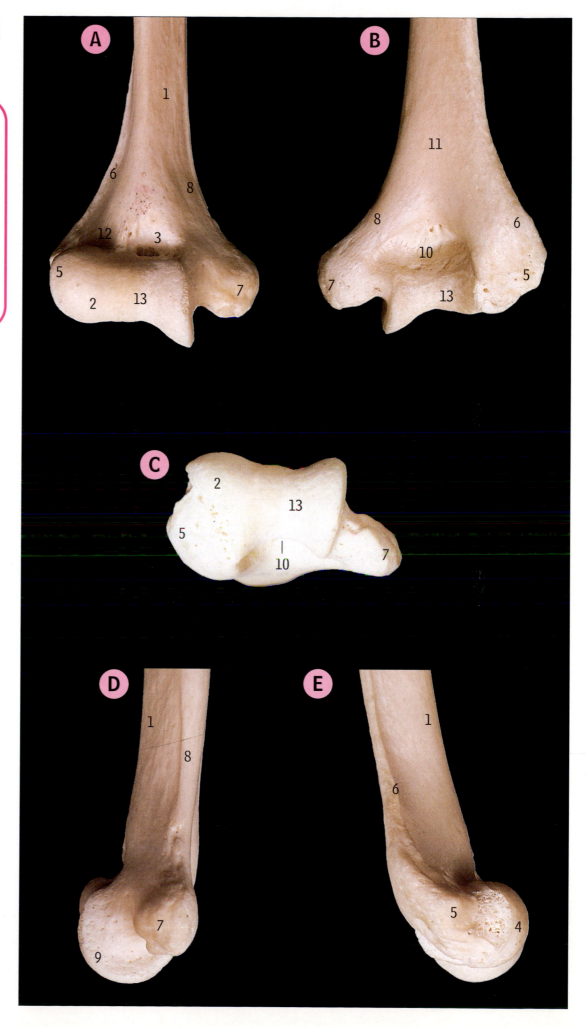

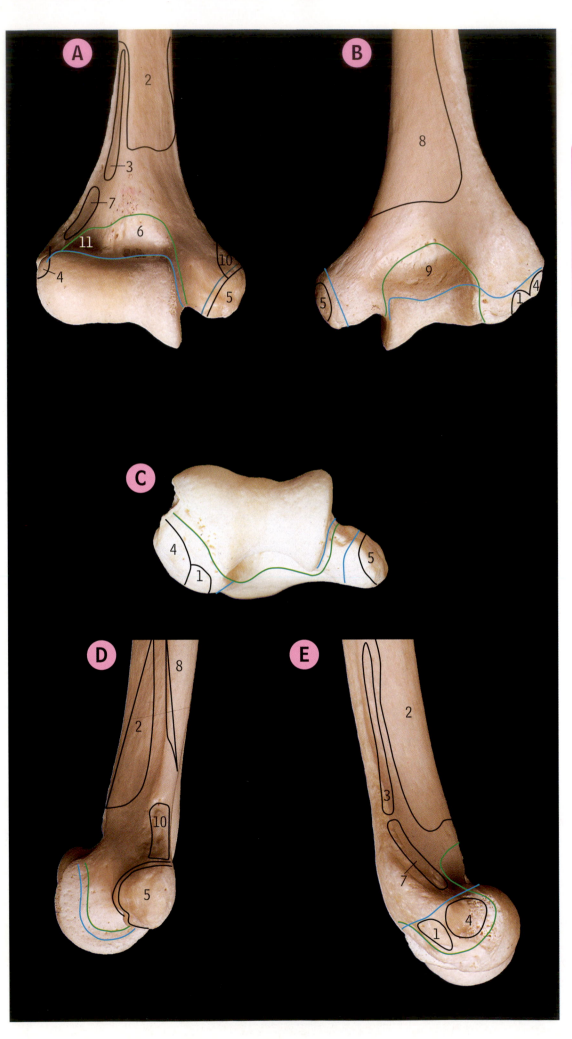

Right humerus, lower end, **A** from the front, **B** from behind, **C** from below, **D** from the medial side, **E** from the lateral side. Attachments
Blue lines = epiphysial lines; green lines = capsular attachment of elbow joint

1 Anconeus
2 Brachialis
3 Brachioradialis
4 Common extensor origin
5 Common flexor origin
6 Coronoid fossa
7 Extensor carpi radialis longus
8 Medial head of triceps
9 Olecranon fossa
10 Pronator teres, humeral head
11 Radial fossa

- The ulnar and radial collateral ligaments of the elbow joint are attached to the medial and lateral epicondyles respectively (beneath the common flexor and extensor origins, 5 and 4).
- Pain from a tear of the common extensor origin at 4 is known as 'tennis elbow'. Note that extensor carpi radialis longus (7) and brachioradialis (3) arise from the humeral shaft above the common extensor origin.
- Note that the capsular attachment runs above the radial, coronoid and olecranon fossae (11, 6 and 9) which therefore lie within the joint. The fossae house fat pads which are intracapsular and extrasynovial in position.

Right radius, upper end, **A** from the front, **B** from behind, **C** from the medial side, **D** from the lateral side

 1 Anterior border
 2 Anterior oblique line
 3 Anterior surface
 4 Head
 5 Interosseous border
 6 Lateral surface
 7 Neck
 8 Posterior border
 9 Posterior surface
10 Rough area for pronator teres
11 Tuberosity

- The head of the radius (4) is at its upper end; the head of the ulna is at its lower end (page 98, E3).
- The tuberosity (11) is rough posteriorly for the attachment of the biceps tendon, and smooth anteriorly where it is covered by the intervening bursa.
- The shaft is triangular in cross-section, and its surfaces are anterior (3), posterior (9) and lateral (6); its borders are interosseous (5), anterior (1) and posterior (8) (compare with the ulna, page 98).

Right radius, lower end, **E** from the front, **F** from behind, **G** from the medial side, **H** from the lateral side

 1 Anterior surface
 2 Dorsal tubercle
 3 Groove for abductor pollicis longus
 4 Groove for extensor carpi radialis brevis
 5 Groove for extensor carpi radialis longus
 6 Groove for extensor digitorum and extensor indicis
 7 Groove for extensor pollicis brevis
 8 Groove for extensor pollicis longus
 9 Interosseous border
10 Lateral surface
11 Posterior surface
12 Styloid process
13 Ulnar notch

- The lower end of the radius is concave anteriorly (at the lower label 1 in E), with the ulnar notch medially (G13) and the dorsal tubercle on the posterior surface (F2).

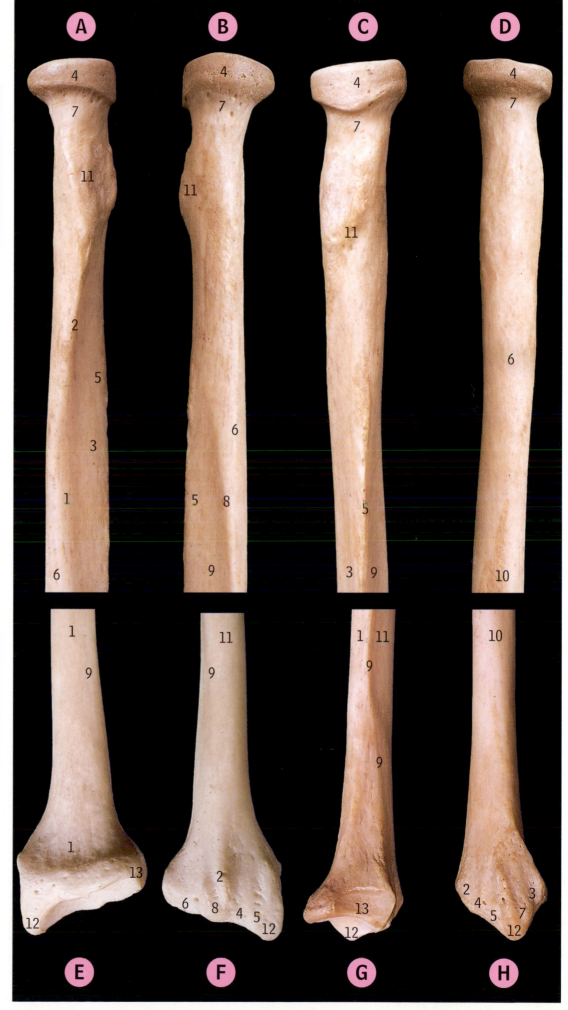

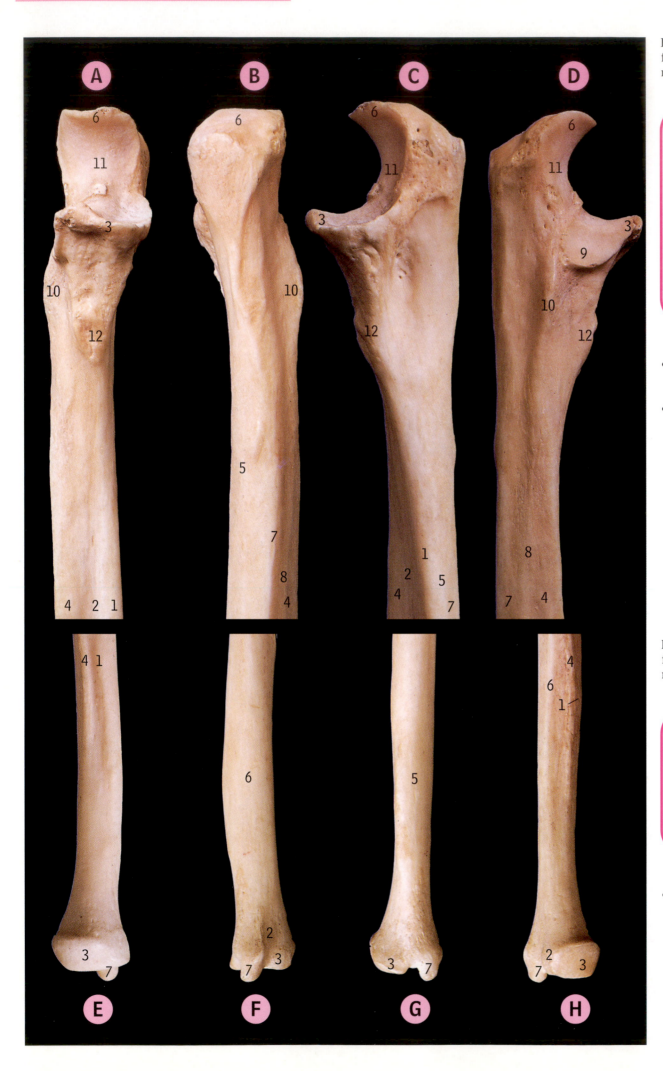

Right ulna, upper end, **A** from the front, **B** from behind, **C** from the medial side, **D** from the lateral side

1 Anterior border
2 Anterior surface
3 Coronoid process
4 Interosseous border
5 Medial surface
6 Olecranon
7 Posterior border
8 Posterior surface
9 Radial notch
10 Supinator crest
11 Trochlear notch
12 Tuberosity

- The trochlear notch (11) faces forwards, with the radial notch (9) on the lateral side.
- The upper part of the shaft is triangular in cross-section but the lower quarter is almost cylindrical. The surfaces of the shaft are anterior (2), posterior (8) and medial (5); the borders are interosseous (4), anterior (1) and posterior (7) (compare with the radius, page 97).

Right ulna, lower end, **E** from the front, **F** from behind, **G** from the medial side, **H** from the lateral side

1 Anterior surface
2 Groove for extensor carpi ulnaris
3 Head
4 Interosseous border
5 Medial surface
6 Posterior surface
7 Styloid process

- The head of the ulna (3) is at its lower end, with the styloid process (7) situated posteromedially. The head of the radius is at its upper end (page 97).

A Right radius and ulna, upper ends, from above and in front

1 Coronoid process of ulna
2 Head of radius
3 Neck of radius
4 Olecranon of ulna
5 Trochlear notch of ulna
6 Tuberosity of radius
7 Tuberosity of ulna

B Right radius and ulna, lower ends, from below

1 Attachment of articular disc
2 Dorsal tubercle
3 Groove for extensor carpi radialis brevis
4 Groove for extensor carpi radialis longus
5 Groove for extensor carpi ulnaris
6 Groove for extensor digitorum and extensor indicis
7 Groove for extensor pollicis longus
8 Styloid process of radius
9 Styloid process of ulna
10 Surface for disc
11 Surface for lunate
12 Surface for scaphoid

Articulation of right humerus, radius and ulna, **C** from the front, **D** from behind

1 Capitulum of humerus
2 Coronoid process of ulna
3 Head of radius
4 Lateral epicondyle of humerus
5 Medial epicondyle of humerus
6 Olecranon of ulna
7 Radial notch of ulna
8 Trochlea of humerus

• The elbow joint and the proximal radio-ulnar joint share a common synovial cavity.

Supracondylar fracture of the humerus is usually seen in young children following a fall on the outstretched hand, producing a posterior displacement of the fragment. The structures at risk are the brachial artery and the median nerve and in a postero-lateral displacement, the radial nerve. If the artery has a fragment of bone against it, ischaemic changes may occur in the forearm and hand which will lead to a Volkmann's ischaemic contracture.

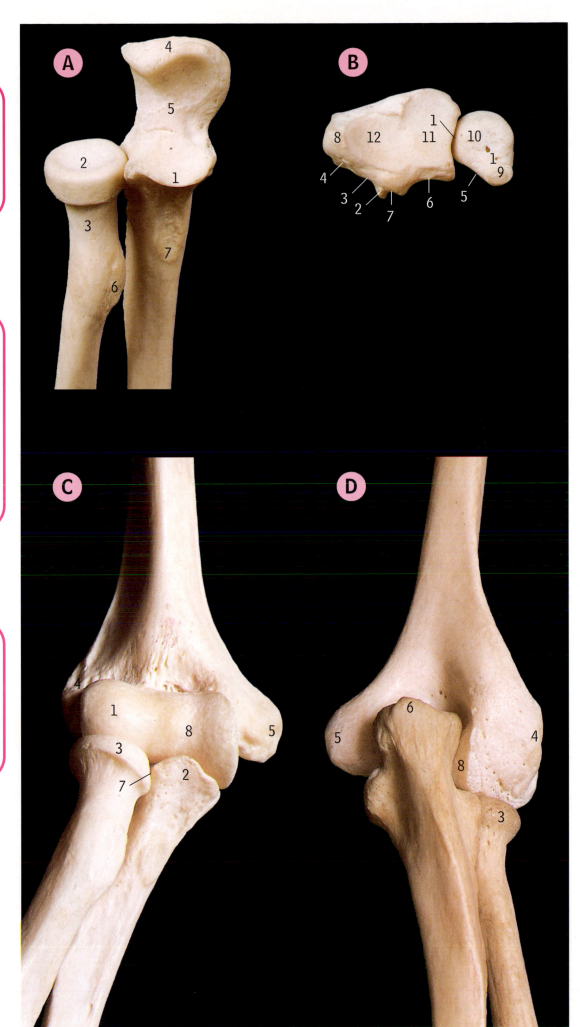

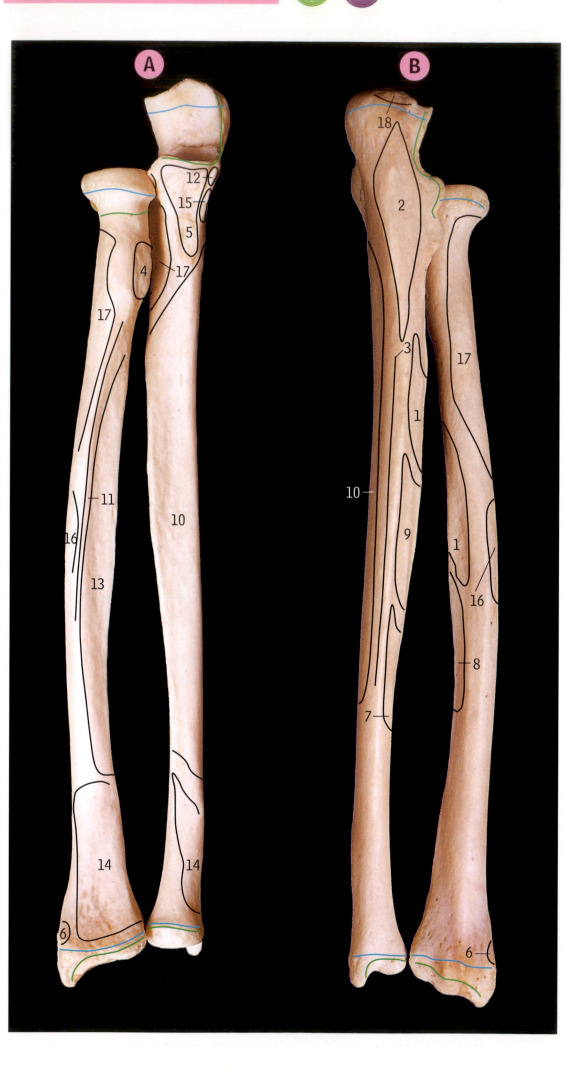

Right radius and ulna, **A** from the front, **B** from behind. Attachments
Blue lines = epiphysial lines; green lines = capsular attachments of elbow and wrist joints

1 Abductor pollicis longus
2 Anconeus
3 Aponeurotic attachment of flexor digitorum profundus, flexor carpi ulnaris and extensor carpi ulnaris
4 Biceps
5 Brachialis
6 Brachioradialis
7 Extensor indicis
8 Extensor pollicis brevis
9 Extensor pollicis longus
10 Flexor digitorum profundus
11 Flexor digitorum superficialis, radial head
12 Flexor digitorum superficialis, ulnar head
13 Flexor pollicis longus
14 Pronator quadratus
15 Pronator teres, ulnar head
16 Pronator teres
17 Supinator
18 Triceps

- Abductor pollicis longus (1) and extensor pollicis brevis (8) are the only two muscles to have an origin from the posterior surface of the radius (although both extend on to the interosseous membrane and the abductor also has an origin from the posterior surface of the ulna). These muscles remain companions as they wind round the lateral side of the radius (page 127, C2 and 11) and form the radial boundary of the anatomical snuffbox (page 128, B1 and 5).

- Flexor pollicis longus has an occasional small additional origin from the lateral (or rarely the medial) side of the coronoid process of the ulna (beside the lower part of the brachialis attachment).

- Note the position of the epiphysial and capsular lines at the lower ends of the bones; the epiphyses are extracapsular.

- In the young subject the radius sometimes fractures across the lower epiphysis following an injury to the wrist. In the adult the term 'Colles' fracture' refers to a transverse break across the lower radius within about 2.5 cm of the lower end of the bone. The ulnar styloid process is also often fractured.

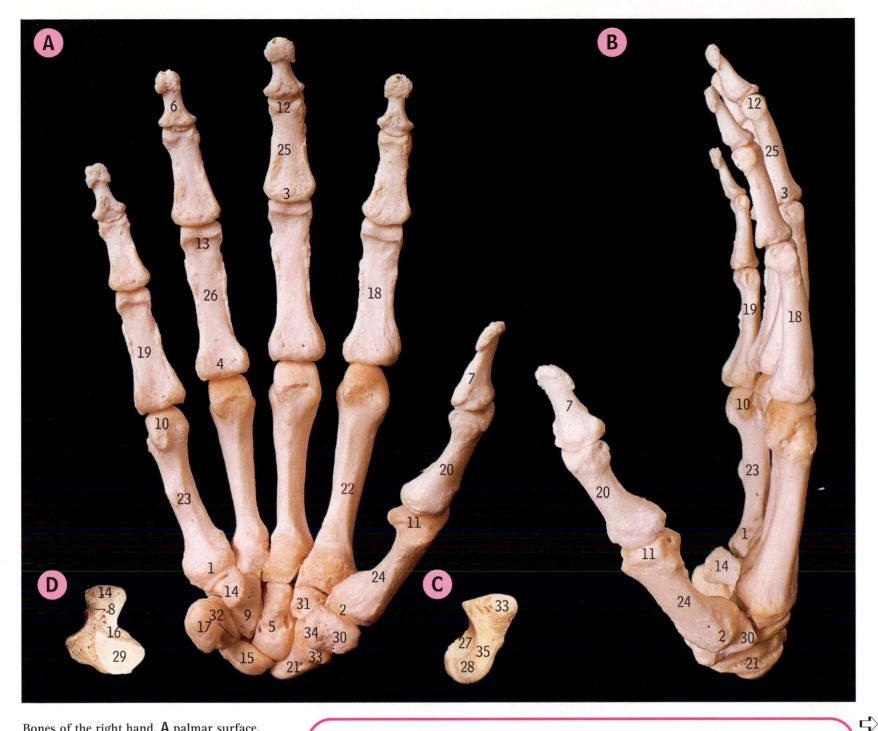

Bones of the right hand, **A** palmar surface, **B** from the lateral side, **C** scaphoid, palmar surface, **D** hamate from the medial side

- The scaphoid, lunate, triquetral and pisiform bones form the proximal row of carpal bones.
- The trapezium, trapezoid, capitate and hamate bones form the distal row of carpal bones.
- The tubercle (33) and waist (35) are the non-articular parts of the scaphoid and therefore contain nutrient foramina. A fracture across the waist may therefore interfere with the blood supply of the proximal pole of the bone and lead to avascular necrosis. The waist of the scaphoid lies in the anatomical snuffbox; the tubercle may be palpated in front of the radial boundary of the snuffbox.

1 Base of fifth metacarpal	19 Proximal phalanx of little finger
2 Base of first metacarpal	20 Proximal phalanx of thumb
3 Base of middle phalanx of middle finger	21 Scaphoid
4 Base of proximal phalanx of ring finger	22 Shaft of second metacarpal
5 Capitate	23 Shaft of fifth metacarpal
6 Distal phalanx of ring finger	24 Shaft of first metacarpal
7 Distal phalanx of thumb	25 Shaft of middle phalanx of middle finger
8 Groove for deep branch of ulnar nerve	26 Shaft of proximal phalanx of ring finger
9 Hamate	27 Surface for capitate
10 Head of fifth metacarpal	28 Surface for lunate
11 Head of first metacarpal	29 Surface for triquetral
12 Head of middle phalanx of middle finger	30 Trapezium
13 Head of proximal phalanx of ring finger	31 Trapezoid
14 Hook of hamate	32 Triquetral
15 Lunate	33 Tubercle of scaphoid
16 Palmar surface, hamate	34 Tubercle of trapezium
17 Pisiform	35 Waist of scaphoid
18 Proximal phalanx of index finger	

Upper limb bones

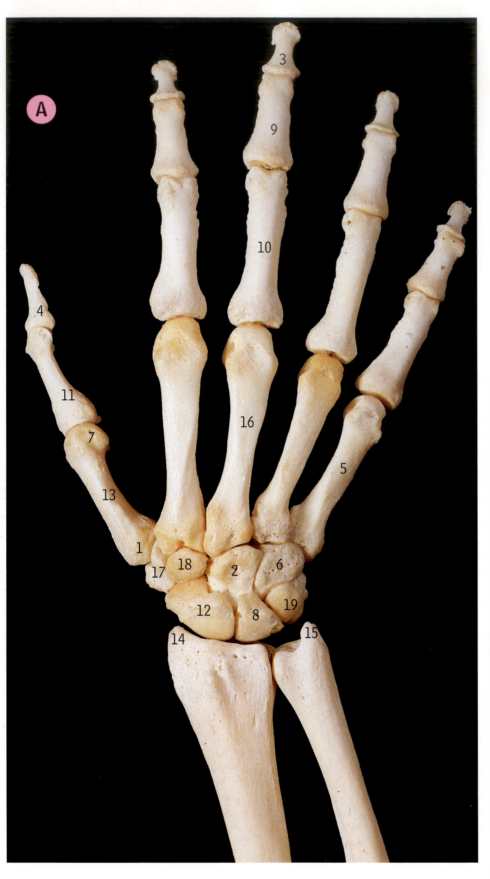

Bones of the right hand, **A** dorsal surface

1	Base of first metacarpal	11	Proximal phalanx of thumb
2	Capitate	12	Scaphoid
3	Distal phalanx of middle finger	13	Shaft of first metacarpal
4	Distal phalanx of thumb	14	Styloid process of radius
5	Fifth metacarpal	15	Styloid process of ulna
6	Hamate	16	Third metacarpal
7	Head of first metacarpal	17	Trapezium
8	Lunate	18	Trapezoid
9	Middle phalanx of middle finger	19	Triquetral
10	Proximal phalanx of middle finger		

Bones of the right hand, **B** palmar surface, **C** dorsal surface. Attachments

Pale green lines = ligament attachments

1	Abductor digiti minimi
2	Abductor pollicis brevis
3	Abductor pollicis longus
4	Extensor carpi radialis brevis
5	Extensor carpi radialis longus
6	Extensor carpi ulnaris
7	Extensor expansion
8	Extensor pollicis brevis
9	Extensor pollicis longus
10	First dorsal interosseous
11	First palmar interosseous
12	Flexor carpi radialis
13	Flexor carpi ulnaris
14	Flexor digiti minimi brevis
15	Flexor digitorum profundus
16	Flexor digitorum superficialis
17	Flexor pollicis brevis
18	Flexor pollicis longus
19	Fourth dorsal interosseous
20	Fourth palmar interosseous
21	Oblique head of adductor pollicis
22	Opponens digiti minimi
23	Opponens pollicis
24	Pisohamate ligament
25	Pisometacarpal ligament
26	Second dorsal interosseous
27	Second palmar interosseous
28	Third dorsal interosseous
29	Third palmar interosseous
30	Transverse head of adductor pollicis

- The wrist joint (properly called the radiocarpal joint) is the joint between (proximally) the lower end of the radius and the interarticular disc which holds the lower ends of the radius and the ulna together, and (distally) the scaphoid, lunate and triquetral bones.
- The midcarpal joint is the joint between the proximal and distal rows of carpal bones (see the note on page 101).
- The carpometacarpal joint of the thumb is the joint between the trapezium and the base of the first metacarpal.

Colles' fracture is named after Abraham Colles, a Dublin surgeon. This fracture of the lower end of the radius with posterior displacement is commonly due to a fall on to an outstretched hand. Complications include median nerve irritation, rupture of the extensor pollicis longus tendon, and subluxation of the distal radio-ulnar joint.

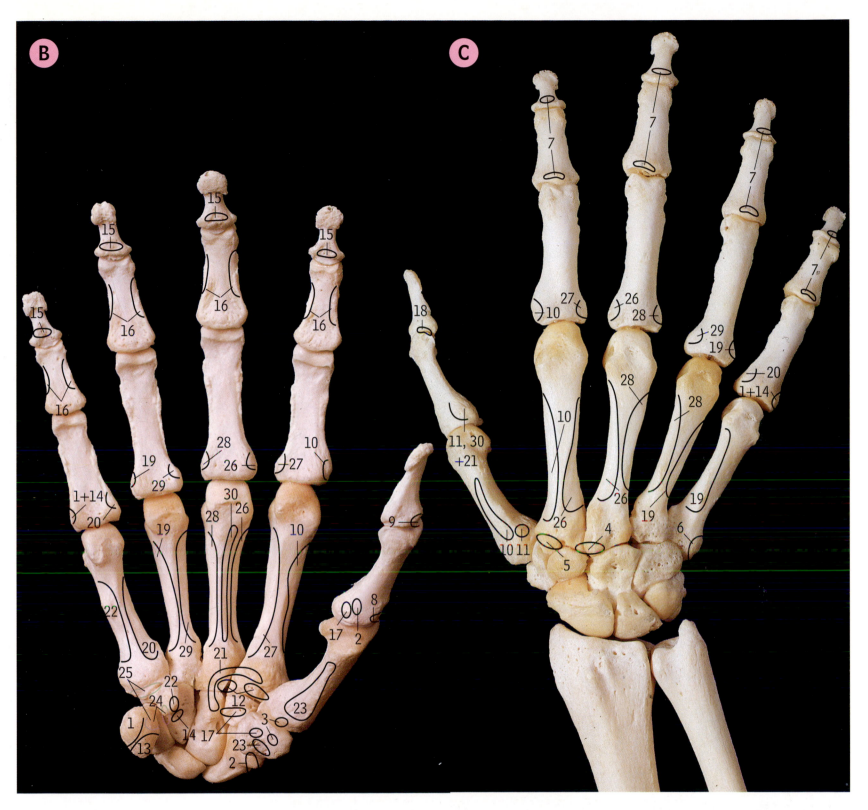

- The metacarpophalangeal joints are the joints between the heads of the metacarpals and the bases of the proximal phalanges.
- The interphalangeal joints are the joints between the head of one phalanx and the base of the adjoining phalanx.
- The pisiform is a sesamoid bone in the tendon of flexor carpi ulnaris and is anchored by the pisohamate and pisometacarpal ligaments (B24 and B25).
- In official anatomical terminology, the origin of flexor pollicis brevis (B17) from the trapezium (and flexor retinaculum) is referred to as the superficial head, and that from the trapezoid and capitate as the deep head (often small or even absent,

and to be distinguished from the first palmar interosseous (C11) which is sometimes considered to be synonymous with the deep head).
- Dorsal interossei arise from the sides of two adjacent metacarpal bones (as at C26, from the sides of the second and third metacarpals); palmar interossei arise only from the metacarpal of their own finger (as at B27, from the second metacarpal). Compare with dissection B on page 136 and note that when looking at the palm, parts of the dorsal interossei can be seen as well as the palmar interossei, but when looking at the dorsum of the hand (as on page 140) only dorsal interossei are seen.

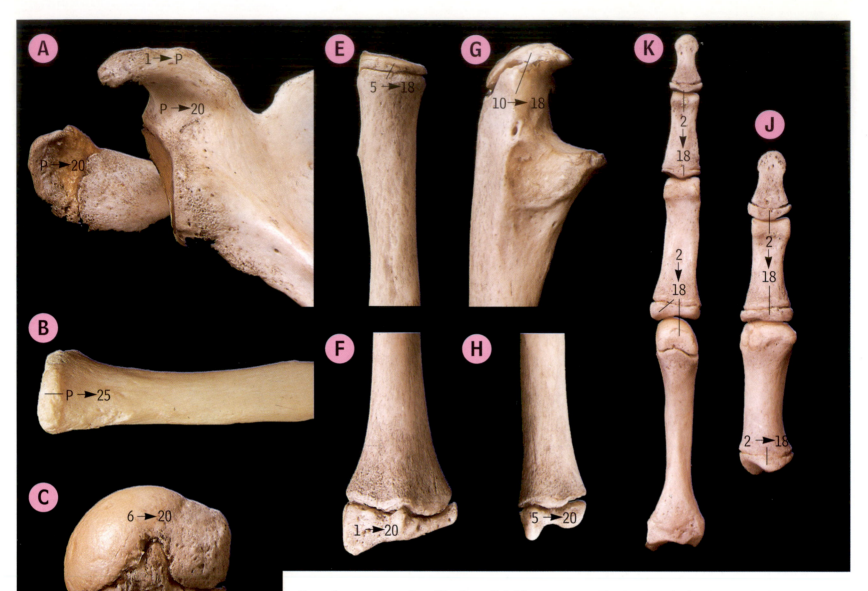

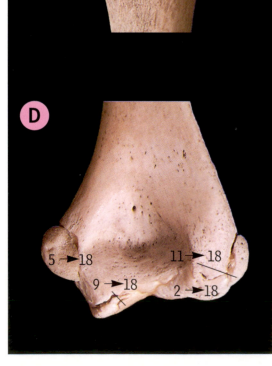

Secondary centres of ossification of right upper limb bones

A Scapula, upper lateral part
B Clavicle, sternal end
C and **D** Humerus, upper and lower ends
E and **F** Radius, upper and lower ends
G and **H** Ulna, upper and lower ends
J First metacarpal and phalanges of thumb
K Second metacarpal and phalanges of index finger
Figures in years after birth, commencement of ossification → fusion.
P = puberty
The first figure indicates the approximate date when ossification begins in the secondary centre, and the second figure (beyond the arrowhead) when the centre finally becomes fused with the rest of the bone. Single average dates have been given (both here and for the lower limb bone centres on pages 274 and 275) and although there may be considerable individual variations, the 'growing end' of the bone (when fusion occurs last) is constant. The dates in females are often a year or more earlier than in males

- Apart from the acromial, coracoid and subcoracoid centres illustrated (A), the scapula usually has other centres for the inferior angle, medial border, and the lower part of the rim of the glenoid cavity (all P → 20; see pages 90 and 92).

- The clavicle is the first bone in the body to start to ossify (fifth week of gestation). It ossifies in membrane, but the ends of the bone have a cartilaginous phase of ossification; a secondary centre appearing at the sternal end (B) unites with the body at about the twenty-fifth year.

- The centre illustrated at the upper end of the humerus (C) is the result of the union at six years of centres for the head (one year), greater tubercle (three years) and lesser tubercle (five years).

- At the lower end of the humerus (D) the centres for the capitulum, trochlea and lateral epicondyle fuse together before uniting with the shaft.

- All the phalanges (as in K), and the first metacarpal (J) have a secondary centre at their proximal ends; the other metacarpals (as in K) have one at their distal ends.

- All the carpal bones are cartilaginous at birth and none has a secondary centre. The largest, the capitate, is the first to begin to ossify (in the second month after birth), followed in a month or so by the hamate, with the triquetral at three years, lunate at four years, scaphoid, trapezoid and trapezium at five years and the pisiform last at nine years or later. There are often variations in the above common pattern.

- Primary centres for the body or shafts of bones usually begin to ossify about the eighth week of fetal life, but the clavicle is the first bone to commence ossification (see the second note above).

Right shoulder, from the front. Surface markings

The clavicle is subcutaneous throughout its length. Its acromial end (1) at the acromioclavicular joint (2) lies at a slightly higher level than the acromion of the scapula (3). At the most lateral part of the shoulder, the deltoid overlies the humerus; the acromion of the scapula does not extend so far laterally. Compare the positions of the features noted here with the dissection on the next page

- The nipple in the male (11) normally lies at the level of the fourth intercostal space.
- The deltopectoral groove containing the cephalic vein (8) is formed by the adjacent borders of deltoid (4) and pectoralis major (15).
- The lower border of pectoralis major (10) forms the anterior axillary fold.
- Note that the most lateral bony point in the shoulder is the greater tubercle (7).

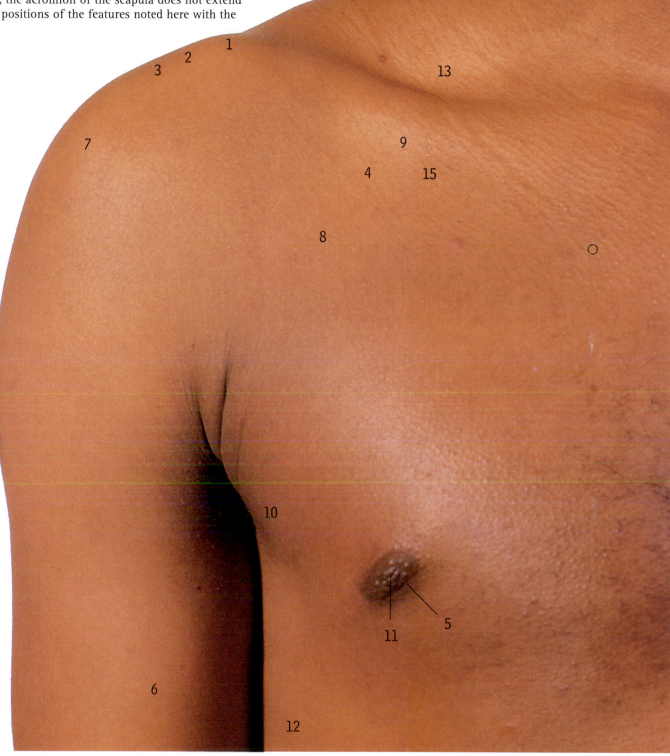

1 Acromial end of clavicle	9 Infraclavicular fossa
2 Acromioclavicular joint	10 Lower margin of pectoralis major
3 Acromion	11 Nipple
4 Anterior margin of deltoid	12 Serratus anterior
5 Areola	13 Supraclavicular fossa
6 Biceps	14 Trapezius
7 Deltoid overlying greater tubercle of humerus	15 Upper margin of pectoralis major
8 Deltopectoral groove and cephalic vein	

Reducing shoulder dislocations. Usually under general anaesthetic the arm is fully externally rotated, adducted across the body and, still in adduction, swung into internal rotation (Kocher's method). In another method (Hippocratic) an unbooted foot is placed in the patient's axilla and, with slight traction on the hand, the humerus is gently levered back into the glenoid fossa. Both before and after reduction it is important to check for axillary nerve damage.

A Right shoulder, from the front. Superficial dissection

Removal of skin and fascia displays branches of the supraclavicular nerve (6) crossing the clavicle (9), and the cephalic vein (7) lying in the deltopectoral groove between deltoid (13) and pectoralis major (11)

1 A superficial venous plexus
2 Accessory nerve
3 Acromial end of clavicle
4 Acromioclavicular joint
5 Acromion of scapula
6 Branches of supraclavicular nerve
7 Cephalic vein
8 Cervical nerve to trapezius
9 Clavicle
10 Clavicular head of sternocleidomastoid
11 Clavicular part of pectoralis major
12 Clavipectoral fascia
13 Deltoid
14 Sternal head of sternocleidomastoid
15 Sternocostal part of pectoralis major
16 Tip of shoulder
17 Trapezius

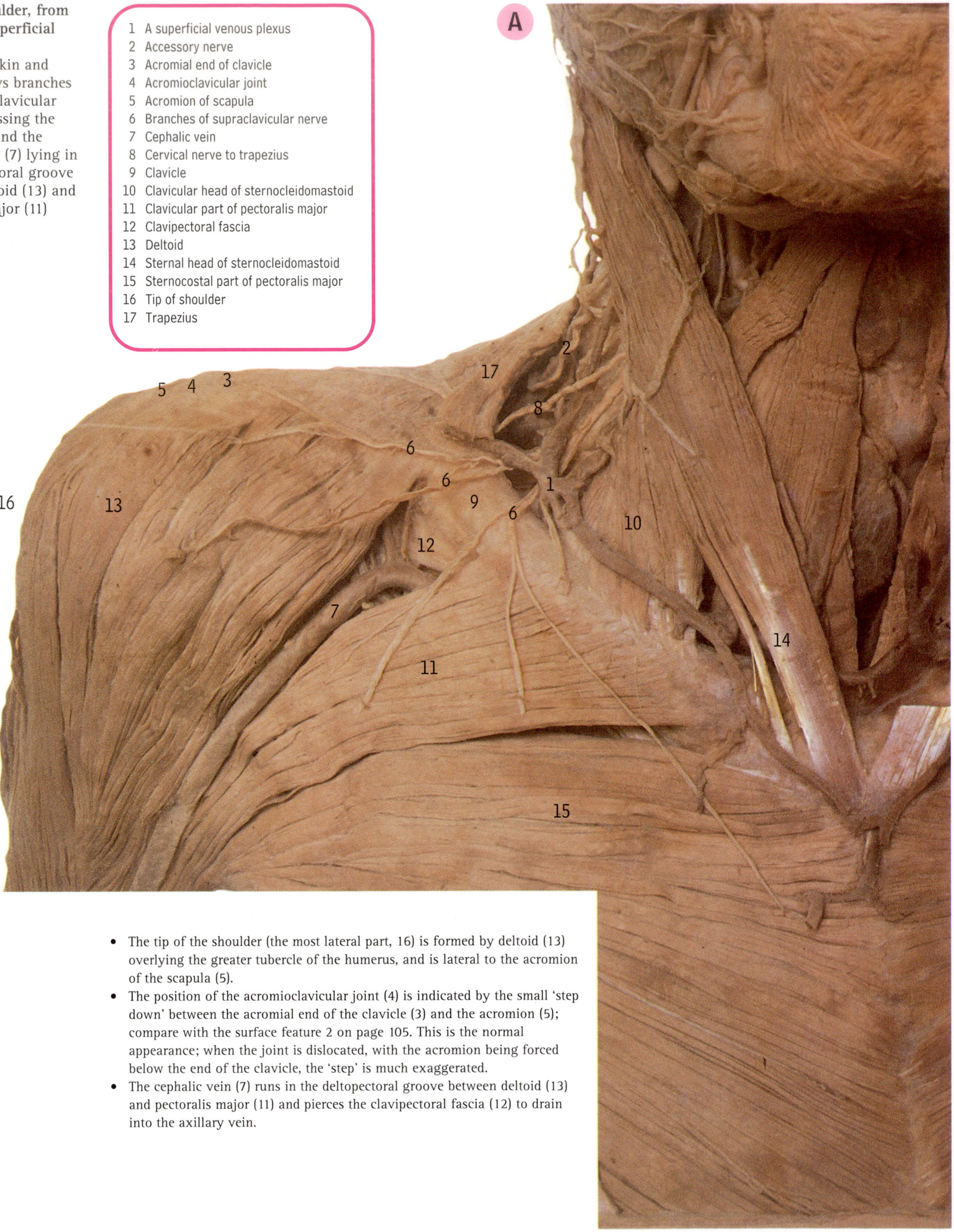

- The tip of the shoulder (the most lateral part, 16) is formed by deltoid (13) overlying the greater tubercle of the humerus, and is lateral to the acromion of the scapula (5).
- The position of the acromioclavicular joint (4) is indicated by the small 'step down' between the acromial end of the clavicle (3) and the acromion (5); compare with the surface feature 2 on page 105. This is the normal appearance; when the joint is dislocated, with the acromion being forced below the end of the clavicle, the 'step' is much exaggerated.
- The cephalic vein (7) runs in the deltopectoral groove between deltoid (13) and pectoralis major (11) and pierces the clavipectoral fascia (12) to drain into the axillary vein.

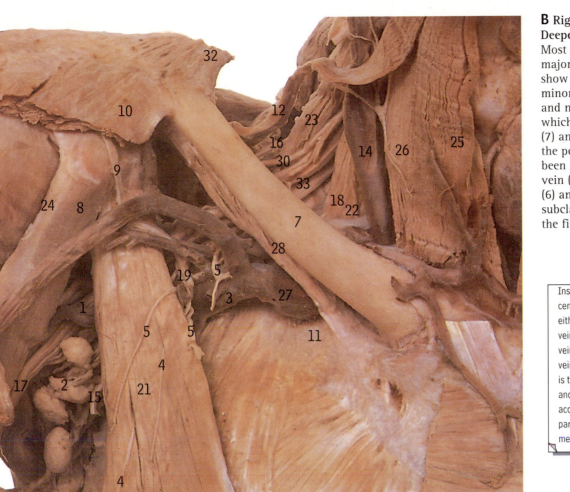

B Right shoulder, from the front. Deeper dissection
Most of deltoid (10) and pectoralis major (20) have been removed to show the underlying pectoralis minor (21) and its associated vessels and nerves. The clavipectoral fascia which passes between the clavicle (7) and the upper (medial) border of the pectoralis minor (21) has also been removed to show the axillary vein (3) receiving the cephalic vein (6) and continuing as the subclavian vein (27) as it crosses the first rib (11).

> Insertion of a central venous line. A central venous line may be introduced via either the subclavian or internal jugular veins. Occasionally, the brachiocephalic vein on the right is used to enter the great veins of the neck. From here the catheter is threaded into the superior vena cava and the right atrium. Central venous access is used to deliver chemotherapy, parenteral nutrition or antibiotics, or to measure the central venous pressure.

- The clavipectoral fascia, a small part of which is seen at A12 and which passes between the clavicle (B7) and pectoralis minor (B21), is pierced by the cephalic vein (A7, B6), branches of the thoraco-acromial vessels (as at B19), the lateral pectoral nerve (B5) and lymph vessels.

> Klumpke's paralysis is a brachial plexus lesion most commonly due to an abduction strain at the time of childbirth causing injury to the C8 and T1 roots of the brachial plexus, resulting in a claw hand deformity due to the paralysis of the intrinsic hand muscles (T1 myotome). The sympathetic trunk may also be affected, causing constriction of the pupil and a Horner's syndrome. The injury may not be diagnosed at birth, becoming obvious only when the baby fails to grasp objects normally.

1 Anterior circumflex humeral artery and musculocutaneous nerve
2 Axillary lymph nodes
3 Axillary vein
4 Branch of medial pectoral nerve
5 Branches of lateral pectoral nerve
6 Cephalic vein
7 Clavicle
8 Coracobrachialis
9 Coracoid process and acromial branch of thoraco-acromial artery
10 Deltoid
11 First rib
12 Inferior belly of omohyoid (displaced upwards)
13 Intercostobrachial nerve
14 Internal jugular vein
15 Lateral thoracic artery
16 Long thoracic nerve (to serratus anterior)
17 Median nerve
18 Nerve to subclavius
19 Pectoral branch of thoraco-acromial artery
20 Pectoralis major
21 Pectoralis minor
22 Phrenic nerve overlying scalenus anterior
23 Scalenus medius
24 Short head of biceps
25 Sternohyoid
26 Sternothyroid
27 Subclavian vein
28 Subclavius
29 Subscapularis
30 Suprascapular nerve
31 Tendon of long head of biceps
32 Trapezius
33 Trunks of brachial plexus

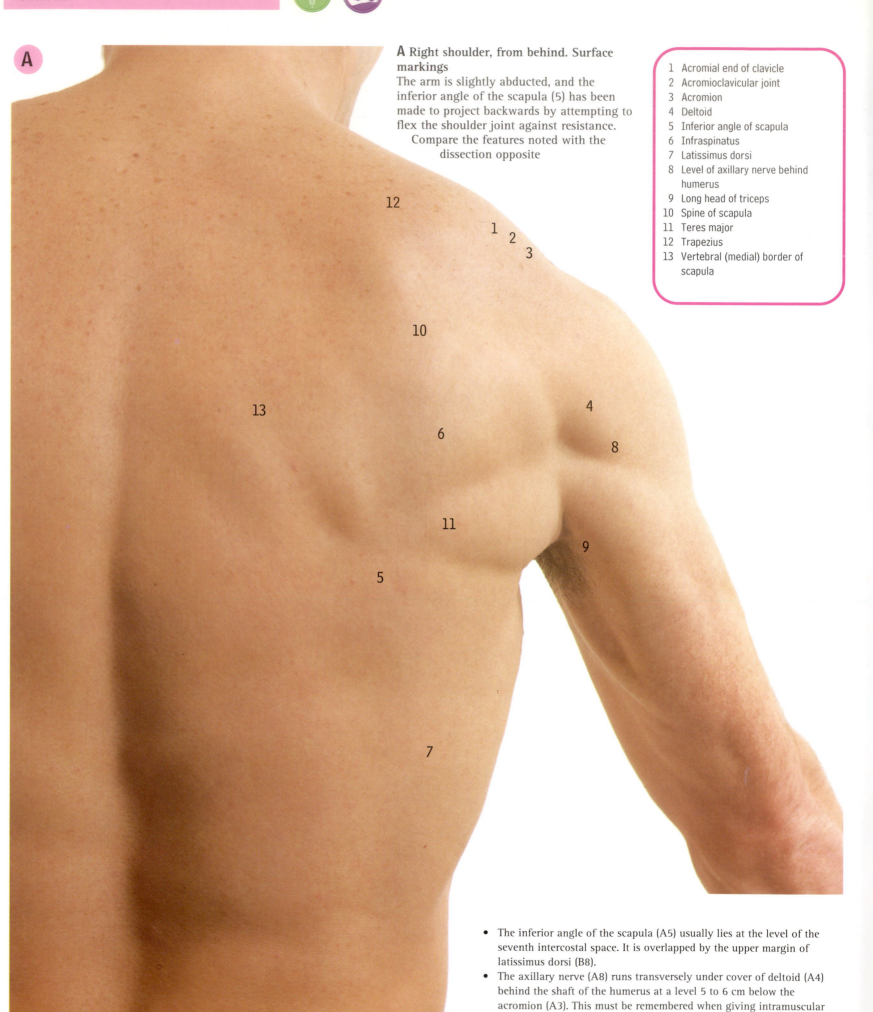

A Right shoulder, from behind. Surface markings

The arm is slightly abducted, and the inferior angle of the scapula (5) has been made to project backwards by attempting to flex the shoulder joint against resistance. Compare the features noted with the dissection opposite

1 Acromial end of clavicle
2 Acromioclavicular joint
3 Acromion
4 Deltoid
5 Inferior angle of scapula
6 Infraspinatus
7 Latissimus dorsi
8 Level of axillary nerve behind humerus
9 Long head of triceps
10 Spine of scapula
11 Teres major
12 Trapezius
13 Vertebral (medial) border of scapula

- The inferior angle of the scapula (A5) usually lies at the level of the seventh intercostal space. It is overlapped by the upper margin of latissimus dorsi (B8).
- The axillary nerve (A8) runs transversely under cover of deltoid (A4) behind the shaft of the humerus at a level 5 to 6 cm below the acromion (A3). This must be remembered when giving intramuscular injections into the deltoid.
- Latissimus dorsi (A7, B8) and teres major (A11, B14) form the lower boundary of the posterior wall of the axilla.

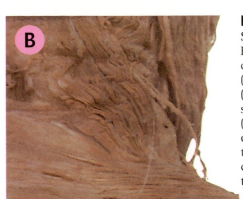

B Right shoulder, from behind.
Superficial dissection
From above and below, trapezius (16) converges on to the spine of the scapula (13). The upper margin of latissimus dorsi (8) overlaps the inferior angle of the scapula and the lowest part of teres major (14). The long head of triceps (9) is seen emerging between teres minor (15) and teres major (14). As at the front, deltoid (5) covers the shoulder joint and upper part of the humerus

1 Acromial end of clavicle	10 Medial cutaneous branches of dorsal rami of thoracic nerves
2 Acromioclavicular joint	11 Posterior cutaneous nerve of arm
3 Acromion	
4 Branches of upper lateral cutaneous nerve of arm	12 Rhomboid major
5 Deltoid	13 Spine of scapula
6 Fascia over infraspinatus	14 Teres major
7 Intercostobrachial nerve	15 Teres minor
8 Latissimus dorsi	16 Trapezius
9 Long head of triceps	17 Triangle of auscultation

- The triangle of auscultation (17) is bounded by the trapezius, latissimus dorsi, and the medial border of the scapula; its floor is partly formed by rhomboid major. If the arms are brought forwards, the sixth intercostal space becomes available for auscultation.

Intramuscular injections are most commonly performed in the centre or anterior part of the deltoid muscle because injecting the posterior deltoid may endanger the axillary nerve as it exits the quadrangular (quadrilateral) space.

A Right shoulder, from behind

Interrupted line = outline of scapula

Most of trapezius (20) and deltoid (5) have been removed to show the underlying muscles. The medial cut edge of trapezius remains near the line of the thoracic spines (18). Levator scapulae (9), rhomboid minor (13) and rhomboid major (12) are seen converging on to the vertebral (medial) border of the scapula, and supraspinatus (15) lies above the spine of the scapula (14)

- Muscles producing movements at the shoulder joint:

Abduction: supraspinatus and deltoid (middle fibres) for about 120°; further abduction requires scapular rotation produced by serratus anterior and trapezius.

Adduction: pectoralis major, latissimus dorsi, teres major, teres minor.

Flexion: deltoid (anterior fibres), pectoralis major, biceps, coracobrachialis.

Extension: deltoid (posterior fibres), latissimus dorsi, teres major.

Lateral rotation: infraspinatus, teres minor.

Medial rotation: pectoralis major, subscapularis, latissimus dorsi, teres major.

1 Acromial end of clavicle	12 Rhomboid major
2 Acromioclavicular joint	13 Rhomboid minor
3 Acromion	14 Spine of scapula
4 Branch of dorsal ramus of a thoracic nerve	15 Supraspinatus
5 Deltoid	16 Teres major
6 Erector spinae	17 Teres minor
7 Infraspinatus	18 Third thoracic spinous process
8 Latissimus dorsi	19 Thoracic part of thoracolumbar fascia
9 Levator scapulae	20 Trapezius
10 Long head of triceps	
11 Posterior circumflex humeral vessels and axillary nerve	

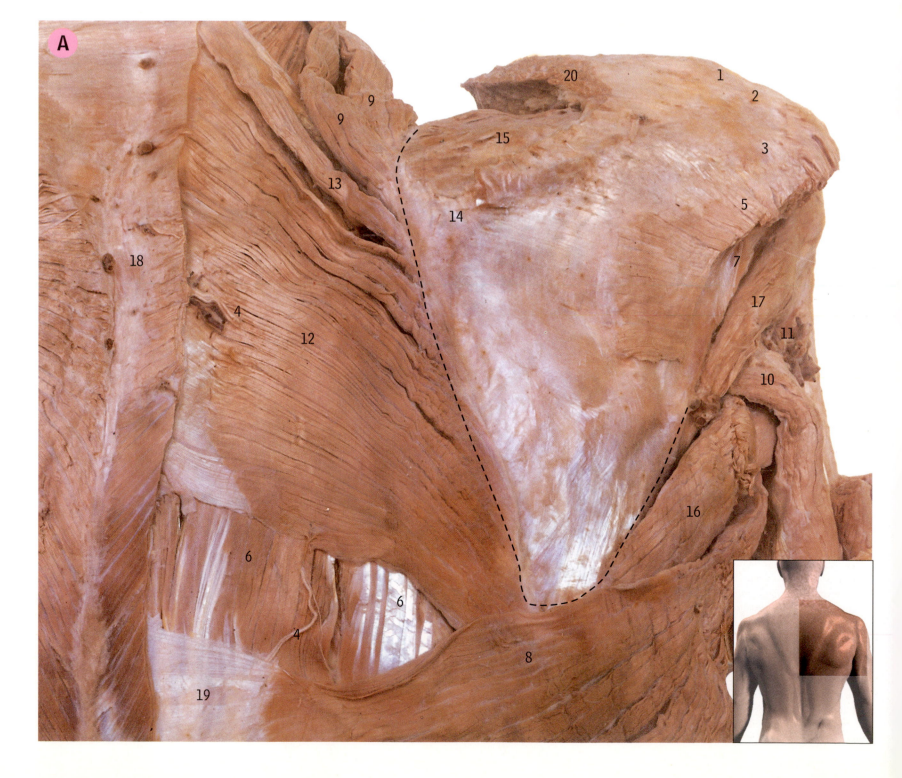

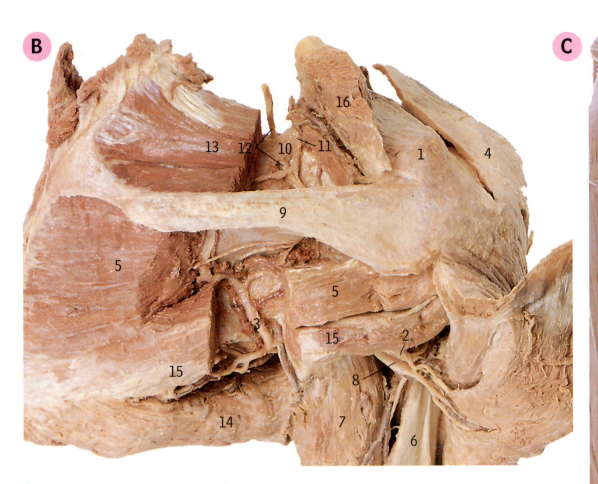

B Right shoulder, from the right and behind
The central parts of supraspinatus (13) and infraspinatus (5) have been removed to show the suprascapular nerve (12) which supplies both muscles. The removal of parts of infraspinatus and teres minor displays the anastomosis between the circumflex scapular branch of the subscapular artery and the suprascapular artery. Deltoid (4) has been reflected laterally to show the axillary nerve (2) and the posterior circumflex humeral vessels passing backwards through the quadrilateral space (see note below)

1	Acromioclavicular joint	
2	Axillary nerve	
3	Circumflex scapular artery	
4	Deltoid	
5	Infraspinatus	
6	Lateral head of triceps	
7	Long head of triceps	
8	Posterior circumflex humeral artery	
9	Spine of scapula	
10	Superior transverse scapular (suprascapular) ligament	
11	Suprascapular artery	
12	Suprascapular nerve	
13	Supraspinatus	
14	Teres major	
15	Teres minor	
16	Trapezius	

- As it lies just beneath the capsule of the shoulder joint, the axillary nerve may be injured by dislocation of the joint.
- The suprascapular artery (B11) passes into the supraspinous fossa superficial to the superior transverse scapular ligament (B10); the suprascapular nerve (B12) passes deep to the ligament.
- The axillary nerve (B2) and posterior circumflex humeral vessels (B8) pass backwards through the quadrilateral space which (viewed from behind) is bounded by teres minor (B15), below by teres major (B14), medially by the long head of triceps (B7), and laterally by the humerus. (Viewed from the front, the upper boundary of the space is subscapularis – see page 116, A23.)

C Right shoulder and upper arm, from the right
Deltoid (5) extends over the tip of the shoulder to its attachment halfway down the lateral side of the shaft of the humerus. Biceps (2) is on the front of the arm below pectoralis major (8) and triceps (7 and 6) is at the back

1	Acromion
2	Biceps
3	Brachialis
4	Brachioradialis
5	Deltoid
6	Lateral head of triceps
7	Long head of triceps
8	Pectoralis major
9	Trapezius

A

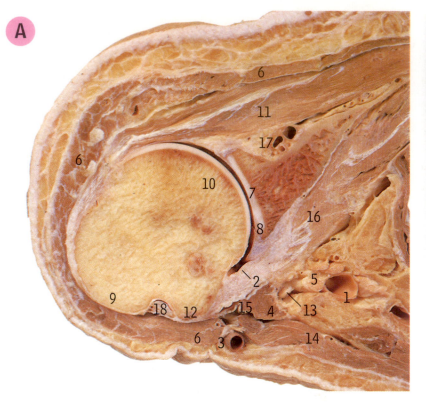

B

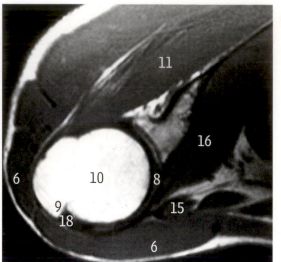

1	Axillary artery	11	Infraspinatus
2	Capsule	12	Lesser tubercle
3	Cephalic vein	13	Musculocutaneous nerve
4	Coracobrachialis	14	Pectoralis major
5	Cords of brachial plexus	15	Short head of biceps
6	Deltoid	16	Subscapularis
7	Glenoid cavity	17	Suprascapular nerve and vessels
8	Glenoid labrum	18	Tendon of long head of biceps in
9	Greater tubercle		intertubercular groove
10	Head of humerus		

Right shoulder joint, A horizontal section, B axial MR image
Viewed from above, this section shows the articulation of the head of the humerus (10) with the glenoid cavity of the scapula (7). The tendon of the long head of biceps (18) lies in the groove between the greater and lesser tubercles of the humerus (9 and 12). Subscapularis (16) passes immediately in front of the joint, and infraspinatus (11) behind it. Compare the MR image in B with features in A

C

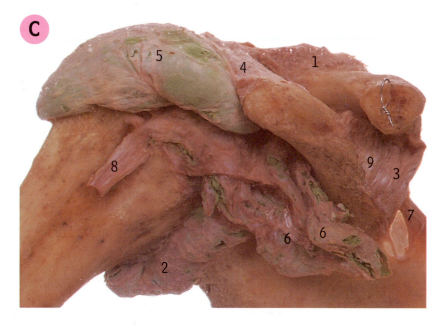

D

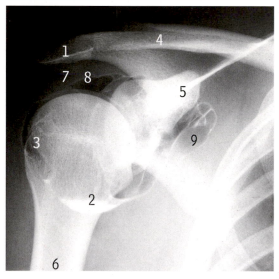

D Double contrast arthrogram of shoulder

C Right shoulder joint, from the front
The synovial joint cavity inside the capsule (2) and the subacromial bursa (5) have been injected separately with green resin

1	Acromioclavicular joint	6	Subscapularis bursa
2	Capsule of shoulder joint	7	Superior transverse scapular
3	Conoid ligament		(suprascapular) ligament
4	Coraco-acromial ligament	8	Tendon of long head of biceps
5	Subacromial bursa	9	Trapezoid ligament

1	Acromion
2	Axillary pouch
3	Bicipital tendon sheath
4	Clavicle
5	Coracoid process
6	Humerus
7	Region of subacromial bursa
8	Site of rotator cuff muscles
9	Subscapularis bursa

- The subscapularis bursa (C6 and D9) normally communicates with the synovial cavity of the shoulder joint.
- The subacromial bursa (C5 and D7) does not normally communicate with the shoulder joint; it is separated from the joint by the supraspinatus tendon. Only if the tendon is ruptured can the two cavities become continuous with one another.

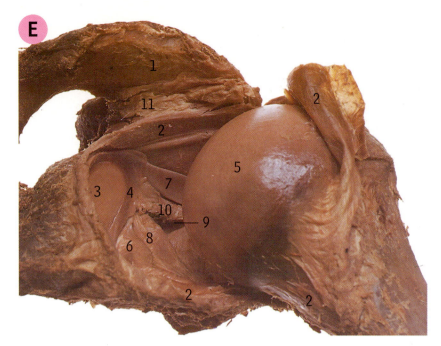

E Right shoulder joint, opened from behind
In this view, after removing all the posterior part of the capsule, the inner surface of the front of the capsule (2) is seen, with its reinforcing glenohumeral ligaments (6, 8 and 10)

1	Acromion	7	Long head of biceps
2	Capsule	8	Middle glenohumeral ligament
3	Glenoid cavity	9	Opening into subscapularis bursa
4	Glenoid labrum	10	Superior glenohumeral ligament
5	Head of humerus	11	Supraspinatus
6	Inferior glenohumeral ligament		

- The joint cavity communicates with the subscapularis bursa through an opening (9) between the superior (10) and middle (8) glenohumeral ligaments.
- The tendon of the long head of biceps (7) is continuous with the glenoid labrum (4).

Supraspinatus tendon calcification, calcium deposits within the supraspinatus tendon near its insertion, is of unknown aetiology and may cause eventual attrition and rupture of the tendon.

Rotator cuff tears often follow dislocation of the shoulder and can be in any of the rotator cuff muscles (supraspinatus, infraspinatus, teres minor and subscapularis).

Supraspinatus tendinitis. Deep to the acromion and the deltoid muscle, the subacromial bursa allows the supraspinatus tendon below it to glide freely in abduction. Injury or inflammatory change in the bursa can cause a painful arc on abduction, particularly when the greater tuberosity comes in contact with the acromion. On abduction the pain will be relieved when the acromion no longer touches the supraspinatus muscle. This painful 'arc' is fairly diagnostic of the condition.

Bicipital tendinitis. The long head of biceps muscle in the bicipital groove may become frayed and rupture. The presenting symptoms include pain on supination against resistance and, if there is rupture, a noticeable bulge appears on the anterior surface of the arm when the elbow is flexed.

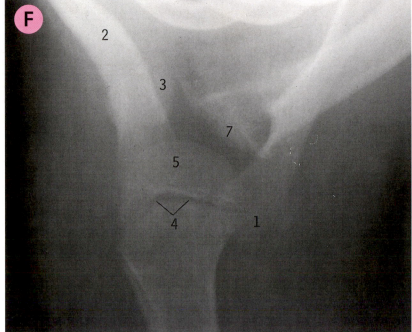

Shoulder radiographs, F axial projection in a six-year-old child, G anteroposterior projection in a nine-year-old child
In F the head of the humerus (5) lies against the glenoid cavity of the scapula (7), whose coracoid process (3) is seen end-on. In G note the epiphysial line (4) at the upper end of the humerus

- The upper humeral epiphysis is a compound structure made up of epiphyses for the head, and greater and lesser tubercles. It rests on the spike-like upper humeral shaft, giving the appearance of an inverted V to the epiphysial line on the radiograph (G4). Compare with page 104, C.

1	Acromion
2	Clavicle
3	Coracoid process
4	Epiphysial line
5	Head of humerus
6	Lateral border of scapula
7	Rim of glenoid cavity
8	Spine of scapula

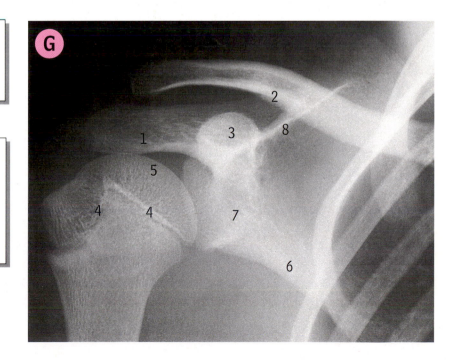

A Right axilla, anterior wall
Pectoralis major (7) has been reflected upwards and laterally, and the clavipectoral fascia which passes from subclavius (10) to pectoralis minor (8) has been removed

1 Axillary sheath
2 Branches of medial pectoral nerve
3 Cephalic vein
4 Clavicle
5 First rib
6 Lateral pectoral nerve
7 Pectoralis major
8 Pectoralis minor
9 Subclavian vein
10 Subclavius
11 Thoraco-acromial vessels

- The clavipectoral fascia (here removed, between subclavius, 10, and pectoralis minor, 8) is pierced by the cephalic vein (3), thoraco-acromial vessels (11), lateral pectoral nerve (6) and lymphatics.
- The axillary sheath (1) is the downward continuation of the prevertebral fascia of the neck and forms a dense covering for the axillary artery and the surrounding parts of the brachial plexus.
- The lateral pectoral nerve (6) is related to the medial (upper) border of pectoralis minor (8). The medial pectoral nerve (2) is related to the lateral (lower) border of pectoralis minor.

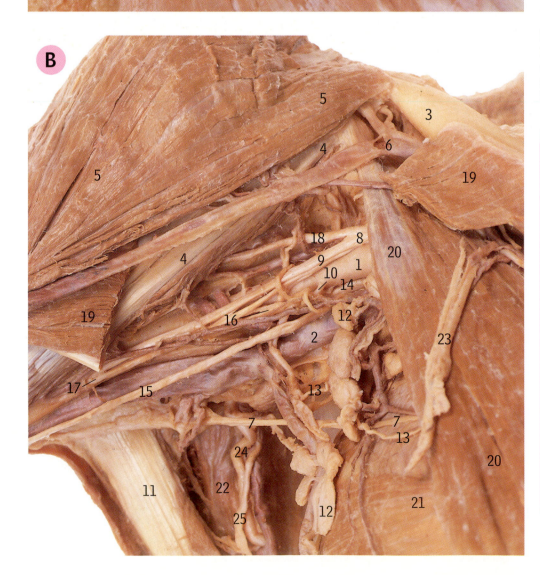

B Right axilla and brachial plexus, from the front
Pectoralis major (19) has been reflected and the clavipectoral fascia removed, together with the axillary sheath (A1) which surrounded the axillary artery and brachial plexus

1 Axillary artery
2 Axillary vein
3 Clavicle
4 Coracobrachialis
5 Deltoid
6 Entry of cephalic vein into deltoid vein
7 Intercostobrachial nerve
8 Lateral cord of brachial plexus
9 Lateral root of median nerve
10 Lateral thoracic artery
11 Latissimus dorsi
12 Lymph nodes
13 Lymph vessels
14 Medial cord brachial plexus
15 Medial cutaneous nerve of arm
16 Medial root of median nerve
17 Median nerve
18 Musculocutaneous nerve
19 Pectoralis major
20 Pectoralis minor
21 Serratus anterior
22 Subscapularis
23 Thoraco-acromial vessels and lateral pectoral nerve
24 Thoracodorsal artery
25 Thoracodorsal nerve

C Right brachial plexus, from the front

Pectoralis major and minor (28 and 29) have been reflected and the axillary sheath (A1) removed, together with most of the axillary vein (4) and its tributaries

1 Anterior circumflex humeral artery
2 Axillary artery
3 Axillary nerve
4 Axillary vein
5 Branch from first thoracic nerve to intercostobrachial nerve
6 Circumflex scapular artery
7 Clavicle
8 Communication between 23 and 13
9 Coracobrachialis and short head of biceps
10 Deltoid
11 Entry of cephalic vein
12 First rib
13 Intercostobrachial nerve (cut end)
14 Lateral cord
15 Lateral pectoral nerve
16 Lateral root of median nerve
17 Lateral thoracic artery
18 Latissimus dorsi
19 Long head of biceps
20 Long thoracic nerve
21 Loop between medial and lateral pectoral nerves
22 Lower subscapular nerve
23 Medial cutaneous nerve of arm
24 Medial cutaneous nerve of forearm
25 Medial root of median nerve
26 Median nerve
27 Musculocutaneous nerve
28 Pectoralis major
29 Pectoralis minor
30 Radial nerve
31 Serratus anterior
32 Subclavian vein
33 Subclavius
34 Subscapularis
35 Teres major
36 Thoraco-acromial artery
37 Thoracodorsal artery
38 Thoracodorsal nerve
39 Ulnar nerve

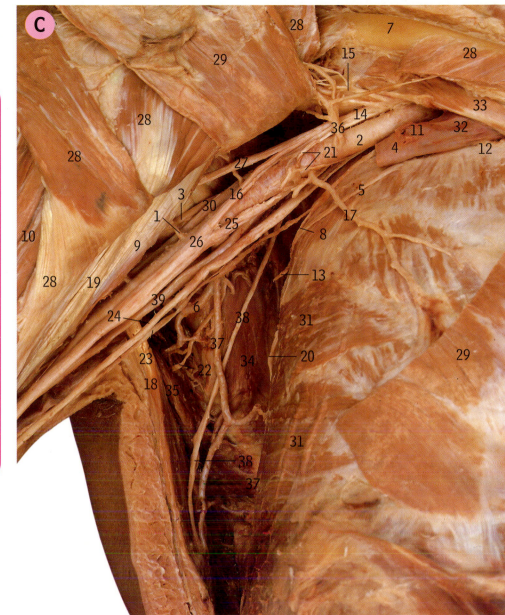

Winging of the scapula, a scapula that sticks out from the posterior chest like an angel's wing, is noticeable when the hand is pushed against a wall or used to open a door. It is most commonly caused by a weakness of the serratus anterior or sometimes latissimus dorsi muscles. This may be due to muscular disease or nerve palsy.

Erb's paralysis (Erb–Duchenne palsy) is a brachial plexus injury to the upper roots, often an obstetric injury during a difficult delivery when traction is applied to the baby's head when it is in lateral flexion and the shoulder of the opposite side is pulled downwards. This most common of obstetric brachial plexus injuries involves the fifth and sixth cervical nerve roots. After delivery the baby's arm may be in a 'waiter's tip' position with the shoulder adducted, the elbow extended and the wrist pronated and flexed. It is usually evident at birth and may improve during the first year of the child's life.

• Boundaries of the axilla:
Anterior wall: pectoralis major and minor (B19 and 20) and the clavipectoral fascia with subclavius.
Posterior wall: subscapularis (C34), teres major (C35) and latissimus dorsi (C18).
Medial wall: serratus anterior (C31) overlying the first four ribs and intercostal muscles.
Lateral wall: the intertubercular groove of the humerus, coracobrachialis and biceps (C9).
Apex: the space between the clavicle (C7), first rib (C12) and upper border of the scapula.
Base: the concavity of skin and axillary fascia between the lower borders of pectoralis major and latissimus dorsi.

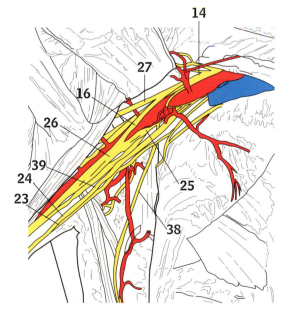

A Right brachial plexus and branches

In this front view of the plexus, all the blood vessels have been removed to show the cords of the plexus and their branches more clearly. Note the 'capital M' pattern formed by the musculocutaneous nerve (18), the lateral root of the median nerve (8), the median nerve itself (17), the medial root of the median nerve (16) and the ulnar nerve (26). In this specimen the tendon of latissimus dorsi (9) is unusually broad and has become blended with the long head of triceps (10)

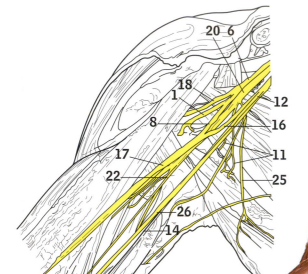

Axillary nerve paralysis often occurs after shoulder dislocations or fractures of the humeral neck. The sensory loss is over the lateral upper part of the arm and the motor loss of the deltoid muscle makes it extremely difficult to abduct the arm or put one's hand into the trouser pocket.

1	Axillary nerve	15	Medial head of triceps
2	Biceps	16	Medial root of median nerve
3	Coracobrachialis	17	Median nerve
4	Deltoid	18	Musculocutaneous nerve
5	Intercostobrachial nerve	19	Pectoralis minor and lateral pectoral nerve
6	Lateral cord	20	Posterior cord
7	Lateral head of triceps	21	Radial nerve
8	Lateral root of median nerve	22	Radial nerve branches to triceps
9	Latissimus dorsi	23	Subscapularis
10	Long head of triceps	24	Teres major
11	Lower subscapular nerve	25	Thoracodorsal nerve
12	Medial cord	26	Ulnar nerve
13	Medial cutaneous nerve of arm	27	Upper subscapular nerves
14	Medial cutaneous nerve of forearm		

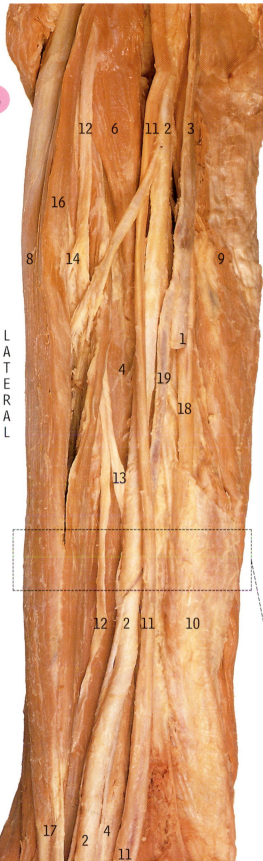

B Right arm. Vessels and nerves, from the front

Biceps (16 and 8) has been turned laterally to show the musculocutaneous nerve (12) emerging from coracobrachialis (6), giving branches to biceps and brachialis (14 and 13) and becoming the lateral cutaneous nerve of the forearm (7) on the lateral side of the biceps tendon (17). The median nerve (11) gradually crosses over in front of the brachial artery (2) from the lateral to the medial side. The ulnar nerve (18) passes behind the medial intermuscular septum (10), and the end of the basilic vein (1) is seen joining a vena comitans (19) of the brachial artery to form the brachial vein (3)

1 Basilic vein (cut end)	11 Median nerve
2 Brachial artery	12 Musculocutaneous nerve
3 Brachial vein	13 Nerve to brachialis
4 Brachialis	14 Nerve to short head of biceps
5 Brachioradialis	15 Pronator teres
6 Coracobrachialis	16 Short head of biceps
7 Lateral cutaneous nerve of forearm	17 Tendon of biceps
8 Long head of biceps	18 Ulnar nerve
9 Long head of triceps	19 Vena comitans of brachial artery
10 Medial intermuscular septum	

- The musculocutaneous nerve (B12) supplies coracobrachialis (B6), biceps (B16 and 8) and brachialis (B4), and at the level where the muscle fibres of biceps become tendinous (B17) it pierces the deep fascia to become the lateral cutaneous nerve of the forearm (B7).
- The median nerve does not give off any muscular branches in the arm.
- The ulnar nerve (B18) leaves the anterior compartment of the arm by piercing the medial intermuscular septum (B10), and does not give off any muscular branches in the arm.

C Cross section of the right arm, from below

Looking from the elbow towards the shoulder, the section is taken through the middle of the arm. The musculocutaneous nerve (9) lies between brachialis (4) and biceps (2), and the median nerve (8) is on the medial side of the brachial artery (3) which has several venae comitantes adjacent (unlabelled). The ulnar nerve (13), with the superior ulnar collateral artery (11) beside it, is behind the median nerve (8) and the basilic vein (1). The radial nerve and the profunda brachii vessels (10) are in the posterior compartment at the lateral side of the humerus (6)

FRONT

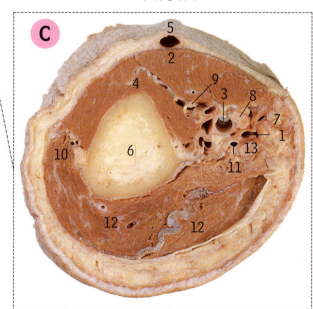

1 Basilic vein
2 Biceps
3 Brachial artery
4 Brachialis
5 Cephalic vein
6 Humerus
7 Medial cutaneous nerve of forearm
8 Median nerve
9 Musculocutaneous nerve
10 Radial nerve and profunda brachii vessels
11 Superior ulnar collateral artery
12 Triceps
13 Ulnar nerve

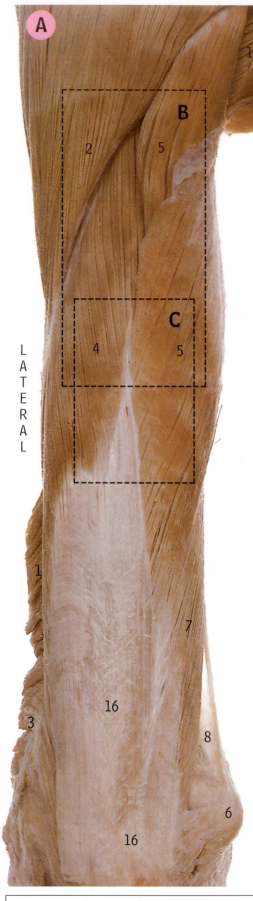

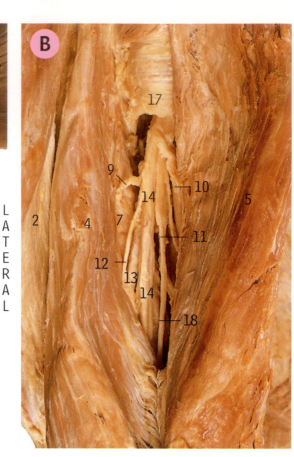

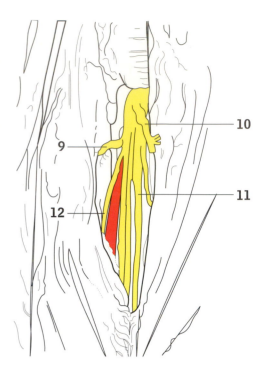

Left arm, from behind, **A** triceps, **B** and **C**, radial nerve

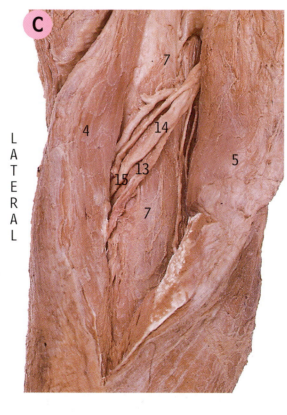

1	Brachioradialis
2	Deltoid
3	Extensor carpi radialis longus
4	Lateral head of triceps
5	Long head of triceps
6	Medial epicondyle
7	Medial head of triceps
8	Medial intermuscular septum
9	Nerve to lateral head of triceps
10	Nerve to long head of triceps
11	Nerve to medial head of triceps
12	Nerve to medial head of triceps
13	Profunda brachii artery
14	Radial nerve
15	Radial nerve at radial groove
16	Tendon of triceps
17	Teres major
18	Ulnar nerve

Radial nerve paralysis, commonly known as a 'crutch' palsy or 'Saturday night' palsy, is usually caused by compression of the radial nerve in the axilla (crutches fitting poorly into the armpit or a drunken stupor in which arms are flopped over an armchair are common causes). The injury affects both elbow and wrist extension but the sensory loss is minimal: normally only a small area of skin superficial to the first dorsal interosseous muscle on the back of the hand.

- Triceps has three heads but four nerves – the medial head receives two branches. All the muscular branches (B10, 11, 12 and 9) arise from the radial nerve (B14) high up, well before the nerve has reached the radial groove (C15) at the lateral side of the humerus (page 93, B4). The usual order of origin of the branches from above downwards is: nerve to the long head (B10), medial head (B11), lateral head (B9), and medial head (B12). The first branch to the medial head runs for part of its course close to the ulnar nerve (B18), and is therefore sometimes called the ulnar collateral nerve.

- The medial head (7) of triceps would be better known as the deep head, since most of it is under cover of the long and lateral heads (5 and 4 in A and B), but on the lower medial side of the arm part of the medial head (A7) does project below the long head (A5).

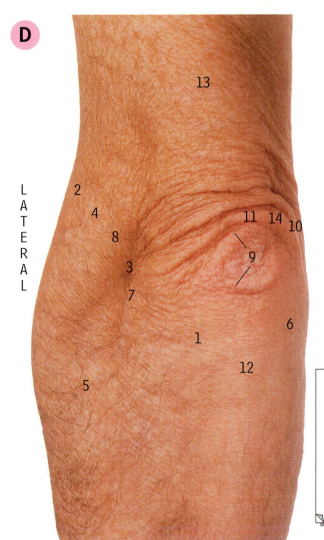

D Left elbow, from behind. Surface markings
With the elbow fully extended, the extensor muscles (5, 4) form a bulge on the lateral side. In the adjacent hollow can be felt the head of the radius (7) and the capitulum of the humerus (3) which indicate the line of the humeroradial part of the elbow joint. The lateral and medial epicondyles of the humerus (8 and 10) are palpable on each side. Wrinkled skin lies at the back of the prominent olecranon of the ulna (11), and in this arm the margin of the olecranon bursa (9) is outlined. The most important structure in this region is the ulnar nerve (14) which is palpable as it lies in contact with the humerus behind the medial epicondyle (10). The posterior border of the ulna (12) is subcutaneous throughout its whole length

1   Anconeus
2   Brachioradialis
3   Capitulum of humerus
4   Extensor carpi radialis longus
5   Extensor muscles
6   Flexor carpi ulnaris
7   Head of radius
8   Lateral epicondyle of humerus
9   Margin of olecranon bursa
10 Medial epicondyle of humerus
11 Olecranon of ulna
12 Posterior border of ulna
13 Triceps
14 Ulnar nerve

Ulnar nerve paralysis commonly results from damage to the nerve behind the medial epicondyle of the humerus, producing impaired power of adduction of the wrist, an inability to spread the fingers and claw hand (paralysis of the interossei). There may also be an inability to adduct the thumb. The hypothenar muscles become wasted and sensation may be impaired in the ulnar one and a half fingers on both palmar and dorsal surfaces.

Triceps tendon reflex, elicited by tapping with a hammer the triceps tendon just above the olecranon, extends the elbow. This pathway is mainly through C7. Diminished or exaggerated reflex responses indicate abnormality of function. A cut nerve, for example, will give no reflex, whereas exaggerated reflexes are characteristic of an upper motor neurone lesion because the responsiveness of the anterior horn cells has been enhanced by elimination of the normal inhibition from higher centres.

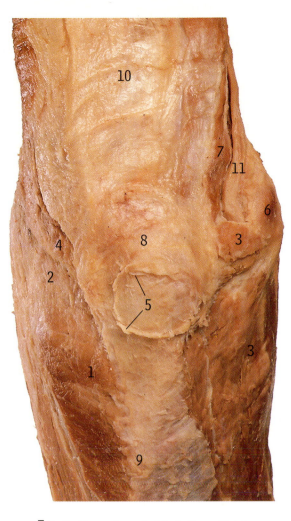

E Left elbow, from behind. Superficial dissection
Skin and subcutaneous tissue and some deep fascia have been removed, but the margin of the olecranon bursa (5) has been preserved. The ulnar nerve (11) is behind the medial epicondyle (6) and passes downwards under cover of flexor carpi ulnaris (3)

1   Anconeus
2   Common extensor origin
3   Flexor carpi ulnaris
4   Lateral epicondyle of humerus
5   Margin of olecranon bursa
6   Medial epicondyle of humerus
7   Medial head of triceps
8   Olecranon of ulna
9   Posterior border of ulna
10 Triceps tendon
11 Ulnar nerve

Olecranon bursitis is inflammation of the bursa overlying the olecranon process of the ulna, associated with prolonged pressure at this point.

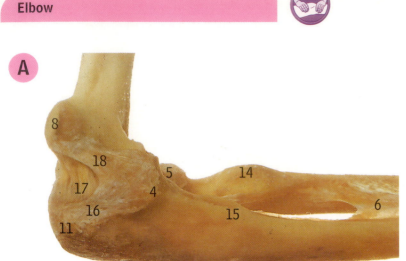

A

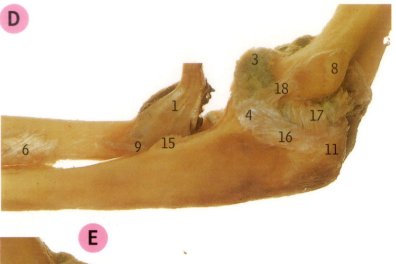

D

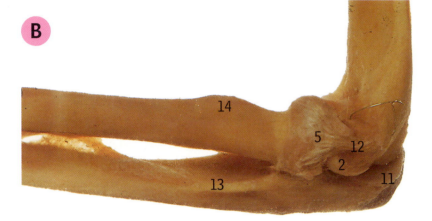

B

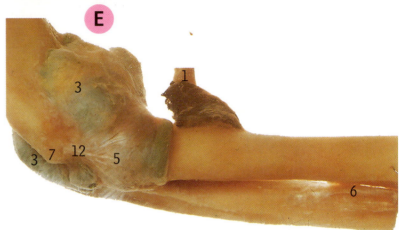

E

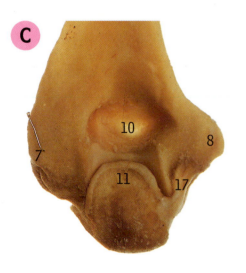

C

Left elbow joint and proximal radio-ulnar joint, **A** from the medial side, **B** from the lateral side, **C** from behind

Right elbow joint and proximal radio-ulnar joint, **D** from the medial side, **E** from the lateral side, **F** from behind

In A, B and C the forearm is flexed to a right angle. In D, E and F the forearm is partially flexed, and the synovial cavity within the capsule (3) and the bursa beneath the biceps tendon (1) have been injected with green resin

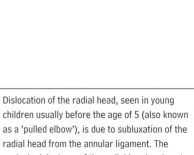

F

- The synovial cavity of the proximal radio-ulnar joint is continuous with that of the elbow joint (the synovial cavity of the distal radio-ulnar joint is *not* continuous with that of the wrist joint).
- Posteriorly and above, the capsule of the elbow joint is attached to the upper part of the floor of the olecranon fossa, not to the upper margin of the fossa (page 96, B).
- The main ligaments of the elbow and proximal radio-ulnar joints are the capsule (3), the ulnar and radial collateral ligaments (16 to 18 and 12) and the annular ligament (5).

1 Biceps tendon and underlying bursa
2 Capitulum
3 Capsule (distended)
4 Coronoid process of ulna
5 Head and neck of radius covered by annular ligament
6 Interosseous membrane
7 Lateral epicondyle
8 Medial epicondyle
9 Oblique cord
10 Olecranon fossa
11 Olecranon of ulna
12 Radial collateral ligament
13 Supinator crest of ulna
14 Tuberosity of radius
15 Tuberosity of ulna
16 Ulnar collateral ligament: oblique band
17 Ulnar collateral ligament: posterior band
18 Ulnar collateral ligament: upper band

Dislocation of the radial head, seen in young children usually before the age of 5 (also known as a 'pulled elbow'), is due to subluxation of the radial head from the annular ligament. The conical adult shape of the radial head and neck develop late and their cylindrical shape in childhood reduces retention by the annular ligament.

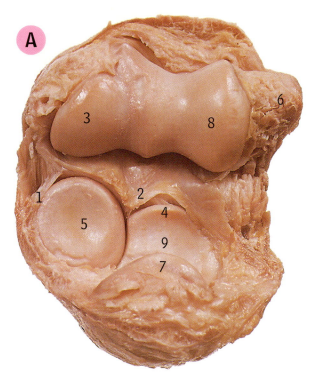

A Left elbow joint, opened from behind
The joint has been 'forced open' from behind: the capitulum (3) and trochlea (8) of the lower end of the humerus are seen from below with the forearm in forced flexion to show the upper ends of the radius and ulna (5 and 9) from above

1 Annular ligament
2 Anterior part of capsule
3 Capitulum of humerus
4 Coronoid process of ulna
5 Head of radius
6 Medial epicondyle of humerus
7 Olecranon of ulna
8 Trochlea of humerus
9 Trochlear notch of ulna

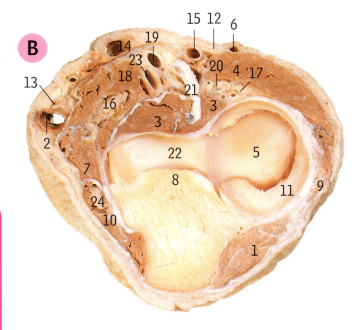

B Cross-section of the left elbow
The section is viewed from below, looking towards the shoulder, and is just below the point where the brachial artery has divided into radial and ulnar arteries (19 and 23). The cut has passed immediately below the trochlea (22) and capitulum (5) of the humerus, and has gone through the coronoid process of the ulna (8). The radial nerve (20) and its posterior interosseous branch (17) lie between brachioradialis (4) and brachialis (3). The median nerve (16) is under the main part of pronator teres (18), and the ulnar nerve (24) is passing under flexor carpi ulnaris (10)

1 Anconeus
2 Basilic vein
3 Brachialis
4 Brachioradialis
5 Capitulum of humerus
6 Cephalic vein
7 Common flexor origin
8 Coronoid process of ulna
9 Extensor carpi radialis longus and brevis
10 Flexor carpi ulnaris
11 Fringe of synovial membrane
12 Lateral cutaneous nerve of forearm
13 Medial cutaneous nerve of forearm
14 Median basilic vein
15 Median cephalic vein
16 Median nerve
17 Posterior interosseous nerve
18 Pronator teres
19 Radial artery
20 Radial nerve
21 Tendon of biceps
22 Trochlea of humerus
23 Ulnar artery
24 Ulnar nerve

Radiograph of the elbow, C anterior projection, D lateral projection

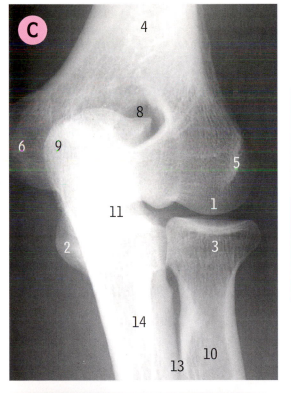

1 Capitulum of humerus
2 Coronoid process of ulna
3 Head of radius
4 Humerus
5 Lateral epicondyle of humerus
6 Medial epicondyle of humerus
7 Neck of radius
8 Olecranon fossa of humerus
9 Olecranon of ulna
10 Radius
11 Trochlea of humerus
12 Trochlear notch of ulna
13 Tuberosity of radius
14 Ulna

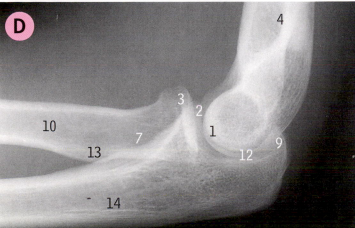

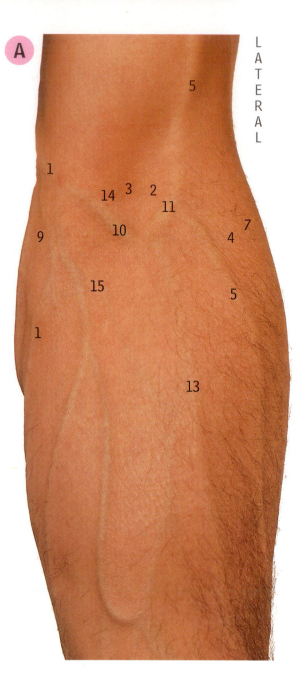

LATERAL

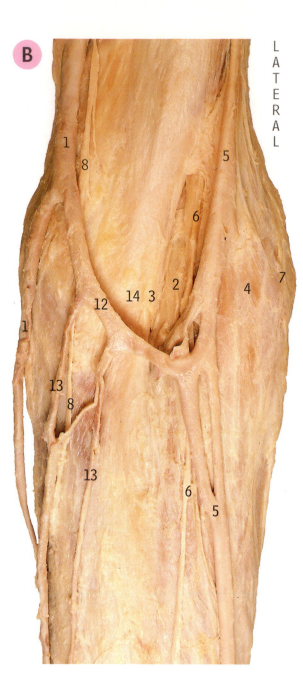

LATERAL

Left cubital fossa, A surface markings, B superficial veins

In A there is an M-shaped pattern of superficial veins (see notes). In B the cephalic (5) and basilic (1) veins are joined by a median cubital vein (12) into which drain two small median forearm veins (13). In C (page 123) the deep fascia has been removed but the bicipital aponeurosis (C2) is preserved; it runs downwards and medially from the biceps tendon (B2), crossing the brachial artery (C3) and the median nerve (C9). The musculocutaneous nerve becomes the lateral cutaneous nerve of the forearm (B6, C8) at the lateral border of biceps where the muscle becomes tendinous. Brachioradialis (A4, B4, C5) forms the lateral boundary and pronator teres (A15, C12) the medial boundary of the cubital fossa. The brachial arteriogram in D (page 123) outlines the main arteries (D2, 5 and 7)

1	Basilic vein
2	Biceps tendon
3	Brachial artery
4	Brachioradialis
5	Cephalic vein
6	Lateral cutaneous nerve of forearm
7	Lateral epicondyle
8	Medial cutaneous nerve of forearm
9	Medial epicondyle
10	Median basilic vein
11	Median cephalic vein
12	Median cubital vein
13	Median forearm vein
14	Median nerve
15	Pronator teres

Auscultation of the brachial pulse. The arterial blood pressure is usually taken from the brachial artery using a sphygmomanometer. The artery is first palpated in the cubital fossa anterior to the medial epicondyle and just superomedial to the bicipital aponeurosis, which may be felt as a strong band across the elbow. A cuff with pressure gauge placed proximally on the arm is inflated above the suspected systolic blood pressure and slowly deflated until the arterial pressure waves are heard through the stethoscope at the cubital fossa; this is the systolic blood pressure. Further release of the air from the cuff makes the beats disappear; this happens at the diastolic blood pressure. The pressure waves are normally heard at around the levels of 120 mmHg for systolic pressure and 70–80 mmHg for diastolic pressure.

- The superficial veins on the front of the elbow such as the cephalic (5) and basilic (1) and their intercommunicating tributaries are those most commonly used for intravenous injections and obtaining specimens of venous blood. The pattern of veins is typically M-shaped (as in A) or H-shaped (as in B), but there is much variation and it is not always possible or necessary to name every vessel.
- The order of the structures in the cubital fossa from lateral to medial is: biceps tendon (2), brachial artery (3) and median nerve (14).

Biceps tendon reflex is elicited by tapping with a hammer the end of the thumb held against the biceps tendon. The reflex (mainly through C6) causes the biceps muscle to contract and the elbow to flex.

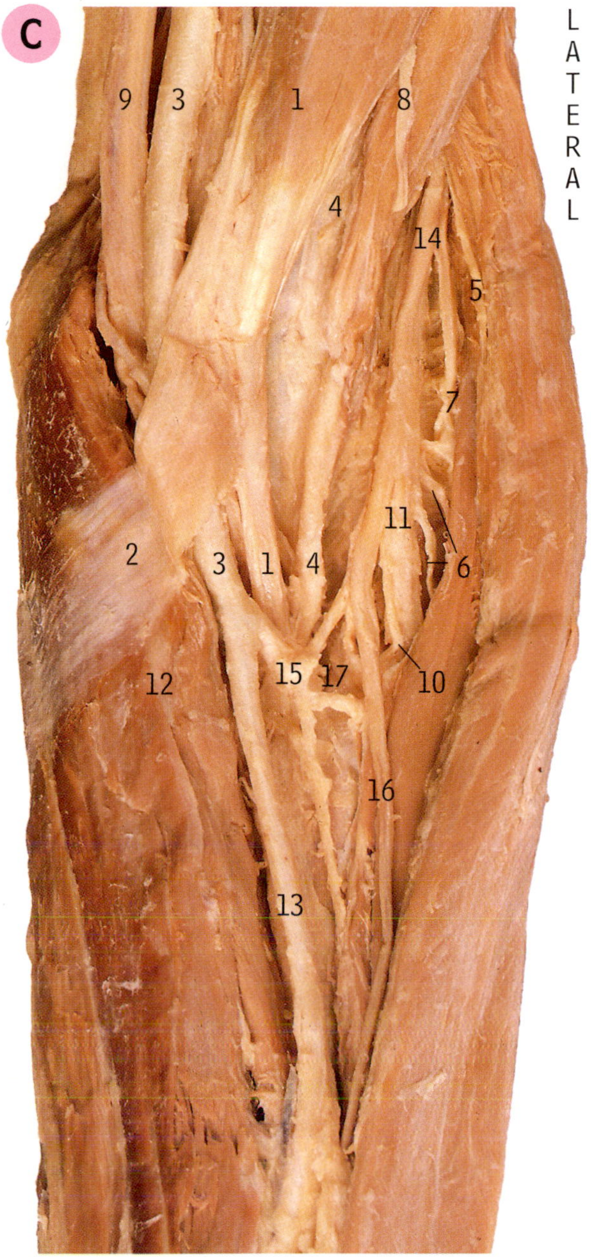

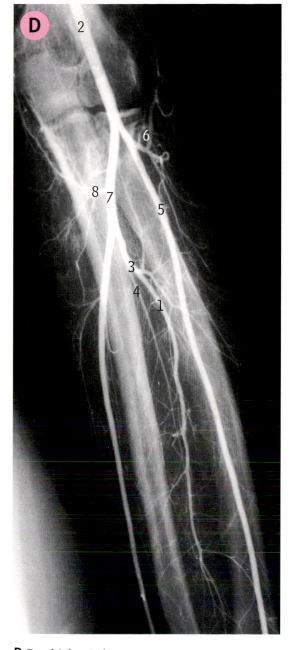

C

LATERAL

C Left elbow and upper forearm, from the front

Brachioradialis (5) and extensor carpi radialis longus (7) have been displaced laterally to show the radial nerve (14) giving off branches to those muscles and then dividing into the superficial (cutaneous) branch (16) and the deep (posterior interosseous) branch (11) which enters supinator

1 Biceps
2 Bicipital aponeurosis
3 Brachial artery
4 Brachialis
5 Brachioradialis and nerve
6 Branches to extensor carpi radialis brevis
7 Extensor carpi radialis longus and nerve
8 Lateral cutaneous nerve of forearm
9 Median nerve
10 Nerve to supinator
11 Posterior interosseous nerve
12 Pronator teres
13 Radial artery
14 Radial nerve
15 Radial recurrent artery
16 Superficial branch of radial nerve
17 Supinator

D Brachial arteriogram

1 Anterior interosseous artery
2 Brachial artery
3 Common interosseous artery
4 Posterior interosseous artery
5 Radial artery
6 Radial recurrent artery
7 Ulnar artery
8 Ulnar recurrent artery

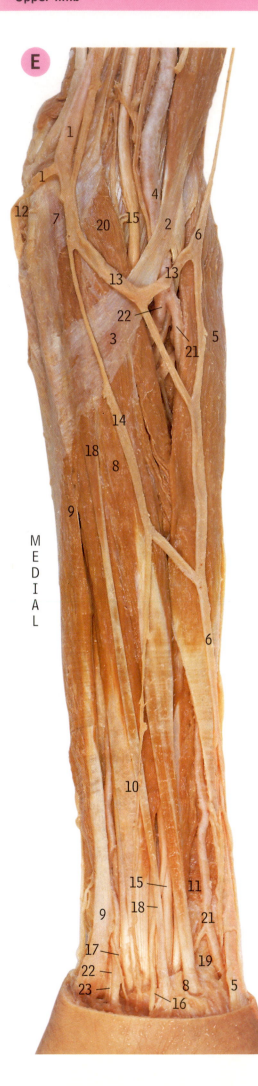

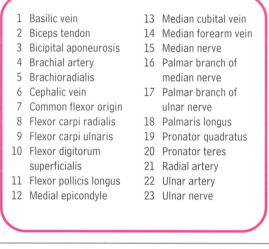

E Left forearm, from the front. Superficial muscles

Skin and fascia have been removed, but the larger superficial veins (1, 6 and 13) have been preserved. On the lateral side the radial artery (21) is largely covered by brachioradialis (5). At the wrist the tendon of flexor carpi radialis (8) has the radial artery (21) on its lateral side; on its medial side is the median nerve (15), slightly overlapped from the medial side by the tendon of palmaris longus (18) (if present; it is absent in 13 per cent of arms)

1	Basilic vein	13	Median cubital vein
2	Biceps tendon	14	Median forearm vein
3	Bicipital aponeurosis	15	Median nerve
4	Brachial artery	16	Palmar branch of
5	Brachioradialis		median nerve
6	Cephalic vein	17	Palmar branch of
7	Common flexor origin		ulnar nerve
8	Flexor carpi radialis	18	Palmaris longus
9	Flexor carpi ulnaris	19	Pronator quadratus
10	Flexor digitorum	20	Pronator teres
	superficialis	21	Radial artery
11	Flexor pollicis longus	22	Ulnar artery
12	Medial epicondyle	23	Ulnar nerve

Venous cutdown. If central venous cannulation is not available, the two major cutdown sites in the upper limb are the median cubital vein in the cubital fossa and the cephalic vein at the wrist over the radiocarpal joint. Both procedures involve aseptic technique and knowing the precise positions of these superficial veins.

F Left forearm, from the front, deep muscles

All vessels and nerves have been removed, together with the superficial muscles, to show the deep flexor group – flexor digitorum profundus (10), flexor pollicis longus (11) and pronator quadratus (13)

1	Abductor pollicis	8	Flexor carpi radialis
	longus	9	Flexor carpi ulnaris
2	Biceps	10	Flexor digitorum
3	Brachialis		profundus
4	Brachioradialis	11	Flexor pollicis longus
5	Common flexor origin	12	Flexor retinaculum
6	Extensor carpi	13	Pronator quadratus
	radialis brevis	14	Pronator teres
7	Extensor carpi	15	Supinator
	radialis longus		

Venepuncture of the upper limb is a very common procedure carried out on most hospital admissions. The most common sites are the median cubital vein in the antecubital fossa and the cephalic vein just proximal to the wrist joint. In difficult patients the dorsal venous arch of the hand may also be used. A useful tip is that veins are more fixed at a branching 'tree of a tributary'.

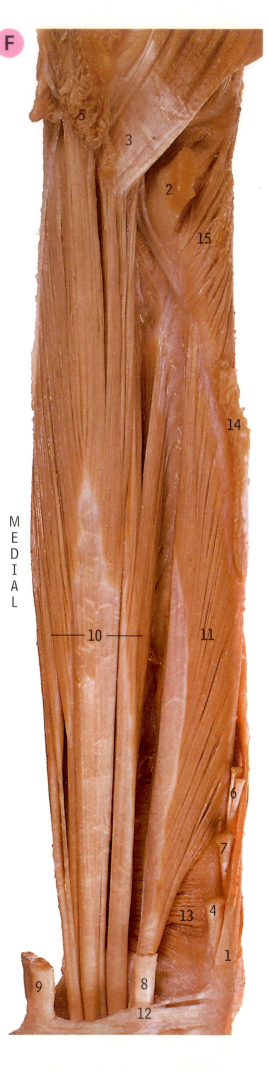

A Right cubital fossa and forearm. Arteries

The arteries have been injected, and after removal of most of the superficial muscles the brachial artery (4) is seen dividing into the radial artery (18) and the ulnar artery (20). The radial artery gives off the radial recurrent (19) which runs upwards in front of supinator, giving branches to the carpal extensor muscles (10 and 9). The ulnar artery gives off the anterior and posterior ulnar recurrent vessels (2 and 15), and its common interosseous branch (8) is seen giving off the anterior interosseous (1) which passes down in front of the interosseous membrane between flexor pollicis longus (13) and flexor digitorum profundus (12)

1 Anterior interosseous artery overlying interosseous membrane	8 Common interosseous artery	16 Pronator quadratus
2 Anterior ulnar recurrent artery	9 Extensor carpi radialis brevis	17 Pronator teres
3 Biceps tendon	10 Extensor carpi radialis longus	18 Radial artery
4 Brachial artery	11 Flexor carpi ulnaris	19 Radial recurrent artery overlying supinator
5 Brachialis	12 Flexor digitorum profundus	20 Ulnar artery
6 Brachioradialis	13 Flexor pollicis longus	
7 Common flexor origin	14 Medial epicondyle of humerus	
	15 Posterior ulnar recurrent artery	

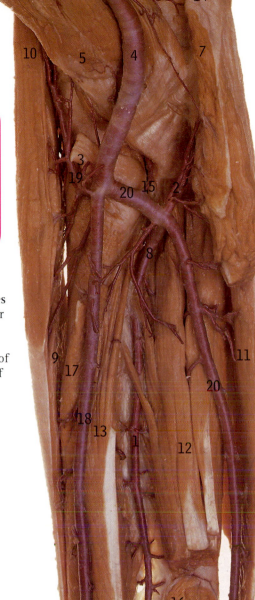

B Right cubital fossa and forearm. Arteries and nerves

Most of the humeral origins of pronator teres and flexor carpi radialis (from the common flexor origin, 9 and 7) and palmaris longus have been removed to show the median nerve (12) passing superficial to the deep head of pronator teres (18) and then deep to the upper border of the radial head of flexor digitorum superficialis (14)

> Volkmann's contracture is a deformity of the upper limb due to muscular ischaemia following injury to the brachial artery (supracondylar fracture of the humerus). An interruption of the blood supply of the upper limb musculature leads to necrosis and eventual fibrosis. Clinically, the finger flexors are usually the most severely affected, the patient being unable to extend the fingers when the wrist is flexed.

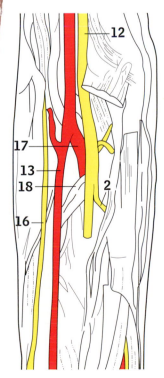

- In the lower part of B flexor carpi ulnaris (B8) has been displaced medially to show the underlying ulnar nerve and artery (B19).

> Anterior interosseous nerve entrapment. The deep branch of the median nerve, the anterior interosseous, may be trapped around the elbow following a fracture. The result is weakness in the flexor pollicis longus or flexor profundus muscles of the index and middle finger, making it impossible to flex the distal phalanx.

1 A muscular branch of median nerve	11 Lateral cutaneous nerve of forearm
2 Anterior interosseous nerve	12 Median nerve
3 Biceps	13 Radial artery
4 Brachial artery	14 Radial head of flexor digitorum superficialis
5 Brachialis	15 Radial recurrent artery
6 Brachioradialis (displaced laterally)	16 Superficial terminal branch of radial nerve overlying extensor carpi radialis longus
7 Common flexor origin	17 Ulnar artery
8 Flexor carpi ulnaris (displaced medially)	18 Ulnar head of pronator teres
9 Humeral head of pronator teres	19 Ulnar nerve and artery
10 Humero-ulnar head of flexor digitorum superficialis	

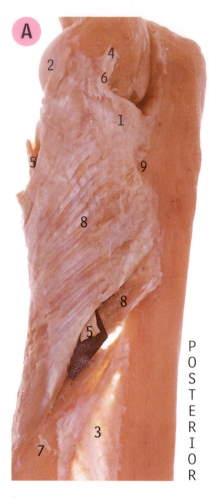

A

A Left elbow from the lateral side

With the forearm in midpronation and seen from the lateral side so that the radius (7) lies in front of the ulna, all muscles have been removed except supinator (8) to show its humeral and ulnar origins (see notes)

1 Annular ligament
2 Capitulum of humerus
3 Interosseous membrane
4 Lateral epicondyle
5 Posterior interosseous nerve
6 Radial collateral ligament
7 Radius
8 Supinator
9 Supinator crest of ulna

B Left forearm, from the lateral side. Deep muscles

1 Abductor pollicis longus
2 Biceps
3 Extensor carpi radialis brevis
4 Extensor carpi radialis longus (double)
5 Extensor indicis
6 Extensor pollicis brevis
7 Extensor pollicis longus
8 Extensor retinaculum
9 Flexor pollicis longus
10 Pronator teres
11 Supinator

C Left forearm, from behind. Posterior interosseous nerve

1 Abductor pollicis longus
2 Branch of posterior interosseous artery
3 Extensor carpi radialis brevis
4 Extensor carpi radialis longus
5 Extensor carpi ulnaris
6 Extensor digitorum
7 Extensor indicis
8 Extensor pollicis brevis
9 Extensor pollicis longus
10 Extensor retinaculum
11 Posterior interosseous nerve
12 Supinator

Posterior interosseous nerve entrapment occurs where this branch of the radial nerve passes through two planes of fibres within the supinator muscle, often following elbow trauma or a fibrous band within the supinator. There is no sensory loss if the posterior interosseous nerve alone is damaged but there is an inability to extend the fingers using the extensor digitorum muscle. Extension of the distal interphalangeal joints of the fingers is possible through the action of the ulnar-innervated interossei. If the radial nerve itself is damaged there is a more pronounced wrist drop.

- The fibres of the interosseous membrane (3) pass obliquely downwards from the radius (7) to the ulna, so transmitting weight from the hand and radius to the ulna.
- The supinator muscle (8) arises from the lateral epicondyle of the humerus (4), radial collateral ligament (6), annular ligament (1), supinator crest of the ulna (9) and bone in front of the crest (page 98, D10), and an aponeurosis overlying the muscle. From these origins the fibres wrap themselves round the upper end of the radius above the pronator teres attachment, to be attached to the lateral surface of the radius and extending anteriorly and posteriorly as far as the tuberosity of the radius.

B

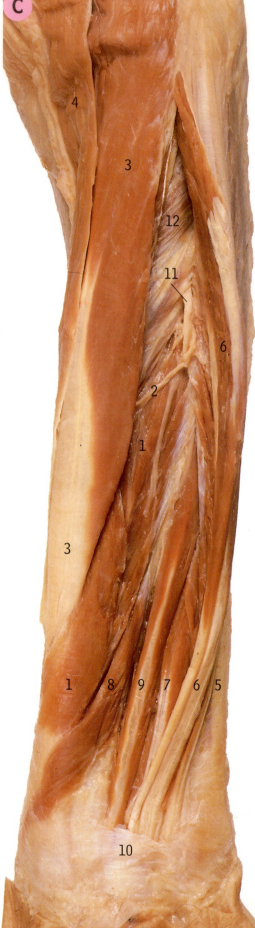

C

D

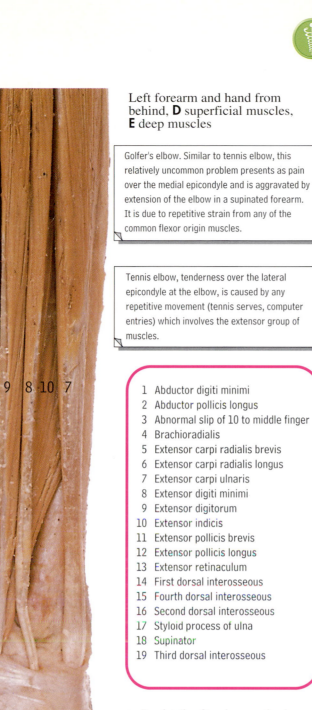

E

Left forearm and hand from behind, **D** superficial muscles, **E** deep muscles

Golfer's elbow. Similar to tennis elbow, this relatively uncommon problem presents as pain over the medial epicondyle and is aggravated by extension of the elbow in a supinated forearm. It is due to repetitive strain from any of the common flexor origin muscles.

Tennis elbow, tenderness over the lateral epicondyle at the elbow, is caused by any repetitive movement (tennis serves, computer entries) which involves the extensor group of muscles.

1 Abductor digiti minimi
2 Abductor pollicis longus
3 Abnormal slip of 10 to middle finger
4 Brachioradialis
5 Extensor carpi radialis brevis
6 Extensor carpi radialis longus
7 Extensor carpi ulnaris
8 Extensor digiti minimi
9 Extensor digitorum
10 Extensor indicis
11 Extensor pollicis brevis
12 Extensor pollicis longus
13 Extensor retinaculum
14 First dorsal interosseous
15 Fourth dorsal interosseous
16 Second dorsal interosseous
17 Styloid process of ulna
18 Supinator
19 Third dorsal interosseous

De Quervain's disease is a chronic inflammatory thickening of the tendon sheath usually seen in the abductor pollicis longus and extensor pollicis brevis muscles as they run across the lower end of the radius near the radial styloid. There is a palpable thickening of the tendon sheath and pain on movements of the thumb.

- For details of tendons on the dorsum of the hand, see page 138.

A Palm of the left hand

Interrupted lines = radial and ulnar arteries and palmar arches
The surface markings of various structures within the wrist and hand are indicated; not all of them are palpable, e.g. the superficial and deep palmar arches (13 and 12), but their relative positions are important

1	Abductor digiti minimi
2	Abductor pollicis brevis
3	Adductor pollicis
4	Distal transverse crease
5	Distal wrist crease
6	Flexor carpi radialis
7	Flexor carpi ulnaris
8	Flexor digiti minimi brevis
9	Flexor pollicis brevis
10	Head of metacarpal
11	Hook of hamate
12	Level of deep palmar arch
13	Level of superficial palmar arch
14	Longitudinal crease
15	Median nerve
16	Middle wrist crease
17	Palmaris brevis
18	Palmaris longus
19	Pisiform
20	Proximal transverse crease
21	Proximal wrist crease
22	Radial artery
23	Thenar eminence
24	Ulnar artery and nerve

B Dorsum of the left hand

The fingers are extended at the metacarpophalangeal joints, causing the extensor tendons of the fingers (2, 3 and 4) to stand out, and partially flexed at the interphalangeal joints. The thumb is extended at the carpometacarpal joint and partially flexed at the metacarpophalangeal and interphalangeal joints. The lines proximal to the bases of the fingers indicate the ends of the heads of the metacarpals and the level of the metacarpophalangeal joints. The anatomical snuffbox (1) is the hollow between the tendons of abductor pollicis longus and extensor pollicis brevis (5) laterally and extensor pollicis longus (6) medially

1	Anatomical snuffbox
2	Extensor digiti minimi
3	Extensor digitorum
4	Extensor indicis
5	Extensor pollicis brevis and abductor pollicis longus
6	Extensor pollicis longus
7	Extensor retinaculum
8	First dorsal interosseous
9	Head of ulna
10	Styloid process of radius

- The curved lines (10) proximal to the bases of the fingers indicate the ends of the heads of the metacarpals and the level of the metacarpophalangeal joints.
- The creases on the fingers indicate the level of the interphalangeal joints.

- The middle crease at the wrist indicates the level of the wrist joint.
- The radial artery at the wrist (22) is the commonest site for feeling the pulse. The vessel is on the radial side of the tendon of flexor carpi radialis (6) and can be compressed against the lower end of the radius.

- The median nerve at the wrist (15) lies on the ulnar side of the tendon of flexor carpi radialis (6) and is slightly overlapped from the ulnar side by the tendon of palmaris longus (18) (although this muscle is absent in 13 per cent of limbs).
- The ulnar nerve and artery at the wrist (24) are on the radial side of the tendon of flexor carpi ulnaris (7) and the pisiform bone (19). The artery is on the radial side of the nerve and its pulsation can be felt, though less easily than that of the radial artery (22).
- Abductor pollicis brevis (2) and flexor pollicis brevis (9), together with the underlying opponens pollicis, are the muscles which form the thenar eminence, the 'bulge' at the base of the thumb. Abductor digiti minimi (1) and flexor digiti minimi brevis (8), together with the underlying opponens digiti minimi, form the muscles of the hypothenar eminence, the less prominent bulge on the ulnar side of the palm where palmaris brevis (17) lies subcutaneously.

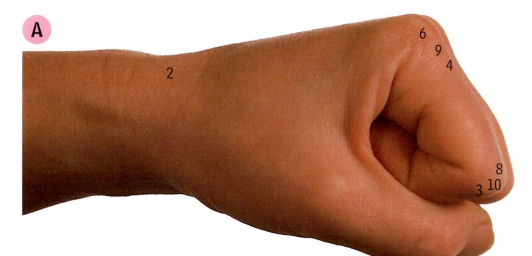

Movements of the fingers, **A** flexion of the metacarpophalangeal joints and flexion of the interphalangeal joints, **B** extension of the metacarpophalangeal joints and flexion of the interphalangeal joints, **C** extension of the metacarpophalangeal and interphalangeal joints

When 'making a fist' with all finger joints flexed (A), the heads of the metacarpals (6) form the knuckles. To extend the metacarpophalangeal joints (B9) requires the activity of the long extensor tendons of the fingers, but to extend the interphalangeal joints (C10 and 5) as well requires the activity of the interossei and lumbricals, pulling on the dorsal extensor expansions (page 140). Only if the metacarpophalangeal joints remain flexed can the long extensors extend the interphalangeal joints

1	Base of distal phalanx
2	Base of metacarpal
3	Base of middle phalanx
4	Base of proximal phalanx
5	Distal interphalangeal joint
6	Head of metacarpal
7	Head of middle phalanx
8	Head of proximal phalanx
9	Metacarpophalangeal joint
10	Proximal interphalangeal joint

• Muscles producing movements at the metacarpophalangeal joints:

Flexion: flexor digitorum profundus, flexor digitorum superficialis, lumbricals, interossei, with flexor digiti minimi brevis for the little finger and flexor pollicis longus, flexor pollicis brevis and the first palmar interosseous for the thumb.

Extension: extensor digitorum, extensor indicis (index finger) and extensor digiti minimi (little finger), with extensor pollicis longus and extensor pollicis brevis for the thumb.

Adduction: palmar interossei; when flexed, the long flexors assist.

Abduction: dorsal interossei and the long extensors, with abductor digiti minimi for the little finger.

• Muscles producing movements at the interphalangeal joints:

Flexion: at the proximal joints, flexor digitorum superficialis and flexor digitorum profundus; at the distal joints, flexor digitorum profundus. For the thumb, flexor pollicis longus.

Extension: with the metacarpophalangeal joints flexed, extensor digitorum, extensor indicis and extensor digiti minimi; with the metacarpophalangeal joints extended, interossei and lumbricals. For the thumb, extensor pollicis longus.

• Muscles producing movements at the wrist joint:

Flexion: flexor carpi radialis, flexor carpi ulnaris, palmaris longus, with assistance from flexor digitorum superficialis, flexor digitorum profundus, flexor pollicis longus and abductor pollicis longus.

Extension: extensor carpi radialis longus and brevis, extensor carpi ulnaris, assisted by extensor digitorum, extensor indicis, extensor digiti minimi and extensor pollicis longus.

Abduction: flexor carpi radialis, extensor carpi radialis longus and brevis, abductor pollicis longus and extensor pollicis brevis.

Adduction: flexor carpi ulnaris, extensor carpi ulnaris.

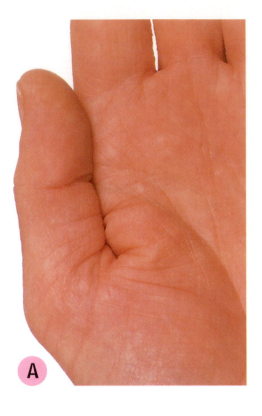

A

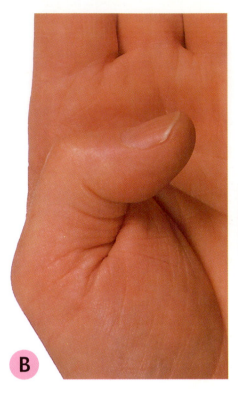

B

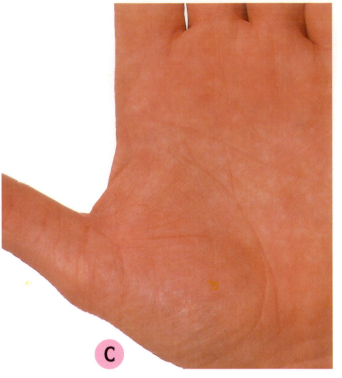

C

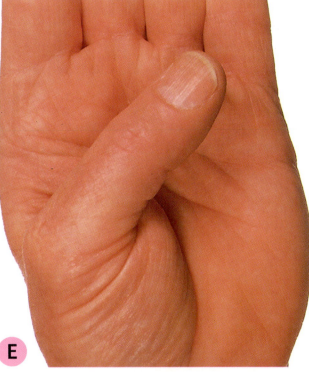

D

E

Movements of the thumb, **A** in the anatomical position, **B** in flexion, **C** in extension, **D** in abduction, **E** in opposition

With the thumb in the anatomical position (A), the thumb nail is at right angles to the fingers because the first metacarpal is at right angles to the others (page 101). This is a rather artificial position; in the normal position of rest the thumb makes an angle of about 60˚ with the plane of the palm (i.e. it is partially abducted). Flexion (B) means bending the thumb across the palm, keeping the phalanges at right angles to the palm. Extension (C) is the opposite movement, away from the palm. In abduction (D) the thumb is lifted forwards from the plane of the palm, and continuation of this movement inevitably leads to opposition (E), with rotation of the first metacarpal, twisting the whole digit so that the pulp of the thumb can be brought towards the palm at the base of the little finger (or more commonly in everyday use, to contact or overlap any of the flexed fingers). Opposition is a combination of abduction with flexion and medial rotation at the carpometacarpal joint; it is not necessarily accompanied by flexion at the other thumb joints

• Muscles producing movements at the carpometacarpal joint of the thumb:
Flexion: flexor pollicis brevis, opponens pollicis, and (when the other thumb joints are flexed) flexor pollicis longus.
Extension: abductor pollicis longus, extensor pollicis longus, extensor pollicis brevis.
Abduction: abductor pollicis brevis, abductor pollicis longus.
Adduction: adductor pollicis.
Opposition: opponens pollicis, flexor pollicis brevis, reinforced by adductor pollicis and flexor pollicis longus.

Wrist drop is a complication of radial nerve palsy due to loss of the forearm extensor muscles and may be associated with a loss of sensation over the first dorsal web space between the thumb and index finger.

A Palm of the left hand. Palmar aponeurosis
Removal of skin reveals the palmar aponeurosis with its thick central part (1) and the thin lateral and medial parts (3 and 4) which overlie the thenar and hypothenar muscles, respectively. The central part divides into slips (2) for each finger; in the intervals between adjacent slips (as at 5) digital vessels and nerves are seen

1 Central part of aponeurosis
2 Digital slips of aponeurosis
3 Lateral part of aponeurosis overlying thenar muscles
4 Medial part of aponeurosis overlying hypothenar muscles
5 Palmar digital vessels and nerves in interval between slips
6 Superficial transverse metacarpal ligaments

• The palmar aponeurosis is continuous with the distal edge of the flexor retinaculum; the palmaris longus tendon is attached to the aponeurosis and the distal part of the retinaculum.

1 Abductor digiti minimi
2 Abductor pollicis brevis
3 Adductor pollicis and digital branches of median nerve
4 Central part of palmar aponeurosis and filaments of palmar branch of median nerve
5 Fibrous sheath (partly removed)
6 First lumbrical
7 Flexor carpi radialis
8 Flexor carpi ulnaris
9 Flexor digiti minimi brevis
10 Flexor digitorum profundus tendon overlying superficialis tendon
11 Flexor digitorum superficialis
12 Flexor pollicis brevis and muscular (recurrent) branch of median nerve
13 Flexor retinaculum and palmar branch of median nerve
14 Median nerve and overlying palmar branch
15 Palmar digital vessels and nerves
16 Palmaris brevis and filament of palmar branch of ulnar nerve
17 Radial artery
18 Ulnar nerve and artery passing beneath superficial part of flexor retinaculum

B Palm of the left hand. Superficial dissection
The central part of the aponeurosis remains (4) but the medial and lateral parts have been removed to show the underlying muscles. At the base of the thumb, abductor pollicis brevis (2) lies lateral to flexor pollicis brevis (12). On the ulnar side, palmaris brevis (16) overlies the proximal part of abductor digiti minimi (1) and flexor digiti minimi brevis (9). At the wrist the median nerve (14) is on the ulnar side of the tendon of flexor carpi radialis (7)

Dupuytren's contracture is a deformity of the hand due to thickening of the palmar aponeurosis (of unknown aetiology) with resultant fibrosis and eventual contracture of the fingers. Presenting as a small hard nodule in the base of the ring finger, it tends to affect the ring and little finger as puckering and adherence of the palmar aponeurosis to the skin. Eventually the metacarpophalangeal and proximal interphalangeal joints become permanently flexed.

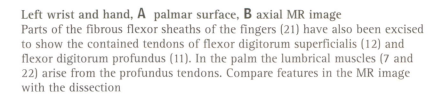

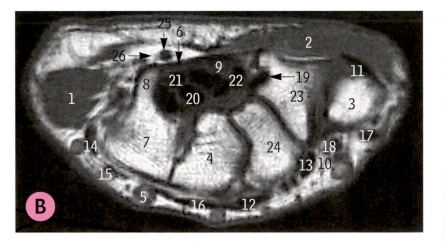

Left wrist and hand, A palmar surface, B axial MR image

Parts of the fibrous flexor sheaths of the fingers (21) have also been excised to show the contained tendons of flexor digitorum superficialis (12) and flexor digitorum profundus (11). In the palm the lumbrical muscles (7 and 22) arise from the profundus tendons. Compare features in the MR image with the dissection

1	Abductor digiti minimi	16	Median nerve
2	Abductor pollicis brevis	17	Median nerve, digital branch
3	Abductor pollicis longus	18	Median nerve, palmar cutaneous
4	Adductor pollicis		branch
5	Brachioradialis	19	Median nerve, recurrent branch
6	First dorsal interosseous	20	Palmaris brevis
7	First lumbrical	21	Remaining parts of fibrous flexor
8	Flexor carpi radialis		sheath
9	Flexor carpi ulnaris	22	Second lumbrical
10	Flexor digiti minimi brevis	23	Ulnar artery
11	Flexor digitorum profundus	24	Ulnar artery, deep branch
12	Flexor digitorum superficialis	25	Ulnar nerve
13	Flexor pollicis brevis	26	Ulnar nerve, deep branch
14	Flexor pollicis longus	27	Ulnar nerve, digital branch
15	Flexor retinaculum	28	Ulnar nerve, muscular branch

- The lumbrical muscles have no bony attachments. They arise from the tendons of flexor digitorum profundus (A11) – the first and second (A7 and A22) from the tendons of the index and middle fingers respectively, and the third and fourth from adjacent sides of the middle and ring, and ring and little fingers respectively. Each is attached distally to the radial side of the dorsal digital expansion of each finger (page 140).

1	Abductor digiti minimi muscle	14	Tendon of extensor carpi ulnaris muscle
2	Abductor pollicis brevis muscle	15	Tendon of extensor digiti minimi muscle
3	Base of first metacarpal	16	Tendon of extensor digitorum muscle
4	Capitate	17	Tendon of extensor pollicis brevis muscle
5	Dorsal venous arch	18	Tendon of extensor pollicis longus muscle
6	Flexor retinaculum	19	Tendon of flexor carpi radialis muscle
7	Hamate	20	Tendon of flexor digitorum profundus muscle
8	Hook of hamate	21	Tendon of flexor digitorum superficialis muscle
9	Median nerve	22	Tendon of flexor pollicis longus muscle
10	Radial artery	23	Trapezium
11	Tendon of abductor pollicis longus muscle	24	Trapezoid
12	Tendon of extensor carpi radialis brevis muscle	25	Ulnar artery
13	Tendon of extensor carpi radialis longus muscle	26	Ulnar nerve

Superficial palmar arch, **A** incomplete in the left hand, **B** complete in the right hand

1 A common palmar digital artery
2 A palmar digital artery
3 A palmar digital nerve
4 Abductor digiti minimi
5 Abductor pollicis brevis
6 Abductor pollicis longus
7 Common palmar digital branch of ulnar nerve
8 Common stem of 28 and 26
9 Deep branch of ulnar artery
10 Deep branch of ulnar nerve
11 Deep palmar arch
12 First lumbrical
13 Flexor carpi radialis
14 Flexor carpi ulnaris and pisiform
15 Flexor digitorum profundus
16 Flexor digitorum superficialis
17 Flexor pollicis brevis
18 Flexor pollicis longus
19 Flexor retinaculum
20 Fourth lumbrical
21 Median nerve
22 Median nerve dividing into common palmar digital branches
23 Muscular (recurrent) branch of median nerve
24 Opponens digiti minimi
25 Palmaris brevis
26 Princeps pollicis artery
27 Radial artery
28 Radialis indicis artery
29 Superficial palmar arch
30 Superficial palmar branch of radial artery
31 Ulnar artery
32 Ulnar nerve

Arterial cutdown, performed when repeated arterial blood sampling is required in a seriously ill patient, is performed on the radial artery in the non-dominant hand. Before proceeding it is important to confirm the patency of the deep palmar arch by occluding the radial artery and checking the ulnar arterial pulse (Allen's test).

Carpal tunnel syndrome is a constellation of symptoms due to compression of the median nerve in the carpal tunnel beneath the flexor retinaculum. The sensory loss is usually in the radial two and a half fingers and the thumb, and the presenting symptom may be pain along the radial side of the hand. The muscles affected are the three small thenar muscles (abductor pollicis brevis, flexor pollicis brevis and opponens pollicis). If the condition is of long standing there may be a noticeable loss of muscle bulk in the thenar eminence. Treatment usually involves division of the flexor retinaculum distal to the wrist.

- In two-thirds of hands the superficial palmar arch is not complete (as in A29). In the other third it is usually completed by the superficial palmar branch of the radial artery (B30).
- In the palm the superficial arterial arch (29) and its branches (as at 1) lie superficial to the common palmar digital nerves (22 and 7), but on the fingers the palmar digital nerves (as at 3) lie superficial (anterior) to the palmar digital arteries (as at 2).

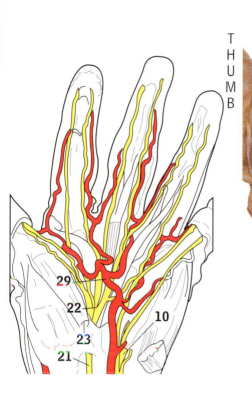

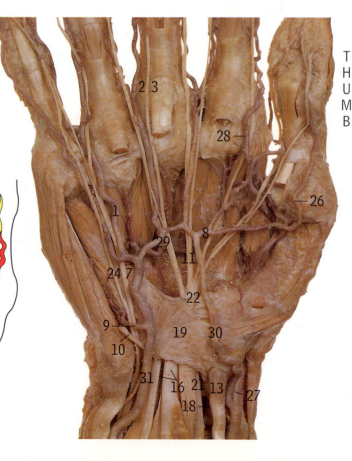

A Palm of the right hand, with synovial sheaths
The synovial sheaths of the wrist and fingers have been emphasized by blue tissue. On the middle finger the fibrous flexor sheath has been removed (but retained on the other fingers, as at 3) to show the whole length of the synovial sheath (22). On the index and ring fingers the synovial sheath projects slightly proximal to the fibrous sheath. The synovial sheath of the little finger is continuous with the sheath surrounding the finger flexor tendons under the flexor retinaculum (the ulnar bursa, 24), and the sheath of flexor pollicis longus is the radial bursa (20), which also continues under the retinaculum (9)

1 Abductor digiti minimi	13 Palmar branch of ulnar nerve
2 Abductor pollicis brevis	14 Palmar digital artery
3 Fibrous flexor sheath	15 Palmar digital nerve
4 Flexor carpi radialis	16 Palmaris brevis
5 Flexor carpi ulnaris	17 Palmaris longus
6 Flexor digiti minimi brevis	18 Pisiform bone
7 Flexor digitorum superficialis	19 Radial artery
8 Flexor pollicis brevis	20 Radial bursa and flexor pollicis longus
9 Flexor retinaculum	21 Superficial palmar arch
10 Median nerve	22 Synovial sheath
11 Muscular (recurrent) branch of median nerve	23 Ulnar artery
12 Palmar branch of median nerve	24 Ulnar bursa
	25 Ulnar nerve

- In the carpal tunnel (beneath the flexor retinaculum), one synovial sheath envelops the eight tendons of flexor digitorum superficialis and profundus (A24), another envelops the flexor pollicis longus tendon (A20), and flexor carpi radialis (in its own compartment of the flexor retinaculum) has its own sheath also (A4). The synovial sheaths for flexor carpi radialis and flexor pollicis longus extend as far as the tendon insertions.
- The sheath of the long finger flexors is continuous with the digital synovial sheath of the little finger, but is *not* continuous with the digital synovial sheaths of the ring, middle or index fingers; these fingers have their own synovial sheaths whose proximal ends project slightly beyond the *fibrous* sheaths within which the digital *synovial* sheaths lie.
- The muscular (recurrent) branch (A11) of the median nerve usually supplies abductor pollicis brevis, flexor pollicis brevis and opponens pollicis, but of all the muscles in the body flexor pollicis brevis (A8) is the one most likely to have an anomalous supply: in about one-third of hands by the median nerve, in another third by the ulnar nerve, and in the rest by both the median and ulnar nerves.

B Long flexor tendons and vincula of the right middle finger
The fibrous and synovial sheaths have been removed, and the flexor tendons (1 and 2) have been pulled anteriorly to show the vincula (3 and 6), which are small fibrous bands carrying blood vessels from the sheaths to the tendons

1 Flexor digitorum profundus
2 Flexor digitorum superficialis
3 Long vinculum of superficialis tendon
4 Lumbrical muscles
5 Metacarpal bone
6 Short vinculum of profundus tendon

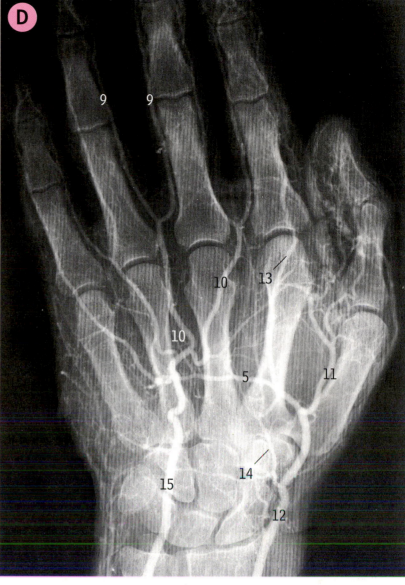

Palm of the right hand, C deep palmar arch, D arteriogram of palmar arteries

Most muscles and tendons have been removed and the arteries have been distended by injection. The deep palmar arch (5) is seen giving off the palmar metacarpal arteries (10) which join the common palmar digital arteries (3) from the superficial arch. Compare C with the vessels in the arteriogram

1   Abductor pollicis longus
2   Branch of anterior interosseous artery to anterior carpal arch
3   Common palmar digital arteries (from superficial arch)
4   Deep branch of ulnar artery
5   Deep palmar arch
6   Flexor carpi radialis
7   Flexor carpi ulnaris and pisiform
8   Head of ulna
9   Palmar digital arteries
10 Palmar metacarpal arteries
11 Princeps pollicis artery
12 Radial artery
13 Radialis indicis artery (anomalous origin)
14 Superficial palmar branch of radial artery
15 Ulnar artery

Arterial punctures. Two common sites for obtaining arterial blood in the upper limb are the brachial artery in the cubital fossa and the radial artery at the wrist. The site for the brachial artery is just medial and superior to the bicipital aponeurosis (i.e. the same point at which the blood pressure is taken). The radial artery is easily palpable just proximal to the wrist joint and lateral to the tendon of flexor carpi radialis. It is important when doing a radial arterial puncture to check that there is an ulnar artery pulse to allow for anastomosis should any damage or spasm follow the radial puncture (Allen's test).

Trigger finger, a form of digital tenosynovitis (a chronic inflammatory thickening of the tendon sheath), often affects the synovial sheaths of the fingers at the level of the metacarpal head. The patient presents with a clicking sensation and tenderness on the affected digit. The condition is known as trigger thumb when it affects the thumb.

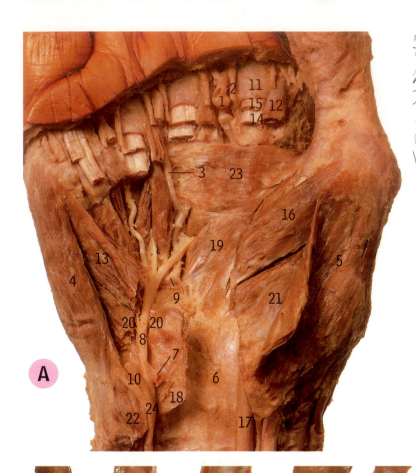

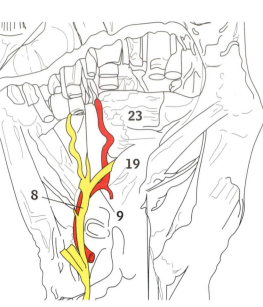

A Palm of the right hand. Deep branch of the ulnar nerve

The long flexor tendons (15 and 14) and lumbricals (12) have been cut off near the heads of the metacarpals, and parts of the hypothenar muscles removed to show the deep branches of the ulnar nerve and artery (8 and 7) running into the palm and curling laterally to pass between the transverse and oblique heads of adductor pollicis (23 and 19)

1 A common palmar digital artery	13 Flexor digiti minimi brevis
2 A palmar digital nerve	14 Flexor digitorum profundus
3 A palmar metacarpal artery	15 Flexor digitorum superficialis
4 Abductor digiti minimi	16 Flexor pollicis brevis
5 Abductor pollicis brevis	17 Flexor pollicis longus
6 Carpal tunnel	18 Flexor retinaculum (cut edge)
7 Deep branch of ulnar artery	19 Oblique head of adductor pollicis
8 Deep branch of ulnar nerve	20 Opponens digiti minimi
9 Deep palmar arch	21 Opponens pollicis
10 Digital branches of ulnar nerve	22 Pisiform
11 Fibrous flexor sheath	23 Transverse head of adductor pollicis
12 First lumbrical	24 Ulnar nerve

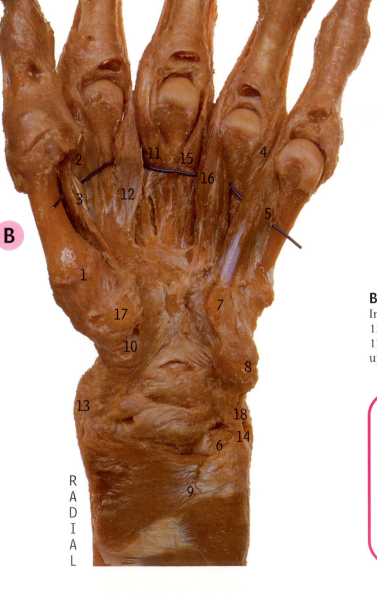

B Left wrist and palm. Interosseous muscles

In this deep dissection with most other muscles removed, the palmar interossei (3, 12, 16 and 5) are shown superficial to the blue marker and the dorsal interossei (2, 11, 15 and 4) deep to it. The capsule of the distal radio-ulnar joint has been opened up to show the head of the ulna (6)

1 Capsule of carpometacarpal joint of thumb	10 Scaphoid
2 First dorsal interosseous	11 Second dorsal interosseous
3 First palmar interosseous	12 Second palmar interosseous
4 Fourth dorsal interosseous	13 Styloid process of radius
5 Fourth palmar interosseous	14 Styloid process of ulna
6 Head of ulna	15 Third dorsal interosseous
7 Hook of hamate	16 Third palmar interosseous
8 Pisiform	17 Trapezium
9 Pronator quadratus	18 Ulnar collateral ligament

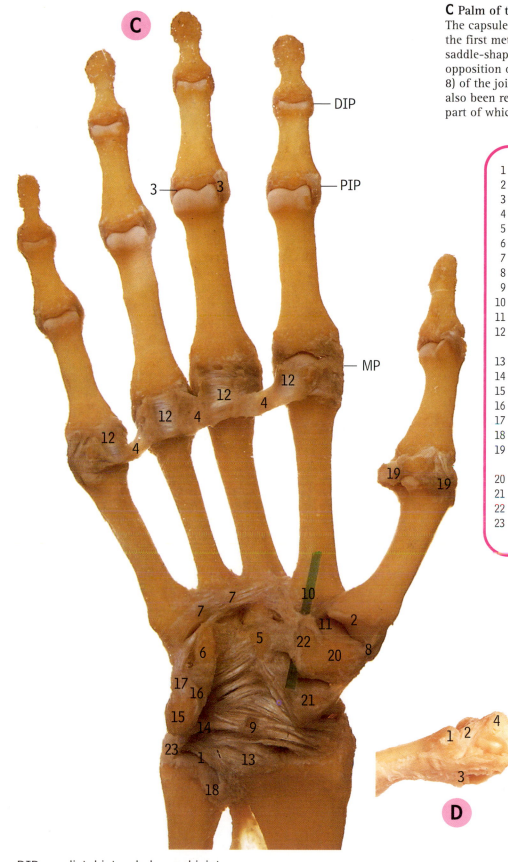

C on image

C Palm of the right hand. Ligaments and joints
The capsule of the carpometacarpal joint of the thumb (between the base of the first metacarpal and the trapezium) has been removed, to show the saddle-shaped joint surfaces which allow the unique movement of opposition of the thumb to occur. The palmar and lateral ligaments (11 and 8) of the joint remain intact. The capsule of the distal radio-ulnar joint has also been removed to show the articular disc, but the wrist joint, the ulnar part of which lies distal to the disc, has not been opened

DIP, PIP, MP labels on image

1	Articular disc of distal radio-ulnar joint
2	Base of first metacarpal
3	Collateral ligament of interphalangeal joint
4	Deep transverse metacarpal ligament
5	Head of capitate
6	Hook of hamate
7	Interosseous metacarpal ligament
8	Lateral ligament of carpometacarpal joint of thumb
9	Lunate
10	Marker in groove on trapezium for flexor carpi radialis tendon
11	Palmar ligament of carpometacarpal joint of thumb
12	Palmar ligament of metacarpophalangeal joint with groove for flexor tendon
13	Palmar radiocarpal ligament
14	Palmar ulnocarpal ligament
15	Pisiform
16	Pisohamate ligament
17	Pisometacarpal ligament
18	Sacciform recess of capsule of distal radio-ulnar joint
19	Sesamoid bones of flexor pollicis brevis tendons (with adductor pollicis on ulnar side)
20	Trapezium
21	Tubercle of scaphoid
22	Tubercle of trapezium
23	Ulnar collateral ligament of wrist joint

- The collateral ligaments of the metacarpophalangeal and interphalangeal joints (C3, D2) pass obliquely forwards from the posterior part of the side of the head of the proximal bone to the anterior part of the side of the base of the distal bone. They become tightest in flexion.
- Opposition of the thumb is a combination of flexion and abduction with medial rotation of the first metacarpal (page 130). The saddle-shape of the joint between the base of the first metacarpal and the trapezium, together with the way that the capsule and its reinforcing ligaments are attached to the bones, ensures that when flexor pollicis brevis and opponens pollicis contract they produce the necessary metacarpal rotation.
- The articular disc (1) holds the lower ends of the the radius and ulna together, and separates the distal radio-ulnar joint from the wrist joint, so that the cavities of these joints are not continuous (unlike those of the elbow and proximal radio-ulnar joints, which have one continuous cavity – page 120).

DIP   distal interphalangeal joint
PIP   proximal interphalangeal joint
MP   metacarpophalangeal joint

D Metacarpophalangeal (MP) joint of the right index finger, from the radial side
Part of the capsule has been removed to define the collateral ligament (2)

1	Base of proximal phalanx
2	Collateral ligament
3	Fibrous flexor sheath
4	Head of second metacarpal

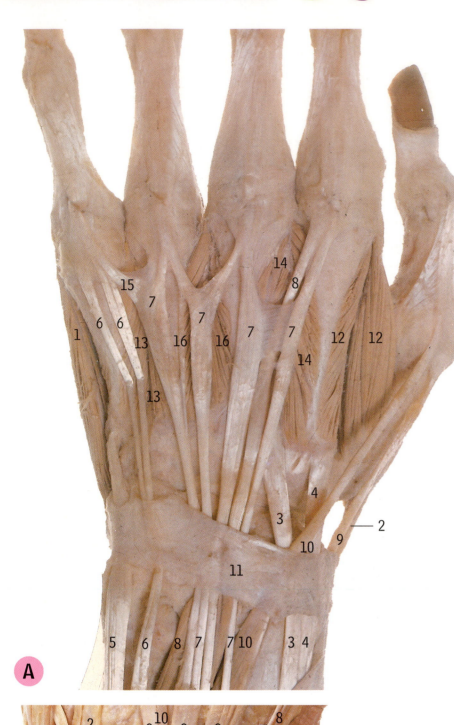

A Dorsum of the left hand. Muscles and tendons
All vessels, nerves and fascia have been removed to show the long tendons passing under the extensor retinaculum (11). See the notes below for the identification of finger tendons

1 Abductor digiti minimi
2 Abductor pollicis longus
3 Extensor carpi radialis brevis
4 Extensor carpi radialis longus
5 Extensor carpi ulnaris
6 Extensor digiti minimi
7 Extensor digitorum
8 Extensor indicis
9 Extensor pollicis brevis
10 Extensor pollicis longus
11 Extensor retinaculum
12 First dorsal interosseous
13 Fourth dorsal interosseous
14 Second dorsal interosseous
15 Slip from extensor digitorum to little finger
16 Third dorsal interosseous

- On the dorsum of the hand the tendon of extensor indicis (A8) lies on the ulnar side of the extensor digitorum tendon (A7) to the index finger.
- It is normal for the tendon of extensor digiti minimi (A6) to be double. In this specimen the extensor digitorum tendon (A7) to the ring finger is also double, with a slip passing to the middle finger.
- The 'tendon' of extensor digitorum (A15) to the little finger normally consists, as here, of a slip from the digitorum tendon (A7) to the ring finger, joining the digiti minimi tendon (A6) just proximal to the metacarpophalangeal joint. Similar slips may join adjacent digitorum tendons on other fingers, as here between the ring and middle fingers.

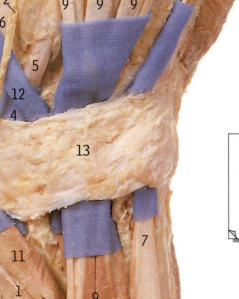

B Dorsum of the right wrist and hand. Synovial sheaths
Fascia and cutaneous branches of the ulnar nerve have been removed; the extensor reticulum (13) and the radial nerve (2) have been preserved and the synovial sheaths have been emphasized by blue tissue. From the radial to the ulnar side, the six compartments of the extensor retinaculum contain the tendons of: a, abductor pollicis longus and extensor pollicis brevis (1 and 11); b, extensor carpi radialis longus and brevis (6 and 5); c, extensor pollicis longus (12); d, extensor digitorum and extensor indicis (9 and 10); e, extensor digiti minimi (8); f, extensor carpi ulnaris (7)

Wrist ganglion is a cystic swelling, usually arising from a herniation of synovial fluid, found at the wrist near the lower end of the radius and ulna. Although the condition normally requires no treatment, the treatment of old was to hit the ganglion with a heavy book to rupture the hernial sac.

1 Abductor pollicis longus
2 Branches of radial nerve
3 Cephalic vein
4 Common sheath for 5 and 6
5 Extensor carpi radialis brevis
6 Extensor carpi radialis longus
7 Extensor carpi ulnaris
8 Extensor digiti minimi
9 Extensor digitorum
10 Extensor indicis
11 Extensor pollicis brevis
12 Extensor pollicis longus
13 Extensor retinaculum

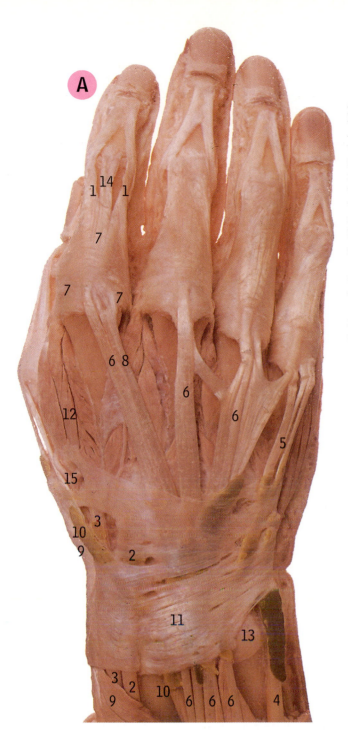

A Dorsum of the right hand. Muscles and tendons
All vessels and nerves (except the radial artery, 15) have been removed; the extensor retinaculum (11) is preserved, together with some fascia distal to it to give it some support to synovial sheaths which have been partially injected with green resin (compare with page 138, B). The margins of the distal parts of the extensor digital expansions (as at 7 and 1) have been emphasized by removal of the intervening connective tissue

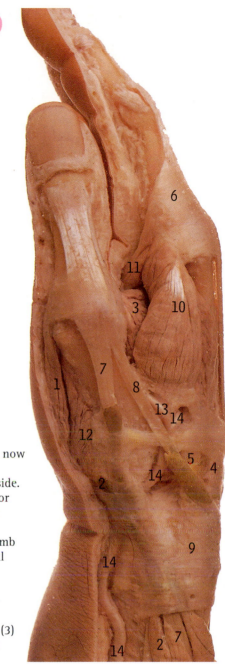

B Right hand. Muscles and tendons, from the radial side
This is the specimen seen in A, now rotated to show muscles and tendons on the radial (lateral) side. The synovial sheaths of extensor pollicis brevis (7) and extensor pollicis longus (8) show some injected resin. Between the thumb and index finger the first dorsal interosseous (10) passes to the expansion (6), with the first lumbrical (11) running into the expansion just beyond the interosseous. Adductor pollicis (3) passes through to the proximal phalanx of the thumb

1 Collateral slip of expansion to distal phalanx	9 Extensor pollicis brevis
2 Extensor carpi radialis brevis	10 Extensor pollicis longus
3 Extensor carpi radialis longus	11 Extensor retinaculum
4 Extensor carpi ulnaris	12 First dorsal interosseous
5 Extensor digiti minimi	13 Head of ulna
6 Extensor digitorum	14 Intermediate part of expansion to middle phalanx
7 Extensor expansion	15 Radial artery
8 Extensor indicis	

1 Abductor pollicis brevis	9 Extensor retinaculum
2 Abductor pollicis longus	10 First dorsal interosseous
3 Adductor pollicis	11 First lumbrical
4 Extensor carpi radialis brevis	12 Opponens pollicis
5 Extensor carpi radialis longus	13 Princeps pollicis artery (unusual origin)
6 Extensor expansion	14 Radial artery
7 Extensor pollicis brevis	
8 Extensor pollicis longus	

- In A the extensor digitorum tendon to the ring finger (6) is double, as well as giving a slip to the digiti minimi tendon (5), and to the extensor tendon of the middle finger. Some fascia distal to the extensor retinaculum (11) is preserved.
- At the lateral side of the wrist the radial artery (B14) lies in the 'anatomical snuffbox', which is bounded laterally by the tendons of abductor pollicis longus and extensor pollicis brevis (B2 and B7) and medially by the tendon of extensor pollicis longus (B8).
- The princeps pollicis artery (B13) has a more proximal origin than usual; it normally arises from the radial artery after that artery has passed through the first dorsal interosseous muscle (B10) to enter the palm.
- The radial artery (B14, and A15 on page 140) enters the palm by passing through the first dorsal interosseous muscle (B10, and A13 on page 140), in B just after giving off the princeps pollicis artery (B13), and in A on page 140 after giving off the first dorsal metacarpal artery (A13).
- The anterior interosseous artery (A4 on page 140) pierces the interosseous membrane above pronator quadratus (here removed) to anastomose with the posterior interosseous artery (A14 on page 140) and join the dorsal carpal arch (A7 on page 140).

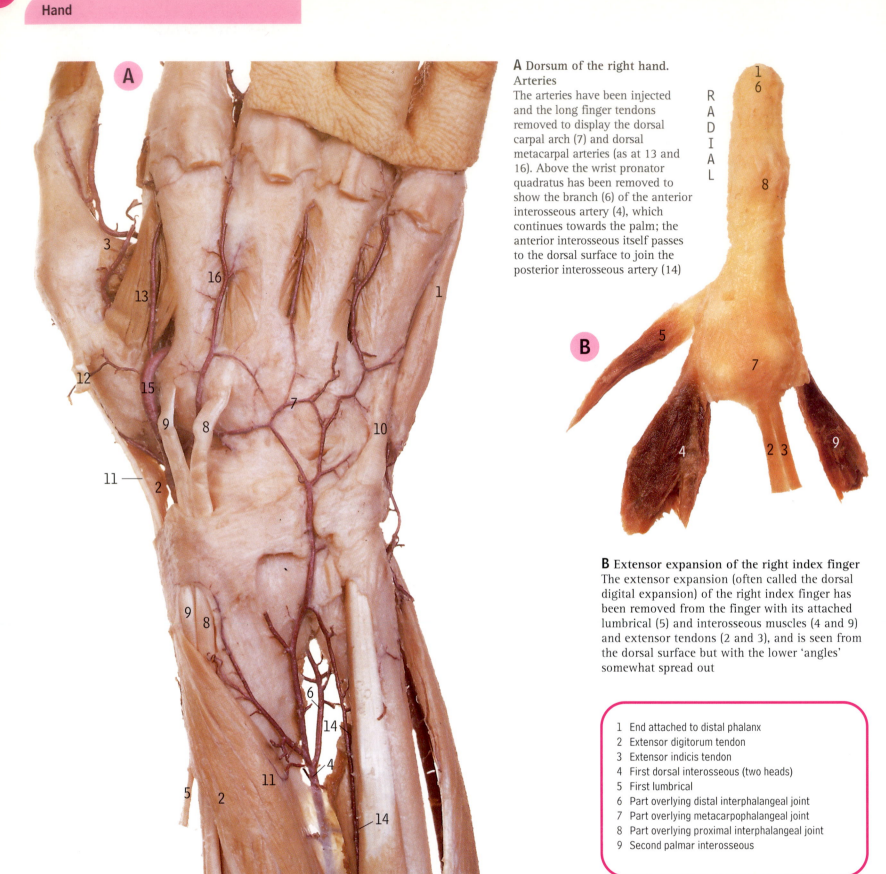

A Dorsum of the right hand. Arteries
The arteries have been injected and the long finger tendons removed to display the dorsal carpal arch (7) and dorsal metacarpal arteries (as at 13 and 16). Above the wrist pronator quadratus has been removed to show the branch (6) of the anterior interosseous artery (4), which continues towards the palm; the anterior interosseous itself passes to the dorsal surface to join the posterior interosseous artery (14)

R
A
D
I
A
L

B Extensor expansion of the right index finger
The extensor expansion (often called the dorsal digital expansion) of the right index finger has been removed from the finger with its attached lumbrical (5) and interosseous muscles (4 and 9) and extensor tendons (2 and 3), and is seen from the dorsal surface but with the lower 'angles' somewhat spread out

1 End attached to distal phalanx
2 Extensor digitorum tendon
3 Extensor indicis tendon
4 First dorsal interosseous (two heads)
5 First lumbrical
6 Part overlying distal interphalangeal joint
7 Part overlying metacarpophalangeal joint
8 Part overlying proximal interphalangeal joint
9 Second palmar interosseous

1 Abductor digiti minimi
2 Abductor pollicis longus
3 Adductor pollicis and branch of princeps pollicis artery
4 Anterior interosseous artery
5 Brachioradialis
6 Branch of anterior interosseous artery to anterior carpal arch

7 Dorsal carpal arch
8 Extensor carpi radialis brevis
9 Extensor carpi radialis longus
10 Extensor carpi ulnaris
11 Extensor pollicis brevis
12 Extensor pollicis longus
13 First dorsal interosseous and first dorsal metacarpal artery

14 Posterior interosseous artery
15 Radial artery
16 Second dorsal interosseous and second dorsal metacarpal artery

• Three tendons pass to different levels of the thumb: abductor pollicis longus (A2) to the base of the first metacarpal, extensor pollicis brevis (A11) to the base of the proximal phalanx, and extensor pollicis longus (A12) to the base of the distal phalanx.

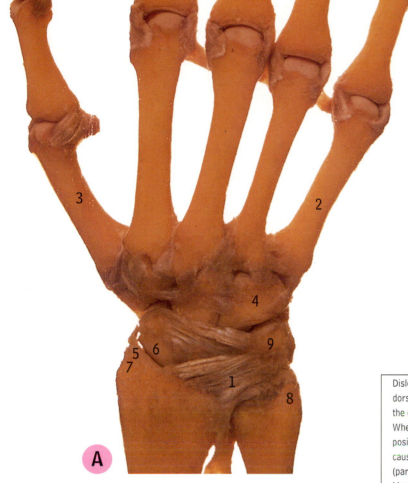

A Dorsum of the right hand. Ligaments and joints

Most joint capsules have been removed, including the radial parts of the wrist joint capsule, so showing the articulation between the scaphoid (6) and the lower end of the radius (7)

1 Dorsal radiocarpal ligament	6 Scaphoid
2 Fifth metacarpal	7 Styloid process of radius
3 First metacarpal	8 Styloid process of ulna
4 Hamate	9 Triquetral
5 Radial collateral ligament of wrist joint	

- The wrist (radiocarpal) joint is the joint between the lower end of the radius and the articular disc of the distal radio-ulnar joint proximally, and the scaphoid, lunate and triquetral bones distally.
- The midcarpal joint is the joint between the scaphoid, lunate and triquetral proximally, and the trapezium, trapezoid, capitate and hamate distally.
- Extension of the wrist occurs at the wrist and midcarpal joints, but most of the movement takes place at the wrist joint.
- Flexion of the wrist occurs at the wrist and midcarpal joints, but most of the movement takes place at the midcarpal joint.

Dislocation of the lunate. Severe forced dorsiflexion of the wrist may occasionally cause the capitate or lunate to become dislocated. When the wrist then returns to the neutral position the lunate is displaced anteriorly, causing pressure on the median nerve (paraesthesia in the thumb and index fingers). Manipulative reduction consists of pulling on the wrist, which is then palmar flexed slowly while pressure is applied on the lunate itself.

Avascular necrosis of the scaphoid. Following a fracture of the waist of the scaphoid, avascular necrosis may occur in the proximal fragment. This is because the arterial supply to this bone goes from distal to proximal and, therefore, the proximal fragment becomes avascular. The condition is often not diagnosed until 6 to 8 weeks after injury; the X-ray may then show an apparent increase in the density of the avascular fragment. Later there may be osseous collapse and removal of this segment may be necessary if immobilization has not occurred at the time of injury.

Coronal section of the right wrist, **B** dissection, **C** MR image

Viewed from the dorsal surface, the section has passed through the wrist near this surface, and the first and fifth metacarpals have not been included in the cut. The arrows between the two rows of carpal bones indicate the line of the midcarpal joint. Compare the MR image with the section

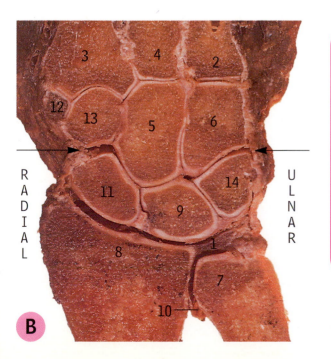

1 Articular disc	
2 Base of fourth metacarpal	
3 Base of second metacarpal	
4 Base of third metacarpal	
5 Capitate	
6 Hamate	
7 Head of ulna	
8 Lower end of radius	
9 Lunate	
10 Sacciform recess of distal radio-ulnar joint	
11 Scaphoid	
12 Trapezium	
13 Trapezoid	
14 Triquetral	

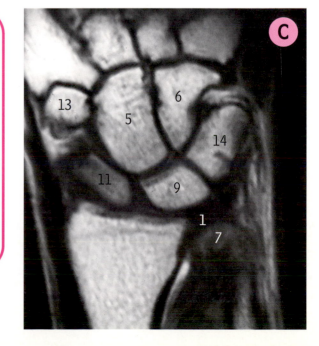

Right wrist and midcarpal joints, A wrist joint, opened up in forced flexion, B midcarpal joint, opened up in forced extension
Both joints have been opened up (far beyond the normal range of movement) in order to demonstrate the bones of the joint surfaces. The wrist joint in A has been forced open in flexion, since flexion takes place mostly at this joint, and the midcarpal joint in B has been forced open in extension, since extension takes place mostly at this joint. The proximal (wrist joint) surfaces of the scaphoid (21), lunate (13) and triquetral (24) are seen in A, and their distal (midcarpal joint) surfaces in B

1 Articular disc	8 Flexor carpi radialis tendon	15 Palmar arch vein	22 Styloid process of radius
2 Capitate	9 Flexor carpi ulnaris tendon	16 Palmaris longus tendon	23 Styloid process ulna
3 Extensor carpi radialis brevis	10 Flexor digitorum profundus tendon	17 Radial artery	24 Triquetral
4 Extensor carpi radialis longus	11 Flexor digitorum superficialis tendon	18 Radial artery, palmar arch branch	25 Ulnar artery
5 Extensor carpi ulnaris	12 Hamate	19 Radial surface for lunate	
6 Extensor digiti minimi	13 Lunate	20 Radial surface for scaphoid	
7 Extensor digitorum	14 Median nerve	21 Scaphoid	

BACK OF RIGHT
THUMB EDGE   D O R S A L

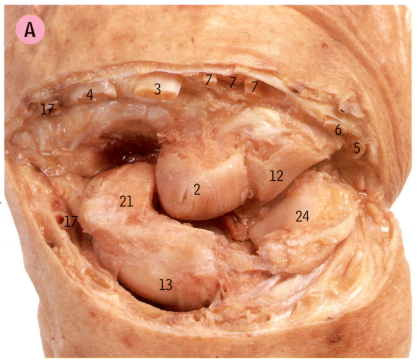

R A D I A L

D O R S A L

P A L M A R   FRONT OF RIG
THUMB ED

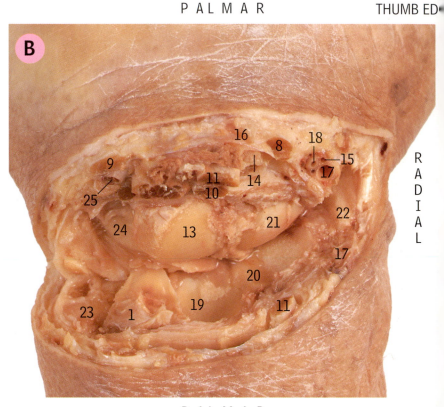

R A D I A L

P A L M A R

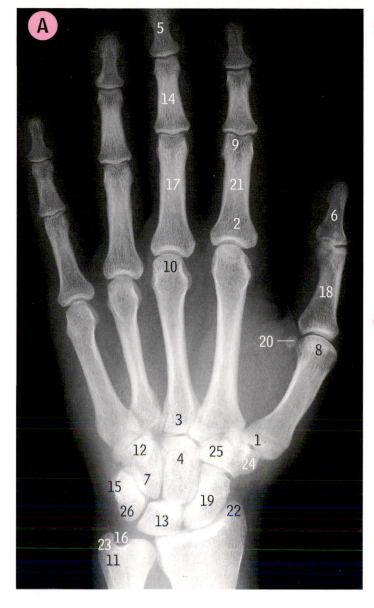

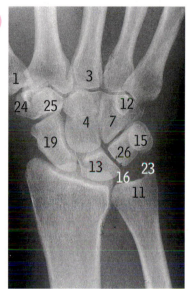

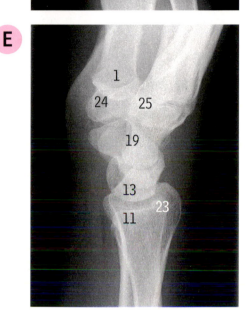

Wrist and hand radiographs, **A** dorsopalmar projection, **B** of a four-year-old child, **C** oblique projection, **D** posteroanterior projection, **E** lateral projection

Compare the epiphyses of the metacarpals and phalanges seen in B with the bony specimens in K and J on page 104

- The epiphysis at the lower end of the radius appears on a radiograph at 2 years and in the ulna at 6 years. The first carpal bone to appear is the capitate at 1 year.

1 Base of first metacarpal	15 Pisiform
2 Base of phalanx	16 Position of articular disc of distal radio-ulnar joint
3 Base of third metacarpal	
4 Capitate	17 Proximal phalanx of middle finger
5 Distal phalanx of middle finger	18 Proximal phalanx of thumb
6 Distal phalanx of thumb	19 Scaphoid
7 Hamate	20 Sesamoid bone in flexor pollicis brevis
8 Head of first metacarpal	
9 Head of phalanx	21 Shaft of phalanx
10 Head of third metacarpal	22 Styloid process at lower end of radius
11 Head of ulna	23 Styloid process of ulna
12 Hook of hamate	24 Trapezium
13 Lunate	25 Trapezoid
14 Middle phalanx of middle finger	26 Triquetral

Scaphoid fracture occurs across the waist of the scaphoid and initially may be extremely difficult to diagnose. There is acute tenderness in the anatomical snuffbox which is bounded by the tendons of extensor pollicis longus and abductor pollicis longus. Owing to the proximal segment's loss of blood supply, avascular necrosis is a known complication of this fracture and osteoarthritis may be severe if not treated.

Chapter 4

Thorax

Typical ribs, from behind, **A** the left fifth rib (a typical upper rib), **B** the left seventh rib (a typical lower rib)

1 Angle
2 Articular facet of tubercle
3 Articular facets of head
4 Costal groove
5 Crest of head
6 Neck
7 Non-articular part of tubercle
8 Shaft

- The typical ribs are the third to the ninth.
- Typical ribs have a head with two facets (3), and a tubercle with articular and non-articular parts (2 and 7) at the junction of the neck (6) and shaft (8). The shaft has external and internal surfaces, an angle (1) and a costal groove (4).
- In typical upper ribs the articular facet of the tubercle is curved (A2) but becomes increasingly flattened in lower ribs (B2).

C Typical rib and vertebra articulated, from above

1 Articular facet of transverse process
2 Articular part of tubercle
3 Neck of rib
4 Non-articular part of tubercle
5 Upper costal facet of head of rib
6 Upper costal facet of vertebral body

- The lower of the two facets on the head of a typical rib articulates with the upper costal facet (6) on the vertebral body having the same number as the rib. The upper facet on the head of the rib (5) articulates with the vertebral body above. These form the joints of the heads of ribs.
- The articular facet of the tubercle of a rib (C2) articulates with the costal facet of the transverse process of a vertebra. These are the costotransverse joints.
- The joints of the heads of the ribs and the costotransverse joints collectively form the costovertebral joints.

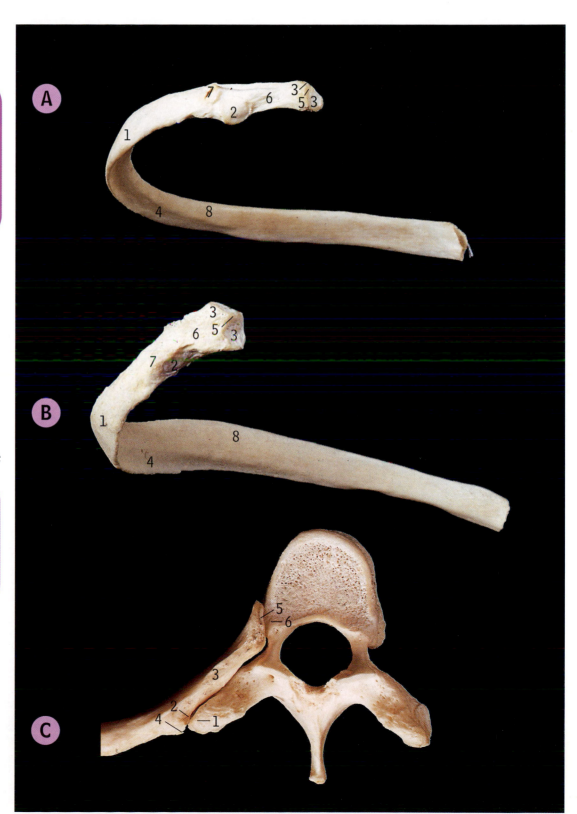

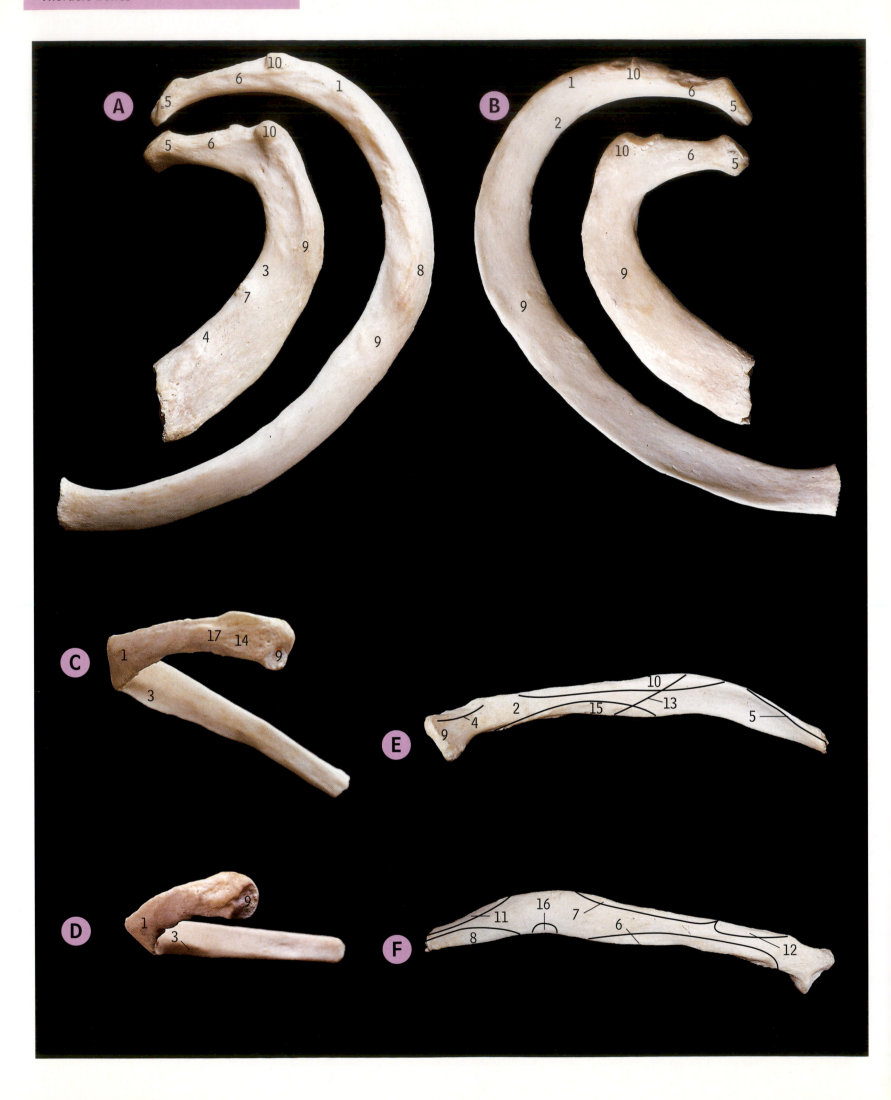

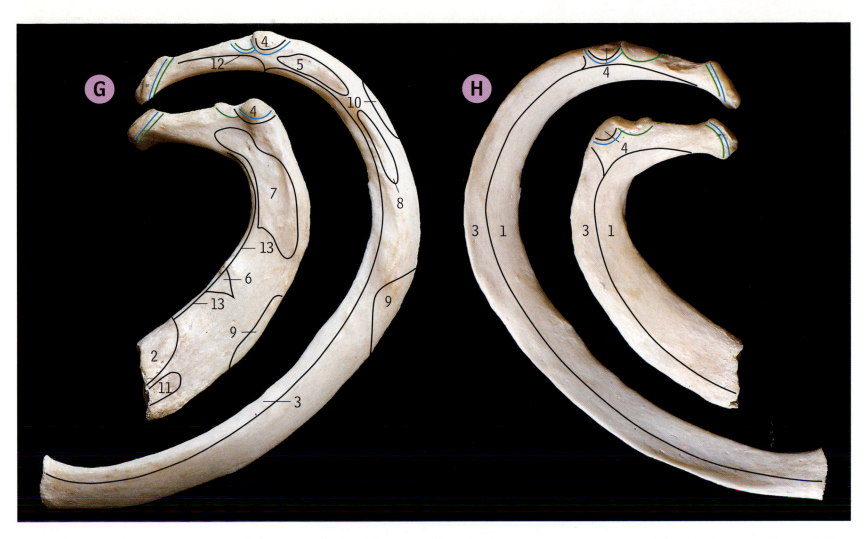

Left first rib (inner) and second rib (outer), **A** from above, **B** from below

1 Angle
2 Costal groove
3 Groove for subclavian artery and first thoracic nerve
4 Groove for subclavian vein
5 Head
6 Neck
7 Scalene tubercle
8 Serratus anterior tuberosity
9 Shaft
10 Tubercle

- The second rib gives origin to part of the first, and the whole of the second, digitation of serratus anterior.
- The atypical ribs are the first, second, tenth, eleventh and twelfth.
- The first rib has a head with one facet (A5), a prominent tubercle (A10), no angle and no costal groove. The shaft has superior and inferior surfaces.
- The second rib has a head with two facets (B5), an angle (B1) near the tubercle (B10), a broad costal groove (B2) posteriorly, and an external surface facing upwards and outwards with the inner surface facing correspondingly downwards and inwards.
- The tenth rib has a head with one or two facets (C9), a tubercle with or without an articular facet (C17), and a costal groove (C3).

Atypical left lower ribs, **C** tenth rib from behind, **D** eleventh rib from behind, **E** twelfth rib from the front, with attachments, **F** twelfth rib from behind, with attachments

1 Angle
2 Area covered by pleura
3 Costal groove
4 Costotransverse ligament
5 Diaphragm
6 Erector spinae
7 External intercostal
8 External oblique
9 Head
10 Internal intercostal
11 Latissimus dorsi
12 Levator costae
13 Line of pleural reflexion
14 Neck
15 Quadratus lumborum
16 Serratus posterior inferior
17 Tubercle

- The eleventh rib has a head with one facet (D9), no tubercle but there is an angle (D1) and a slight costal groove (D3).
- The twelfth rib has a head with one facet (E9) but there is no tubercle, no angle and no costal groove. The shaft tapers at its end (the ends of all other ribs widen slightly).

Left first rib (inner) and second rib (outer), **G** from above, **H** from below. Attachments
Blue lines = epiphysial lines; green lines = capsule attachments of costovertebral joints

1 Area covered by pleura
2 Costoclavicular ligament
3 Intercostal muscles and membranes
4 Lateral costotransverse ligament
5 Levator costae
6 Scalenus anterior
7 Scalenus medius
8 Scalenus posterior
9 Serratus anterior
10 Serratus posterior superior
11 Subclavius
12 Superior costotransverse ligament
13 Suprapleural membrane

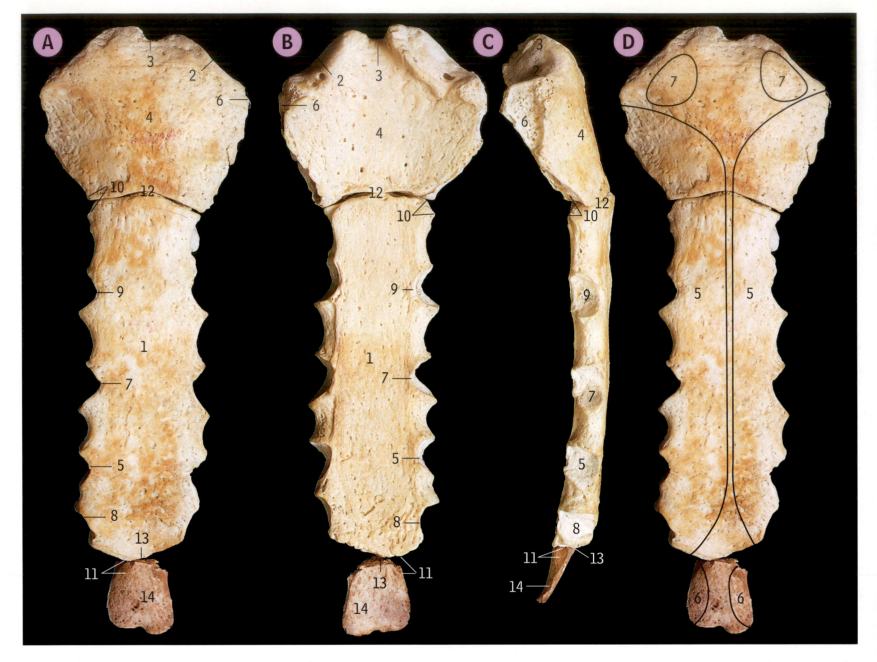

The sternum, A from the front, B from behind, C from the right

1	Body
2	Clavicular notch
3	Jugular notch
4	Manubrium
5	Notch for fifth costal cartilage
6	Notch for first costal cartilage
7	Notch for fourth costal cartilage
8	Notch for sixth costal cartilage
9	Notch for third costal cartilage
10	Notches for second costal cartilage
11	Notches for seventh costal cartilage
12	Sternal angle and manubriosternal joint
13	Xiphisternal joint
14	Xiphoid process

- The sternum consists of the manubrium (4), body (1) and xiphoid process (14).
- The body of the sternum (1) is formed by the fusion of four sternebrae, the sites of the fusion sometimes being indicated by three slight transverse ridges.
- The manubrium (4) and body (1) are bony but the xiphoid process (14), which varies considerably in size and shape, is cartilaginous although it frequently shows some degree of ossification.
- The manubriosternal and xiphisternal joints (12 and 13) are both symphyses, the surfaces being covered by hyaline cartilage and united by a fibrocartilaginous disc.

Bone marrow aspiration . The sternum is a site of red marrow, even in adulthood, and is a convenient site to obtain a bone marrow biopsy. Another common site is the posterior iliac crest.

The sternum, D from the front, E from behind. Attachments

1	Area covered by left pleura
2	Area covered by right pleura
3	Area in contact with pericardium
4	Diaphragm
5	Pectoralis major
6	Rectus abdominis
7	Sternocleidomastoid
8	Sternohyoid
9	Sternothyroid
10	Transversus thoracis

- The two pleural sacs are in contact from the levels of the second to fourth costal cartilages (E2 and 1).

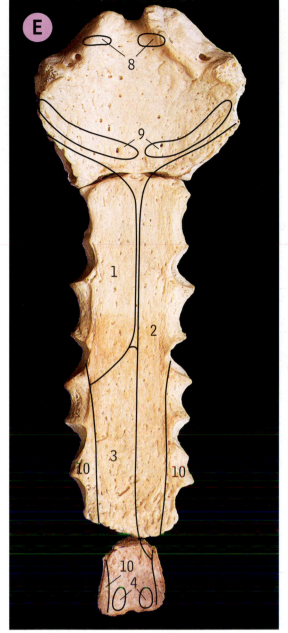

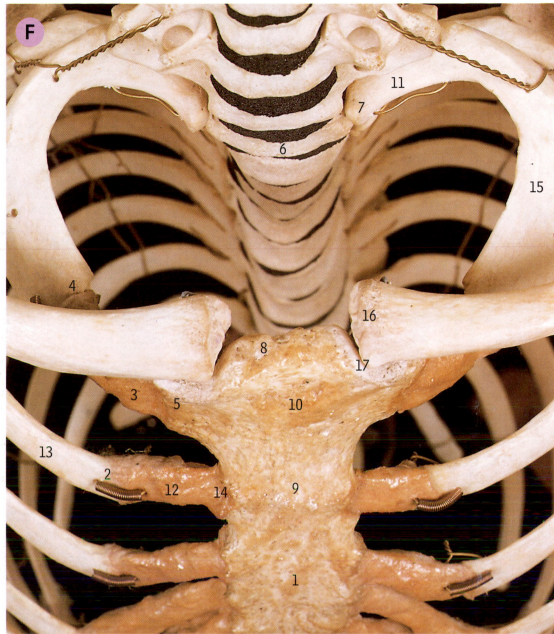

Rib/sternal fractures are most commonly seen following motor vehicle accidents (where the chest is crushed against the steering wheel) or during contact sports. In both situations a common complication is pneumothorax, the broken rib edge having torn the visceral pleura, allowing air to enter the pleural cavity, collapsing the lung.

Costochondritis, sometimes known as Tietze's syndrome, is inflammation of the costochondral and chondrosternal joints. It is commonly misdiagnosed as cardiac disease.

F Thoracic inlet, from above and in front, in an articulated skeleton

- The thoracic inlet (upper aperture of the thorax) is approximately the same size and shape as the outline of the kidney, and is bounded by the first thoracic vertebra (6), first ribs (15), and costal cartilages (3), and the upper border of the manubrium of the sternum (jugular notch, 8). It does not lie in a horizontal plane but slopes downwards and forwards.

- The second costal cartilage (12) joins the manubrium and body of the sternum (10 and 1) at the level of the manubriosternal joint (9). This is an important landmark, since the joint line is palpable as a ridge at the slight angle between the manubrium and body, and the second costal cartilage and rib can be identified lateral to it. Other ribs can be identified by counting down from the second; the first costal cartilage and the end of the first rib are under cover of the clavicle and not easily felt.

- The thoracic 'inlet' is sometimes known clinically as the thoracic 'outlet'.

1 Body of sternum
2 Costochondral joint
3 First costal cartilage
4 First costochondral joint
5 First sternocostal joint
6 First thoracic vertebra
7 Head of first rib
8 Jugular notch
9 Manubriosternal joint (angle of Louis)
10 Manubrium of sternum
11 Neck of first rib
12 Second costal cartilage
13 Second rib
14 Second sternocostal joint
15 Shaft of first rib
16 Sternal end of clavicle
17 Sternoclavicular joint

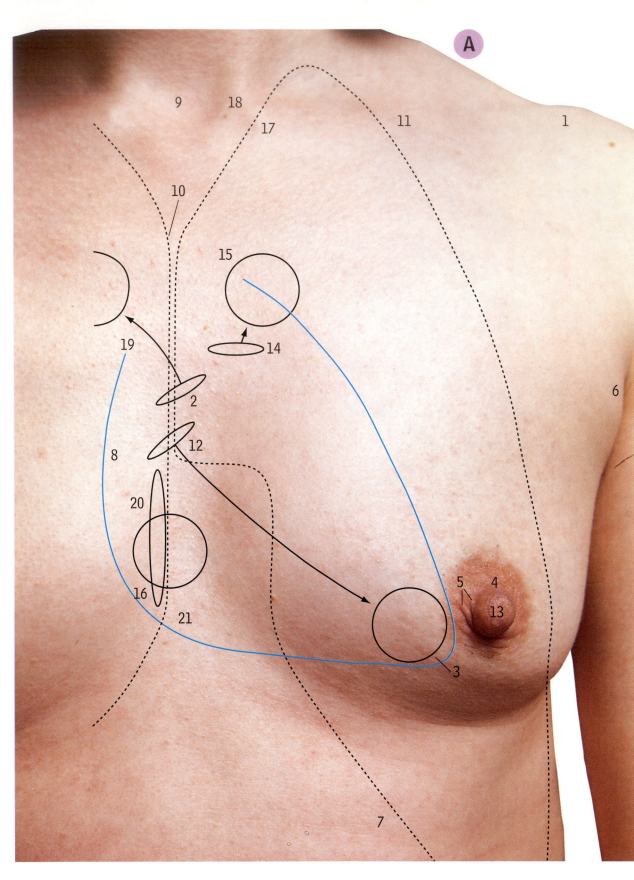

A Surface markings of the heart, left pleura and lung, in the female
Blue line = heart; dotted lines = pleura
The positions of the four heart valves are indicated by ellipses, and the sites where the sounds of the corresponding valves are best heard with the stethoscope are indicated by the circles

1   Acromioclavicular joint
2   Aortic valve
3   Apex of heart
4   Areola of breast
5   Areolar gland of breast
6   Axillary tail of breast
7   Costal margin (at eighth costal cartilage)
8   Fourth costal cartilage
9   Jugular notch
10 Manubriosternal joint
11 Midpoint of clavicle
12 Mitral valve
13 Nipple of breast
14 Pulmonary valve
15 Second costal cartilage
16 Sixth costal cartilage
17 Sternoclavicular joint
18 Sternocleidomastoid
19 Third costal cartilage
20 Tricuspid valve
21 Xiphisternal joint

Heart sounds (sites of auscultation). The four heart valves are best heard in specific places on the anterior chest wall related to the direction of blood flow through that valve – the aortic over the second right intercostal space just lateral to the sternum; the pulmonary in the second left intercostal space just lateral to the sternum; the tricuspid valve in the midline at the lower end of the sternum; the mitral valve towards the apex at the level of the fourth or fifth left costal cartilage usually just medial to the nipple.

- The manubriosternal joint (10) is palpable and a guide to identifying the second costal cartilage (15) which joins the sternum at this level (see page 149, F9, 14 and 12).
- The pleura and lung extend into the neck for 2.5 cm above the medial third of the clavicle.
- In the midclavicular line the lower limit of the *pleura* reaches the eighth costal cartilage, in the midaxillary line it reaches the tenth rib, and at the lateral border of the erector spinae muscle it crosses the twelfth rib. The lower border of the *lung* is about two ribs higher than the pleural reflexion.
- Behind the sternum the pleural sacs are adjacent to one another in the midline from the level of the second to fourth costal cartilages, but then diverge owing to the mass of the heart on the left.

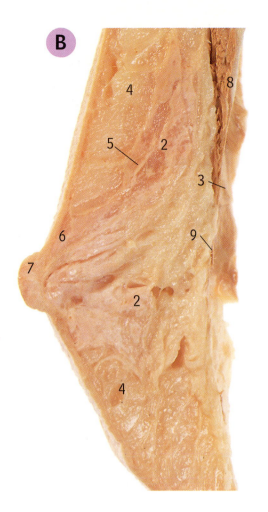

B

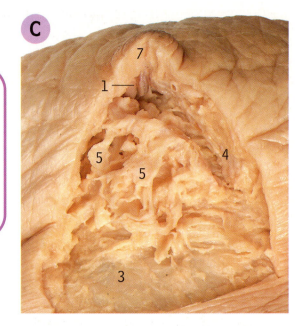

C

Female breast (mammary gland), **B** median sagittal section, **C** dissection of lower part, from the front and below, **D** xeromammogram

1 Ampulla of lactiferous duct
2 Condensed glandular tissue
3 Fascia over pectoralis major
4 Fat
5 Fibrous septum
6 Lactiferous duct
7 Nipple
8 Pectoralis major
9 Retromammary space

Breast examination. A retracted nipple or a dimpling of the skin on breast examination are signs that should be taken seriously and may indicate an underlying pathology. Examination of the axillary lymph nodes is an important part of all breast examinations.

Orange-peel texture of the skin, also known as 'peau d'orange', is a pitting oedema of the breast often seen with an underlying malignancy due to blocked lymphatics.

Dimpling of the breast, particularly associated with fixation of a lump, may be due to contraction of the suspensory ligaments of the breast caused by an underlying pathology.

Retraction of the nipple. Many women have congenitally inverted nipples and this should not be a cause for concern except when preparing to breast feed. As a new occurrence this should have a high index of suspicion for malignancy.

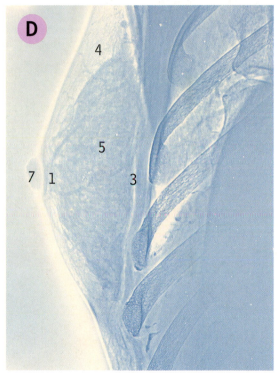

D

Carcinoma of the breast is one of the most common causes of death in young women. Most commonly found in the upper outer quadrant of the breast, metastases from these cancers often spread to the axillary lymph nodes. Surgeons may remove nodes from the axilla to 'stage' the disease.

Mastectomy is the removal of a breast. It is now a less popular operation, the most common procedure being a lumpectomy or removal of the tumour and possibly surrounding lymph nodes in the axilla. This more conservative surgical procedure can be followed by chemotherapy and/or radiotherapy, depending on the underlying pathology.

E

E Diagram of lymph drainage of the breast
There is a diffuse network of anastomosing lymphatic channels within the breast, including the overlying skin, and *lymph in any part may travel to any other part*. Larger channels drain most of the lymph to axillary nodes, but some from the medial part pass through the thoracic wall near the sternum to parasternal nodes adjacent to the internal thoracic vessels. These are the commonest and initial sites for cancerous spread, but other nodes may be involved (especially in the later spread of disease); these include infraclavicular and supraclavicular (deep cervical) nodes, nodes in the mediastinum, and nodes in the abdomen (via the diaphragm and rectus sheath). Spread to the opposite breast may also occur

A Right side of the thorax, from behind with the arm abducted

With the arm fully abducted, the medial (vertebral) border of the scapula (5) comes to lie at an angle of about 60° to the vertical, and indicates approximately the line of the oblique fissure of the lung (interrupted line)

1 Deltoid
2 Fifth intercostal space
3 Inferior angle of scapula
4 Latissimus dorsi
5 Medial border of scapula
6 Spine of scapula
7 Spinous process of third thoracic vertebra
8 Teres major
9 Trapezius

- The line of the oblique fissure of the lung runs from the level of the spine of the third thoracic vertebra (7) to the sixth costal cartilage at the lateral border of the sternum (see B). With the arm fully abducted the vertebral border of the scapula (5) is a good guide to the direction of this fissure.

B Surface markings of the right side of the thorax, from the right, with the arm abducted

The black line indicates the extent of the pleura, and the solid green line the lower limit of the lung; note the gap between the two at the lower part of the thorax, indicating the costodiaphragmatic recess of pleura, which does not contain any lung. The transverse and oblique fissures of the lung are represented by the interrupted green lines

1 Costal margin	4 Floor of axilla
2 Digitations of serratus anterior	5 Latissimus dorsi
3 External oblique	6 Pectoralis major

- The transverse fissure of the right lung is represented by a line drawn horizontally backwards from the fourth costal cartilage until it meets the line of the oblique fissure (described in A) running forwards to the sixth costal cartilage. The triangle so outlined indicates the middle lobe of the lung, with the superior lobe above it and the inferior lobe below and behind it.
- On the left side where the lung has only two lobes, superior and inferior, there is no transverse fissure; the surface marking for the oblique fissure is similar to that on the right.
- The asterisks represent the places where the lower edges of the lung and pleura cross the eighth and tenth ribs respectively in the midaxillary line.

C Muscles of the thorax. Left external and internal intercostal muscles, from the front
Pectoral and abdominal muscles have been removed, together with all vessels and nerves and the anterior intercostal membranes, to show the external and internal intercostal muscles (as at 2 and 3)

1 Eighth costal cartilage
2 External intercostal
3 Internal intercostal
4 Ninth costal cartilage
5 Second costal cartilage
6 Second rib
7 Seventh costal cartilage
8 Sternal angle
9 Tenth costal cartilage
10 Xiphoid process

Flail chest is the result of multiple rib fractures, most commonly following a motor vehicle accident. The section of the broken ribs moves paradoxically during respiration, making breathing extremely painful and distressing.

- The fibres of the external intercostal muscles (2) run downwards and medially, and near the costochondral junctions (as between 5 and 6) give place to the anterior intercostal membrane (here removed); these are thin sheets of connective tissue through which the underlying internal intercostal muscles (3) can be seen.
- The fibres of the internal intercostal muscles (3) run downwards and laterally. At the front they are covered by the anterior intercostal membranes, and at the back of the thorax they give place to the posterior intercostal membranes. The different directions of the muscle fibres enable the two muscle groups to be distinguished – down and medially for the externals (2), and down and laterally for the internals (3).
- The seventh costal cartilage (7) is the lowest to join the sternum and together with the eighth, ninth and tenth cartilages (1, 4 and 9) forms the costal margin.

Muscles of the thorax. Right intercostal muscles, **A** from the outside, **B** from the inside

In A each intercostal space has been dissected to a different depth, showing from above downwards an external intercostal muscle (2), internal intercostal (9), innermost intercostal (8) and pleura (10). The main intercostal vessels and nerve lie between the internal and innermost muscles; the nerve (12) is seen in the sixth interspace immediately below the sixth rib (13) and lying on the outer surface of the innermost intercostal (8), but the artery and vein are under cover of the costal groove. The vessels as well as the nerve are seen in the fifth intercostal space when this is dissected from the inside of the thorax, as in B; here the pleura and innermost intercostal muscle have been removed, and the vessels (5 and 4) and fifth intercostal nerve (3) lie against the inner surface of the internal intercostal (9)

 1   Eighth rib
 2   External intercostal
 3   Fifth intercostal nerve
 4   Fifth posterior intercostal artery
 5   Fifth posterior intercostal vein
 6   Fifth rib
 7   Fourth rib
 8   Innermost intercostal
 9   Internal intercostal
10   Pleura
11   Seventh rib
12   Sixth intercostal nerve
13   Sixth rib

- The internal intercostal muscles are continuous posteriorly with the posterior intercostal membranes which are covered up by the medial ends of the external intercostals (as at 2).

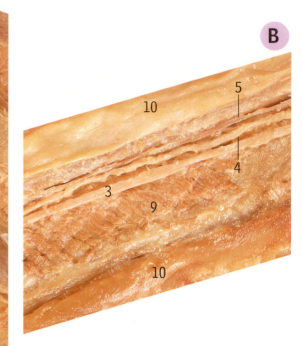

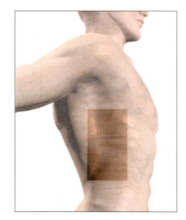

Intercostal nerve block. Using an anaesthetic agent this may be performed at any point along the course of an intercostal nerve in the subcostal groove, just below each rib.

Clicking-rib syndrome is a form of costochondritis along the costal margin due to subluxation of a rib causing irritation of the intercostal nerves. This may be confused clinically with upper abdominal problems.

A Muscles of the thorax. **Right transversus thoracis, from behind (inside view)**

This view of the internal surface of the thoracic wall shows the posterior surface of the right half of the sternum and adjacent wall, with the pleura removed. The internal thoracic artery (4) is seen passing deep to the slips of transversus thoracis (7, previously called sternocostalis)

- Transversus thoracis (A7) is in the same plane as the innermost intercostal muscles at the lateral side of the thoracic wall (B3) and the subcostal muscles on the posterior part (B4).
- The subcostal muscles (B4) span more than one rib. They and the innermost intercostals (B3, intercostales intimi) are often poorly developed or absent in the upper part of the thorax.

1 Body of sternum
2 Diaphragm
3 Internal intercostal
4 Internal thoracic artery
5 Second rib
6 Sixth rib
7 Slips of transversus thoracis muscle
8 Sternal angle
9 Xiphoid process

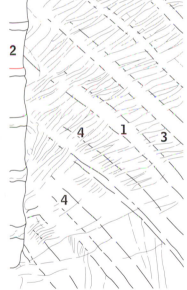

B Muscles of the thorax. **Left lower subcostal and innermost intercostal muscles**

This view of the lower left hemithorax is seen from the right and in front, with vertebral bodies (as at 2) sectioned and the pleura, vessels and nerves removed, and shows part of the innermost layer of thoracic wall muscles (3 and 4)

1 Eighth rib
2 Eighth thoracic vertebra
3 Innermost intercostal
4 Subcostal

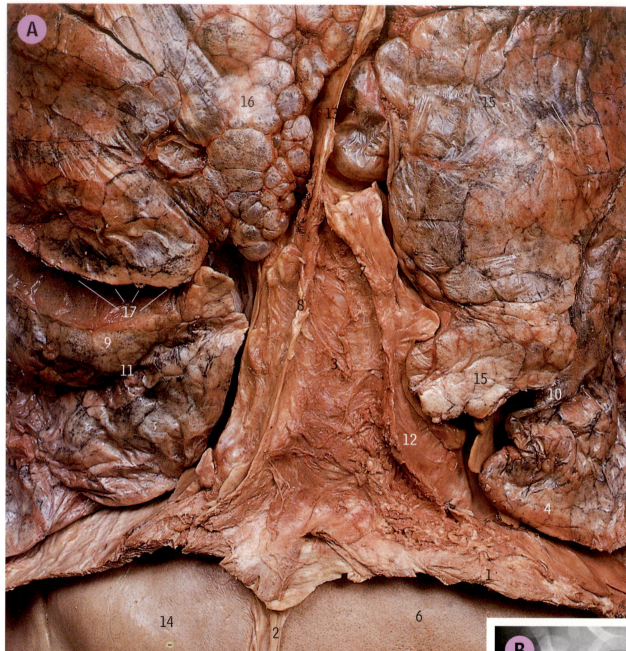

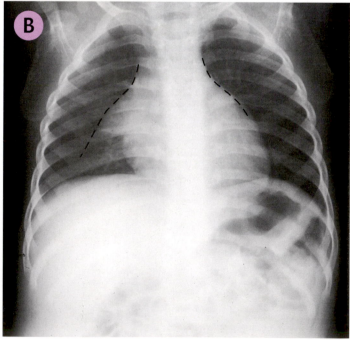

A Lungs and pericardium, from the front

The anterior thoracic and abdominal walls have been removed. The cut edges of the two pleural sacs are seen lying adjacent to one another (13), but lower down over the front of the pericardium (3) they become separated (8 and 7)

- The pleurae become separated at the level of the fourth costal cartilage (junction of 13, 8 and 7) owing to the leftward bulge of the heart, and therefore the central part of the fibrous pericardium (3) is not covered by pleura.

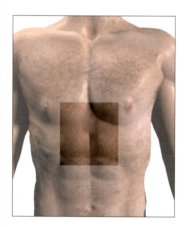

B Chest radiograph of a child

The child's thymus can be seen on a plain chest radiograph, appearing as a spinnaker sail (sail sign), as outlined by the interrupted line

1 Diaphragm	10 Oblique fissure
2 Falciform ligament	11 Oblique fissure of right lung
3 Fibrous pericardium	12 Pleura overlying pericardium
4 Inferior lobe of left lung	13 Right and left parietal pleurae in contact
5 Inferior lobe of right lung	14 Right lobe of liver
6 Left lobe of liver	15 Superior lobe of left lung
7 Line of reflexion of left pleura	16 Superior lobe of right lung
8 Line of reflexion of right pleura	17 Transverse fissure of right lung
9 Middle lobe of right lung	

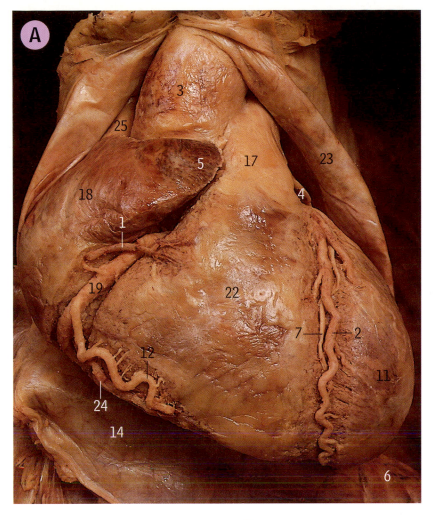

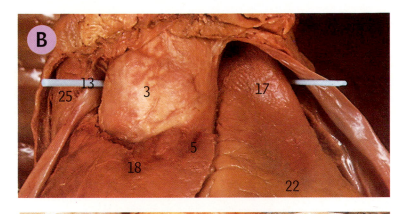

Heart and pericardium, A from the front, B with marker in the transverse sinus, C oblique sinus after removal of the heart

In A the pericardium has been incised and turned back (23) to display the anterior surface of the heart. The pulmonary trunk (17) leaves the right ventricle (22) in front and to the left of the ascending aorta (3), which is overlapped by the auricle (5) of the right atrium (18). The superior vena cava (25) is to the right of the aorta and still largely covered by pericardium. The anterior interventricular branch (2) of the left coronary artery and the great cardiac vein (7) lie in the interventricular groove between the right and left ventricles (22 and 11), and the right coronary artery (19) is in the atrioventricular groove between the right ventricle (22) and right atrium (18). In B only the upper part of another heart is shown, with a marker in the transverse sinus, the space behind the aorta (3) and pulmonary trunk (17). In C the heart has been removed from the pericardium, leaving the orifices of the great vessels. The dotted line indicates the attachment of the single sleeve of serous pericardium surrounding the aorta (3) and pulmonary trunk (17). The interrupted line indicates the attachment of another more complicated but still single sleeve of serous pericardium surrounding all the other six great vessels (the four pulmonary veins, 10, 9, 20 and 21, and the superior and inferior venae cavae, 25 and 8). The narrow interval between the two sleeves is the transverse sinus; the solid line in C indicates the path of the marker in B. The area of the pericardium (16) between the pulmonary veins and limited above by the reflexion of the serous pericardium on to the back of the heart is the oblique sinus

- The right border of the heart is formed by the right atrium (A18).
- The left border is formed mostly by the left ventricle (A11) with at the top the uppermost part (infundibulum) of the right ventricle (A22) and the tip of the left auricle (A4).
- The inferior border is formed by the right ventricle (A22) with a small part of the left ventricle at the apex (page 158, A2).

1 Anterior cardiac vein	15 Pericardium turned laterally over lung
2 Anterior interventricular branch of left coronary artery	16 Posterior wall of pericardial cavity and oblique sinus
3 Ascending aorta	17 Pulmonary trunk
4 Auricle of left atrium	18 Right atrium
5 Auricle of right atrium	19 Right coronary artery
6 Diaphragm	20 Right inferior pulmonary vein
7 Great cardiac vein	21 Right superior pulmonary vein
8 Inferior vena cava	22 Right ventricle
9 Left inferior pulmonary vein	23 Serous pericardium overlying fibrous pericardium (turned laterally)
10 Left superior pulmonary vein	24 Small cardiac vein
11 Left ventricle	25 Superior vena cava
12 Marginal branch of right coronary artery	
13 Marker in transverse sinus	
14 Pericardium fused with tendon of diaphragm	

Pericardial effusion is an accumulation of fluid within the pericardial sac; i.e. between the visceral and parietal serous layers of the pericardium. If a large amount of fluid accumulates this can lead to cardiac tamponade.

Cardiac tamponade occurs when sufficient fluid, usually blood, accumulates within the pericardial cavity to restrict filling of the heart during diastole, leading to reduced blood pressure, tachycardia and eventually cardiac arrest if the tamponade is not relieved by a pericardial puncture.

Myocardial infarction, better known as a 'heart attack', is death of a section of heart muscle due to prolonged, progressive ischaemia. Occasionally the dead heart muscle may then thin and expand, causing an aneurysm.

A

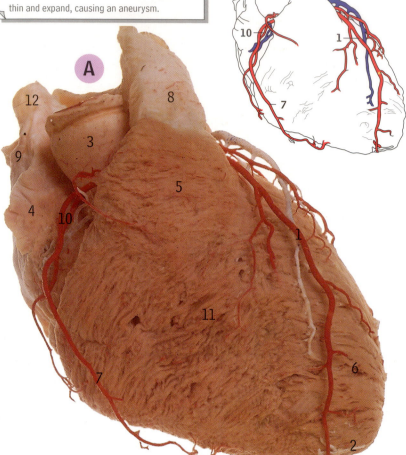

A Heart, from the front, with blood vessels injected

The coronary arteries have been injected with red latex and the cardiac veins with grey latex. The pulmonary trunk (8) passes upwards from the infundibulum (5) of the right ventricle (11), and at its commencement it is just in front and to the left of the ascending aorta (3)

1 Anterior interventricular branch of left coronary artery and great cardiac vein in interventricular groove
2 Apex
3 Ascending aorta
4 Auricle of right atrium (displaced laterally)
5 Infundibulum of right ventricle
6 Left ventricle
7 Marginal branch of right coronary artery
8 Pulmonary trunk
9 Right atrium
10 Right coronary artery in anterior atrioventricular groove
11 Right ventricle
12 Superior vena cava

- The *sternocostal* surface of the heart is the *anterior* surface (as seen in A on page 157 and A above) formed mainly by the right ventricle (A11, D7), with parts of the left ventricle (A6) and right atrium (A9 and D10).
- The *apex* of the heart (A2) is formed by the left ventricle.
- The *base* of the heart is the *posterior* surface, formed mainly by the left atrium (B8) with a small part of the right atrium (B13).
- The *inferior* surface is the *diaphragmatic* surface, formed by the two ventricles (mainly the left) (B10 and B15).

Coronary bypass surgery is performed to provide a reliable arterial supply to cardiac muscle distal to stenosis (narrowing) of a coronary artery. Many different techniques have been used over the last thirty years but most commonly the internal thoracic arteries are freed from the anterior thoracic wall and anastomosed to coronary artery segments distal to the obstructions.

B

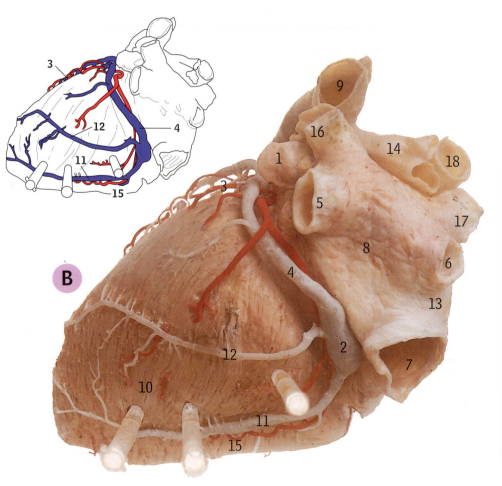

B Heart, from behind, with blood vessels injected

1 Auricle of left atrium
2 Coronary sinus in posterior atrioventricular groove
3 Great cardiac vein and anterior interventricular branch of left coronary artery
4 Great cardiac vein and circumflex branch of left coronary artery
5 Inferior left pulmonary vein
6 Inferior right pulmonary vein
7 Inferior vena cava
8 Left atrium
9 Left pulmonary artery
10 Left ventricle
11 Middle cardiac vein and posterior interventricular branch of right coronary artery in posterior interventricular groove
12 Posterior vein of left ventricle
13 Right atrium
14 Right pulmonary artery
15 Right ventricle
16 Superior left pulmonary vein
17 Superior right pulmonary vein
18 Superior vena cava

C Right atrium, from the front and right

The anterior wall has been incised near its left margin and reflected to the right, showing on its internal surface the vertical crista terminalis (2) and horizontal pectinate muscles (7). The fossa ovalis (3) is on the interatrial septum, and the opening of the coronary sinus (6) is to the left of the inferior vena caval opening (4)

1 Auricle	8 Position of atrioventricular node
2 Crista terminalis	9 Position of intervenous tubercle
3 Fossa ovalis	10 Superior vena cava
4 Inferior vena cava	11 Tricuspid valve
5 Limbus	12 Valve of coronary sinus
6 Opening of coronary sinus	13 Valve of inferior vena cava
7 Pectinate muscles	

- The intervenous tubercle, which is rarely detectable in the human heart (9), may have served in the embryo to direct blood from the superior vena cava towards the tricuspid orifice.
- The fossa ovalis (3) forms part of the interatrial septum, and is part of the embryonic primary septum.
- The limbus (5), which forms the margin of the fossa ovalis (3), represents the lower margin of the embryonic secondary septum. Before the primary and secondary septa fuse (at birth), the gap between them forms the foramen ovale.
- The sinuatrial node (SA node, not illustrated) is embedded in the anterior wall of the atrium at the upper end of the crista terminalis, just below the opening of the superior vena cava.
- The atrioventricular node (AV node, 8) is embedded in the interatrial septum, just above and to the left of the opening of the coronary sinus (6).

D Right ventricle, from the front

1 Anterior cusp of tricuspid valve
2 Anterior papillary muscle
3 Ascending aorta
4 Auricle of right atrium
5 Chordae tendineae
6 Inferior vena cava
7 Infundibulum of right ventricle
8 Posterior papillary muscle
9 Pulmonary trunk
10 Right atrium
11 Septomarginal trabecula
12 Superior vena cava
13 Trabeculae on interventricular septum

A cardiac pacemaker is an electrical impulse-generating device that is inserted under the skin of the anterior chest wall, from which wires pass to the heart tissue to maintain a regular heart beat. The insertion is a relatively minor surgical procedure, with the pacemaker lying on the pectoralis major muscle.

Valvular disease, a common complication of rheumatic fever, produces valvular incompetence most frequently in the mitral and aortic valves. This can be readily seen on video colour Doppler ultrasonography.

- The septomarginal trabecula (11), which conducts part of the right limb of the atrioventricular bundle from the interventricular septum (13) to the anterior papillary muscle (2), was formerly known as the moderator band.
- The chordae tendineae (5) connect the cusps of the tricuspid valve to the papillary muscles. The usual arrangement of the connexions is given in the following notes.
- The anterior papillary muscle (2) is large and connected to the anterior (1) and posterior cusps.
- The posterior papillary muscle (8) is small and connected to the posterior and septal cusps.
- Several small septal papillary muscles are connected to the septal and anterior cusps.
- The posterior papillary muscle (8) is so called because it is *behind* the anterior cusp (1), but it might be better named *inferior* because it is on the *floor* of the ventricle.

Ventricular hypertrophy is the enlargement of the ventricular heart muscle. Ischaemia, cardiomyopathy and valvular disease are common causes.

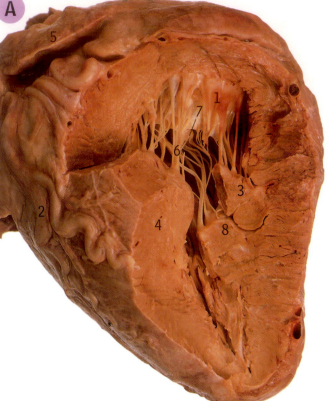

A

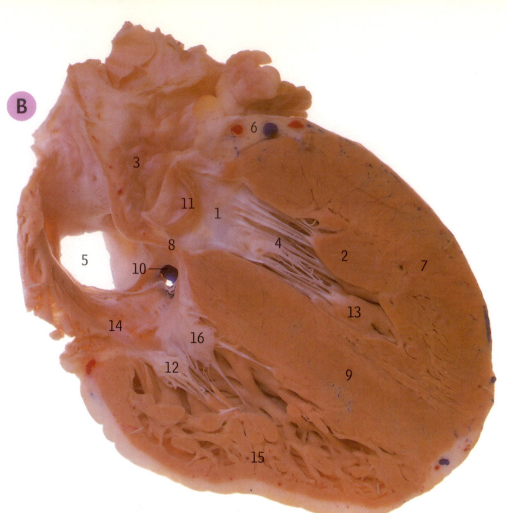

B

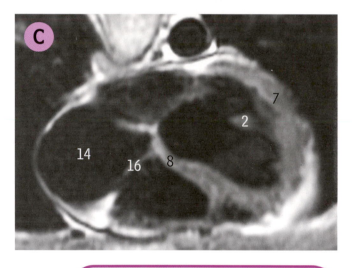

C

A Left ventricle, from the left and below
The ventricle has been opened by removing much of the left, anterior and posterior walls, and is viewed from below, looking upwards to the under-surface of the cusps of the mitral valve (1 and 7) which are anchored to the anterior and posterior papillary muscles (3 and 8) by chordae tendineae (6). The posterior cusp (7) is largely hidden by the anterior cusp (1) in this view

Heart, B coronal section of the ventricles, C axial MR image
In B the heart has been cut in two in the coronal plane, and this is the posterior section seen from the front, looking towards the back of both ventricles. The section has passed immediately in front of the anterior cusp of the mitral valve (1) and the posterior cusp of the aortic valve (11). Compare features seen in the MR image with the section

1 Anterior cusp of mitral valve
2 Anterior interventricular branch of left coronary artery
3 Anterior papillary muscle
4 Anterior ventricular wall
5 Auricle of left atrium
6 Chordae tendineae
7 Posterior cusp of mitral valve
8 Posterior papillary muscle

• The cusps of the aortic and pulmonary valves are here given their official names but some English texts use slightly different alternatives, as follows:

	Official	English
	Right	Anterior
Aortic	Left	Left posterior
	Posterior	Right posterior
	Left	Posterior
Pulmonary	Anterior	Left anterior
	Right	Right anterior

1 Anterior cusp of mitral valve
2 Anterior papillary muscle
3 Ascending aorta
4 Chordae tendineae
5 Inferior vena cava
6 Left coronary artery branches and great cardiac vein
7 Left ventricular wall
8 Membranous part of interventricular septum
9 Muscular part of interventricular septum
10 Opening of coronary sinus
11 Posterior cusp of aortic valve
12 Posterior cusp of tricuspid valve
13 Posterior papillary muscle
14 Right atrium
15 Right ventricular wall
16 Septal cusp of tricuspid valve

Intracardiac injections. In acute emergencies, when drugs need to be delivered directly into the heart, a long needle may be inserted through the anterior chest wall to the right of the sternum.

D Tricuspid valve, from the right atrium

The atrium has been opened by incising the anterior wall (2) and turning the flap outwards so that the atrial surface of the atrioventricular orifice is seen, guarded by the three cusps of the tricuspid valve – anterior (1), posterior (7) and septal (8)

1 Anterior cusp of tricuspid valve	6 Pectinate muscles
2 Anterior wall of right atrium	7 Posterior cusp of tricuspid valve
3 Auricle of right atrium	8 Septal cusp of tricuspid valve
4 Crista terminalis	9 Superior vena cava
5 Interatrial septum	

- The posterior cusp (7) of the tricuspid valve is the smallest.

E Pulmonary, aortic and mitral valves, from above

The pulmonary trunk (12) and ascending aorta (3) have been cut off immediately above the three cusps of the pulmonary and aortic valves (7, 2 and 15, and 14, 10 and 6). The upper part of the left atrium (5) has been removed to show the upper surface of the mitral valve cusps (11 and 1)

1 Anterior cusp of mitral valve	9 Ostium of left coronary artery
2 Anterior cusp of pulmonary valve	10 Posterior cusp of aortic valve
3 Ascending aorta	11 Posterior cusp of mitral valve
4 Auricle of right atrium	12 Pulmonary trunk
5 Left atrium	13 Right atrium
6 Left cusp of aortic valve	14 Right cusp of aortic valve
7 Left cusp of pulmonary valve	15 Right cusp of pulmonary valve
8 Marker in ostium of right coronary artery	16 Superior vena cava

F Fibrous framework of the heart

The heart is seen from the right and behind after removing both atria, looking down on to the fibrous rings (4) that surround the mitral and tricuspid orifices and form the attachments for the bases of the valve cusps. The cusps of the pulmonary valve (7, 2 and 13) are seen at the top of the infundibulum of the right ventricle (5), and the aortic valve cusps (12, 9 and 6) have been dissected out from the beginning of the ascending aorta

1 Anterior cusp of mitral valve	9 Posterior cusp of aortic valve
2 Anterior cusp of pulmonary valve	10 Posterior cusp of mitral valve
3 Anterior cusp of tricuspid valve	11 Posterior cusp of tricuspid valve
4 Fibrous ring	12 Right cusp of aortic valve
5 Infundibulum of right ventricle	13 Right cusp of pulmonary valve
6 Left cusp of aortic valve	14 Right fibrous trigone
7 Left cusp of pulmonary valve	15 Septal cusp of tricuspid valve
8 Left fibrous trigone	

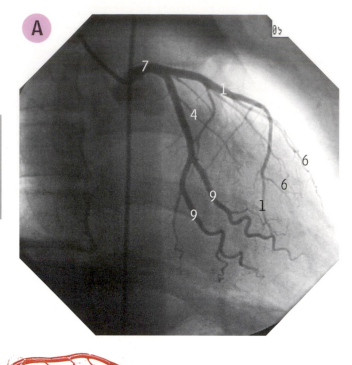

Balloon angioplasty is a procedure, similar to that for cardiac angiography, in which a catheter is inserted, usually via the femoral artery, and the tip of the catheter advanced into the narrowed coronary artery segment. A balloon at the end of the catheter is then blown up to stretch the coronary artery at this site.

Cardiac angiography is a radiographic record of the functional cardiac circulation. It is carried out by means of a long catheter passed through the ascending aorta via the femoral or brachial arteries. The tip of the catheter is placed just inside the opening of a coronary artery and a small injection of radio-opaque contrast material will then show a moving coronary angiogram. This technique is used to find sites of coronary occlusion that may present clinically as angina or myocardial infarction.

A Left coronary arteriogram, right anterior oblique projection. **B** Cast of the coronary arteries, from the front. **C** Right coronary arteriogram, left anterior oblique projection

1 Anterior interventricular branch of left coronary artery
2 Ascending aorta
3 Atrioventricular nodal artery
4 Circumflex branch of left coronary artery
5 Conus artery
6 Diagonal artery
7 Left coronary artery
8 Marginal branch of right coronary artery
9 Obtuse marginal artery
10 Posterior interventricular branch of right coronary artery
11 Right coronary artery
12 Sinuatrial nodal artery

• The interventricular branches are often called by clinicians the descending branches (anterior interventricular = left anterior descending; posterior interventricular = posterior descending).

Angina pectoris, an intermittent, relatively transient constricting central chest pain usually brought on by effort or exercise, is commonly due to coronary artery insufficiency secondary to atherosclerosis.

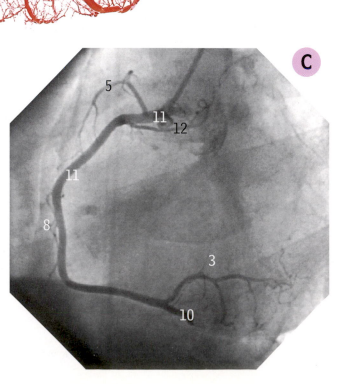

Colour Doppler echocardiographs, **D** apical long axis – left ventricle,
E suprasternal notch view of aortic arch

1	Aortic arch	6	Left common carotid artery
2	Ascending aorta	7	Left subclavian artery
3	Brachiocephalic artery	8	Left ventricle
4	Descending aorta	9	Mitral valve
5	Left atrium	10	Site of transducer

- Colour code: red = blood coming towards transducer; blue = away from the transducer
- This technique enables both anatomy and pathology of blood vessels to be visualized in the living subject without exposure to radiation.

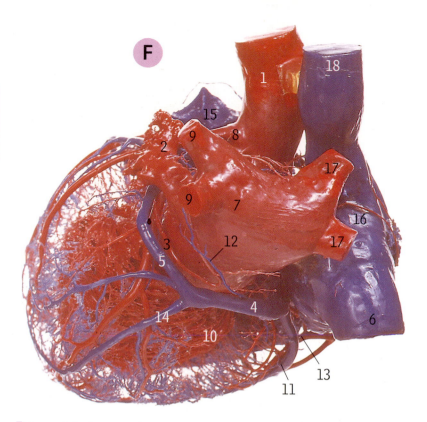

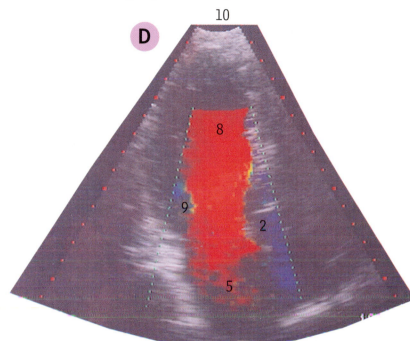

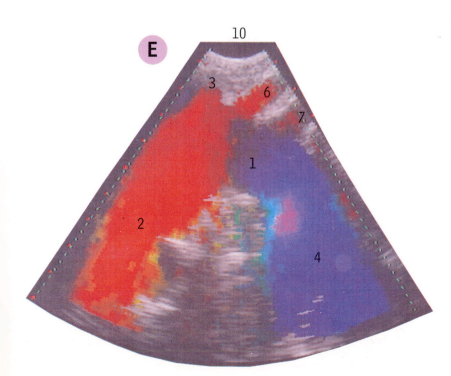

F Cast of the heart and great vessels, from below and behind
This specimen shows the coronary sinus (4) in the atrioventricular groove, and various tributaries (see notes)

1 Ascending aorta
2 Auricle of left atrium
3 Circumflex branch of left coronary artery
4 Coronary sinus
5 Great cardiac vein
6 Inferior vena cava
7 Left atrium
8 Left coronary artery
9 Left pulmonary veins
10 Left ventricle
11 Middle cardiac vein
12 Oblique vein of left atrium
13 Posterior interventricular branch of right coronary artery
14 Posterior vein of left ventricle
15 Pulmonary trunk
16 Right atrium
17 Right pulmonary veins
18 Superior vena cava

- The base of the heart (like the base of the prostate) is its posterior surface, formed largely by the left atrium (F7). Note that the base is not the part of the heart which joins the superior vena cava, aorta and pulmonary trunk; this part has no special name.
- The very small oblique vein of the left atrium (F12) marks the point where the great cardiac vein (F5) becomes the coronary sinus (F4), but in F the junction is unusually far to the right so that the posterior vein of the left ventricle (F14) joins the great cardiac vein (F5) instead of the coronary sinus itself.
- The coronary sinus (F4), which receives most of the venous blood from the heart, lies in the posterior part of the atrioventricular groove between the left atrium and left ventricle (page 158, B2), and opens into the right atrium (page 159, C6).
- The coronary sinus normally receives as tributaries the great cardiac vein (F5), middle cardiac vein (F11), and the small cardiac vein, the posterior vein of the left ventricle (F14) and the oblique vein of the left atrium (F12).
- The small cardiac vein frequently drains directly into the right atrium and not into the coronary sinus.

A

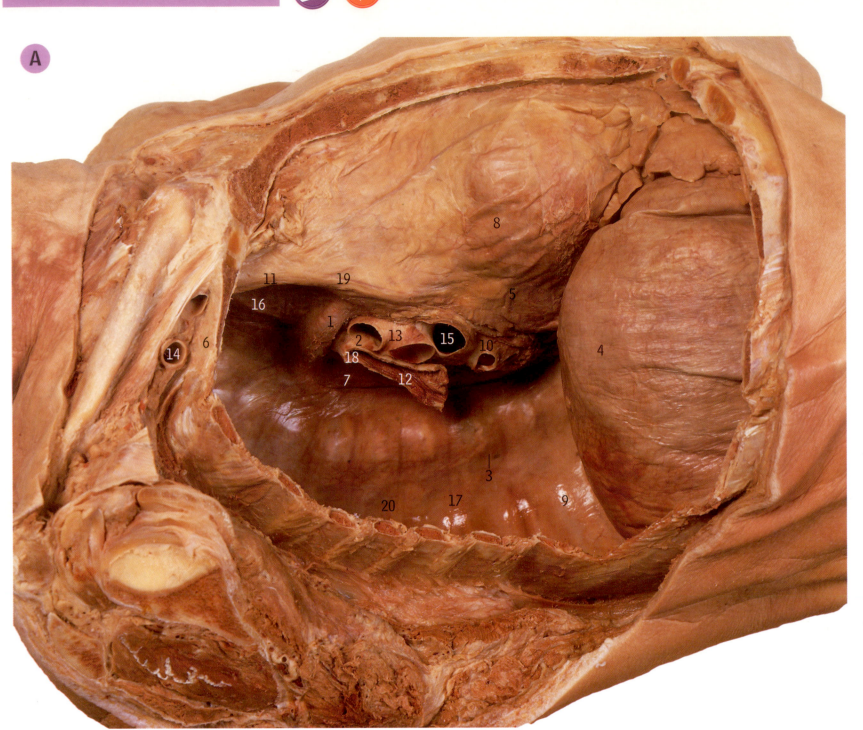

A Right lung root and mediastinal pleura
This is the view of the right side of the mediastinum after removing the lung but with the parietal pleura still intact (with the body lying on its back, head towards the left).

1 Azygos vein	11 Right phrenic nerve
2 Branch of right pulmonary artery to superior lobe	12 Right principal bronchus
	13 Right pulmonary artery
3 Branches of sympathetic trunk to greater splanchnic nerve	14 Right subclavian artery
	15 Right superior pulmonary vein
4 Diaphragm	16 Right vagus nerve
5 Inferior vena cava	17 Sixth right posterior intercostal vessels under parietal pleura
6 Neck of first rib	
7 Oesophagus	18 Superior lobe bronchus
8 Pericardium over right atrium	19 Superior vena cava
9 Pleura, costal	20 Sympathetic trunk and ganglion
10 Right inferior pulmonary vein	

B Right lung root and mediastinum
In a similar specimen to A, most of the pleura has been removed to display the underlying structures. The azygos vein (1) arches over the structures forming the lung root to enter the superior vena cava (23). The highest structures in the lung root are the artery (2) and bronchus (22) to the superior lobe of the lung. The right superior pulmonary vein (16) is in front of the right pulmonary artery, with the right inferior pulmonary vein (10) the lowest structure in the root. Above the arch of the azygos vein the trachea (26), with the right vagus nerve (17) in contact with it, lies in front of the oesophagus (7). Part of the first rib has been cut away to show the structures lying in front of its neck (6) – the sympathetic trunk (25), supreme intercostal vein (24), superior intercostal artery (20) and the ventral ramus of the first thoracic nerve (27). The right recurrent laryngeal nerve (14) hooks underneath the right subclavian artery (15). The right phrenic nerve (11) runs down over the superior vena cava (23) and the pericardium overlying the right atrium (8), and pierces the diaphragm (4) beside the inferior vena cava (5). Contributions from the sympathetic trunk (3) pass over the sides of vertebral bodies superficial to posterior intercostal arteries and veins (as at 19 and 18) to form the greater splanchnic nerve. The lower part of the oesophagus (7) behind the lung root and heart has the azygos vein (1) on its right side

B

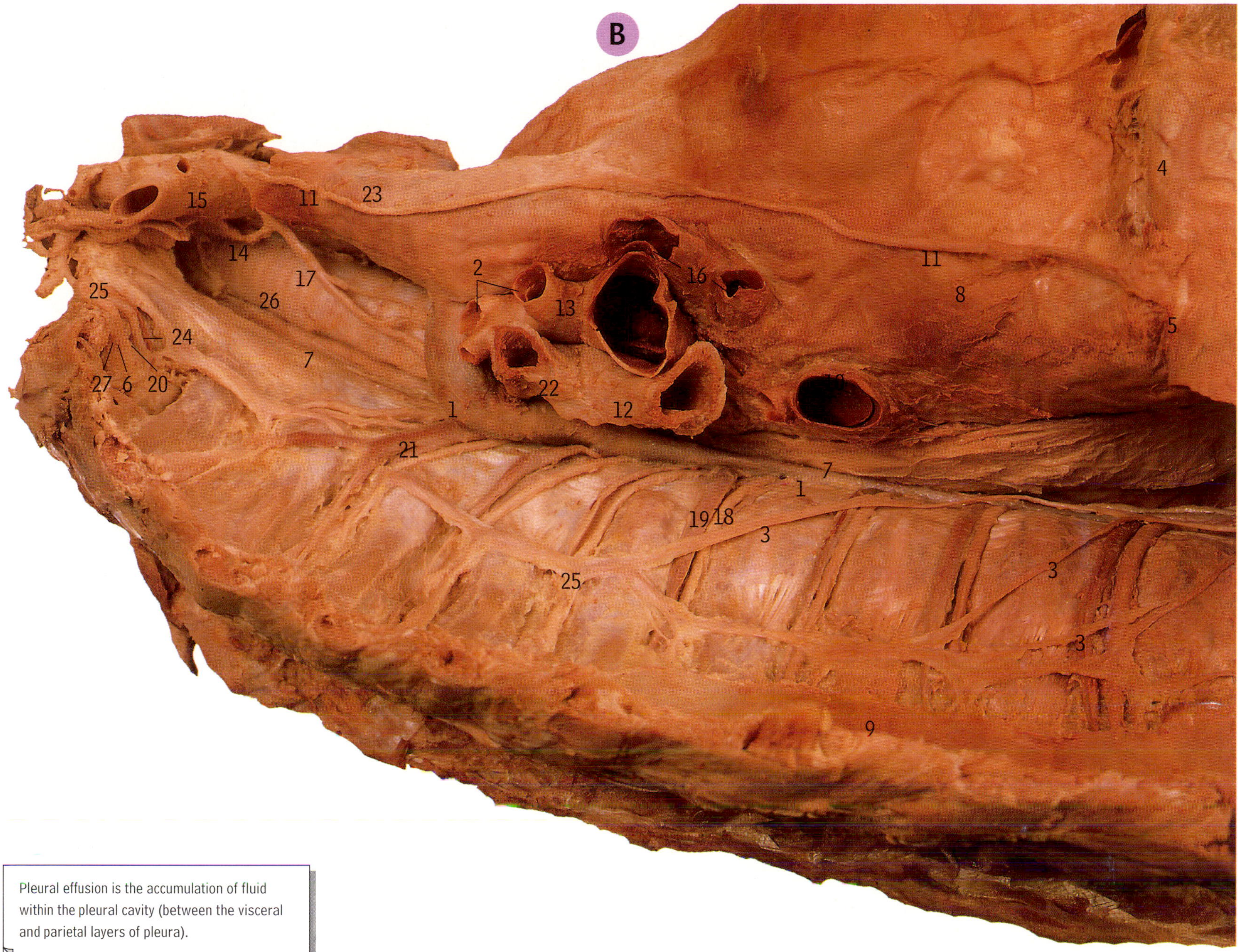

Pleural effusion is the accumulation of fluid within the pleural cavity (between the visceral and parietal layers of pleura).

1 Azygos vein	14 Right recurrent laryngeal nerve
2 Branch of right pulmonary artery to superior lobe	15 Right subclavian artery
	16 Right superior pulmonary vein
3 Branches of sympathetic trunk to greater splanchnic nerve	17 Right vagus nerve
	18 Sixth right posterior intercostal artery
4 Diaphragm	
5 Inferior vena cava	19 Sixth right posterior intercostal vein
6 Neck of first rib	20 Superior intercostal artery
7 Oesphagus	21 Superior intercostal vein
8 Pericardium over right atrium	22 Superior lobe bronchus
9 Pleura (cut edge)	23 Superior vena cava
10 Right inferior pulmonary vein	24 Supreme intercostal vein
11 Right phrenic nerve	25 Sympathetic trunk and ganglion
12 Right principal bronchus	26 Trachea
13 Right pulmonary artery	27 Ventral ramus of first thoracic nerve

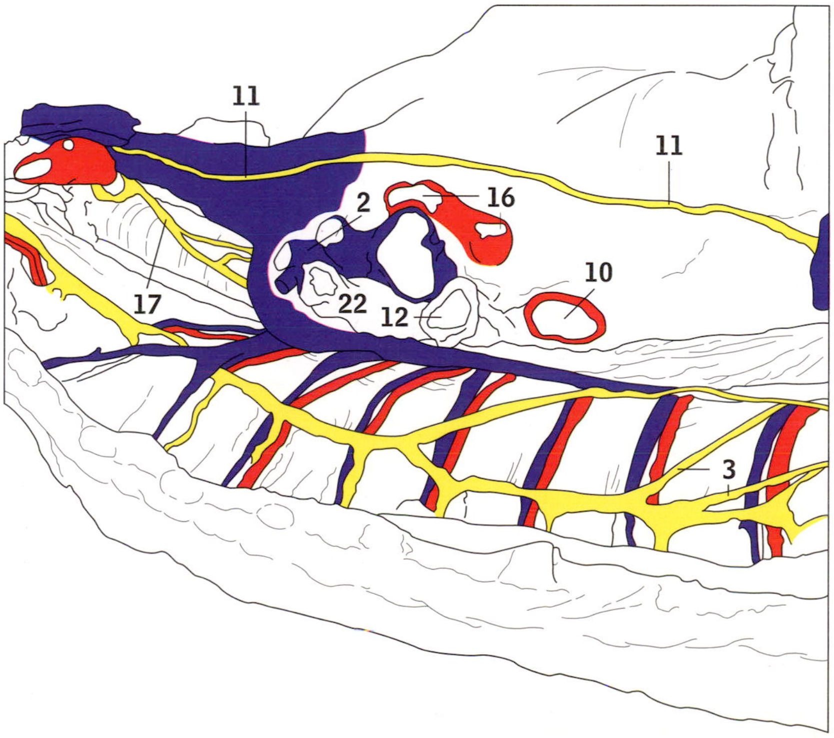

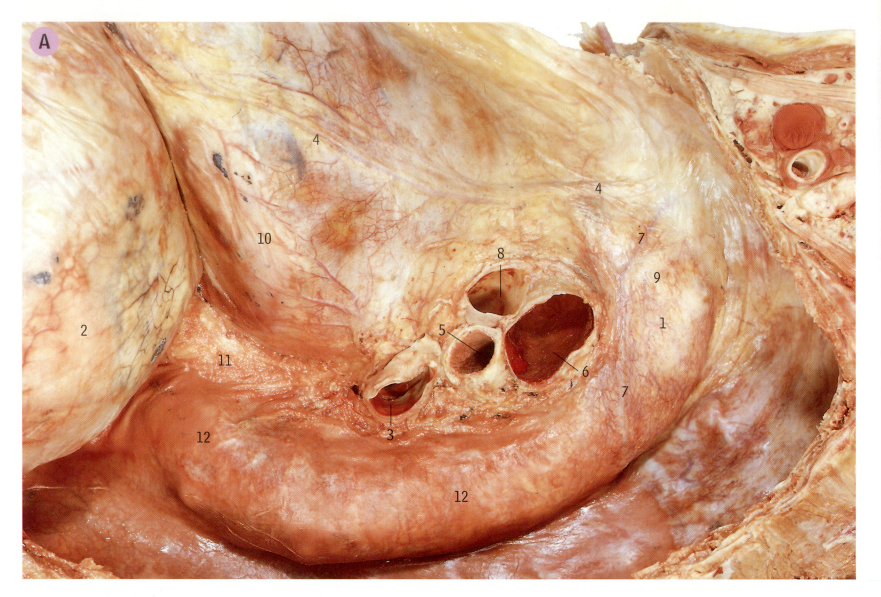

A

A Left lung root and mediastinal pleura
This is the view of the left side of the mediastinum after removing the lung
but with the parietal pleura still intact (with the body lying on its back,
head towards the right). Compare the features seen here with those in the
dissection opposite (a different specimen), from which the pleura has been
removed

- On the left side above the diaphragm, the lower end of the oesophagus lies in a
triangle bounded by the diaphragm below (A2), the heart in front (A10 and B24)
and the descending aorta behind (A12 and B27).

1 Arch of aorta	7 Left superior intercostal vein
2 Diaphragm	8 Left superior pulmonary vein
3 Left inferior pulmonary vein	9 Left vagus nerve
4 Left phrenic nerve and	10 Mediastinal pleura and pericardium
pericardiacophrenic vessels	overlying left ventricle
5 Left principal bronchus	11 Oesophagus
6 Left pulmonary artery	12 Thoracic aorta

Pleurisy is a painful inflammation of the pleura.
The pain is referred from the parietal pleura to
the cutaneous distribution of the intercostal
nerve.

Pneumothorax is the introduction of air
between the visceral and parietal pleurae. This
may result from a stab wound damaging the
parietal pleura or, more commonly, from the
spontaneous bursting of an air sac tearing the
visceral pleura.

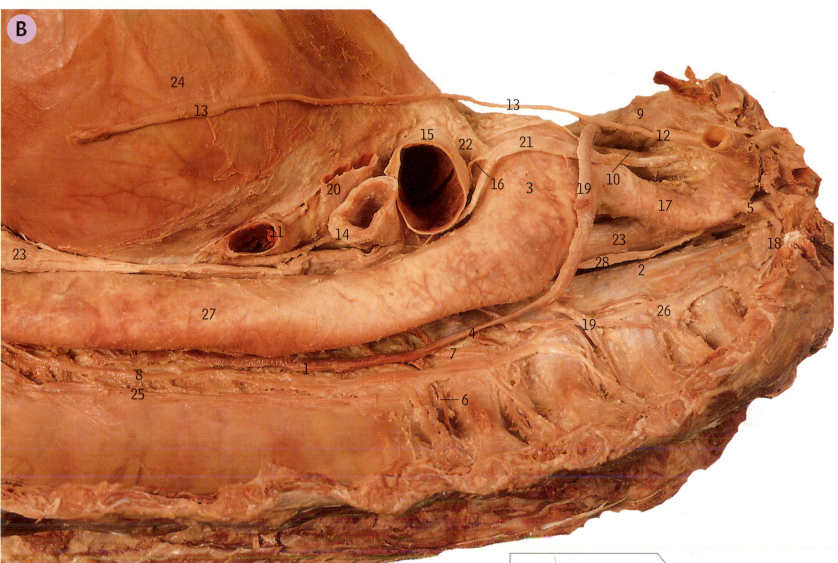

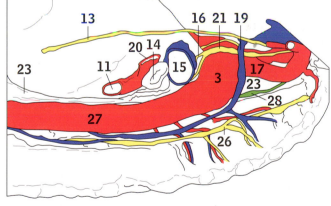

B Left lung root and mediastinum

In a similar specimen to that on the opposite page, most of the pleura has been removed to show the underlying structures seen in A. The left vagus nerve (21) crosses the arch of the aorta (3) with the left phrenic nerve (13) anterior to it; the superior intercostal vein (19) runs over the vagus and under the phrenic. The left recurrent laryngeal nerve (16) hooks round the ligamentum arteriosum (22) while the vagus nerve continues behind the structures forming the lung root. The left pulmonary artery (15) is the highest structure in the root, and the inferior pulmonary vein (11) the lowest. The left superior pulmonary vein (20) is in front of the principal bronchus. The thoracic duct (28) is seen behind the left edge of the oesophagus, and the origin of the left superior intercostal artery (18) from the costocervical trunk (5) of the subclavian artery (17) is shown. In this specimen there is an uncommon communication (4) between the left superior intercostal vein (19) and the accessory hemi-azygos vein (1). Above the diaphragm (not shown, having been pushed beyond the edge of the picture with the lower end of the phrenic nerve, 13) the oesophagus (23) bulges towards the left between the heart and pericardium (24) in front and the descending aorta (27) behind

Coarctation of the aorta is a narrowing of the aorta that is usually congenital, causing reversal of flow through the intercostal arteries. This may present as the radiological sign of 'rib-notching'.

1 Accessory hemi-azygos vein
2 Anterior longitudinal ligament
3 Arch of aorta
4 Communication between 19 and 1
5 Costocervical trunk
6 Fifth left posterior intercostal vein
7 Fourth left posterior intercostal artery
8 Hemi-azygos vein
9 Left brachiocephalic vein
10 Left common carotid artery
11 Left inferior pulmonary vein
12 Left internal thoracic artery
13 Left phrenic nerve
14 Left principal bronchus
15 Left pulmonary artery
16 Left recurrent laryngeal nerve
17 Left subclavian artery
18 Left superior intercostal artery
19 Left superior intercostal vein
20 Left superior pulmonary vein
21 Left vagus nerve
22 Ligamentum arteriosum
23 Oesophagus
24 Pericardium overlying left ventricle
25 Pleura (cut edge)
26 Sympathetic trunk and ganglion
27 Thoracic aorta
28 Thoracic duct

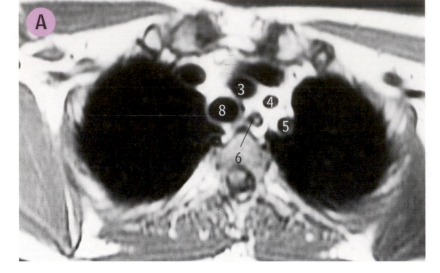

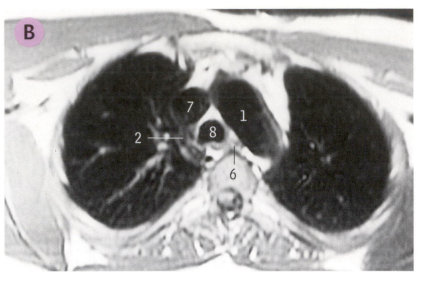

Mediastinum, A and B axial MR images
The section in B is at the level of the arch of the aorta (1), while that in A is higher and shows the three large branches of the arch (3, 4 and 5)

1	Arch of aorta	5	Left subclavian artery
2	Azygos vein	6	Oesophagus
3	Brachiocephalic trunk	7	Superior vena cava
4	Left common carotid artery	8	Trachea

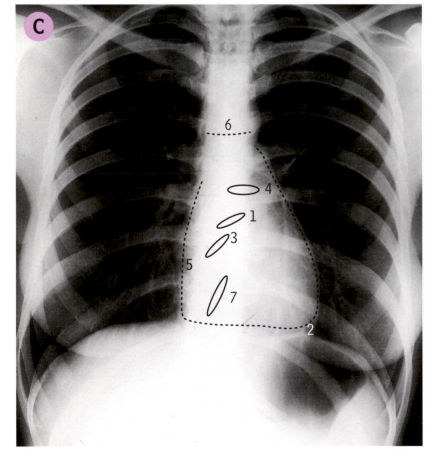

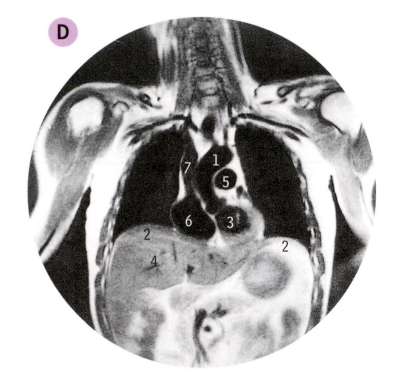

D Thorax, coronal MR image
The section shows the heart and great vessels in the mediastinum, above the domes of the diaphragm (2) and liver (4). The plane of the image is through the left ventricle (3) and right atrium (6)

C Chest radiograph
The surface markings of the heart valves are outlined by the dotted lines

1	Aortic valve	5	Right atrium
2	Apex of heart	6	Site of manubriosternal joint
3	Mitral valve	7	Tricuspid valve
4	Pulmonary valve		

1	Arch of aorta
2	Dome of diaphragm
3	Left ventricle
4	Liver
5	Pulmonary trunk
6	Right atrium
7	Superior vena cava

Phrenic nerve palsy will cause paralysis of the ipsilateral dome of the diaphragm. This can be seen radiographically by noticing paradoxical movement; i.e. instead of descending on inspiration, the diaphragm is pushed upwards by pressure from the underlying abdominal viscera.

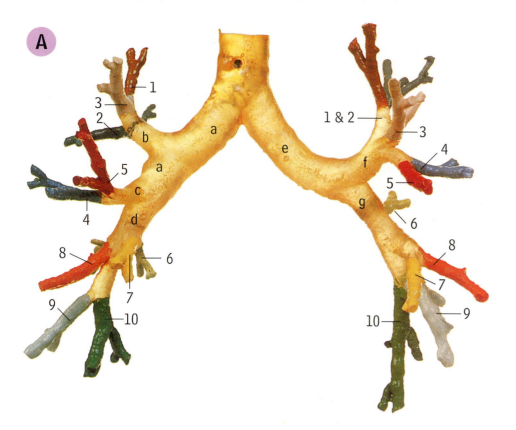

A

Cast of the lower trachea and bronchi, A vertical from the front, B oblique from the left

The main bronchi and lobar bronchi are labelled with letters; the segmental bronchi are labelled with their conventional numbers. In the side view in B the cast has been tilted to avoid overlap, and the right side is more anterior than the left

- The trachea divides into right and left principal bronchi (a and e).
- The right principal bronchus (a) is shorter, wider and more vertical than the left (e).
- The left principal bronchus (e) is longer and narrower and lies more transversely than the right. Foreign bodies are therefore more likely to enter the right principal bronchus than the left.
- The right principal bronchus (a) gives off a superior lobe bronchus (b) and then enters the hilum of the right lung before dividing into middle and inferior lobe bronchi (c and d).
- The left principal bronchus (e) enters the hilum of the lung before dividing into superior and inferior lobe bronchi (f and g).
- The branches of the lobar bronchi are called segmental bronchi and each supplies a segment of lung tissue—bronchopulmonary segment. The segmental bronchi and the bronchopulmonary segments have similar names, and the ten segments of each lung are officially numbered (as here and page 170) as well as being named.
- The segmental bronchi of the left and right lungs are essentially similar except that the apical and posterior bronchi of the superior lobe of the left lung arise from a common stem, thus called the apicoposterior bronchus and labelled here as 1 and 2; also there is no middle lobe of the left lung, and so the corresponding segments bear similar numbers; and the medial basal bronchus (7) of the left lung usually arises in common with the anterior basal (8).
- The apical (superior) bronchus of the inferior lobe (6) of both lungs is the first or highest bronchus to arise from the posterior surface of the bronchial tree, as illustrated in B. When lying on the back fluid may therefore gravitate into this bronchus.

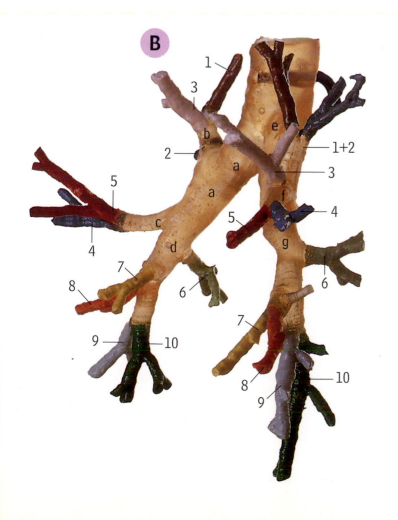

B

RIGHT LUNG	**LEFT LUNG**
Lobar bronchi	
a Principal	e Principal
b Superior lobe	f Superior lobe
c Middle lobe	g Inferior lobe
d Inferior lobe	
Segmental bronchi	
Superior lobe	**Superior lobe**
1 Apical	1 & 2 Apicoposterior
2 Posterior	3 Anterior
3 Anterior	4 Superior lingular
	5 Inferior lingular
Middle lobe	
4 Lateral	
5 Medial	
Inferior lobe	**Inferior lobe**
6 Apical (superior)	6 Apical (superior)
7 Medial basal	7 Medial basal
8 Anterior basal	8 Anterior basal
9 Lateral basal	9 Lateral basal
10 Posterior basal	10 Posterior basal

Cast of the bronchial tree
The bronchus and bronchopulmonary segments have been coloured and labelled with their conventional numbers

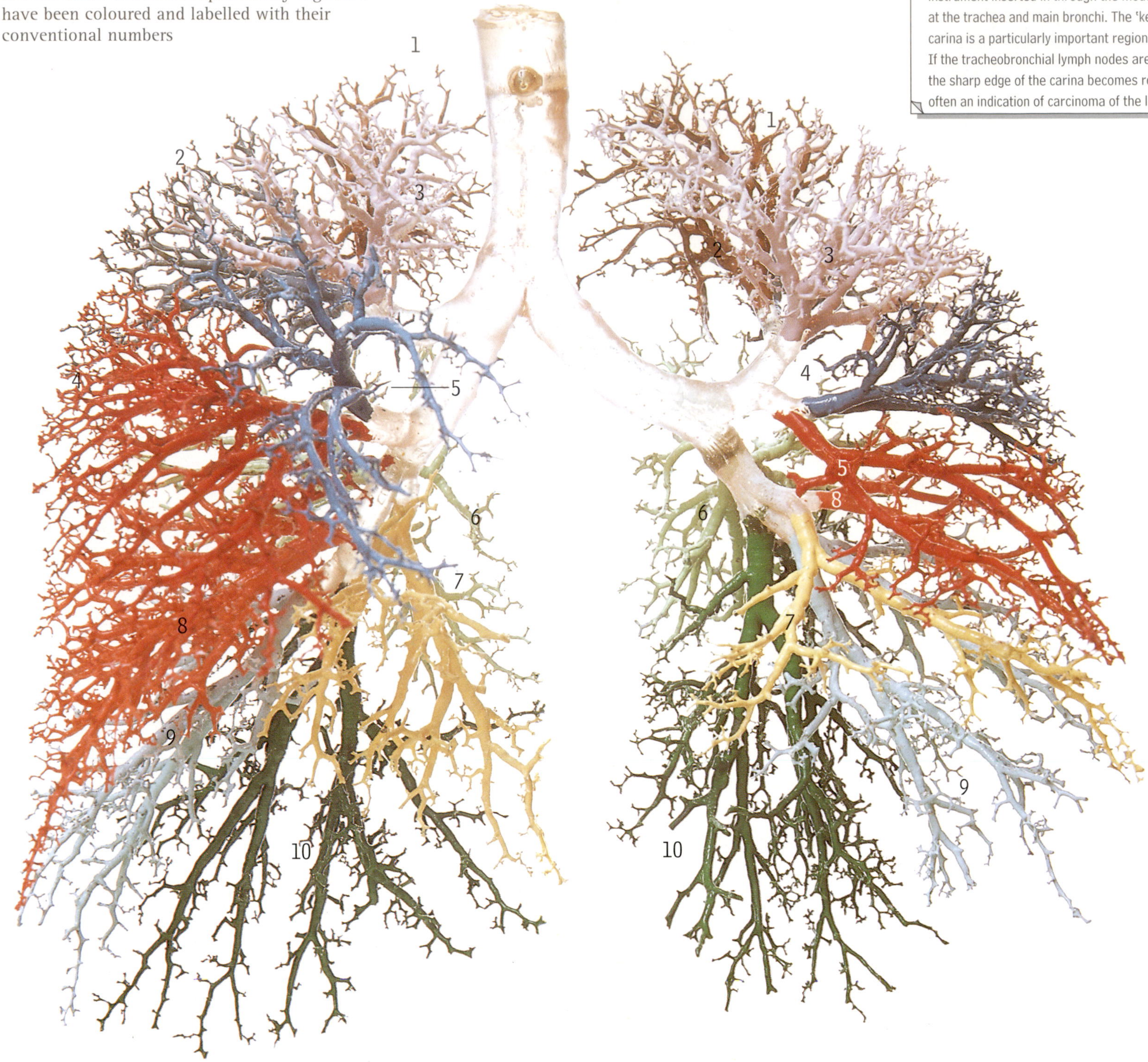

Bronchoscopy is a technique that uses an instrument inserted in through the mouth to look at the trachea and main bronchi. The 'keel-like' carina is a particularly important region to note. If the tracheobronchial lymph nodes are enlarged the sharp edge of the carina becomes rounded, often an indication of carcinoma of the lung.

RIGHT LUNG

Superior lobe
1 Apical
2 Posterior
3 Anterior
Middle lobe
4 Lateral
5 Medial

Inferior lobe
6 Apical (superior)
7 Medial basal
8 Anterior basal
9 Lateral basal
10 Posterior basal

LEFT LUNG

Superior lobe
1 Apical
2 Posterior
3 Anterior
4 Superior lingular
5 Inferior lingular

Inferior lobe
6 Apical (superior)
7 Medial basal (cardiac)
8 Anterior basal
9 Lateral basal
10 Posterior basal

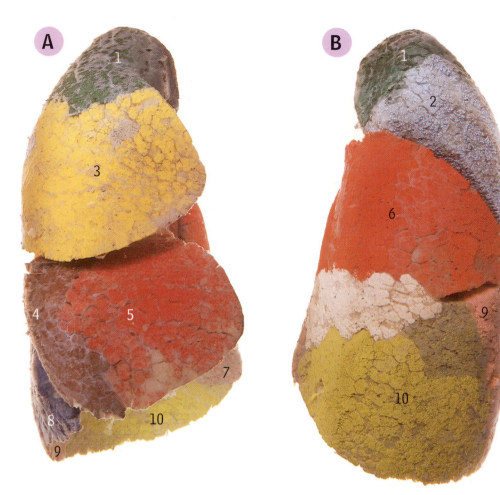

Bronchopulmonary segments of the right lung,
A from the front, **B** from behind

Superior lobe
1 Apical
2 Posterior
3 Anterior

Middle lobe
4 Lateral
5 Medial

Inferior lobe
6 Apical (superior)
7 Medial basal
8 Anterior basal
9 Lateral basal
10 Posterior basal

- A subapical (subsuperior) segmental bronchus and
  bronchopulmonary segment are present in over 50
  per cent of lungs; in this specimen this additional
  segment is shown in white.
- The posterior basal segment (10) is coloured with
  two different shades of green.

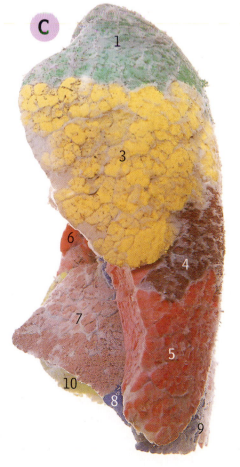

Bronchopulmonary segments of the left lung,
C from the front, **D** from behind

Superior lobe
1 Apical
2 Posterior
3 Anterior
4 Superior lingular
5 Inferior lingular

Inferior lobe
6 Apical (superior)
7 Medial basal (cardiac)
8 Anterior basal
9 Lateral basal
10 Posterior basal

- The apical and posterior segments (1 and 2) are both
  coloured green, having been filled from the
  common apicoposterior bronchus (see page 169).

A Bronchopulmonary segments of the right lung, from the lateral side, **B** Right bronchogram

Superior lobe
1 Apical
2 Posterior
3 Anterior

Middle lobe
4 Lateral
5 Medial

Inferior lobe
6 Apical (superior)
7 Medial basal
8 Anterior basal
9 Lateral basal
10 Posterior basal

Pneumonia is an infection of the lung by either bacteria or viruses and will show up on an X-ray as a white shadow in the normally radiolucent lung tissue.

- The medial basal segment (7) is not seen in the view in A.
- The posterior basal segment in A (10) is coloured with two different shades of green.

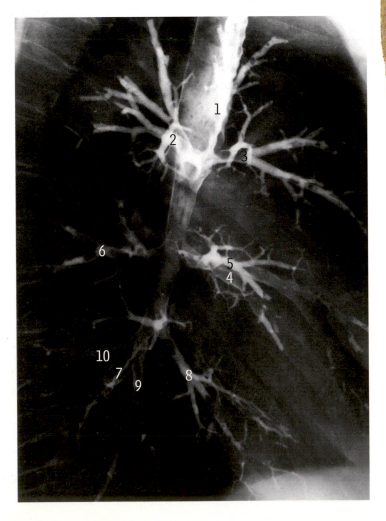

B

(see page 169)

- The apical and posterior segments (1 and 2) are both coloured green, having been filled from the common apicoposterior bronchus (see page 169).

C Bronchopulmonary segments of the left lung, from the lateral side, **D** Left bronchogram

Superior lobe
1 Apical
2 Posterior
3 Anterior
4 Superior lingular
5 Inferior lingular

Inferior lobe
6 Apical (superior)
7 Medial basal (cardiac)
8 Anterior basal
9 Lateral basal
10 Posterior basal

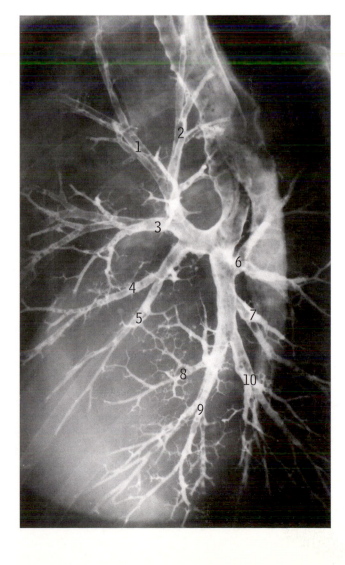

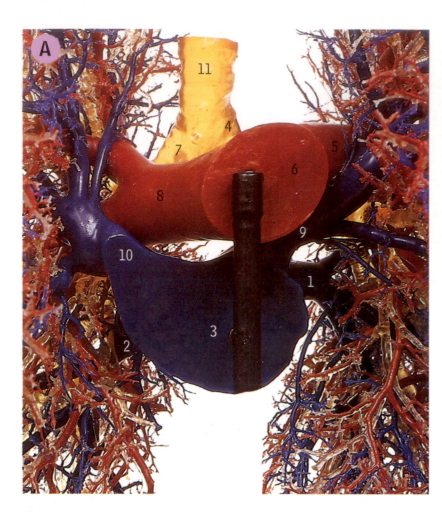

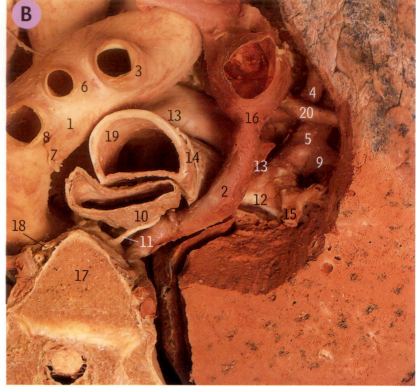

B Lung roots and bronchial arteries, right side from above
In B the thorax has been sectioned transversely at the level of the third thoracic vertebra (17), just above the arch of the aorta (1) whose three larger branches have been removed (8, 6 and 3), and lung tissue at the hilum has been dissected away from above. The oesophagus (10) and trachea (19) have been tilted forwards to show one of the bronchial arteries (11)

A Cast of the bronchial tree and pulmonary vessels, from the front
The pulmonary trunk (6) divides into the left and right pulmonary arteries (5 and 8), and these vessels have been injected with red resin. The four pulmonary veins (9, 1, 2 and 10) which drain into the left atrium (3) have been filled with blue resin. Note that in the living body the pulmonary veins are filled with oxygenated blood from the lungs and would normally be represented by a red colour; similarly the pulmonary arteries contain deoxygenated blood and should be represented by a blue colour

1 Inferior left pulmonary vein	7 Right principal bronchus
2 Inferior right pulmonary vein	8 Right pulmonary artery
3 Left atrium	9 Superior left pulmonary vein
4 Left principal bronchus	10 Superior right pulmonary vein
5 Left pulmonary artery	11 Trachea
6 Pulmonary trunk	

1 Arch of aorta	11 Right bronchial artery
2 Azygos vein	12 Right principal bronchus
3 Brachiocephalic trunk	13 Right pulmonary artery
4 Inferior lobe artery	14 Right vagus nerve
5 Inferior lobe bronchus	15 Superior lobe bronchus
6 Left common carotid artery	16 Superior vena cava
7 Left recurrent laryngeal nerve	17 Third thoracic vertebra
8 Left subclavian artery	18 Thoracic duct
9 Middle lobe bronchus	19 Trachea
10 Oesophagus	20 Tributary of inferior pulmonary vein

> Pulmonary embolism is a serious or fatal condition caused by blood clots, usually from the deep veins of the leg or pelvis, which dislodge and travel through the venous side of the heart to block the pulmonary trunk or its branches.

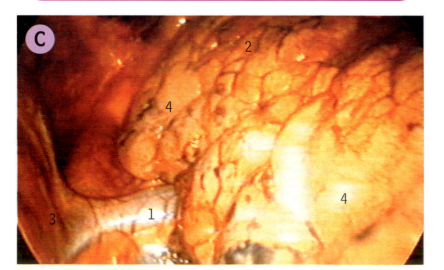

C Laparoscopic view of azygos vein

1 Azygos vein
2 Right lung
3 Superior vena cava
4 Visceral pleura

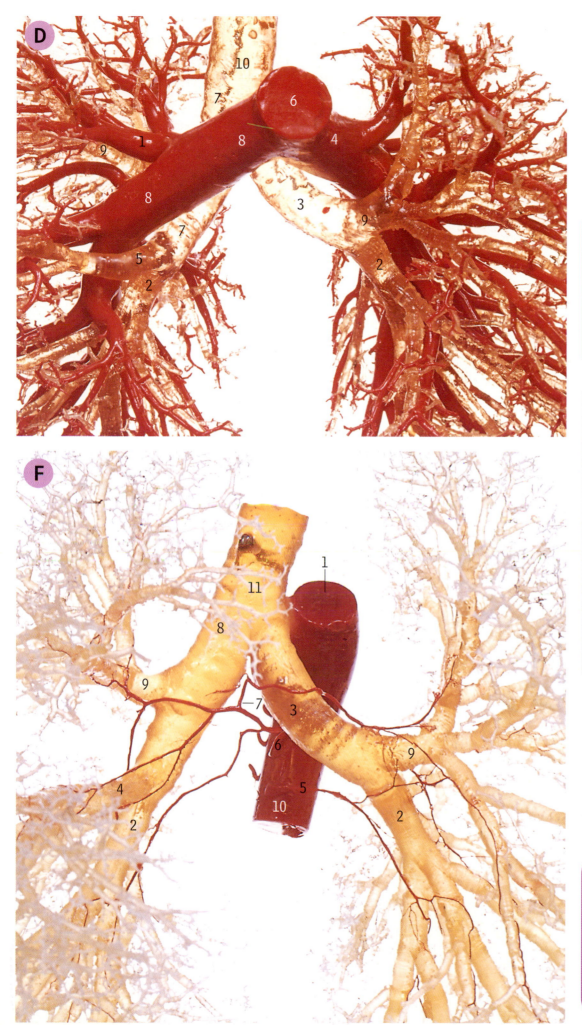

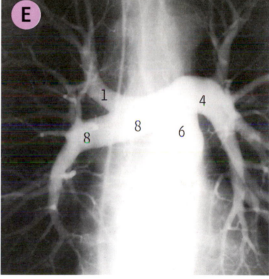

D Cast of the pulmonary arteries and bronchi, from the front, **E** Pulmonary arteriogram

The upper part of the pulmonary trunk (6) is seen end-on after cutting off the lower part, and the bifurcation of the trunk into the left (4) and right (8) pulmonary arteries is in front of the beginning of the left main bronchus (3). In the living body these pulmonary vessels contain deoxygenated blood and would normally be represented by a blue colour, but here they have been filled with red resin. Compare the vessels in the cast with those in the arteriogram E

1 Branch of right pulmonary artery to superior lobe	6 Pulmonary trunk
	7 Right principal bronchus
2 Inferior lobe bronchus	8 Right pulmonary artery
3 Left principal bronchus	9 Superior lobe bronchus
4 Left pulmonary artery	10 Trachea
5 Middle lobe bronchus	

F Cast of the bronchi and bronchial arteries, from the front

Part of the aorta (1 and 10) has been injected with red resin to fill the bronchial arteries. These vessels normally run behind the bronchi and their branches but in this specimen they are in front

1 Arch of aorta	7 Origin of upper left bronchial artery
2 Inferior lobe bronchus	
3 Left principal bronchus	8 Right principal bronchus
4 Middle lobe bronchus	9 Superior lobe bronchus
5 Origin of lower left bronchial artery	10 Thoracic aorta
	11 Trachea
6 Origin of right bronchial artery	

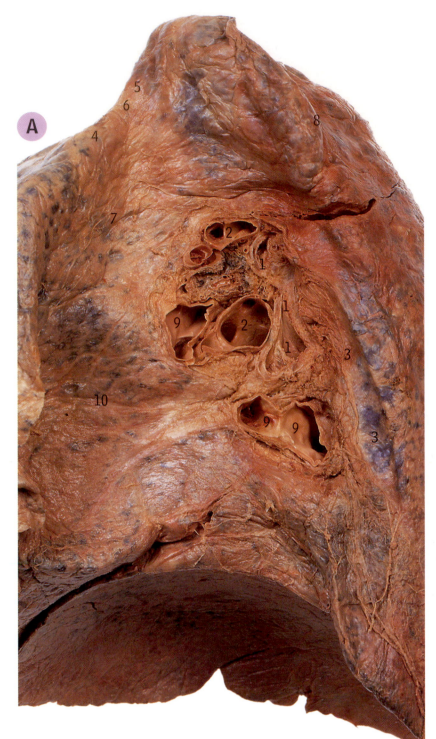

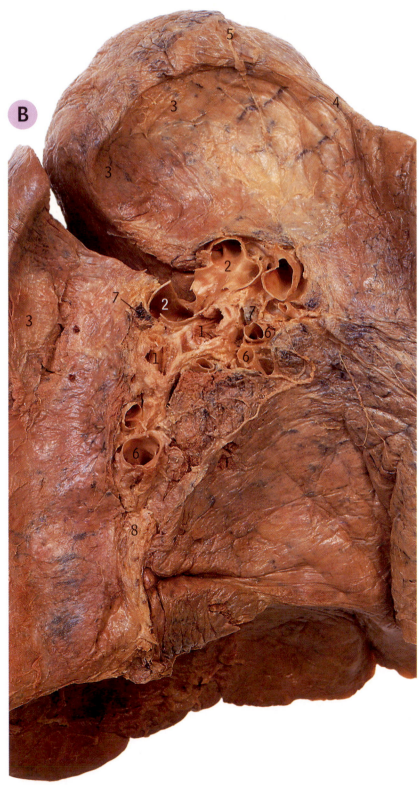

B Medial surface of the left lung

Compare with the right lung in A, and note the large size of the impression made by the aorta on the left lung (B3), in contrast to the smaller azygos groove on the right (A3)

1 Branches of left principal bronchus	5 Groove for left subclavian artery
2 Branches of left pulmonary artery	6 Left pulmonary veins
3 Groove for aorta	7 Lymph node, containing carbon
4 Groove for first rib	8 Pulmonary ligament

A Medial surface of the right lung

In the hardened dissecting room specimen, adjacent structures make impressions on the medial surface of the lung. The most prominent feature on the right side is the groove for the azygos vein (3), above and behind the structures of the lung root (9, 2 and 1)

1 Branches of right principal bronchus	6 Groove for subclavian vein
2 Branches of right pulmonary artery	7 Groove for superior vena cava
3 Groove for azygos vein	8 Oesophageal and tracheal area
4 Groove for first rib	9 Right pulmonary veins
5 Groove for subclavian artery	10 Transverse fissure

- The upper end of the medial surface of the right lung lies against the oesophagus and trachea (A8) with only the pleura intervening, but on the left the subclavian artery (B5) (and the left common carotid in front of it) keep the lung further away from these structures.

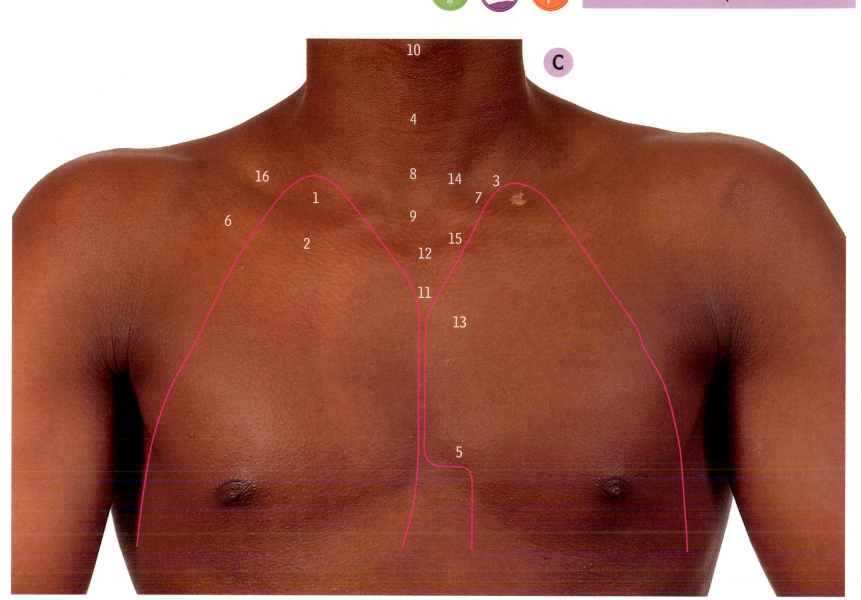

C Lower neck and upper thorax. Surface markings

The magenta line indicates the extent of the pleura and lung on each side; the apices of the pleura and lung (1) rise into the neck for about 3 cm above the medial third of the clavicle. The lower end of the internal jugular vein (7) lies behind the interval between the sternal (14) and clavicular (3) heads of sternocleidomastoid. Behind the sternoclavicular joint (15) the internal jugular and subclavian veins unite to form the brachiocephalic vein. The trachea (8) is felt in the midline above the jugular notch (9), and the arch of the cricoid cartilage (4) is 4 to 5 cm above the notch. The manubriosternal joint is at the level of the second costal cartilage (13) and opposite the lower border of the body of the fourth thoracic vertebra, and the horizontal plane through these points indicates the junction between the superior and inferior parts of the mediastinum. The left brachiocephalic vein passes behind the upper half of the manubrium to unite with the right brachiocephalic at the lower border of the right first costal cartilage (to form the superior vena cava). The midpoint of the manubrium (12) marks the highest level of the arch of the aorta and the origin of the brachiocephalic trunk. Compare many of the features mentioned here with the structures in dissections on pages 178 and 179

1 Apex of pleura and lung
2 Clavicle
3 Clavicular head of sternocleidomastoid
4 Cricoid cartilage
5 Fourth costal cartilage
6 Infraclavicular fossa
7 Internal jugular vein
8 Isthmus of thyroid gland overlying trachea
9 Jugular notch
10 Laryngeal prominence
11 Manubriosternal joint
12 Midpoint of manubrium of sternum
13 Second costal cartilage
14 Sternal head of sternocleidomastoid
15 Sternoclavicular joint
16 Supraclavicular fossa

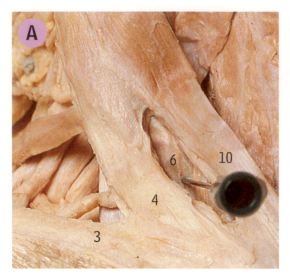

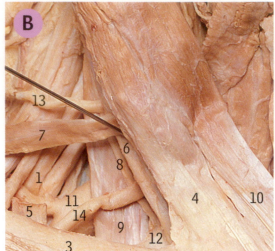

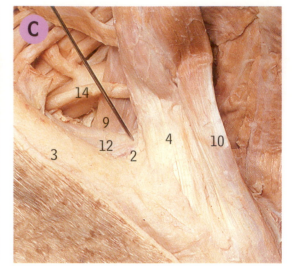

A Dissection of puncture of the right internal jugular vein between the two heads of sternocleidomastoid, with the needle directed backwards and slightly laterally

B Dissection of puncture of the right internal jugular vein at the posterior border of sternocleidomastoid, with the needle directed towards the jugular notch of the sternum

C Dissection of puncture of the right brachiocephalic vein, with the needle directed towards the sternal angle

1	Brachial plexus	8	Phrenic nerve
2	Brachiocephalic vein	9	Scalenus anterior
3	Clavicle	10	Sternal head of sternocleidomastoid
4	Clavicular head of sternocleidomastoid	11	Subclavian artery
5	External jugular vein	12	Subclavian vein
6	Internal jugular vein	13	Superficial cervical artery
7	Omohyoid	14	Suprascapular artery

Subclavian vein catheterization takes advantage of the vascular relations on the superior aspect of the first rib to place a central venous line, normally by an infraclavicular route. The tip of the needle should be pointed as anteriorly as possible towards the jugular notch to avoid injury to posterior structures (apex of the lung, subclavian artery and the brachial plexus). A supraclavicular approach puts the needle into the origin of the brachiocephalic vein.

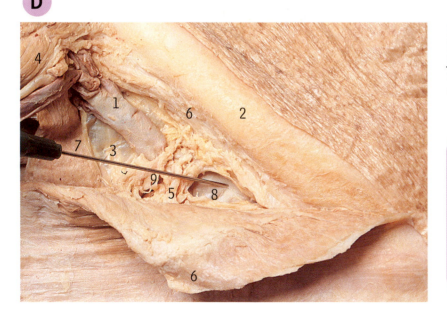

D Dissection of infraclavicular puncture of the subclavian vein, from below the midpoint of the clavicle on a line directed to just above the jugular notch. Part of pectoralis major is detached from the clavicle and the clavipectoral fascia divided to show the subclavian vein deep to it and passing under cover of the clavicle.

1	Cephalic vein
2	Clavicle
3	Clavipectoral fascia
4	Deltoid
5	Lateral pectoral nerve
6	Pectoralis major
7	Pectoralis minor
8	Subclavian vein
9	Thoraco-acromial vessels

E Thoracic inlet and mediastinum, from the front

The anterior thoracic wall and the medial ends of the clavicles have been removed, but part of the parietal pleura (16) remains over the medial part of each lung. The right internal jugular vein has also been removed, displaying the thyrocervical trunk (32) and the origin of the internal thoracic artery (9). Inferior thyroid veins (7) run down over the trachea (33) to enter the left brachiocephalic vein (13). The thymus (31) has been dissected out from mediastinal fat; thymic veins (30) enter the left brachiocephalic vein, and an unusual thymic artery (1) arises from the brachiocephalic trunk (4)

Pancoast's tumour is a tumour of the apex of the lung, usually a bronchogenic carcinoma, which compresses the brachial plexus and the sympathetic trunk with a resultant Horner's syndrome.

Thoracic outlet syndromes, complex in both terminology and pathogenesis, are neural or vascular complications in the upper extremity caused by compression of subclavian vessels and/or nerve roots of the brachial plexus in the scalene triangle above the first rib. Common causes are cervical ribs or constriction bands from cervical outgrowths. Common signs include cold hands, tingling or paralysis of the fourth and fifth digits or weakness of the intrinsic hand muscles.

E

- The remains of the thymus (31) are in front of the pericardium, but in the child, where the thymus is much larger, it may extend upwards in front of the great vessels as high as the lower part of the thyroid gland (12).

1 A thymic artery	13 Left brachiocephalic vein	25 Superficial cervical artery
2 Arch of cricoid cartilage	14 Left common carotid artery	26 Superior vena cava
3 Ascending cervical artery	15 Left vagus nerve	27 Suprascapular artery
4 Brachiocephalic trunk	16 Parietal pleura (cut edge) over lung	28 Sympathetic trunk
5 First rib	17 Phrenic nerve	29 Thoracic duct
6 Inferior thyroid artery	18 Right brachiocephalic vein	30 Thymic veins
7 Inferior thyroid veins	19 Right common carotid artery	31 Thymus
8 Internal jugular vein	20 Right recurrent laryngeal nerve	32 Thyrocervical trunk
9 Internal thoracic artery	21 Right subclavian artery	33 Trachea
10 Internal thoracic vein	22 Right vagus nerve	34 Unusual cervical tributary of 18
11 Isthmus of thyroid gland	23 Scalenus anterior	35 Upper trunk of brachial plexus
12 Lateral lobe of thyroid gland	24 Subclavian vein	36 Vertebral vein

Superior thoracic aperture (thoracic inlet)

A Thoracic inlet. Right upper ribs, from below
This is the view looking upwards into the right side of the thoracic inlet – the region occupied by the cervical pleura, here removed. The under-surface of most of the first rib (5) is seen from below, with the subclavian artery (12) passing over the top of it after giving off the internal thoracic branch (6) which runs towards the top of the picture (to the anterior thoracic wall), and the costocervical trunk whose superior intercostal branch (14) runs down over the neck of the first rib (7). The vertebral vein (22) has come down from the neck and is labelled on its posterior surface before entering the brachiocephalic vein (2, labelled at its opened cut edge). The vertebral vein receives an unusually large supreme intercostal vein (16). On its medial side is the sympathetic trunk (17) with the cervicothoracic ganglion (3). The neck of the first rib (7) has the ventral ramus of the first thoracic nerve (21) below it

1 Brachiocephalic trunk
2 Brachiocephalic vein
3 Cervicothoracic (stellate) ganglion
4 First intercostal nerve
5 First rib
6 Internal thoracic vessels
7 Neck of first rib
8 Recurrent laryngeal nerve
9 Right principal bronchus
10 Second intercostal nerve
11 Second rib
12 Subclavian artery
13 Subclavian vein
14 Superior intercostal artery
15 Superior intercostal vein
16 Supreme intercostal vein (unusually large)
17 Sympathetic trunk
18 Trachea
19 Vagus nerve
20 Ventral ramus of eighth cervical nerve
21 Ventral ramus of first thoracic nerve
22 Vertebral vein

• The neck of the first rib (7) is crossed in order from medial to lateral by the sympathetic trunk (17), supreme intercostal vein (16), superior intercostal artery (14) and the ventral ramus of the first thoracic nerve (21).

ANTERIOR

MEDIAL

Horner's syndrome features ptosis, a constricted pupil (miosis), and a dry and flushed side of the face due to damage or interruption of the sympathetic trunk usually in the upper thoracic or lower cervical region. It is sometimes due to tumours (carcinoma of the lung), but may also be iatrogenic following a stellate ganglion block or cervical sympathectomy.

B Thoracic duct, thoracic part
All viscera and part of the pleura have been removed to show the aorta (14 and 1) in front of the vertebral column and viewed from the right, with the thoracic duct (15) lying between the aorta (14) and azygos vein (2); the lower part of the vein was overlying the duct and has been removed. The cisterna chyli (3), where the thoracic duct begins, is in the abdomen under cover of the right crus of the diaphragm (10)

1 Abdominal aorta
2 Azygos vein
3 Cisterna chyli
4 Coeliac trunk
5 Diaphragm
6 First lumbar artery and first lumbar vertebra
7 Greater splanchnic nerve
8 Medial arcuate ligament
9 Psoas major
10 Right crus of diaphragm
11 Right renal artery
12 Superior mesenteric artery
13 Sympathetic trunk underlying pleura
14 Thoracic aorta
15 Thoracic duct
16 Twelfth thoracic vertebra and subcostal artery

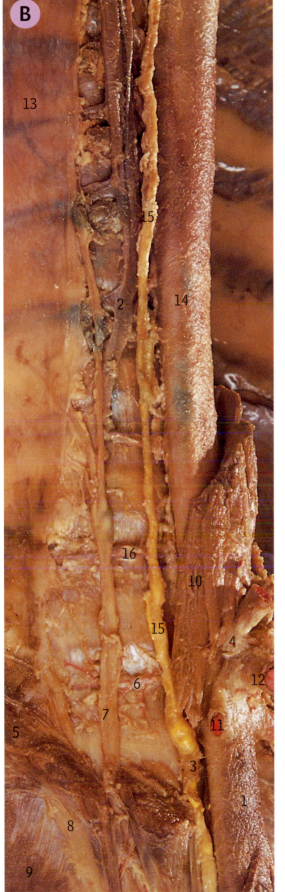

C Thoracic duct, cervical part

In this deep dissection of the left side of the root of the neck and upper thorax, the internal jugular vein (6) joins the subclavian vein (13) to form the left brachiocephalic vein (3). The thoracic duct (15) is double for a short distance just before passing in front of the vertebral artery (9) and behind the common carotid artery (4, whose lower end has been cut away to show the duct). The duct then runs behind the internal jugular vein (6) before draining into the junction of that vein with the subclavian vein (13)

1 Ansa subclavia
2 Arch of aorta
3 Brachiocephalic vein
4 Common carotid artery
5 Inferior thyroid artery
6 Internal jugular vein
7 Internal thoracic artery
8 Longus colli
9 Origin of vertebral artery
10 Phrenic nerve
11 Pleura
12 Subclavian artery
13 Subclavian vein
14 Sympathetic trunk
15 Thoracic duct
16 Vagus nerve

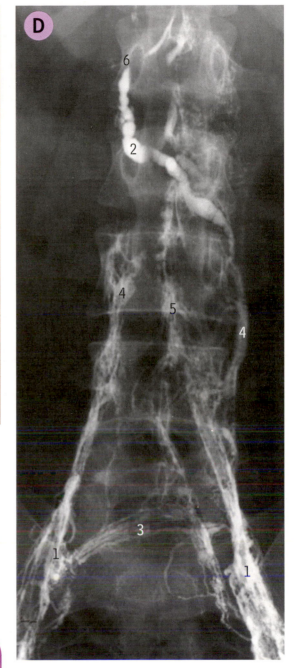

D First day lymphangiogram

1 Common iliac vessels
2 Cysterna chyli
3 Lumbar crossover
4 Para-aortic vessels
5 Pre-aortic vessels
6 Thoracic duct

Chylothorax is the accumulation of lymph within the pleural cavity; i.e. between the visceral and parietal pleurae.

• From the cisterna chyli (B3), situated under cover of the left margin of the right crus of the diaphragm (B10) at the level of the first and second lumbar vertebrae, the thoracic duct passes upwards (through the aortic opening in the diaphragm) on the right side of the front of the thoracic vertebral column between the aorta (B14) and azygos vein (B2), crossing to the left at the level of the fifth to sixth thoracic vertebra and ending by opening into the left side of the union of the left internal jugular (C6) and subclavian (C13) veins after passing between the common carotid artery (in front, C4) and the vertebral artery (behind, C9).

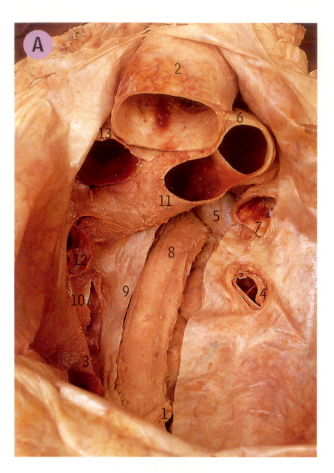

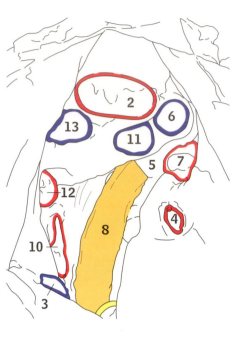

A Oesophagus, lower thoracic part, from the front
The heart has been removed from the pericardial cavity by transecting the great vessels, the pulmonary trunk being cut at the point where it divides into the two pulmonary arteries (11 and 6). Part of the pericardium (9) at the back has been removed to reveal the oesophagus (8). It is seen below the left principal bronchus (5) and is being crossed by the beginning of the right pulmonary artery (11)

1 Anterior vagal trunk
2 Ascending aorta
3 Inferior vena cava
4 Left inferior pulmonary vein
5 Left principal bronchus
6 Left pulmonary artery
7 Left superior pulmonary vein
8 Oesophagus
9 Pericardium (cut edge)
10 Right inferior pulmonary vein
11 Right pulmonary artery
12 Right superior pulmonary vein
13 Superior vena cava

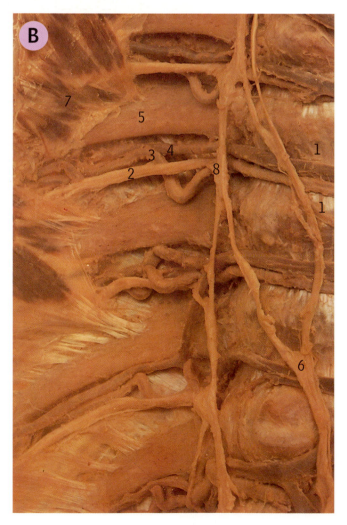

B Intercostal spaces, posterior internal view
This dissection shows the medial ends of some intercostal spaces of the right side, viewed from the front and slightly from the right. The pleura has been removed, revealing subcostal muscles (7) laterally, the nerves and vessels (4, 3 and 2) in the intercostal spaces, and the sympathetic trunk (8) and greater splanchnic nerve (6) on the sides of the vertebral bodies (as at 1)

1 Body of ninth thoracic vertebra
2 Eighth intercostal nerve
3 Eighth posterior intercostal artery
4 Eighth posterior intercostal vein
5 Eighth rib
6 Greater splanchnic nerve
7 Subcostal muscle
8 Sympathetic trunk and ganglia

Intercostal drainage, known also as thoracocentesis, is a procedure in which a needle is inserted through the intercostal muscles into the pleural cavity to remove excess fluid. The needle is inserted along the top of the rib to avoid damage to the neurovascular bundle which lies in the subcostal groove of each intercostal space.

C Joints of the heads of the ribs, from the right
In this part of the right mid-thoracic region, the ribs have been cut short beyond their tubercles, and the joints that the two facets of the head of a rib make with the facets on the sides of adjacent vertebral bodies and the intervening disc are shown, as at 4, 9 and 2, where the radiate ligament (4) covers the capsule of these small synovial joints

1 Greater splanchnic nerve
2 Intervertebral disc
3 Neck of rib
4 Radiate ligament of joint of head of rib
5 Rami communicantes
6 Superior costotransverse ligament
7 Sympathetic trunk
8 Ventral ramus of spinal nerve
9 Vertebral body

D Costotransverse joints, from behind
In this view of the right half of the thoracic vertebral column from behind, costotransverse joints between the transverse processes of vertebrae and the tubercles of ribs are covered by the lateral costotransverse ligaments (as at 4). The dorsal rami of spinal nerves (2) pass medial to the superior costotransverse ligaments (6); ventral rami (8) run in front of these ligaments

1 Costotransverse ligament
2 Dorsal ramus of spinal nerve
3 Lamina
4 Lateral costotransverse ligament
5 Spinous process
6 Superior costotransverse ligament
7 Transverse process
8 Ventral ramus of spinal nerve

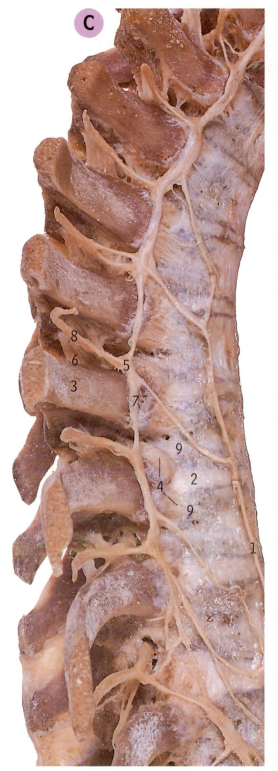

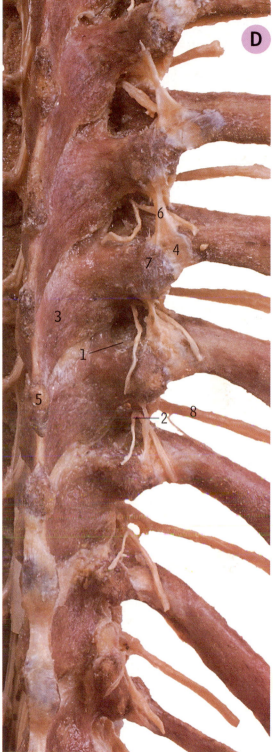

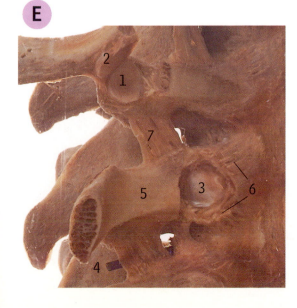

E Costovertebral joints, disarticulated, from the right
In the upper part of the figure, the upper rib has been severed through its neck (5) and the part with the tubercle attached has been turned upwards after cutting through the capsule of the costotransverse joint, to show the articular facet of the tubercle (2) and the transverse process (1). The head of the lower rib has been removed after transecting the radiate ligament (6) and underlying capsule of the joint of the head of the rib (3)

1 Articular facet of transverse process
2 Articular facet of tubercle of rib
3 Cavity of joint of head of rib
4 Marker between anterior and posterior parts of superior costotransverse ligament
5 Neck of rib
6 Radiate ligament
7 Superior costotransverse ligament

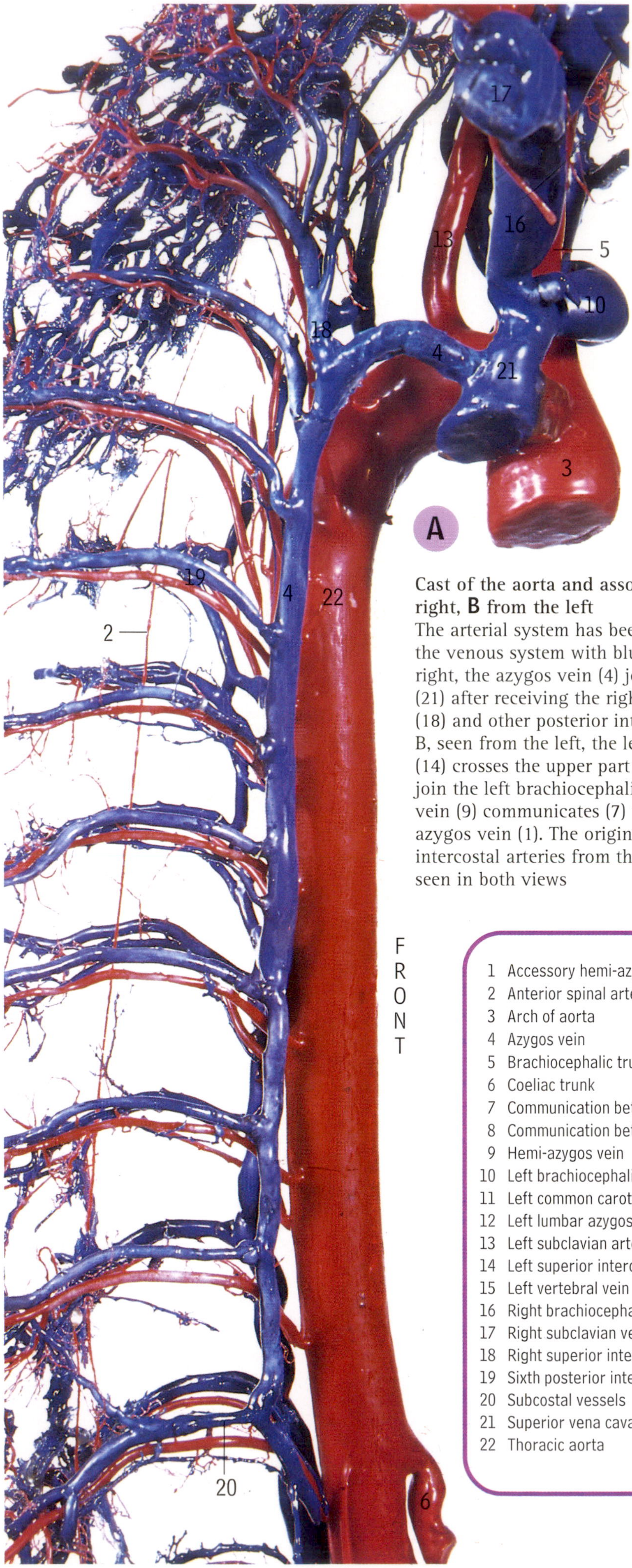

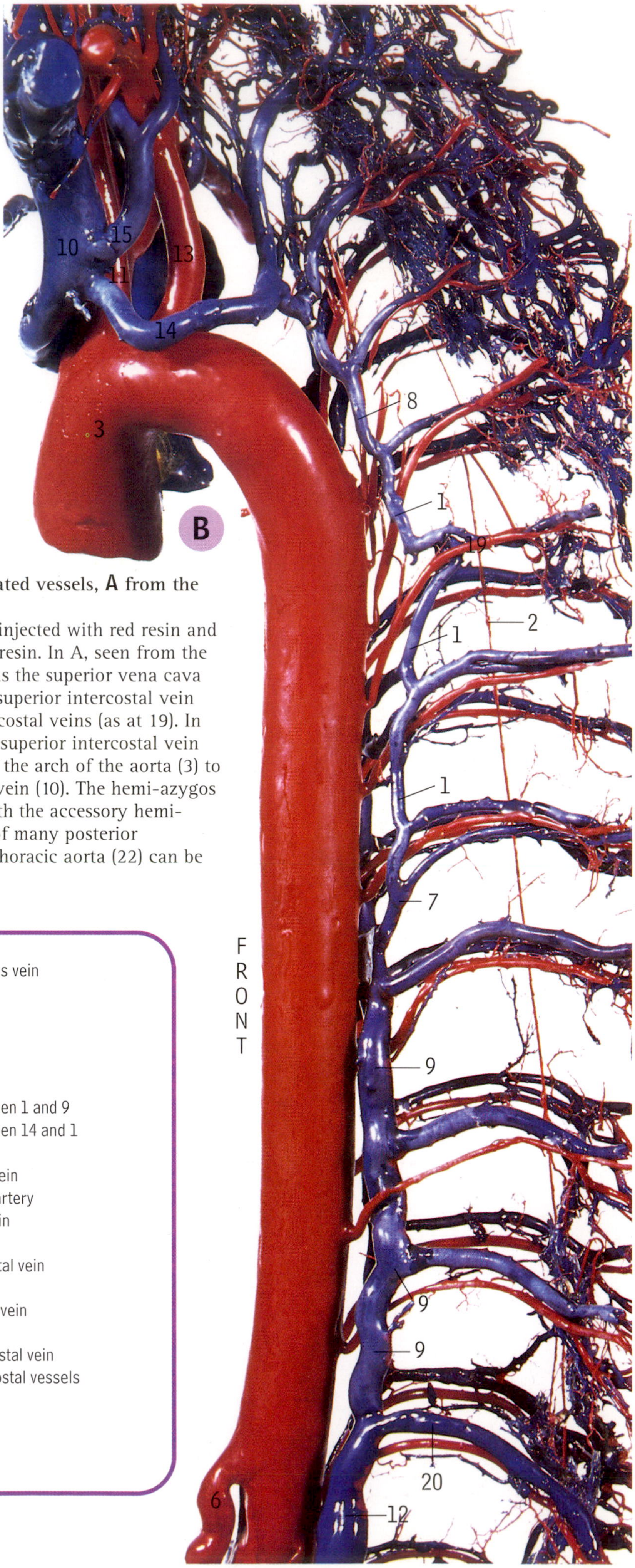

Cast of the aorta and associated vessels, A from the right, B from the left

The arterial system has been injected with red resin and the venous system with blue resin. In A, seen from the right, the azygos vein (4) joins the superior vena cava (21) after receiving the right superior intercostal vein (18) and other posterior intercostal veins (as at 19). In B, seen from the left, the left superior intercostal vein (14) crosses the upper part of the arch of the aorta (3) to join the left brachiocephalic vein (10). The hemi-azygos vein (9) communicates (7) with the accessory hemi-azygos vein (1). The origins of many posterior intercostal arteries from the thoracic aorta (22) can be seen in both views

1	Accessory hemi-azygos vein
2	Anterior spinal artery
3	Arch of aorta
4	Azygos vein
5	Brachiocephalic trunk
6	Coeliac trunk
7	Communication between 1 and 9
8	Communication between 14 and 1
9	Hemi-azygos vein
10	Left brachiocephalic vein
11	Left common carotid artery
12	Left lumbar azygos vein
13	Left subclavian artery
14	Left superior intercostal vein
15	Left vertebral vein
16	Right brachiocephalic vein
17	Right subclavian vein
18	Right superior intercostal vein
19	Sixth posterior intercostal vessels
20	Subcostal vessels
21	Superior vena cava
22	Thoracic aorta

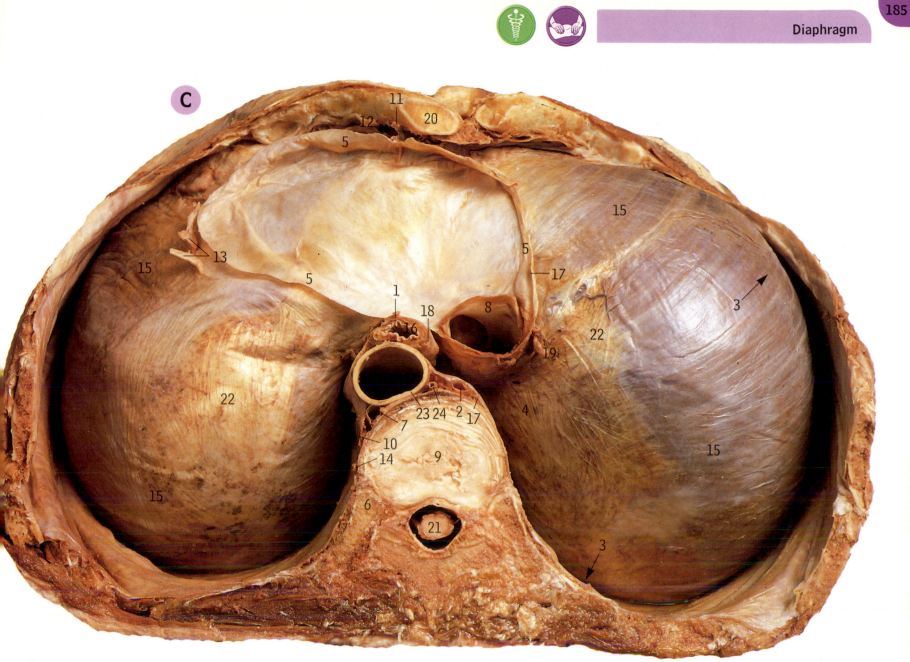

C Diaphragm, from above
The thorax has been transected at the level of the disc between the ninth
and tenth thoracic vertebrae.

> Gastro-oesophageal reflux, also known as
> heartburn, is commonly associated with either a
> hiatus hernia or increased acid production in
> the stomach and peptic ulceration.

1	Anterior vagal trunk	13	Left phrenic nerve
2	Azygos vein	14	Left sympathetic trunk
3	Costodiaphragmatic recess	15	Muscle of diaphragm
4	Costomediastinal recess	16	Oesophagus
5	Fibrous pericardium (cut edge)	17	Pleura (cut edge)
6	Head of left ninth rib	18	Posterior vagal trunk
7	Hemi-azygos vein	19	Right phrenic nerve
8	Inferior vena cava	20	Seventh left costal cartilage
9	Intervertebral disc	21	Spinal cord
10	Left greater splanchnic nerve	22	Tendon of diaphragm
11	Left internal thoracic artery	23	Thoracic aorta
12	Left musculophrenic artery	24	Thoracic duct

- According to the standard textbook description, the foramen for the vena cava is
  at the level of the disc between the eighth and ninth thoracic vertebrae, the
  oesophageal opening at the level of the tenth thoracic vertebra and the aortic
  opening opposite the twelfth thoracic vertebra. However, it is common for the
  oesophageal opening to be nearer the midline, as in this specimen (16), and the
  vena caval foramen (8) is lower than usual.
- The vena caval foramen is in the tendinous part of the diaphragm and the
  oesophageal opening in the muscular part. The so-called aortic opening is not *in*
  the diaphragm but behind it (page 227).
- The central tendon of the diaphragm has the shape of a trefoil leaf and has no
  bony attachment.
- The right phrenic nerve (19) passes through the vena caval foramen, i.e. through
  the tendinous part, but the left phrenic nerve (13) pierces the muscular part in
  front of the central tendon just lateral to the overlying pericardium.
- The phrenic nerves are the *only motor* nerves to the diaphragm, including the
  crura. The supply from lower thoracic (intercostal and subcostal) nerves is purely
  afferent. Damage to one phrenic nerve completely paralyses its own half of the
  diaphragm.

> Hiatus hernia . Commonly of two varieties,
> sliding or rolling, these herniae are protrusions
> of the proximal stomach into the thorax and
> often produce a burning sensation in the
> midsternal region.

Oesophageal imaging

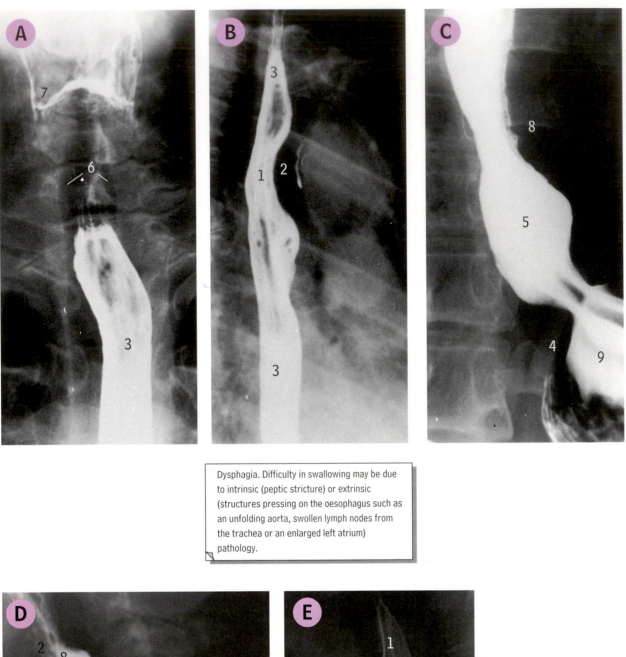

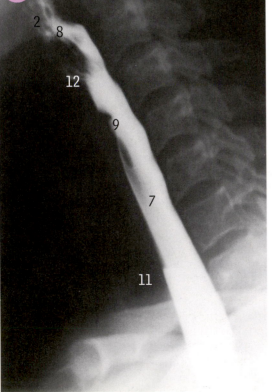

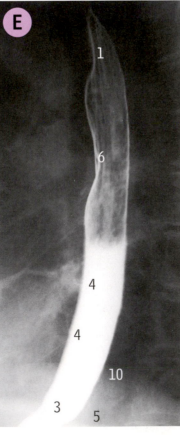

Radiographs of the oesophagus during a barium swallow, **A** lower pharynx and upper oesophagus, **B** middle part, **C** lower end

In A, viewed from the front, some of the barium paste adheres to the pharyngeal wall, outlining the piriform recesses (7), but most of it has passed into the oesophagus (3). In B, viewed obliquely from the left, the oesophagus is identified by the arch of the aorta (2) which shows some calcification in its wall – a useful aid to its identification. In C there is some dilatation at the lower end of the thoracic oesophagus (5) and it is constricted where it passes through the diaphragm (4) to join the stomach (9). The left atrium of the heart (8) lies in front of the lower thoracic oesophagus (page 182, A8), but only when enlarged does the atrium cause an indentation in the oesophagus

1 Aortic impression in oesophagus
2 Arch of aorta with plaque of calcification
3 Barium in oesophagus
4 Diaphragm
5 Lower thoracic oesophagus
6 Margins of trachea (translucent with contained air)
7 Piriform recess in laryngeal part of pharynx
8 Position of left atrium
9 Stomach

Dysphagia. Difficulty in swallowing may be due to intrinsic (peptic stricture) or extrinsic (structures pressing on the oesophagus such as an unfolding aorta, swollen lymph nodes from the trachea or an enlarged left atrium) pathology.

Radiographs of the oesophagus, **D** cervical part, **E** thoracic part

1 Aortic arch impression
2 Base of tongue
3 Gastro-oesophageal junction
4 Left atrium impression
5 Left hemidiaphragm
6 Left principal bronchus impression
7 Oesophagus
8 Oropharynx
9 Postcricoid venous plexus impression
10 Right hemidiaphragm
11 Trachea
12 Vallecula

Chapter 5

Abdomen and pelvis

- The nipple in the male normally lies over the fourth intercostal space.
- The umbilicus normally lies at the level of the disc between the third and fourth lumbar vertebrae.
- The transpyloric plane (10) lies midway between the jugular notch of the sternum and the upper border of the pubic symphysis, or approximately a hand's breadth below the xiphisternal joint (11), and level with the lower part of the body of the first lumbar vertebra.
- The hilum of each kidney is about 5 cm from the midline, that of the left (7) being just above the transpyloric plane and that of the right (8) just below it.
- In life the duodenum and the head of the pancreas (6) may lie at one or more vertebral levels lower than in the standard textbook or cadaveric position, shown here.
- The fundus of the gall bladder (5) lies behind the point where the lateral border of the right rectus sheath meets the costal margin at the ninth costal cartilage.

Abdominal paracentesis is a procedure that drains fluid (ascites) from the abdomen through a cannula inserted through the anterior abdominal wall lateral to the rectus sheath to avoid injury to the epigastric vessels.

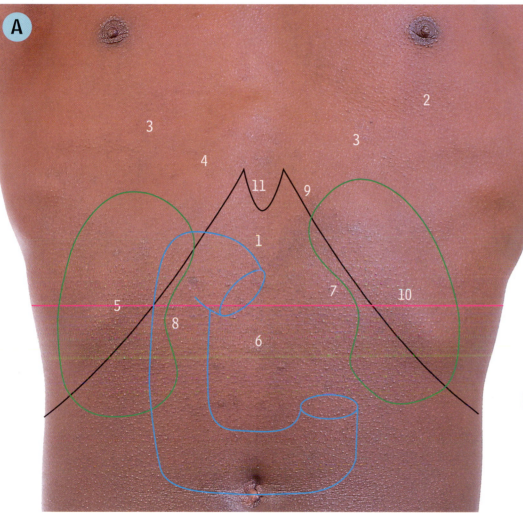

A Anterior abdominal wall above the umbilicus. Surface markings
The solid black line indicates the costal margin. The magenta line indicates the transpyloric plane. The C-shaped duodenum is outlined in blue, and the kidneys in green

1 Aortic opening in diaphragm	6 Head of pancreas and level of second lumbar vertebra
2 Apex of heart in fifth intercostal space	
3 Dome of diaphragm and upper margin of liver	7 Hilum of left kidney
4 Foramen for inferior vena cava in diaphragm	8 Hilum of right kidney
5 Fundus of gall bladder, and junction of ninth costal cartilage and lateral border of rectus sheath	9 Oesophageal opening in diaphragm
	10 Transpyloric plane
	11 Xiphisternal joint

B Regions of the abdomen
The abdomen may be divided into regions by two vertical and two horizontal lines. The vertical lines (VL) pass through the midinguinal points: the upper horizontal line corresponds to the transpyloric plane (TP, A10), the lower line is drawn between the tubercles of the iliac crests (transtubercular plane, TT)

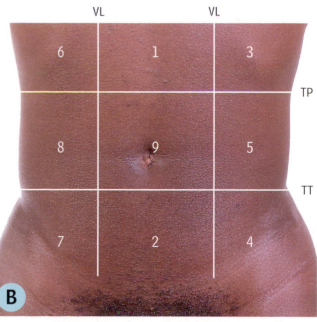

1 Epigastric region	6 Right hypochondrium
2 Hypogastrium or suprapubic	7 Right iliac region or iliac fossa
3 Left hypochondrium	
4 Left iliac region or iliac fossa	8 Right lumbar region
5 Left lumbar region	9 Umbilical region

A Anterior abdominal wall

1 Anterior cutaneous nerve (eighth intercostal)
2 Anterior cutaneous nerve (tenth intercostal)
3 Anterior layer internal oblique aponeurosis
4 External oblique aponeurosis
5 External oblique muscle
6 Ilioinguinal nerve
7 Iliotibial tract
8 Linea alba
9 Linea semilunaris
10 Mons pubis
11 Pectoralis major muscle
12 Posterior layer internal oblique aponeurosis
13 Pyramidalis muscle
14 Rectus abdominis
15 Rectus sheath, anterior
16 Round ligament of uterus
17 Serratus anterior muscle
18 Superficial inguinal lymph node (horizontal group)
19 Superficial inguinal lymph node (vertical group)
20 Superficial inguinal ring
21 Superficial inguinal veins
22 Tendinous intersection
23 Umbilicus

- The rectus sheath (A15) is formed by the internal
  oblique aponeurosis (A3) which splits at the lateral
  border of the rectus muscle (A9) into two layers.
  The posterior (A12) passes behind the muscle to
  blend with the aponeurosis of transversus
  abdominis (B19) to form the posterior wall of the
  sheath (B13), and the anterior layer (A3) passes in
  front of the muscle to blend with the external
  oblique aponeurosis (A4) as the anterior wall (A15).
- The anterior and posterior walls of the sheath unite
  at the medial border of the rectus muscle to form
  the midline linea alba (A8, B11).

Haematoma of the rectus sheath. Direct trauma
to the anterior abdominal wall or a violent
forced expiratory effort, such as at childbirth,
may cause tearing of the inferior or superior
epigastric vessels, extravasating blood into one
or both sheaths of the rectus abdominis
muscles.

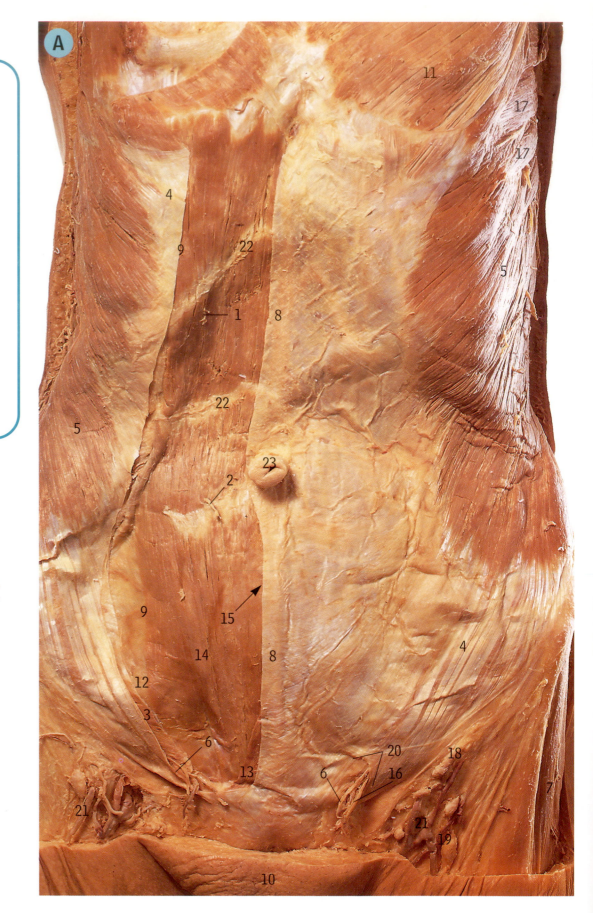

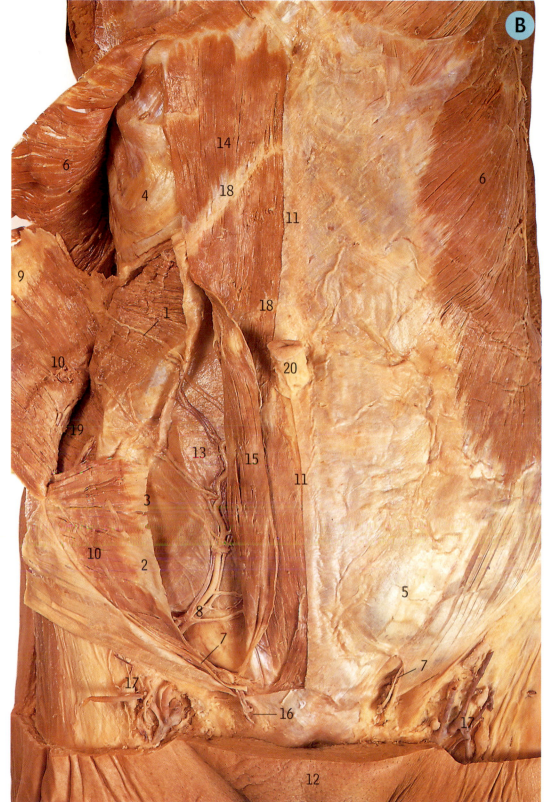

Anterior abdominal wall, **B** rectus sheath,
C axial MR image

1 Anterior cutaneous nerve (tenth intercostal)
2 Anterior layer of internal oblique aponeurosis
3 Anterior wall of rectus sheath
4 Eighth rib
5 External oblique aponeurosis
6 External oblique muscle
7 Ilioinguinal nerve
8 Inferior epigastric vessels
9 Internal oblique aponeurosis
10 Internal oblique muscle
11 Linea alba
12 Mons pubis
13 Posterior wall of rectus sheath
14 Rectus abdominis
15 Rectus abdominis, reflected
16 Round ligament of uterus
17 Superficial inguinal lymph nodes
18 Tendinous intersection
19 Transversus abdominis
20 Umbilicus

- There is no posterior rectus sheath in the lower third of rectus abdominis, below the aruate line (page 193, B1).

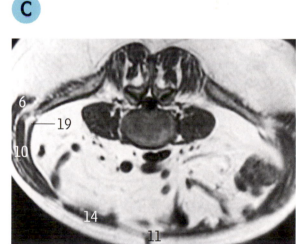

A Anterior abdominal wall, abdominal contents
The anterior abdominal wall, cut along the costal margins (4), has been reflected downwards. The posterior wall of rectus sheath, fascia transversalis (22) and parietal peritoneum (18) have been excised and the abdominal contents exposed *in situ*

 1 Ascending colon
 2 Body of stomach
 3 Caecum
 4 Costal margin
 5 Cut edge of greater omentum
 6 Diaphragm
 7 Duodenum
 8 Falciform ligament
 9 Fundus of stomach
10 Greater omentum
11 Hepatic flexure, colon
12 Ileum
13 Inferior epigastric vessels
14 Left gastro-epiploic vessels
15 Left lobe, liver
16 Lesser omentum
17 Ligamentum teres
18 Parietal peritoneum, cut edge
19 Pyloric sphincter
20 Rectus abdominis muscle
21 Right lobe, liver
22 Transversalis fascia
23 Transverse colon
24 Transversus abdominis muscle
25 Umbilical artery, remnant
26 Umbilicus
27 Urachus

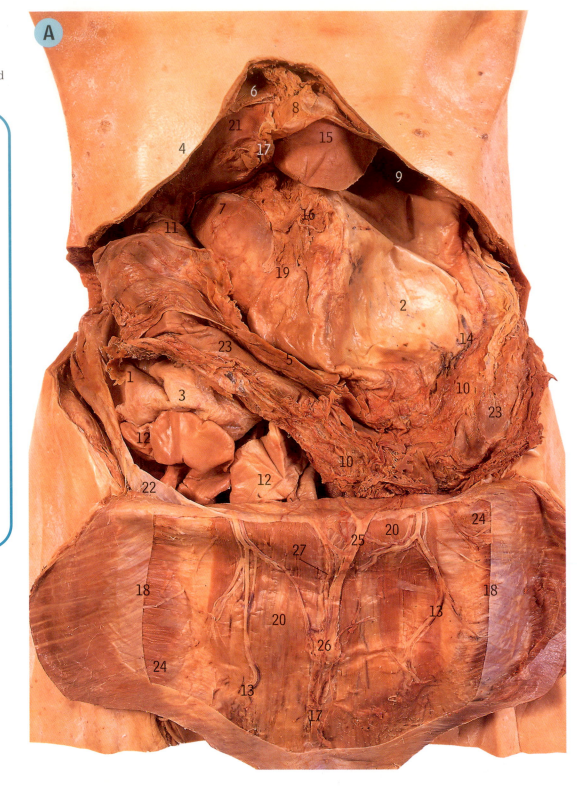

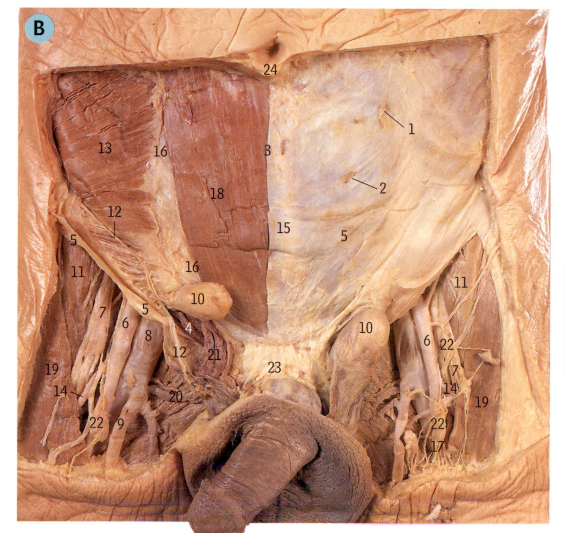

Anterior abdominal wall **B** groin, **C** axial MR image

1 Anterior cutaneous nerve (eleventh intercostal)
2 Anterior cutaneous nerve (twelfth intercostal)
3 Anterior rectus sheath (cut edge)
4 Ductus deferens (vas)
5 External oblique aponeurosis
6 Femoral artery
7 Femoral nerve
8 Femoral vein
9 Great saphenous vein
10 Hernial sac
11 Iliacus muscle
12 Ilioinguinal nerve
13 Internal oblique muscle
14 Lateral circumflex femoral artery
15 Linea alba
16 Linea semilunaris
17 Lymphatic vessels
18 Rectus abdominis muscle
19 Sartorius muscle
20 Scrotal venous connections
21 Spermatic cord
22 Superficial inguinal lymph node
23 Suspensory ligament of penis
24 Umbilicus

• The hernial sac (10), shown here, is not present in normal subjects.

Hernia repair. In the inguinal region, the most common area for herniae, manufactured fibres are often used to patch over large hernia openings. The pectineal and lacunar ligaments are often used as anchor points for sutures.

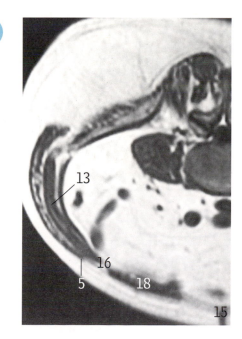

Anterior abdominal wall

- The superficial inguinal ring (14, at the medial end of the inguinal canal) lies 1 cm above the pubic tubercle (13).
- The deep inguinal ring (3, at the lateral end of the inguinal canal) lies 1 cm above the midpoint of the inguinal ligament.
- The femoral artery (4, whose pulsation should normally be palpable) enters the thigh midway between the pubic symphysis (12) and the anterior superior iliac spine (1). This is often referred to as the 'femoral' point or the midinguinal point.
- Note the slight difference between the surface markings of the midpoint of the inguinal ligament (used to find the deep ring, 3) and the midinguinal point (used to locate the femoral artery, 4).

McBurney's point (11) is a site on the surface of the anterior abdominal wall indicating the usual location of the base of the appendix internally. It lies one-third of the way along a line from the right anterior superior iliac spine to the umbilicus (magenta line).

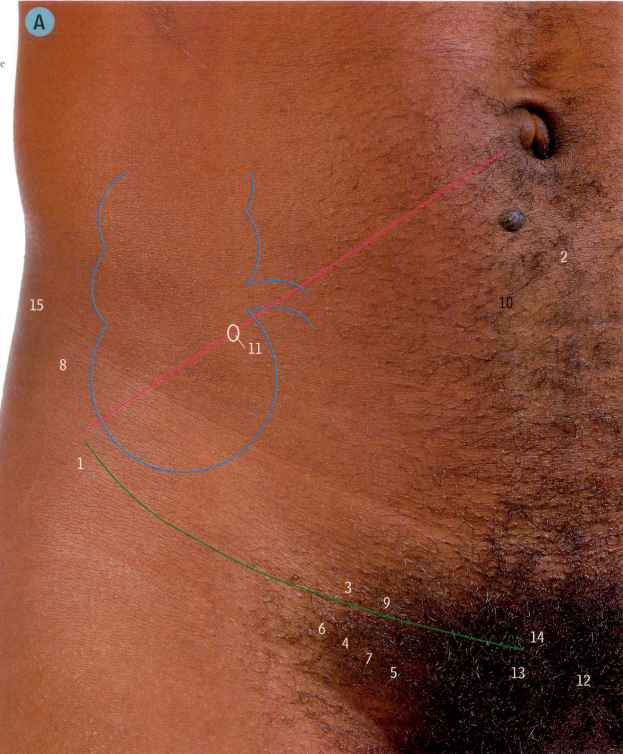

A Anterior abdominal wall, right lower quadrant. Surface markings
The caecum with the ileum opening into it from the left and the ascending colon continuing upwards from it are indicated by the blue line. The inguinal ligament, between the anterior superior iliac spine (1) and the pubic tubercle (13), is indicated by the green line. The femoral artery (4) has the femoral vein (7) on its medial side and the femoral nerve (6) on its lateral side. The femoral canal (5) is on the medial side of the vein. The deep inguinal ring (3) and inferior epigastric vessels (9) are above the artery, while the superficial inguinal ring (14) is above and medial to the pubic tubercle (13)

1 Anterior superior iliac spine	9 Inferior epigastric vessels
2 Bifurcation of aorta (fourth lumbar vertebra)	10 Lower end of inferior vena cava (fifth lumbar vertebra)
3 Deep inguinal ring	11 McBurney's point
4 Femoral artery	12 Pubic symphysis
5 Femoral canal	13 Pubic tubercle
6 Femoral nerve	14 Superficial inguinal ring
7 Femoral vein	15 Tubercle of iliac crest
8 Iliac crest	

B

B Anterior abdominal wall, from behind. Umbilical folds
This view of the peritoneal surface of the central region of
the anterior abdominal wall shows the peritoneal folds
raised by underlying structures. There is one fold above the
umbilicus—the falciform ligament—and there are five below
it: the median umbilical fold (7) in the midline, and a pair of
medial and lateral umbilical folds on each side (6 and 4).
See the notes for the contents of the folds

1	Arcuate line	5	Linea semilunaris
2	Falciform ligament	6	Medial umbilical fold
3	Inguinal triangle	7	Median umbilical fold
	(Hesselbach)	8	Umbilicus
4	Lateral umbilical fold		

- The median umbilical fold (B7) contains the median umbilical
  ligament, which is the obliterated remains of the urachus
  (formed from the allantois, the embryonic connexion between
  the bladder and umbilicus).
- The medial umbilical fold contains the medial umbilical
  ligament, which is the obliterated remains of the umbilical
  artery (C9, B6).
- The lateral umbilical fold (B4) contains the inferior epigastric
  vessels, conducting them from the external iliac vessels to the
  rectus sheath. Although called an umbilical fold, it does not
  extend as far as the umbilicus, since the vessels enter the rectus
  sheath by passing beneath the arcuate line (B1), which is the
  lower border of the posterior wall of the sheath. Below this level
  the three aponeuroses that form the sheath (page 188) all pass in
  front of the rectus muscle.
- The inguinal triangle of Hesselbach is a naturally weak region
  between rectus abdominis and the inferior epigastric vessels.
  Direct inguinal hernia appear through this region.

Peritoneal pain. The visceral peritoneum is
sensitive only to stretch and pressure and these
are experienced as a dull ache. The parietal
peritoneum, however, is sensitive to pain and is
innervated by the spinal nerves.

Umbilical hernia is an anterior abdominal hernia at
the umbilicus which is often congenital and
frequently disappears by the second or third year.

Caput medusae, dilated para-umbilical veins
resembling the hair of Medusa of Greek
mythology, are often a clinical sign of cirrhosis.

C Fetal anterior abdominal wall, from behind
In this full-term fetus the peritoneum and extraperitoneal tissues have been
removed from the anterior abdominal wall to show the umbilical arteries (9)
and left umbilical vein (6) converging at the back of the (unlabelled)
umbilicus

1	Diaphragm	6	Left umbilical vein
2	External oblique	7	Rectus abdominis
3	Falciform ligament	8	Transversus abdominis
4	Inferior epigastric vessels	9	Umbilical artery
5	Internal oblique	10	Urinary bladder

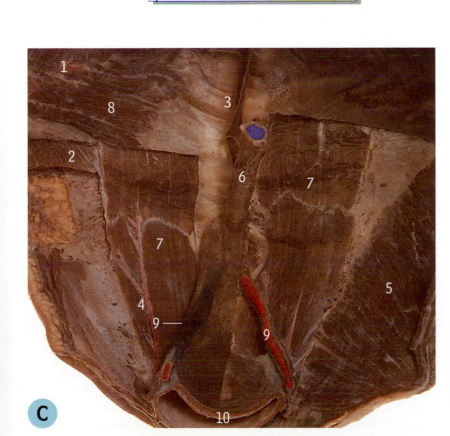

C

Anterior abdominal wall

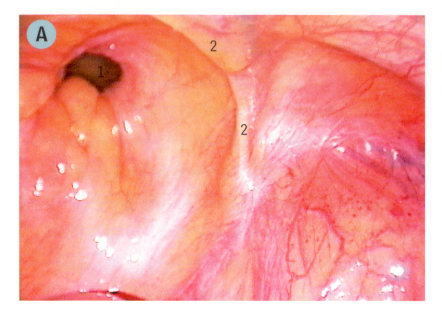

A

Laparoscopic view of direct inguinal hernia, left side

1 Direct inguinal hernia
2 Inferior epigastric vessels

Direct inguinal hernia is a protrusion, often of peritoneum and bowel, through the anterior abdominal wall lying medial to the inferior epigastric vessels, in the inguinal (Hesselbach's) triangle.

B

B Anterior abdominal wall, right side, from behind

Indirect inguinal hernia is a protrusion that follows the course of the vas deferens or round ligament. The neck of the hernia lies in the deep inguinal ring lateral to the inferior epigastric vessels.

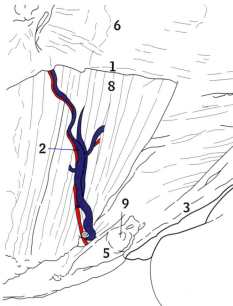

The peritoneum and extraperitoneal tissues have been removed, leaving the inferior epigastric vessels (2) coursing over the back of the rectus muscle (8) to enter the rectus sheath (6) beneath the arcuate line (1)

1 Arcuate line
2 Inferior epigastric vessels
3 Inguinal ligament
4 Linea alba
5 Position of deep inguinal ring
6 Posterior wall of rectus sheath
7 Pubic crest
8 Rectus abdominis
9 Spermatic cord
10 Transversus abdominis
11 Umbilicus

C Upper abdominal viscera, in transverse section

This section through the upper abdomen at the level of the first lumbar vertebra, seen from below looking towards the thorax, shows the general disposition of some of the viscera. The vertebral column (2) bulges forwards into the abdominal cavity, with the kidneys (11 and 20) lying in the trough on either side. The bulk of the liver (21) is on the right side, extending towards the left (12) to overlap part of the stomach (28), and the pancreas (16) lies centrally, also extending towards the left (but on a deeper plane) to overlap part of the left kidney (11). Parts of the colon (32 and 4) are adjacent to the spleen (26), which lies against the part of the diaphragm attached to the thoracic wall in the region of the tenth rib (31)

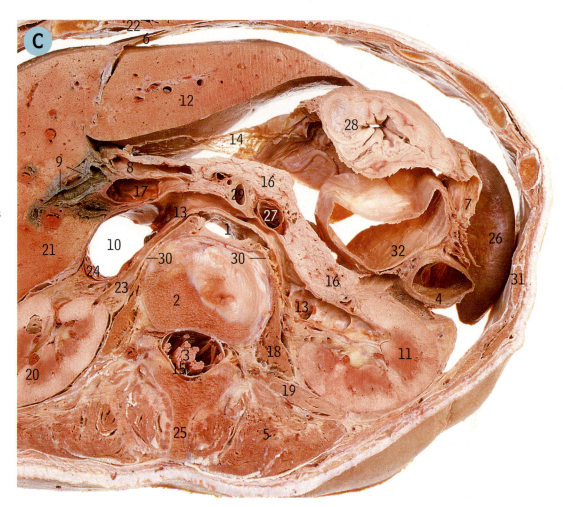

1 Abdominal aorta	12 Left lobe of liver	23 Right renal artery
2 Body of first lumbar vertebra	13 Left renal vein	24 Right renal vein
3 Conus medullaris of spinal cord	14 Lesser omentum	25 Spine of first lumbar vertebra
4 Descending colon	15 Nerve roots of cauda equina	26 Spleen
5 Erector spinae	16 Pancreas	27 Splenic vein
6 Falciform ligament	17 Portal vein	28 Stomach
7 Greater omentum (gastrosplenic ligament)	18 Psoas major	29 Superior mesenteric artery
8 Hepatic artery	19 Quadratus lumborum	30 Sympathetic trunk
9 Hepatic ducts	20 Right kidney	31 Tenth rib
10 Inferior vena cava	21 Right lobe of liver	32 Transverse colon
11 Left kidney	22 Right rectus abdominis	

D CT scan of the upper abdomen, at the level of the coeliac trunk

All CT (computerized tomography) scans of the trunk are, by convention, viewed from below (as with the body lying on the back and the viewer looking towards the head). In D both oral and intravenous contrast media have been used (to emphasize the outlines of the gut and vascular system). To avoid too many labels, only some key features have been numbered, and the various parts of the alimentary tract are unlabelled. The coeliac trunk arising from the aorta (1) is seen to divide as a Y into the splenic artery running towards the left behind the pancreas (6) and the common hepatic artery passing to the right near the portal vein (7). On the left side the spleen (9) and the upper pole of the kidney (4) are shown, but on the right the plane of the scan is too high to show the right kidney

1 Abdominal aorta	6 Pancreas
2 Gall bladder	7 Portal vein
3 Inferior vena cava	8 Right crus of diaphragm
4 Left kidney	9 Spleen
5 Liver	10 Twelfth thoracic vertebra

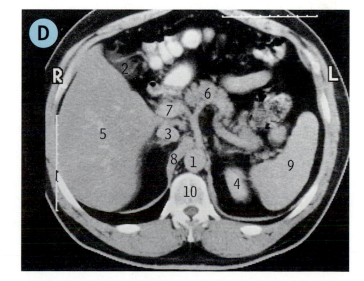

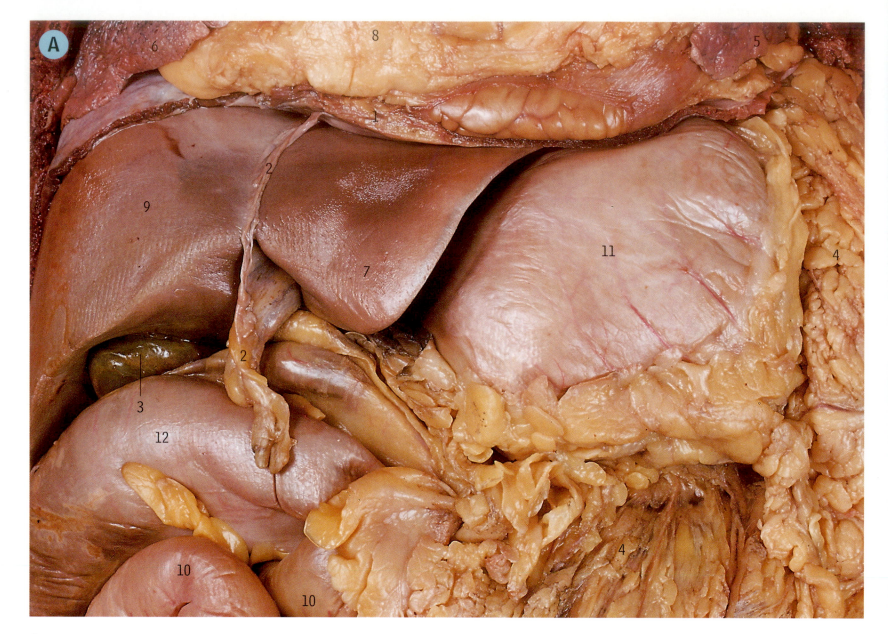

A Upper abdominal viscera, from the front

The thoracic and abdominal walls and the anterior part of the diaphragm have been removed to show the undisturbed viscera. The liver (9 and 7) and stomach (11) are immediately below the diaphragm (1). The greater omentum (4) hangs down from the greater curvature (lower margin) of the stomach (11), overlying much of the small and large intestine but leaving some of the transverse colon (12) and small intestine (10) uncovered. The fundus (tip) of the gall bladder (3) is seen between the right lobe of the liver (9) and transverse colon (12)

Liver biopsy. Liver tissue samples can be obtained using a needle passed through the ninth or tenth right intercostal space in the axillary line during full expiration to reduce the size of the costodiaphragmatic recess and the risk of pneumothorax.

1	Diaphragm	7	Left lobe of liver
2	Falciform ligament	8	Pericardial fat
3	Gall bladder	9	Right lobe of liver
4	Greater omentum	10	Small intestine
5	Inferior lobe of left lung	11	Stomach
6	Inferior lobe of right lung	12	Transverse colon

• For an explanation of peritoneal structures see the diagrams on page 202.

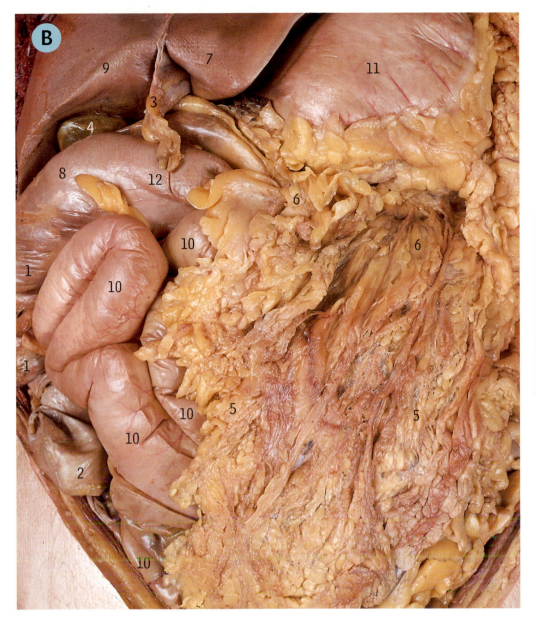

B Upper abdominal viscera, from the front
In this view of the undisturbed abdomen the upper part of the greater omentum (as at 6) overlies much of the transverse colon and mesocolon (with the right part of the transverse colon seen at 12). The lower part of the omentum (5) covers coils of small intestine, some of which (10) are visible beyond the right margin of the omentum. The caecum (2) is at the lower end of the ascending colon (1) which continues upwards into the right colic flexure (hepatic flexure, 8) and then becomes the transverse colon (12)

1 Ascending colon
2 Caecum
3 Falciform ligament
4 Fundus of gall bladder
5 Greater omentum overlying coils of small intestine
6 Greater omentum overlying transverse colon and mesocolon
7 Left lobe of liver
8 Right colic flexure
9 Right lobe of liver
10 Small intestine
11 Stomach
12 Transverse colon

Laparoscopy is a technique of minimal invasive surgery to examine, remove or repair abdominal or pelvic organs/tissue using small tubes inserted through the abdominal wall.

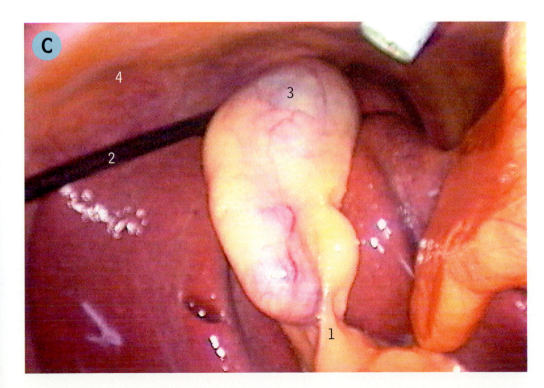

C Laparoscopic view of gall bladder

1 Cystic triangle (Calot)
2 Forceps
3 Fundus of gall bladder
4 Liver

Cholecystectomy. Surgical removal of the gall bladder was classically performed by Kocher's incision parallel to the right costal margin but is now commonly performed laparoscopically.

A Upper abdominal viscera, from the front
In this view of the same specimen as on pages 196 and 197, the greater omentum (5) has been lifted upwards to show its adherence to the transverse colon (8) (see page 202, C)

1 Appendices epiploicae
2 Falciform ligament
3 Gall bladder (fundus)
4 Left lobe of liver
5 Posterior surface of greater omentum
6 Right lobe of liver
7 Small intestine
8 Transverse colon

- The appendices epiploicae (A1) are fat-filled appendages of peritoneum on the various parts of the colon (ascending, transverse, descending and sigmoid). They are not present on the small intestine or the rectum, and may be rudimentary on the caecum and appendix. In abdominal operations they are one feature that helps to distinguish colon from other parts of the intestine.
- In strict anatomical nomenclature the term 'small intestine' includes the duodenum, jejunum and ileum, but clinically it is frequently used to mean jejunum and ileum, with the duodenum being referred to by its own name.

Peritoneal lavage. A procedure that washes peritoneal surfaces by inserting and removing fluid from the abdominal cavity, it is most commonly used to clean infection, or as peritoneal dialysis for renal disease. Diagnostic peritoneal lavage can be performed in cases of abdominal trauma to detect intraperitoneal bleeding.

Peritoneal dialysis. The peritoneum, being a semipermeable membrane, can be used for dialysis in renal disease. Fluid is introduced into the peritoneal cavity, allowed to mix with the contents and then withdrawn, removing with it most circulating toxins. This procedure, frequently repeated, is a satisfactory way of controlling uraemia.

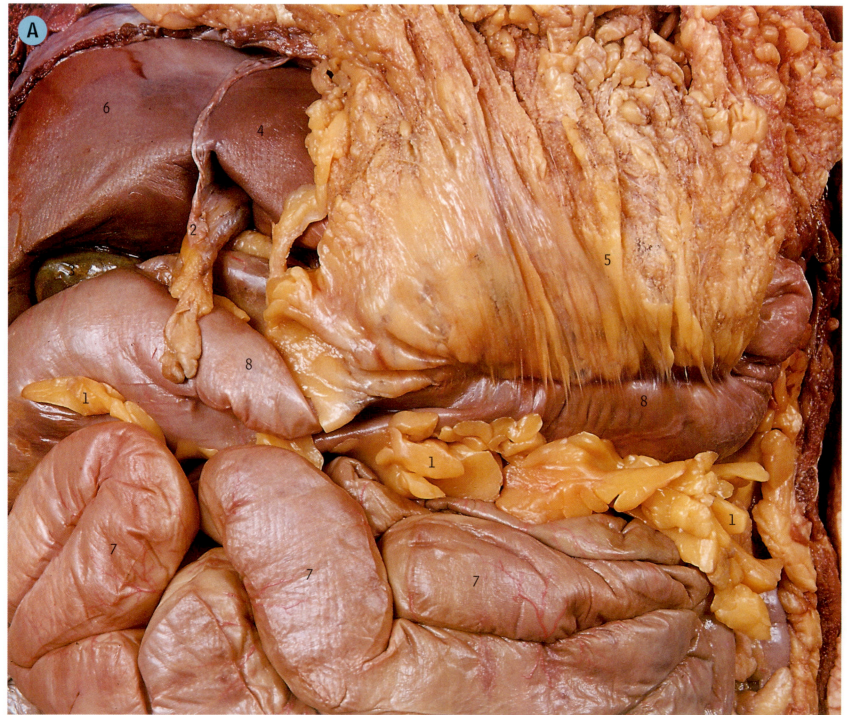

RIGHT

LEFT

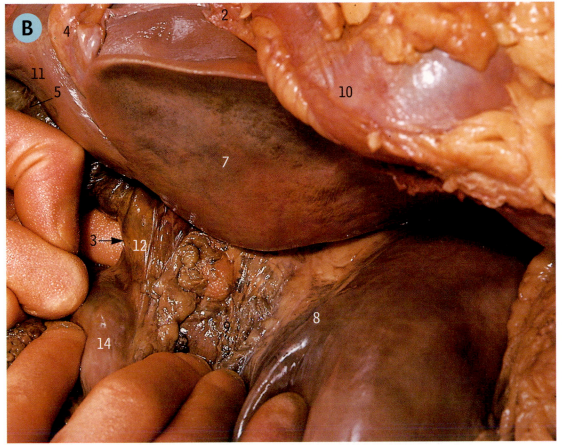

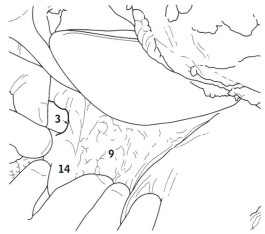

Lesser omentum and epiploic foramen, B from the front, C from the front and the right
In B a finger has been placed in the epiploic foramen (3) behind the right free margin of the lesser omentum (12), and the tip can be seen in the lesser sac, through the transparent lesser omentum (9) which stretches between the liver (7) and the lesser curvature of the stomach (8). In the more lateral view in C, looking into the foramen from the right, the foramen (3) is identified between the right free margin of the lesser omentum (12) in front and the inferior vena cava (6) behind, above the first part of the duodenum (14)

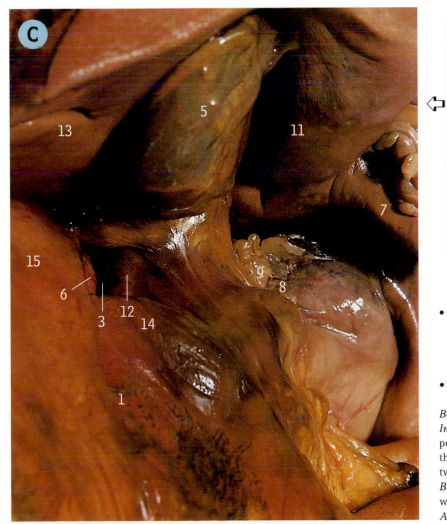

1 Descending (second) part of duodenum
2 Diaphragm
3 Epiploic foramen
4 Falciform ligament
5 Gall bladder
6 Inferior vena cava
7 Left lobe of liver
8 Lesser curvature of stomach
9 Lesser omentum
10 Pericardium
11 Quadrate lobe of liver
12 Right free margin of lesser omentum
13 Right lobe of liver
14 Superior (first) part of duodenum
15 Upper pole of right kidney

- The epiploic foramen (of Winslow, B3 and C3) is the communication between the general peritoneal cavity (sometimes called the greater sac) and the lesser sac (omental bursa), a space lined by peritoneum behind the stomach (B8 and C8) and lesser omentum (B9 and B12) and in front of parts of the pancreas and left kidney.
- The epiploic foramen, the opening into the lesser sac (B3 and C3), has the following boundaries:

Behind – the inferior vena cava (C6).

In front – the right free margin of the lesser omentum (B12) which contains the portal vein, hepatic artery and bile duct (page 203, F27, 17 and 4). The portal vein is the most posterior of these structures, so the foramen may be said to lie between the two great veins – inferior vena cava and portal vein.

Below – the first part of the duodenum (B14 and C14) which passes backwards as well as to the right.

Above – the caudate process of the liver (page 212, A3).

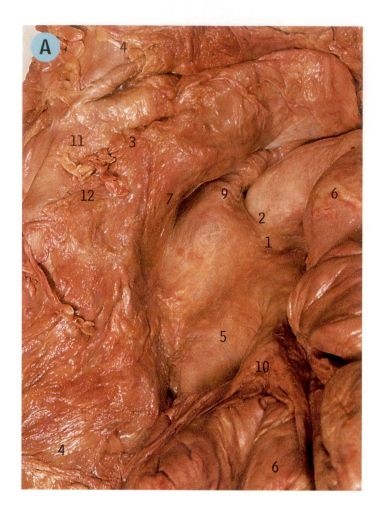

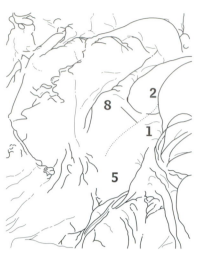

A Upper abdominal viscera, from the front
In this view the stomach (3 and 7), transverse colon (11) and greater omentum (4) have been lifted up to show the region of the duodenojejunal flexure (2). The left end of the horizontal (third) part of the duodenum (5) turns upwards as the ascending (fourth) part (1) which is continuous with the jejunum at the duodenojejunal flexure (2) below the lower border of the pancreas (9)

1 Ascending (fourth) part of duodenum	7 Lesser curvature of stomach
2 Duodenojejunal flexure	8 Line of attachment of root of mesentery
3 Greater curvature of stomach	9 Lower border of pancreas
4 Greater omentum (posterior surface)	10 Mesentery
5 Horizontal (third) part of duodenum	11 Transverse colon (posterior surface)
6 Jejunum	12 Transverse mesocolon (posterior surface)

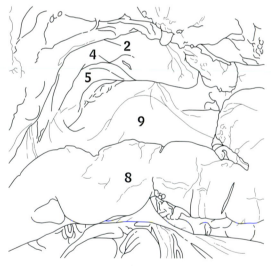

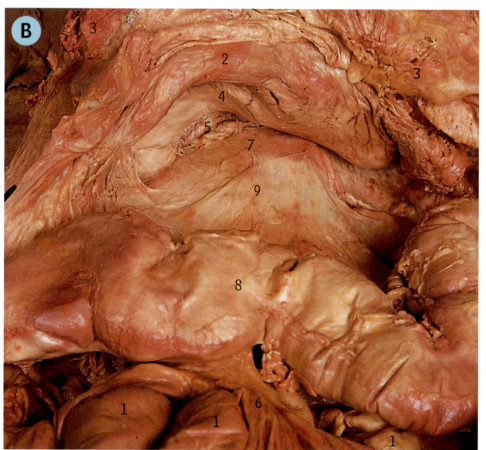

B Lesser sac and transverse mesocolon, from the front
The greater omentum (3) hanging down from the greater curvature of the stomach (2) has been separated from the underlying transverse colon (8) and mesocolon (9) and lifted upwards, and an opening made into the lesser sac (as in D on page 202). This view therefore shows the posterior surface of the greater omentum (3), stomach and lesser omentum (5), and the anterior surface of the transverse mesocolon (9)

Ascites is the accumulation of fluid within the peritoneal cavity due to a variety of causes, including malignancy, inflammation and portal hypertension.

1 Coils of jejunum and ileum	6 Mesentery
2 Greater curvature of stomach	7 Peritoneum of lesser sac overlying pancreas
3 Greater omentum (posterior surface)	8 Transverse colon
4 Lesser curvature of stomach	9 Transverse mesocolon overlying horizontal
5 Lesser omentum (posterior surface)	(third) part of duodenum

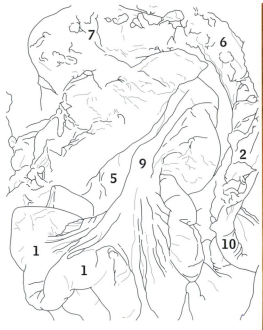

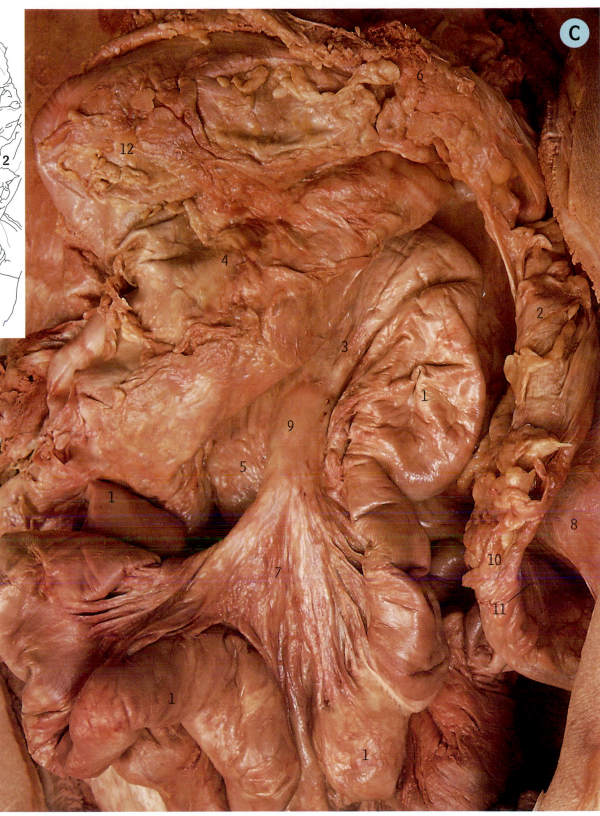

Volvulus, twisting of the bowel on its mesentery, causes ischaemia of the rotated section. It is most commonly seen in the sigmoid colon in those on a high fibre diet, especially in Africa where this condition replaces the acute appendix as the commonest cause of an 'acute abdomen'.

C Mesentery and descending colon, from the front

The stomach (4) and transverse colon (12) have been displaced upwards to show the left end of the root of the mesentery (9) at the duodenojejunal flexure (3). The descending colon (2), which is retroperitoneal, becomes the sigmoid colon (10) when it ceases to be retroperitoneal and acquires a mesocolon (11)

• The root of the mesentery (C9) begins at the duodenojejunal flexure (C3) and passes downwards and to the right, crossing the horizontal (third) part of the duodenum (C5); the superior mesenteric vessels enter the mesentery at this point (see page 204).

1	Coils of jejunum and ileum	7 Mesentery
2	Descending colon	8 Peritoneum overlying
3	Duodenojejunal flexure	external iliac vessels
4	Greater curvature of stomach	9 Root of mesentery
5	Horizontal (third) part of	10 Sigmoid colon
	duodenum	11 Sigmoid mesocolon
6	Left colic (splenic) flexure	12 Transverse colon

Drainage of peritoneal abscesses. Pus within the peritoneal cavity collects in supine individuals in one of the recesses or pouches within the peritoneal cavity. These include the subphrenic spaces, paracolic gutters and the recto-uterine pouch (of Douglas). Different drainage procedures are used for each space.

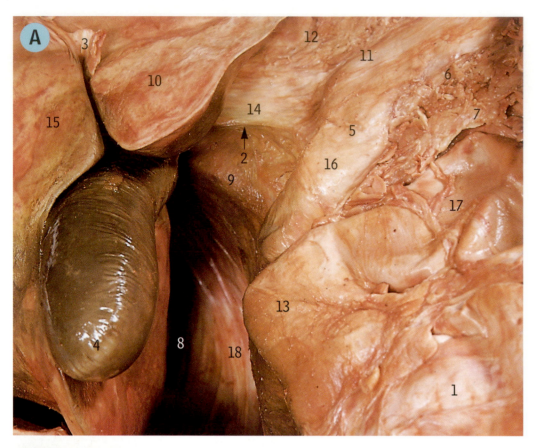

A Hepatorenal pouch of peritoneum, from the right and below

With the body lying on its back and seen from the right (with the head towards the left), the liver (15) has been turned upwards (towards the left) to open up the gap between the liver and upper pole of the right kidney (18) – the hepatorenal pouch of peritoneum (8, Morison's pouch or the right subhepatic compartment of the peritoneal cavity)

1 Ascending colon
2 Epiploic foramen
3 Falciform ligament
4 Gall bladder
5 Gastroduodenal junction
6 Greater curvature of stomach
7 Greater omentum
8 Hepatorenal (Morison's) pouch
9 Inferior vena cava
10 Left lobe of liver
11 Lesser curvature of stomach
12 Lesser omentum overlying pancreas
13 Right colic (hepatic) flexure
14 Right free margin of lesser omentum
15 Right lobe of liver
16 Superior (first) part of duodenum
17 Transverse colon
18 Upper pole of right kidney

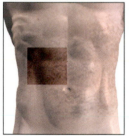

Diagrams of peritoneum. B Normal position, **C** with the lower part of the greater omentum lifted up, **D** with the greater omentum lifted up and separated from the transverse mesocolon and colon, with an opening into the lesser sac, **E** with the greater omentum and transverse mesocolon and colon lifted up, with an opening into the lesser sac

These drawings of a sagittal section through the middle of the abdomen, viewed from the left, illustrate theoretically how the peritoneum forms the lesser omentum (L, passing down to the stomach, S), greater omentum (G), transverse mesocolon (TM) passing to the transverse colon (TC), and the mesentery (M) of the small intestine (SI). The layer in blue represents the peritoneum of the lesser sac. The superior mesenteric artery passes between the head and uncinate process of the pancreas

(P and U), and continues across the duodenum (D) into the mesentery (M) to the small intestine (SI), giving off the middle colic artery which runs in the transverse mesocolon (TM) to the transverse colon (TC). The greater omentum (G) is formed by four layers fused together and also fused with the front of the transverse mesocolon (TM, two layers) and transverse colon. On dissection, no separation between any layers is possible except between the greater omentum and the transverse mesocolon. The six layers between the stomach and transverse colon are sometimes collectively known as the gastrocolic omentum. B corresponds to the dissections on pages 196 and 197, C to page 198, D to page 200B, and E to page 205B. The small arrows in D and E indicate the layers cut to make artificial openings into the lesser sac

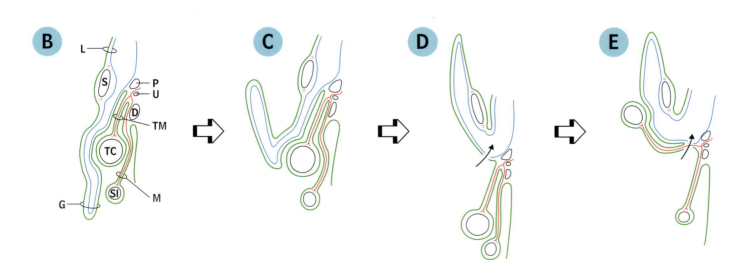

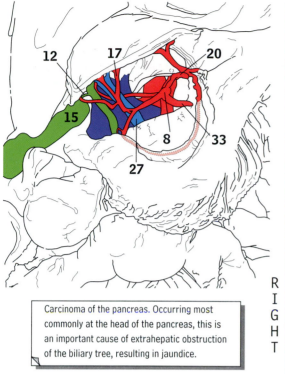

F

Carcinoma of the pancreas. Occurring most commonly at the head of the pancreas, this is an important cause of extrahepatic obstruction of the biliary tree, resulting in jaundice.

F Coeliac trunk and surrounding area
Part of the left lobe of the liver (21), and most of the lesser and greater omentum (24 and 7) have been removed, together with peritoneum of the central part of the posterior abdominal wall (posterior wall of the lesser sac), to show some of the most important structures in the upper abdomen: the coeliac trunk (8) and its branches (20, 33 and 9), the portal vein (27), and the bile duct (4) formed by the union of the cystic duct (12) from the gall bladder (15) with the common hepatic duct (10) from the liver (32 and 21)

1 Abdominal aorta	19 Left crus of diaphragm
2 Abdominal part of oesophagus	20 Left gastric artery
3 Accessory hepatic artery	21 Left lobe of liver
4 Bile duct	22 Left renal vein
5 Body of pancreas	23 Left triangular ligament
6 Body of stomach	24 Lesser omentum containing right and
7 Branches of left and right gastro-	left gastric arteries
epiploic arteries in greater omentum	25 Median arcuate ligament
8 Coeliac trunk	26 Oesophageal branch of left gastric
9 Common hepatic artery	artery
10 Common hepatic duct	27 Portal vein
11 Cystic artery	28 Pyloric part of stomach
12 Cystic duct	29 Right crus of diaphragm
13 Diaphragm	30 Right gastric artery
14 Falciform ligament	31 Right gastro-epiploic artery
15 Gall bladder	32 Right lobe of liver
16 Gastroduodenal artery	33 Splenic artery
17 Hepatic artery and right and left	34 Superior (first) part of duodenum
branches	35 Superior mesenteric artery
18 Inferior vena cava	36 Transverse colon

- The cystic artery (11) is normally derived from the right branch of the hepatic artery and passes behind the common hepatic and cystic ducts. Here it comes from the hepatic artery itself (17) and passes in front of the bile duct (4).
- If an accessory hepatic artery is present (as in this specimen, 3) it passes behind the portal vein (27), not in front as would the normal artery.
- It is normal for the right gastric artery (30) to be much smaller than the left (20).
- The left gastric artery (20) passes upwards and to the left and then turns down to run along the lesser curvature of the stomach between the two layers of peritoneum that form the lesser omentum (24). It gives off an oesophageal branch which passes up through the oesophageal opening in the diaphragm and supplies the lower part of the oesophagus (2). The accompanying veins (not shown here) drain to the left gastric vein and thence to the portal vein, making the lower end of the oesophagus one of the most important sites of portal–systemic anastomosis.

Portocaval shunt. Portal vein obstruction from any cause produces varices at sites of portosystemic venous anastomoses. Direct anastomotic connection of parts of the portal vein into the inferior vena cava reduces portal hypertension and the potential consequences of varicosities.

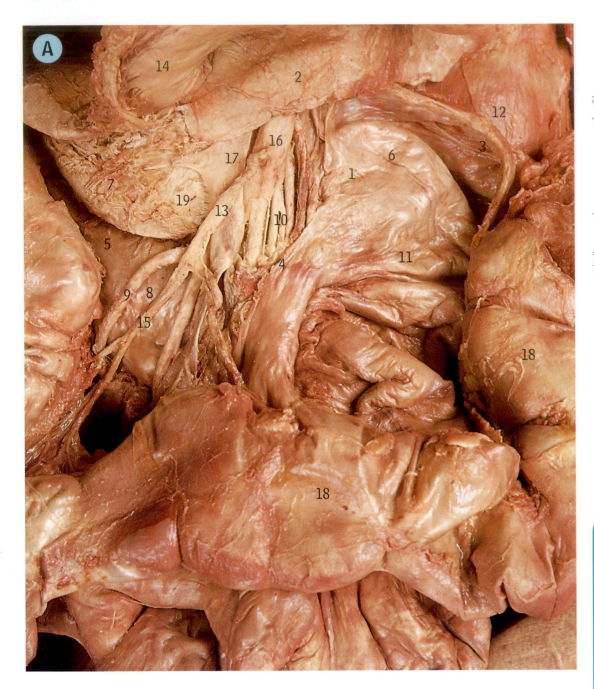

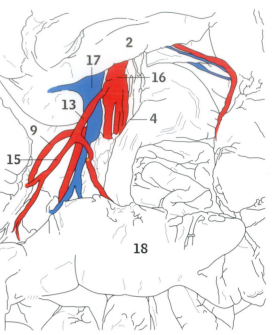

A Superior mesenteric vessels
The stomach (14) has been lifted upwards and transverse mesocolon removed, leaving the transverse colon (18) in its normal position. Part of the peritoneum of the mesentery (4) has been dissected away to show branches of the superior mesenteric artery (16)

- The right colic artery (15) is normally a branch of the superior mesenteric artery (16) but often (as here) arises from its middle colic branch (13).
- The superior mesenteric vein (17) lies on the right side of its companion artery (16). They appear at the lower border of the pancreas (2), crossing the uncinate process (19) of the head of the pancreas (7) and lower down crossing the horizontal (third) part of the duodenum (8) which is where they enter or leave the root of the mesentery (4).

1 Ascending (fourth) part of duodenum
2 Body of pancreas
3 Branches of left colic vessels
4 Cut edge of peritoneum at root of mesentery
5 Descending (second) part of duodenum
6 Duodenojejunal flexure
7 Head of pancreas
8 Horizontal (third) part of duodenum
9 Ileocolic artery
10 Jejunal and ileal arteries
11 Jejunum
12 Lower pole of left kidney
13 Middle colic artery
14 Posterior surface of pyloric part of stomach
15 Right colic artery
16 Superior mesenteric artery
17 Superior mesenteric vein
18 Transverse colon
19 Uncinate process of head of pancreas

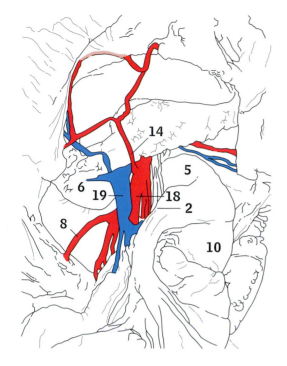

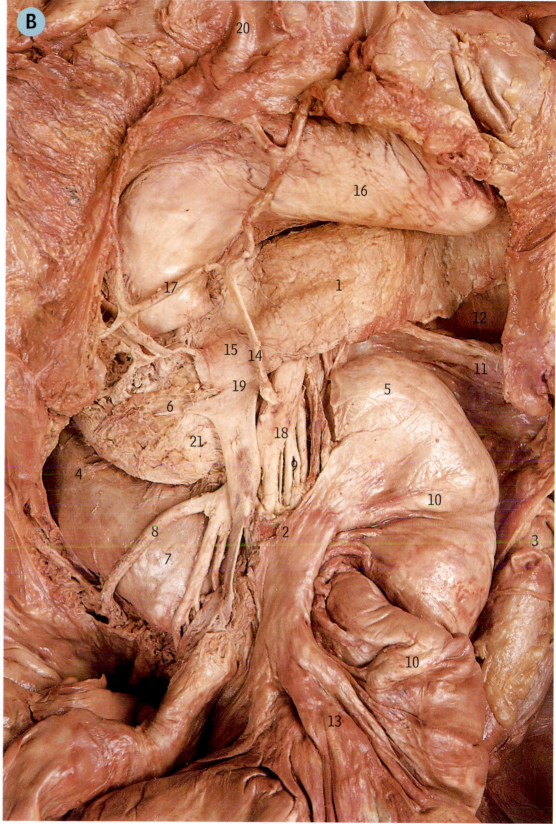

B Superior mesenteric vessels

This dissection is similar to A opposite, but here the stomach (16) and transverse colon (20) have both been lifted upwards, so lifting the middle colic artery (14) upwards also. The root of the mesentery (2) begins at the duodenojejunal flexure (5) and passes obliquely downwards to the right over the horizontal (third) part of the duodenum (7), where the superior mesenteric vessels and their branches (19, 18 and 9) become enclosed between the two layers of the peritoneum that form the mesentery (see B on page 202)

1   Body of pancreas
2   Cut edge of peritoneum at root of mesentery
3   Descending colon
4   Descending (second) part of duodenum
5   Duodenojejunal flexure
6   Head of pancreas
7   Horizontal (third) part of duodenum
8   Ileocolic artery
9   Jejunal and ileal arteries
10 Jejunum
11 Left colic vessels
12 Left kidney
13 Mesentery
14 Middle colic artery
15 Neck of pancreas
16 Posterior surface of body of stomach
17 Right branch of middle colic artery
18 Superior mesenteric artery
19 Superior mesenteric vein
20 Transverse colon
21 Uncinate process of head of pancreas

• In its normal position the middle colic artery runs downwards from its superior mesenteric origin (A13), but obviously when the transverse colon is lifted upwards (as here, B20, and in E on page 202), the vessel (B14) passes upwards also. Textbook drawings of the arteries of the colon often illustrate it in this position, but it must be remembered that with the body in the normal anatomical position it runs downwards.

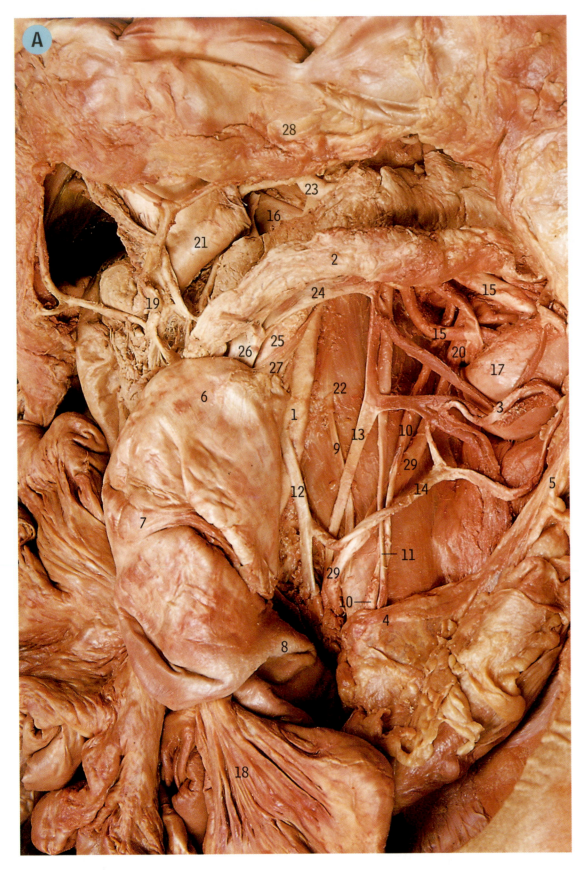

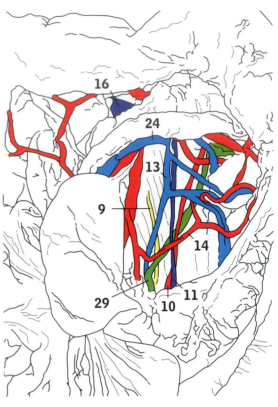

A Inferior mesenteric vessels, from the front

The stomach (21) and transverse colon (28) are lifted upwards. The peritoneum of the posterior abdominal wall has been removed and the left-sided parts of the duodenum (7 and 6) reflected towards the right, to show the origin of the inferior mesenteric artery (12) from the aorta (1). The lower border of the pancreas (2) has been lifted up, revealing the splenic vein (24) with the inferior mesenteric (13) running into it. The ureter (29) has the gonadal vessels (10 and 11) in front of it and the genitofemoral nerve (9) behind it, lying on psoas major (22)

Bowel ischaemia. Areas of bowel predisposed to this condition are the 'watershed area' between the superior and inferior mesenteric arterial supply and, following a twisting (volvulus) of the bowel, in the sigmoid colon.

1 Abdominal aorta
2 Body of pancreas
3 Branches of left colic vessels
4 Cut edge of peritoneum
5 Descending colon
6 Duodenojejunal flexure
7 Duodenum: ascending (fourth)
8 Duodenum: horizontal (third)
9 Genitofemoral nerve
10 Gonadal artery
11 Gonadal vein
12 Inferior mesenteric artery
13 Inferior mesenteric vein
14 Left colic artery
15 Left renal artery
16 Left renal vein
17 Lower pole of left kidney
18 Mesentery
19 Middle colic artery
20 Pelvis of kidney
21 Posterior surface of pyloric part of stomach
22 Psoas major
23 Splenic artery
24 Splenic vein
25 Superior mesenteric artery
26 Superior mesenteric vein
27 Suspensory muscle of duodenum (muscle of Treitz)
28 Transverse colon
29 Ureter

- In this specimen (as in D on page 227) the left gonadal (testicular) artery (10) arises from the renal artery (15) and not from the aorta (1).

B Radiograph of the large intestine

In this double-contrast barium enema (barium and air), the sacculations (haustrations, 9) of the various parts of the colon allow it to be distinguished from the narrower terminal ileum (11), which has become partly filled by barium flowing into it through the ileocaecal junction (5)

1 Ascending colon
2 Caecum
3 Descending colon
4 Hip joint
5 Ileocaecal junction
6 Left colic (splenic) flexure
7 Rectum
8 Right colic (hepatic) flexure
9 Sacculations
10 Sigmoid colon
11 Terminal ileum
12 Transverse colon

Colostomy is the creation of a temporary or permanent exit (stoma) for the colon through the anterior abdominal wall, and is commonly performed following a left colectomy or an abdominoperineal excision of the rectum. Faeces are collected in a disposable bag stuck to the anterior abdominal wall.

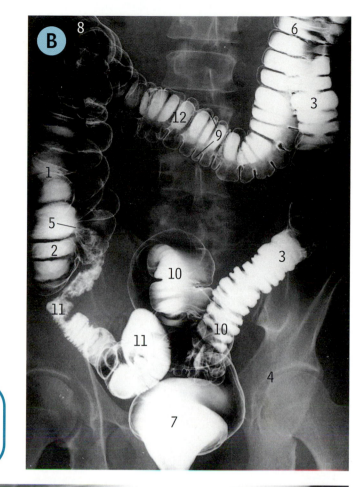

C Small bowel enema via a tube in the duodenum

1 Coils of ileum
2 Coils of jejunum
3 Descending (second) part of duodenum
4 Stomach
5 Valvulae conniventes

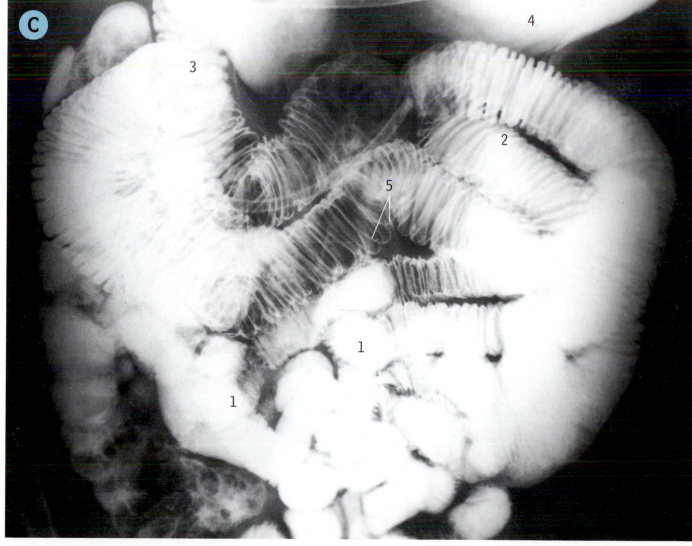

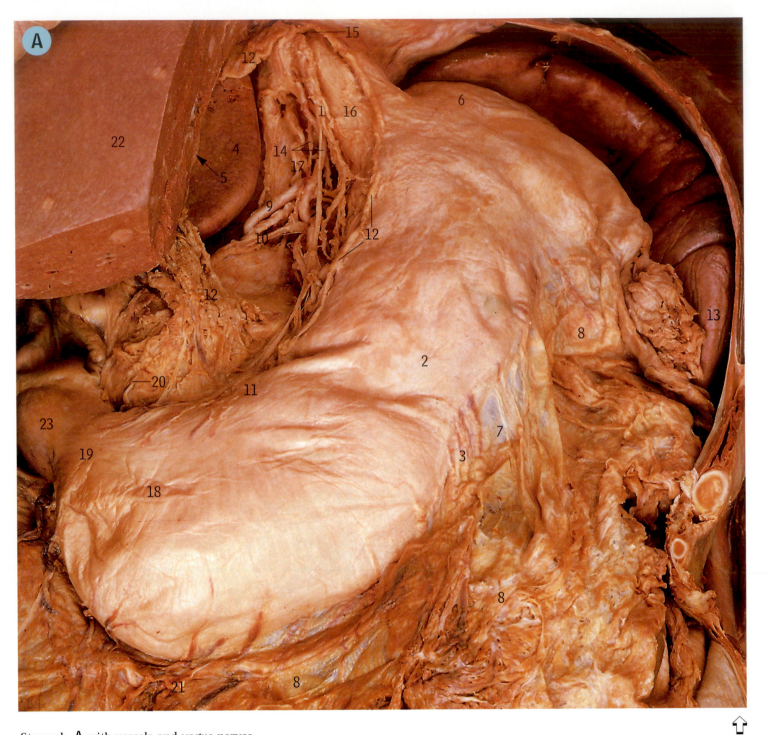

Stomach, A with vessels and vagus nerves, from the front, **B** radiograph after barium meal
The anterior thoracic and abdominal walls and the left lobe of the liver have been removed, with part of the lesser omentum (12), to show the stomach (6, 2, 18 and 19) in its undisturbed position

1 Anterior (left) vagal trunk
2 Body of stomach
3 Branches of left gastro-epiploic vessels
4 Caudate lobe of liver
5 Fissure for ligamentum venosum
6 Fundus of stomach
7 Greater curvature of stomach
8 Greater omentum
9 Left gastric artery
10 Left gastric vein
11 Lesser curvature of stomach
12 Lesser omentum (cut edge)

13 Lower end of spleen
14 Oesophageal branches of left gastric vessels
15 Oesophageal opening in diaphragm
16 Oesophagus
17 Posterior vagal trunk
18 Pyloric antrum
19 Pyloric canal
20 Right gastric artery
21 Right gastro-epiploic vessels and branches
22 Right lobe of liver
23 Superior (first) part of duodenum

Vagotomy is surgical interruption of the vagus nerve performed selectively to reduce acid secretion in the stomach and relieve peptic ulceration. It has been superseded by pharmacological intervention.

Oesophageal varices are dilatations of left gastric vein tributaries that can cause serious bleeding (haematemesis).

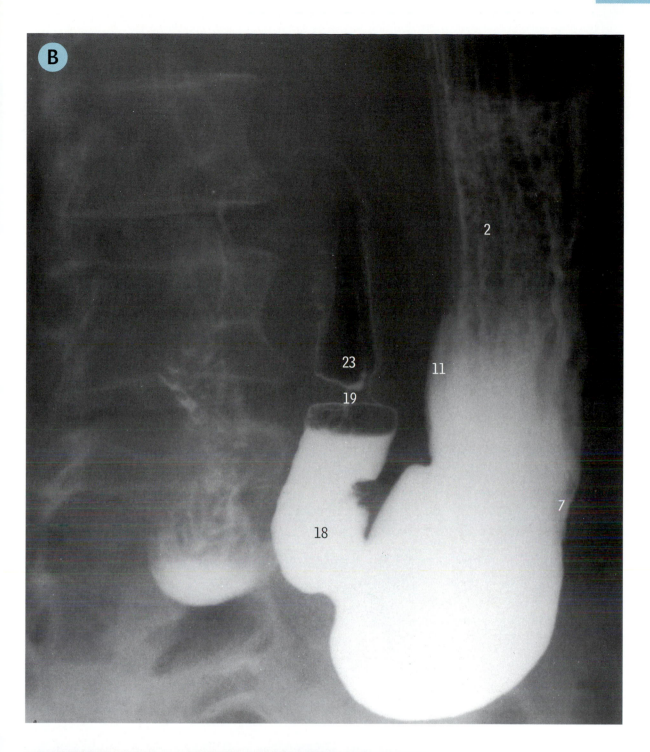

B

2

23

11

19

18

7

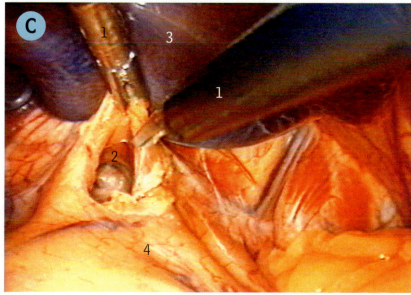

C

1

3

1

2

4

C Laparoscopic view of hiatus hernia

1 Forceps
2 Hiatus hernia
3 Left lobe of liver
4 Stomach

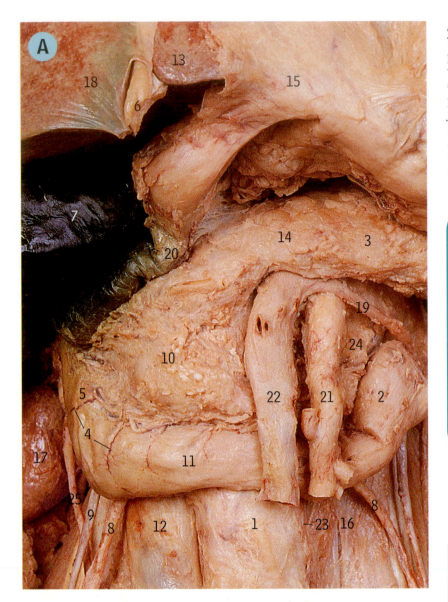

A Duodenum and pancreas

The stomach (15) has been lifted up, the colon and the peritoneum of the posterior abdominal wall removed and branches of the superior mesenteric vessels (21 and 22) cut off. The C-shaped duodenum (20, 5, 11 and 2) is seen embracing the head of the pancreas (10); the neck (14) and body (3) of the pancreas have been displaced slightly upwards to show the splenic vein (19) joining the superior mesenteric vein (22) (to form the portal vein behind the neck of the pancreas). The descending (second) part of the duodenum (5) overlaps the hilum of the right kidney (17). The superior mesenteric artery (21) and vein (22) cross the uncinate process (24) of the head of the pancreas and then the horizontal (third) part of the duodenum (11)

1 Abdominal aorta	13 Left lobe of liver
2 Ascending (fourth) part of duodenum	14 Neck of pancreas
3 Body of pancreas	15 Posterior surface of greater
4 Branches of pancreaticoduodenal	omentum overlying stomach
vessels	16 Psoas major muscle
5 Descending (second) part of	17 Right kidney
duodenum	18 Right lobe of liver
6 Falciform ligament	19 Splenic vein
7 Gall bladder	20 Superior (first) part of duodenum
8 Gonadal artery	21 Superior mesenteric artery
9 Gonadal vein	22 Superior mesenteric vein
10 Head of pancreas	23 Sympathetic trunk
11 Horizontal (third) part of duodenum	24 Uncinate process of head of pancreas
12 Inferior vena cava	25 Ureter

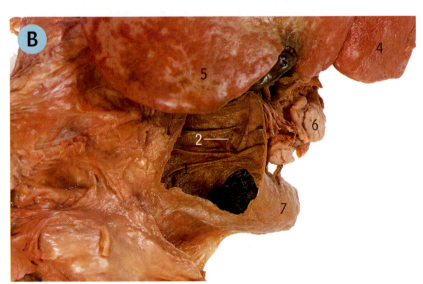

B Duodenal papillae

The anterior wall of the descending (second) part of the duodenum has been removed

1 Circular folds of	5 Liver, right lobe
mucous membrane	6 Pancreas
2 Duodenal papilla	7 Third part of
3 Gall bladder	duodenum
4 Liver, left lobe	

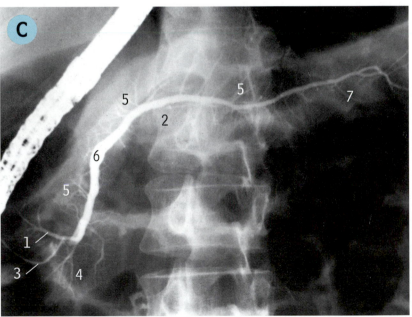

C Endoscopic retrograde cholangiopancreatogram (ERCP)
See page 214 for explanation

1 Accessory pancreatic duct (Santorini)	5 Intralobular ducts
2 Body of pancreas	6 Pancreatic duct (Wirsung)
3 Cannula in ampulla	7 Tail of pancreas
4 Head of pancreas	

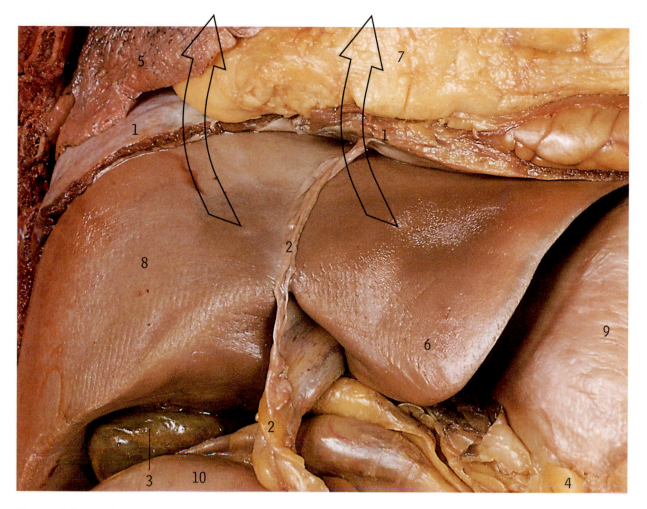

Upper abdominal viscera, from the front
The thoracic and abdominal walls and the anterior part of the diaphragm have been removed to show the undisturbed viscera. The liver (6 and 8) and stomach (9) are immediately below the diaphragm (1). The greater omentum (4) hangs down from the greater curvature (lower margin) of the stomach (9), overlying much of the small and large intestine but leaving some of the transverse colon (10) uncovered. The fundus (tip) of the gall bladder (3) is seen between the right lobe of the liver (8) and transverse colon (10)

Rupture of the liver, commonly caused by trauma, may require removal of a hepatic segment to control bleeding.

1	Diaphragm	6 Left lobe of liver
2	Falciform ligament	7 Pericardial fat
3	Gall bladder	8 Right lobe of liver
4	Greater omentum	9 Stomach
5	Inferior lobe of right lung	10 Transverse colon

• For an explanation of peritoneal structures see the diagrams on page 202.

A Liver, from below and behind

Looking from below and behind with the front edge of the liver lifted, this view shows the posterior and inferior (visceral) surfaces, with no clear demarcation between them. As a general guide, note that the bare area (1) and groove for the inferior vena cava (13) are on the posterior surface, and the fossa for the gall bladder (9) and the structures of the porta hepatis (23, 11, 20 and 5) on the inferior surface. The inferior layer of the coronary ligament is here Z-shaped (at the three key-number 12s); it is normally straight

1 Bare area	11 Hepatic artery	19 Omental tuberosity
2 Caudate lobe	12 Inferior layer of coronary	20 Portal vein
3 Caudate process	ligament	21 Quadrate lobe
4 Colic impression	13 Inferior vena cava	22 Renal impression
5 Common hepatic duct	14 Left lobe	23 Right free margin of lesser
6 Diaphragm	15 Left triangular ligament	omentum in porta hepatis
7 Diaphragm on part of bare area	16 Lesser omentum in fissure for	24 Right lobe
(obstructing view of superior	ligamentum venosum	25 Right triangular ligament
layer of coronary ligament)	17 Ligamentum teres and falciform	26 Suprarenal impression
8 Duodenal impression	ligament in fissure for	
9 Gall bladder	ligamentum teres	
10 Gastric impression	18 Oesophageal groove	

- The caudate (2) and quadrate (21) lobes are classified anatomically as part of the right lobe (24), but functionally they belong to the left lobe (14), since they receive blood from the left branches of the hepatic artery and portal vein, and drain bile to the left hepatic duct.
- The caudate process (3) joins the caudate lobe (2) to the right lobe (24). It is the caudate process (not the caudate lobe) that forms the upper boundary of the epiploic foramen (page 199).
- The posterior surface contains the bare area (1), the groove for the inferior vena cava (13), the caudate lobe (2) and the fissure for the ligamentum venosum (16), the suprarenal impression (26) and most of the right renal impression (22).
- The inferior (visceral) surface contains the porta hepatis where the hepatic artery (11), portal vein (20) and hepatic ducts (5) enter or leave, enclosed within the peritoneum forming the right free margin of the lesser omentum (23). It also contains the quadrate lobe (21), the fossa for the gall bladder (9), the fissure for the ligamentum teres (17), and the gastric (10), duodenal (8) and colic (4) impressions.

Portosystemic anastomoses are connections between the portal venous and systemic venous systems which become clinically important when the portal vein is blocked; they are most commonly seen with liver disease. Varicosities develop in anastomotic regions, especially the oesophagus, anus and bare area of the liver.

Liver abscess. In tropical countries these are often very large and due to amoebic disease. In other areas they are more commonly related to malignancy.

B Cast of the liver, extrahepatic biliary tract and associated vessels, from behind
Yellow = gall bladder and biliary tract
Red = hepatic artery and branches
Light blue = portal vein and tributaries
Dark blue = inferior vena cava, hepatic veins and tributaries
This view, like that of A opposite, shows the inferior and posterior surfaces, as when looking into the abdomen from below with the lower border of the liver pushed up towards the thorax

1 Bile duct	14 Left gastric vein
2 Body of gall bladder	15 Left hepatic duct
3 Caudate lobe	16 Left hepatic vein
4 Caudate process	17 Left lobe
5 Common hepatic duct	18 Neck of gall bladder
6 Cystic artery and veins	19 Portal vein
7 Cystic duct	20 Quadrate lobe
8 Fissure for ligamentum teres	21 Right branch of hepatic artery overlying right branch
9 Fissure for ligamentum venosum	of portal vein
10 Fundus of gall bladder	22 Right gastric vein
11 Hepatic artery	23 Right lobe
12 Inferior vena cava	
13 Left branch of hepatic artery overlying left branch of portal vein	

- The hepatic artery (11) divides like a Y into left (13) and right (21) branches.
- The portal vein (19) divides like a T into left (13) and right (21) branches.
- The common hepatic duct (5) is formed by the union of the left (15) and right (obscured) hepatic ducts, and is joined by the cystic duct (7) to form the bile duct (1).

A Endoscopic retrograde cholangiopancreatogram (ERCP)

In ERCP an endoscope is passed through the mouth, pharynx, oesophagus and stomach into the duodenum, and through it a cannula is introduced into the major duodenal papilla (page 210) and bile duct so that contrast medium can be injected up the biliary tract. (The pancreatic duct can also be cannulated in this way – see C on page 210)

1 Bile duct
2 Common hepatic duct
3 Cystic duct
4 Gall bladder
5 Left hepatic duct
6 Liver shadow and tributaries of hepatic ducts
7 Right hepatic duct

Cholecystitis is acute inflammation of the gall bladder most commonly associated with stones in the biliary system. Pain occurs over the right hypochondrial region near the tip of the ninth rib and the linea semilunaris.

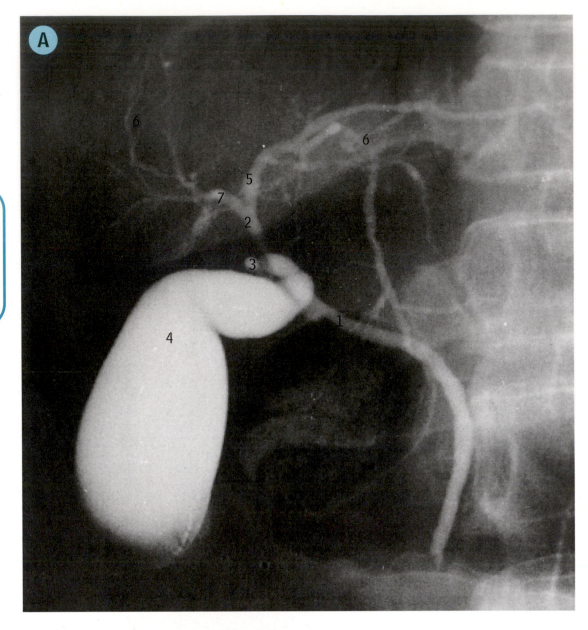

B Ultrasound scan of the gall bladder

To an untrained observer, ultrasound scans are difficult to interpret, but here the gall bladder can be distinguished as a sausage-shaped cavity (3)

1 Cystic duct
2 Fundus of gall bladder
3 Gall bladder
4 Inferior vena cava
5 Portal vein
6 Right dome of diaphragm
7 Right hepatic artery
8 Right lobe of liver
9 Right renal artery

• Ultrasound scans are best interpreted by the operator on a screen.

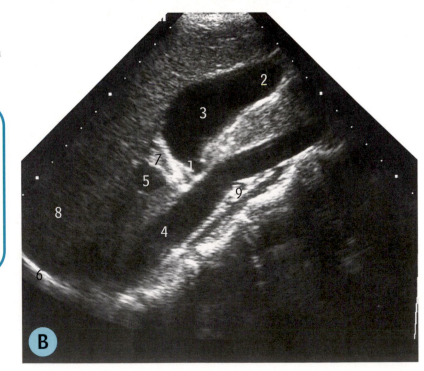

Cast of the portal vein and tributaries, and the mesenteric vessels, from behind
Yellow = biliary tract and pancreatic ducts
Red = arteries
Blue = portal venous system
In this posterior view (chosen in preference to the anterior view, where the many very small vessels to the intestines would have obscured the larger branches), the superior mesenteric vein (22) is seen continuing upwards to become the portal vein (14) after it has been joined by the splenic vein (20). In the porta hepatis the portal vein divides into the left and right branches (8 and 16). Owing to removal of the aorta, the upper part of the inferior mesenteric artery (5) has become displaced slightly to the right and appears to have given origin to the ileocolic artery (4), but this is simply an overlap of the vessels; the origin of the ileocolic from the superior mesenteric is not seen in this view

1   Bile duct
2   Branches of middle colic vessels
3   Coeliac trunk
4   Ileocolic vessels
5   Inferior mesenteric artery
6   Inferior mesenteric vein
7   Left branch of hepatic artery
8   Left branch of portal vein
9   Left colic vessels
10 Left gastric artery and vein
11 Pancreatic duct
12 Pancreatic ducts in head of pancreas
13 Pancreaticoduodenal vessels
14 Portal vein
15 Right branch of hepatic artery
16 Right branch of portal vein
17 Right colic vessels
18 Sigmoid vessels
19 Splenic artery
20 Splenic vein
21 Superior mesenteric artery
22 Superior mesenteric vein

- The inferior mesenteric vein (6) normally drains into the splenic vein (20) behind the body of the pancreas, but it may join the splenic vein nearer the union with the superior mesenteric vein or (as in this specimen) enter the superior mesenteric vein itself (22).
- The colic arteries (2, 17, 4, 9) anastomose with one another near the colonic wall forming what is often called the marginal artery (as at the arrow).

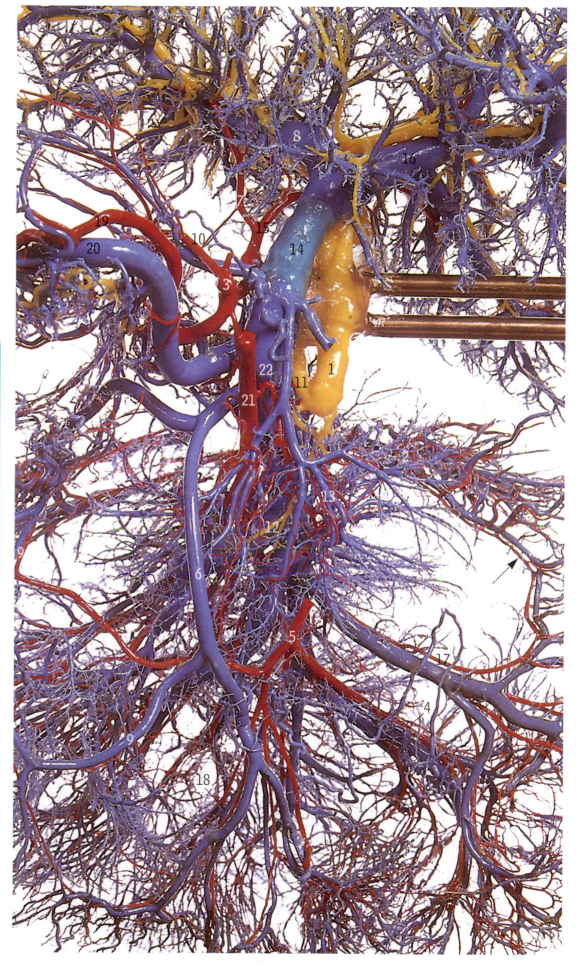

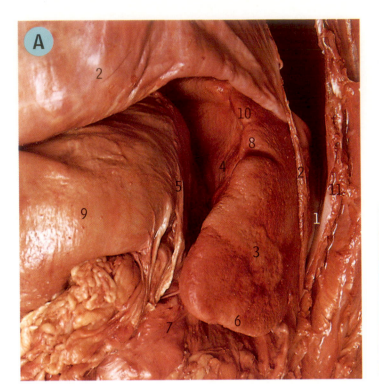

A Spleen, from the front
The left upper anterior abdominal and lower anterior thoracic walls have been removed and part of the diaphragm (2) turned upwards to show the spleen in its normal position, lying adjacent to the stomach (9) and colon (7), with the lower part against the kidney (D16 and 9, opposite)

1 Costodiaphragmatic recess
2 Diaphragm
3 Diaphragmatic surface
4 Gastric impression
5 Gastrosplenic ligament
6 Inferior border
7 Left colic flexure
8 Notch
9 Stomach
10 Superior border
11 Thoracic wall

Splenectomy. Removal of the spleen may be indicated following trauma or in cases of certain blood disorders and is easily accomplished after clamping its pedicle, taking care not to cut through the tail of the pancreas, which lies within the hilum of the spleen.

- The gastrosplenic ligament contains the short gastric and left gastro-epiploic branches of the splenic vessels.
- The lienorenal ligament contains the tail of the pancreas and the splenic vessels.

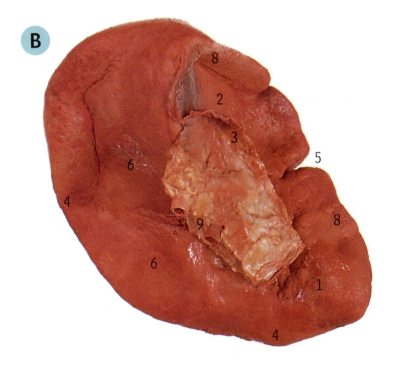

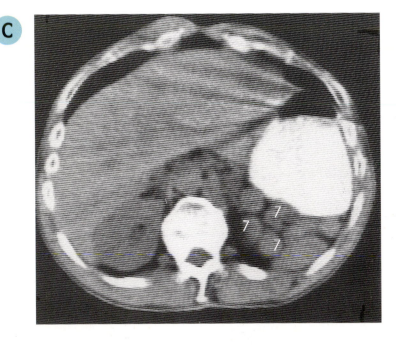

Spleen, **B** visceral surface, **C** CT scan (polysplenia)
In B the spleen has been removed and its visceral or medial surface is shown, with a small part of the gastrosplenic (3) and lienorenal (9) ligaments remaining attached. The scan of the upper abdomen shows a developmental anomaly—several small splenunculi (7) instead of a single organ

Splenomegaly. The most common causes of splenic enlargement are tropical diseases and blood diseases such as haemolytic anaemias. The normal spleen is the size of a clenched fist and is not palpable below the left costal margin.

1 Colic impression
2 Gastric impression
3 Gastrosplenic ligament containing short gastric and left gastro-epiploic vessels
4 Inferior border
5 Notch
6 Renal impression
7 Spleen—multiple splenunculi
8 Superior border
9 Tail of pancreas and splenic vessels in lienorenal ligament

D Spleen, in a transverse section of the left upper abdomen
The section is at the level of the disc (7) between the twelfth thoracic and first lumbar vertebrae, and is viewed from below looking towards the thorax

1 Abdominal aorta
2 Anterior layer of lienorenal ligament
3 Coeliac trunk
4 Costodiaphragmatic recess of pleura
5 Diaphragm
6 Gastrosplenic ligament
7 Intervertebral disc
8 Left gastric artery
9 Left kidney
10 Left lobe of liver
11 Left suprarenal gland
12 Lesser sac
13 Ninth rib
14 Peritoneum of greater sac
15 Posterior layer of lienorenal ligament
16 Spleen
17 Splenic artery
18 Splenic vein
19 Stomach
20 Tail of pancreas
21 Tenth rib

E Interior of the caecum
This is a median sagittal section of the pelvis, right side viewed from the left. The anterior wall has been cut open and reflected to show the lips of the ileocaecal valve (7)

1 Ascending colon
2 Bladder
3 Caecum
4 Cauda equina
5 Coccyx
6 Fibroid in uterine fundus
7 Lips of ileocaecal valve
8 Mesentery of small intestine
9 Pubic symphysis
10 Recto-uterine pouch (of Douglas)
11 Rectum
12 Sacral promontory
13 Sigmoid colon
14 Thecal sac termination
15 Uterine cavity
16 Valvulae conniventes
17 Vesico-uterine pouch

• The three taeniae coli of the ascending colon and caecum converge on the base of the appendix.

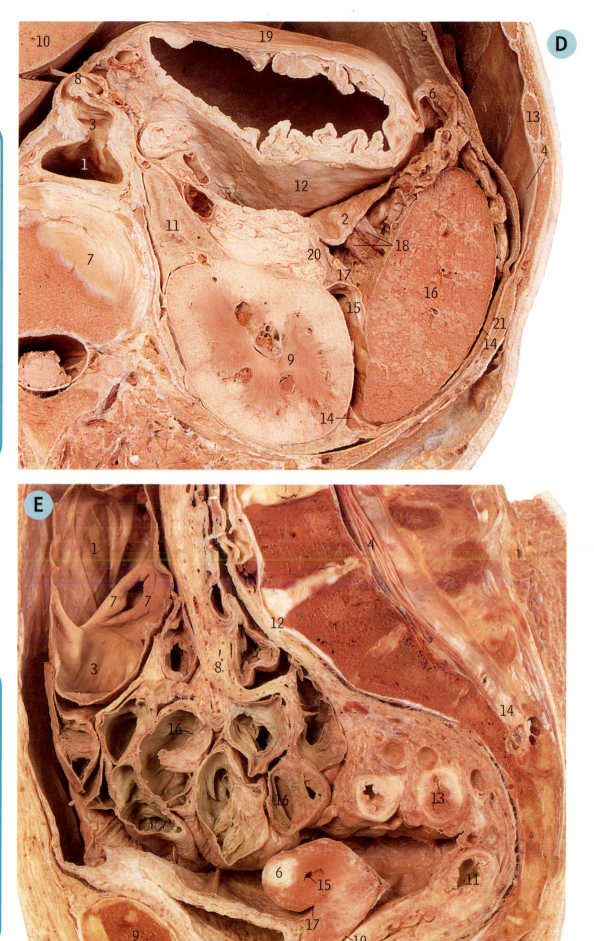

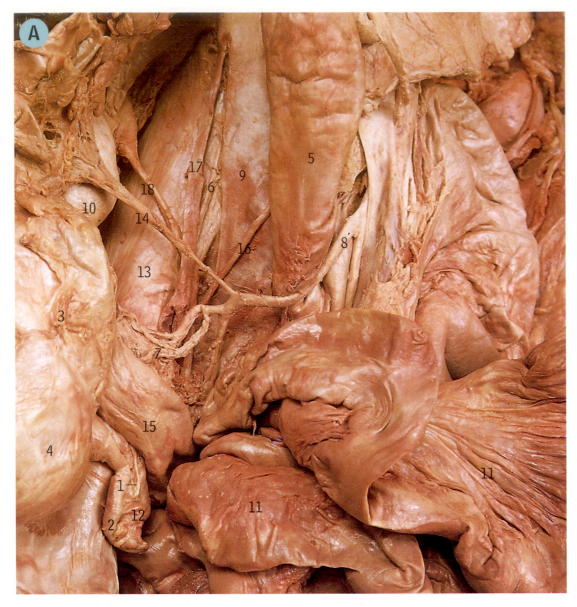

A Appendix, ileocolic artery and related structures, from the front
Most of the peritoneum of the mesentery and posterior abdominal wall have been removed, and coils of small intestine (11) have been displaced to the right of the picture, to show the ileocolic artery (8), terminal ileum (15) and appendix (2) with its appendicular artery (1)

1	Appendicular artery in mesoappendix
2	Appendix
3	Ascending colon
4	Caecum
5	Descending (second) part of duodenum
6	Genitofemoral nerve
7	Ileal and caecal vessels
8	Ileocolic artery
9	Inferior vena cava
10	Lower pole of kidney
11	Mesentery and coils of jejunum and ileum
12	Mesoappendix
13	Psoas major
14	Right colic artery
15	Terminal part of ileum
16	Testicular artery
17	Testicular vein
18	Ureter

- The appendix gets its blood supply from the appendicular artery (1), normally a branch of one of the caecal arteries (7), usually the posterior caecal. The vessel is not at first closely applied to the appendix but approaches it through the mesoappendix (12), the peritoneal fold continuous with the lower part of the mesentery of the terminal ileum (15). If this arterial supply becomes obstructed, the appendix becomes necrotic, as there is no collateral circulation.

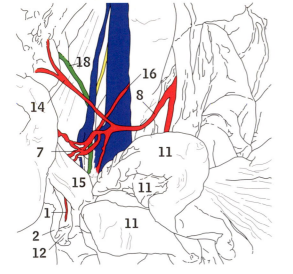

Appendicitis. Classically this condition starts as a para-umbilical pain (visceral peritoneum of midgut) which then moves to the region of McBurney's point in the right iliac fossa owing to parietal peritoneal irritation.

B Caecum and appendix, from the front
The terminal ileum (9) is seen joining the large intestine at the junction of the caecum (4) and ascending colon (2), and the appendix (3) joins the caecum just below the ileocaecal junction

1	Anterior taenia coli
2	Ascending colon
3	Base of appendix
4	Caecum
5	Inferior ileocaecal recess
6	Peritoneum overlying external iliac vessels
7	Retrocaecal recess
8	Superior ileocaecal recess
9	Terminal ileum
10	Tip of appendix

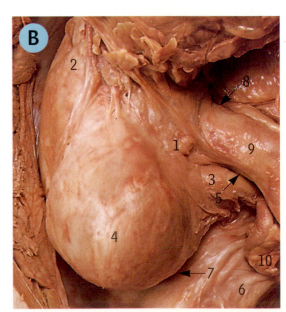

C

D

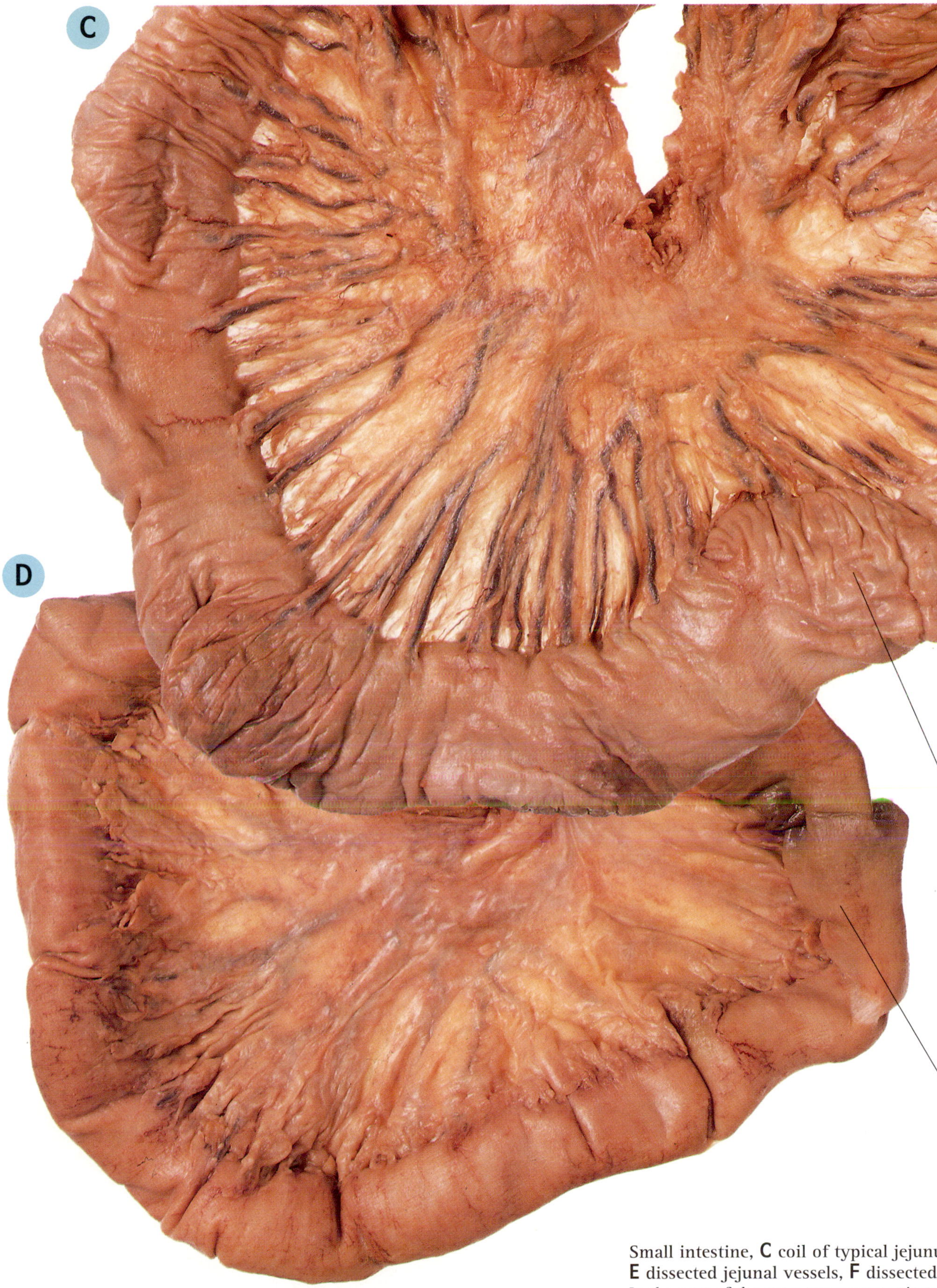

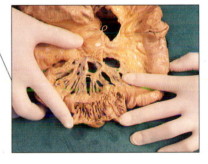

E

F

Small intestine, **C** coil of typical jejunum, **D** coil of typical ileum,
E dissected jejunal vessels, **F** dissected ileal vessels

In the part of the mesentery supporting the jejunum in C, the vessels
anastomose to form one or perhaps two vascular arcades (E) which give off
long straight branches that run to the intestinal wall. The fat in the
mesentery tends to be concentrated near the root, leaving areas or 'windows'
near the gut wall that are devoid of fat. In the mesentery supporting the
ileum in D, the vessels form several arcades with shorter branches (F), and
there are no fat-free areas. The jejunal wall (C) is thicker than that of the
ileum (D) and has a larger lumen. The jejunum also feels thicker, because the
folds of its mucous membrane are more numerous than in the ileum

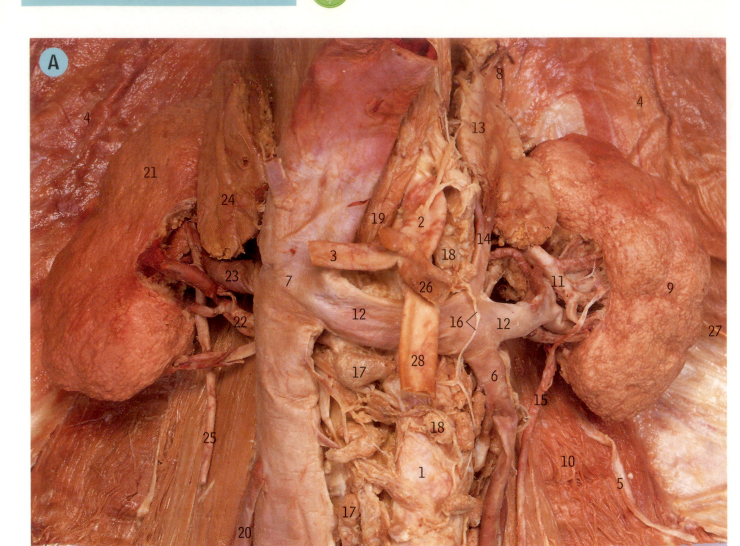

A

Kidneys and suprarenal glands, A dissection, B axial MR image

The kidneys (9 and 21) and suprarenal glands (13 and 24) are displayed on the posterior abdominal wall after the removal of all other viscera. The left renal vein (12) receives the left suprarenal (14) and gonadal (6) vein and then passes over the aorta (1) and deep to the superior mesenteric artery (28) to reach the inferior vena cava (7). In the hilum of the right kidney (21) a large branch of the renal artery (22) passes in front of the renal vein (23). The origins of the renal arteries from the aorta are not seen because they underlie the left renal vein (12) and inferior vena cava (7). The MR image in B passes through the kidneys but is too low to show the suprarenal glands

Aortic aneurysm is a 'ballooning' of either the thoracic or abdominal aorta, occasionally with dissection or rupture causing sudden death. On a plain X-ray an aneurysm may be seen as an enlargement or unfolding of the aortic shadow.

Aortic bruits are audible rhythmic sounds on auscultation of the abdomen, often due to atherosclerotic narrowing (stenosis) of the aorta.

1	Abdominal aorta and aortic plexus	15	Left ureter
2	Coeliac trunk	16	Lymphatic vessels
3	Common hepatic artery	17	Para-aortic lymph nodes
4	Diaphragm	18	Pre-aortic lymph nodes
5	First lumbar spinal nerve	19	Right crus of diaphragm
6	Gonadal vein, left	20	Right gonadal vein
7	Inferior vena cava	21	Right kidney
8	Left inferior phrenic vessels	22	Right renal artery
9	Left kidney	23	Right renal vein
10	Left psoas major	24	Right suprarenal gland
11	Left renal artery	25	Right ureter
12	Left renal vein	26	Splenic artery
13	Left suprarenal gland	27	Subcostal nerve
14	Left suprarenal vein	28	Superior mesenteric artery

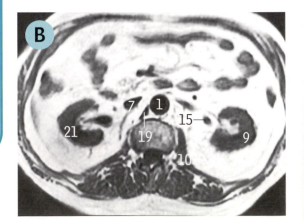

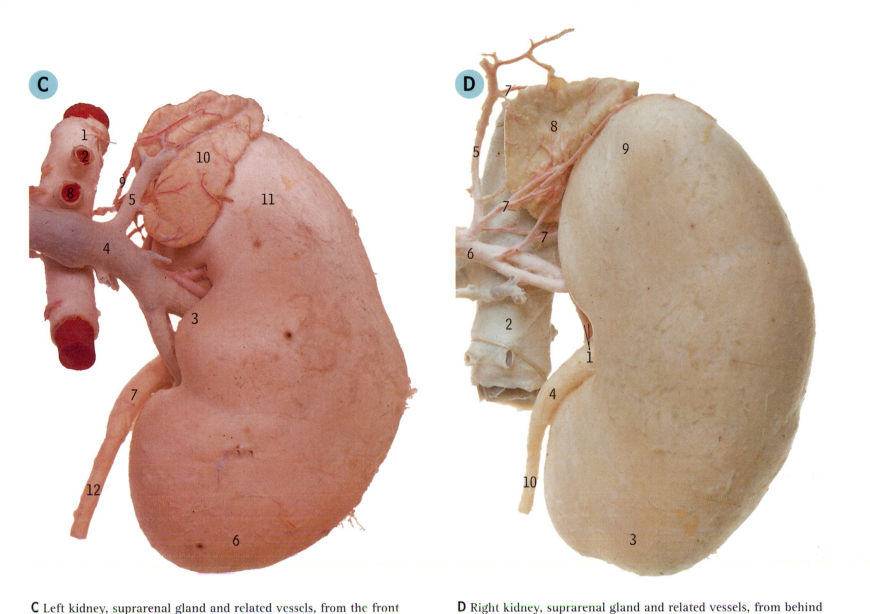

C Left kidney, suprarenal gland and related vessels, from the front
The vessels have been distended by injection of resin, and all fascia has been removed, but the suprarenal gland (10) has been retained in its normal position, lying against the medial side of the upper pole of the kidney (11)

D Right kidney, suprarenal gland and related vessels, from behind
Similar to B, but note that this is the right kidney from behind, not the left; the hilum of each kidney faces medially

1	Abdominal aorta	7	Pelvis of kidney
2	Coeliac trunk	8	Superior mesenteric artery
3	Hilum of kidney	9	Suprarenal arteries
4	Left renal vein overlying renal artery	10	Suprarenal gland
5	Left suprarenal vein	11	Upper pole of kidney
6	Lower pole of kidney	12	Ureter

1	Hilum of kidney	7	Suprarenal arteries
2	Inferior vena cava	8	Suprarenal gland
3	Lower pole of kidney	9	Upper pole of kidney
4	Pelvis of kidney	10	Ureter
5	Right inferior phrenic artery		
6	Right renal artery		

- The ureter (C12, D10) is the constricted downward continuation of the pelvis of the kidney (C7, D4). Note that the correct term is pelvis of the kidney or renal pelvis, not pelvis of the ureter.
- In the hilum of the kidney, the order of the principal structures from front to back is usually remembered as vein, artery, ureter (strictly speaking, pelvis – see note above), although small branches of the vessels may sometimes get out of order. Compare with vein, artery, bronchus in the hilum of the lung (page 166).

- Each suprarenal gland receives arteries from three sources – the inferior phrenic artery, the aorta and the renal artery – but there are not just three arteries: there are several from each source, perhaps up to a total of 20, and only some of the larger ones are shown (as at D7).
- There is usually only one suprarenal vein on each side. On the left (C5) it drains into the renal vein (C4); on the right it is very short and runs directly into the inferior vena cava (in D it is hidden by the gland itself, but is shown in the cast on page 223, D10).
- For details of the renal arteries see pages 222 and 223.

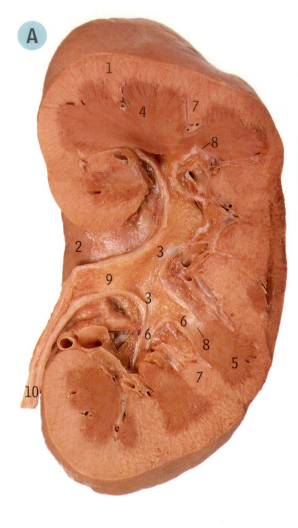

A

A Kidney. Internal structure in longitudinal section
The section is through the centre of the kidney and has included the renal pelvis (9) and beginning of the ureter (10). The major vessels in the hilum (2) have been removed

1	Cortex	6	Minor calyx
2	Hilum	7	Renal column
3	Major calyx	8	Renal papilla
4	Medulla	9	Renal pelvis
5	Medullary pyramid	10	Ureter

- The renal medulla (4) is made up of the medullary pyramids (5), the apices of which form the renal papillae (8) which project into the minor calyces (6).
- The renal columns (7) are the parts of the cortex that intervene between pyramids (5).
- Several minor calyces (6), which receive urine discharged into them from the collecting ducts that open on the renal papillae (8), unite to form a major calyx (3).
- The two or three major calyces (3) unite to form the renal pelvis (9) which passes out through the hilum (2) to become the ureter (10), often with a slight constriction at the junction.
- The hilum is the slit-like space on the medial surface of the kidney where the vessels and renal pelvis enter or leave.

B Cast of the right kidney, from the front
Red = renal artery
Yellow = urinary tract
The posterior division (8) of the renal artery (9) here passes behind the pelvis (7) and upper calyx (upper 5), but all other vessels are in front of the urinary tract; hence this is a right kidney seen from the front (vein, artery, ureter from front to back, and the hilum on the medial side—see page 221), not a left kidney from behind

1	Anterior division	7	Pelvis of kidney
2	Anterior inferior segment artery	8	Posterior division (forming
3	Anterior superior segment artery		posterior segment artery)
	(double)	9	Renal artery
4	Inferior segment artery	10	Superior segment artery
5	Major calyx	11	Ureter
6	Minor calyx		

- The kidney has five arterial segments, named posterior, superior, anterior superior, anterior inferior and inferior. Typically the renal artery (9) divides into anterior (1) and posterior (8) divisions; the posterior supplies the posterior segment and the anterior supplies the remainder. However, the pattern of branching displays many variations.
- This specimen shows a fairly typical pattern, although the superior segment (10) obtains a small additional branch from the posterior division (8), and the anterior superior segment receives two major branches (3).

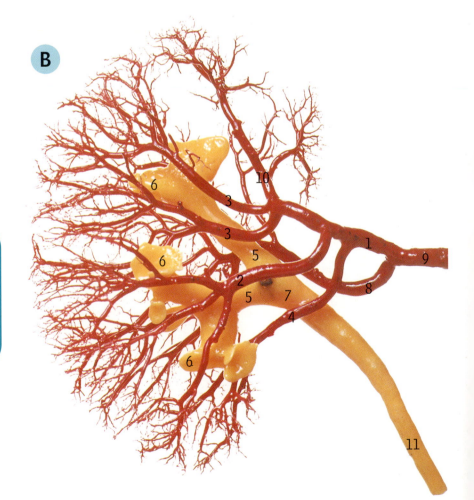

B

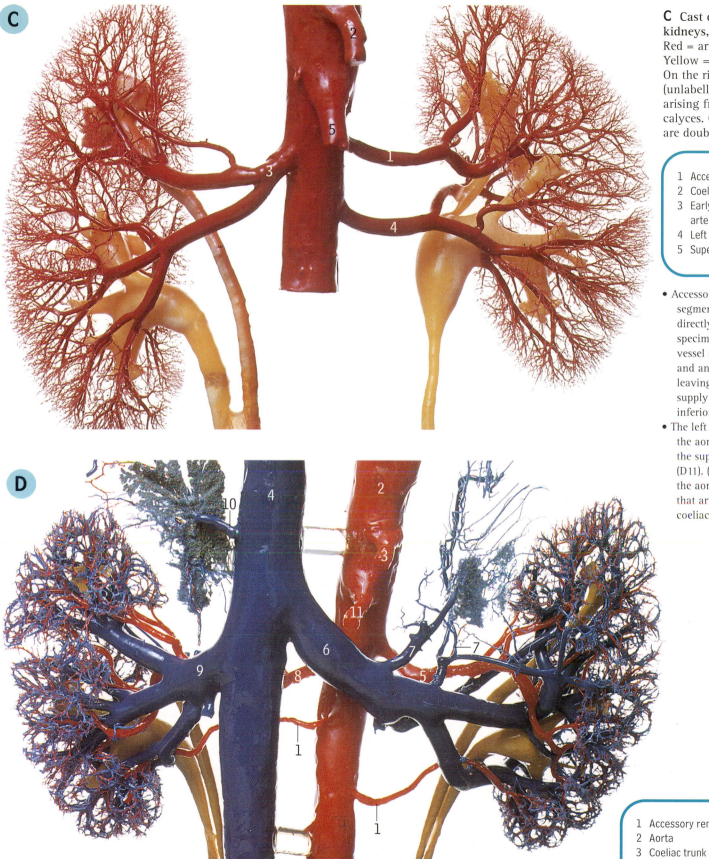

C

D

Red = arteries
Yellow = urinary tracts
On the right side the ureters
(unlabelled) are double, each
arising from a seperate set of
calyces. On the left the arteries
are double (1 and 4)

1 Accessory left renal artery
2 Coeliac trunk
3 Early branching of right renal
artery
4 Left renal artery
5 Superior mesenteric artery

- Accessory renal arteries represent
segmental vessels that arise
directly from the aorta. In this
specimen, the left accessory
vessel (C1) supplies the superior
and anterior superior segments,
leaving the 'normal' vessel to
supply the posterior, anterior
inferior and inferior segments.
- The left renal vein (D6) crosses
the aorta below the origin of
the superior mesenteric artery
(D11). (The splenic vein crosses
the aorta above the origin of
that artery and below the
coeliac trunk, D3.)

D Cast of the kidneys and great vessels, from the front
Red = arteries
Blue = veins
Yellow = urinary tracts
Here both kidneys show double ureters (unlabelled), and there are accessory renal arteries (1) to the
lower poles of both kidneys. The suprarenal glands (also unlabelled) are outlined by their venous
patterns, and the short right suprarenal vein (10) is shown draining directly to the inferior vena
cava (4). On the left there are two suprarenal veins (7), both draining to the left renal vein (6). See
also page 224, A14, A9, A12

1 Accessory renal arteries
2 Aorta
3 Coeliac trunk
4 Inferior vena cava
5 Left renal artery
6 Left renal vein
7 Left suprarenal veins
8 Right renal artery
9 Right renal vein
10 Right suprarenal vein
11 Superior mesenteric artery

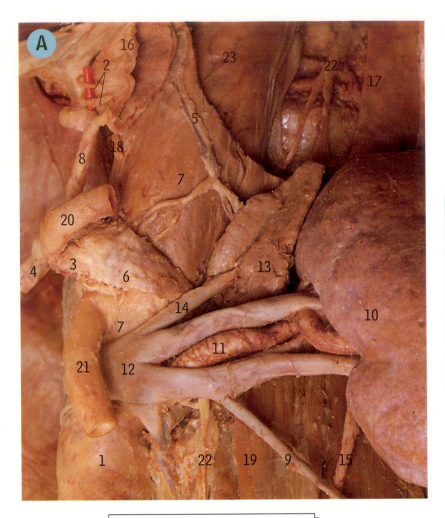

A Left kidney and suprarenal gland, from the front

The left kidney (10) and suprarenal gland (13) are seen on the posterior abdominal wall. Much of the diaphragm has been removed but the oesophageal opening remains, with the end of the oesophagus (16) opening out into the cardiac part of the stomach and a (double) anterior vagal trunk (2) overlying the red marker. The posterior vagal trunk (18) is behind and to the right of the oesophagus. Part of the pleura has been cut away (17) to show the sympathetic trunk (22) on the side of the lower thoracic vertebrae. The left coeliac ganglion and the coeliac plexus (6) are at the root of the coeliac trunk (3)

1	Abdominal aorta	12	Left renal vein
2	Anterior vagal trunk (double, over marker)	13	Left suprarenal gland
		14	Left suprarenal vein
3	Coeliac trunk	15	Left ureter
4	Common hepatic artery	16	Lower end of oesophagus
5	Inferior phrenic vessels	17	Pleura (cut edge)
6	Left coeliac ganglion and coeliac plexus	18	Posterior vagal trunk
		19	Psoas major
7	Left crus of diaphragm	20	Splenic artery
8	Left gastric artery	21	Superior mesenteric artery
9	Left gonadal vein	22	Sympathetic trunk
10	Left kidney	23	Thoracic aorta
11	Left renal artery		

Nephrectomy is the surgical removal of a kidney (for malignancy or polycystic disease) and takes advantage of the renal fasciae for access to, and closure of, the site.

Superior mesenteric artery syndrome causes increased pressure in the left renal vein and potentially renal disease by reducing the lumen of this vein as it is sandwiched between the aorta and the superior mesenteric artery (nutcracker effect).

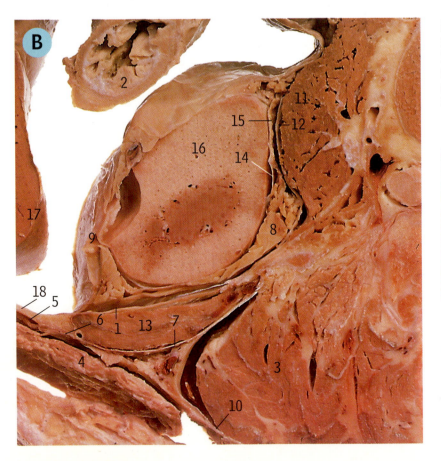

B Right kidney and renal fascia in transverse section from below

In the transverse section of the lower part of the right kidney (B16), seen from below looking towards the thorax, the renal fascia (15) has been dissected out from the perirenal fat (8) and the kidney's own capsule (14). (There was a small cyst on the surface of this kidney.) The section also displays the three layers (10, 7 and 1) of the lumbar fascia (6; see notes on page 94).

1	Anterior layer of lumbar fascia	10	Posterior layer of lumbar fascia
2	Coil of small intestine	11	Psoas major
3	Erector spinae	12	Psoas sheath
4	External oblique	13	Quadratus lumborum
5	Internal oblique	14	Renal capsule
6	Lumbar fascia	15	Renal fascia
7	Middle layer of lumbar fascia	16	Right kidney
8	Perirenal fat	17	Right lobe of liver
9	Peritoneum	18	Transversus abdominis

- Outside the kidney's own capsule (renal capsule, 14), there is variable amount of fat (perirenal fat, 8) and outside this is a condensation of connective tissue forming the renal fascia (15).

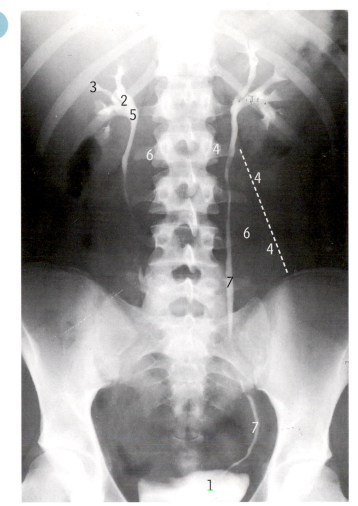

C Intravenous urogram (IVU)
Contrast medium injected intravenously is excreted by the kidneys to outline the calyces (3 and 2), renal pelvis (5) and the ureters (7) which enter the bladder (1) in the pelvis

1	Bladder	5	Renal pelvis
2	Major calyx	6	Transverse processes of lumbar
3	Minor calyx		vertebrae
4	Psoas shadow	7	Ureter

- Radiologically the ureters normally lie near the tips of the transverse processes of the lumbar vertebrae.

Ureteric calculi. Stones within the ureter descend from the kidney towards the bladder and may lodge at the pelvi-ureteric junction, the brim of the pelvis, or at the entrance to the bladder where the ureter traverses the bladder wall, causing excruciating pain.

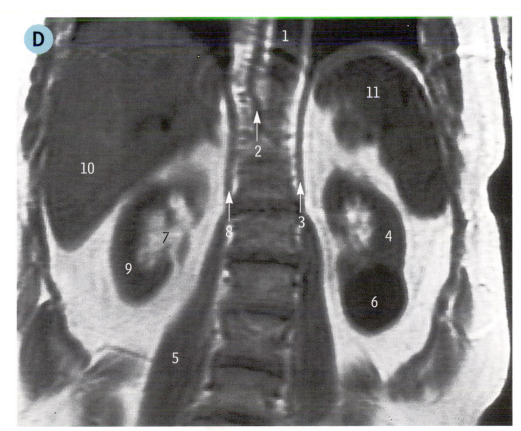

D Upper abdomen, coronal MR image

1	Aorta	7	Renal pelvis
2	Azygos vein	8	Right crus of diaphragm
3	Left crus of diaphragm	9	Right kidney
4	Left kidney	10	Right lobe of liver
5	Psoas major muscle	11	Spleen
6	Renal cyst		

Abdominal aortic aneurysm. This ballooning of the lower abdominal aorta may extend distally to involve both external iliac arteries and proximally as far as the renal arteries causing renal insufficiency. Treatment is surgical replacement of this section of the aorta with a graft or stent.

A Surface markings of the kidneys, from behind
The upper pole of the left kidney rises to the level of the eleventh rib, but the right kidney is slightly lower (due to the bulk of the liver on the right). The hilum of each kidney is 5 cm (2 in) from the midline. The lower edge of the costodiaphragmatic recess of the pleura crosses the twelfth rib; compare with the dissection below (B6)

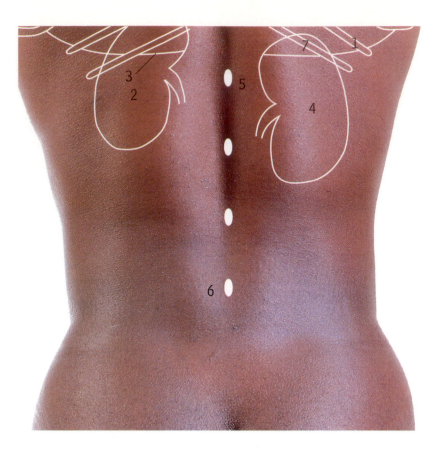

1 Eleventh rib
2 Left kidney
3 Lower edge of pleura
4 Right kidney
5 Spinous process of first lumbar vertebra
6 Spinous process of fourth lumbar vertebra
7 Twelfth rib

Renal biopsy is a procedure that is best performed only at the lower pole of the kidney because upper pole biopsies may damage the pleura (an immediate posterior relation of the upper part of the kidney) and cause a pneumothorax.

B Right kidney, from behind
Most thoracic and abdominal muscles have been removed to show the three nerves (9, 3 and 4) that lie behind the kidney (5). Much more important is the relationship of the upper part of the kidney to the pleura. A window has been cut in the parietal pleura above the twelfth rib (12) to open into the costodiaphragmatic recess (1), whose lower limit (6) runs transversely behind the kidney and in front of the obliquely placed twelfth rib

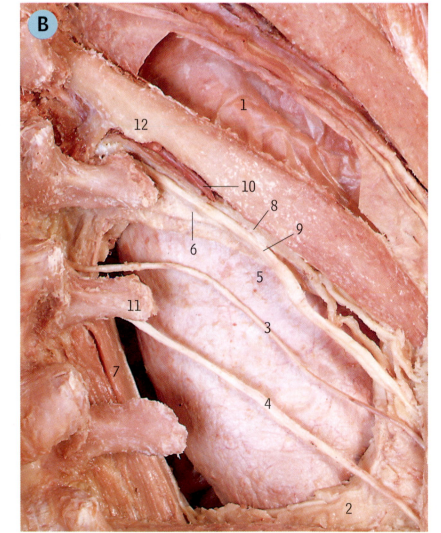

1 Costodiaphragmatic recess of pleura
2 Extraperitoneal tissue
3 Iliohypogastric nerve
4 Ilio-inguinal nerve
5 Kidney
6 Lower edge of pleura
7 Psoas major
8 Subcostal artery
9 Subcostal nerve
10 Subcostal vein
11 Transverse process of second lumbar vertebra
12 Twelfth rib

A Diaphragm from below

1 Aorta	11 Inferior vena caval opening
2 Azygos vein	12 Left crus
3 Cauda equina	13 Lumbar fascia
4 Central tendon of diaphragm	14 Median arcuate ligament
5 Costal margin	15 Oesophageal opening
6 Diaphragm	16 Psoas major
7 Erector spinae muscles	17 Quadratus lumborum
8 First lumbar intervertebral disc	18 Right crus
9 Hemi-azygos vein	19 Spinal cord
10 Inferior phrenic vessels	

- The right crus of the diaphragm (A18) has a more extensive origin (from the upper three lumbar vertebrae and intervening discs) than the left (A12) (from the upper two) because of the greater bulk of the liver on the right; the crura help to pull the liver downwards when the diaphragm contracts.
- Fibres of the right crus (A18) form the right and left boundaries of the oesophageal opening (A15).

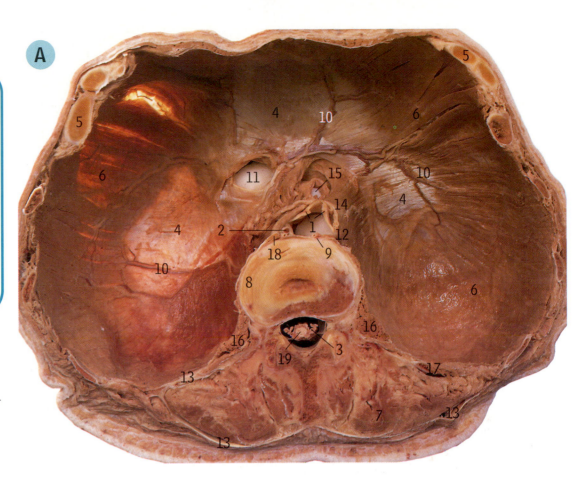

B Posterior abdominal wall, left side

The structures on the posterior abdominal wall are viewed from the front. The body of the pancreas (2) has been turned upwards to expose the splenic vein (21). The suprarenal gland (23) appears detached from the superior pole of the kidney (compared with A13 and 10, page 224)

1 Aorta and aortic plexus	14 Ovarian vein
2 Body of pancreas	15 Para-aortic lymph node
3 First lumbar spinal nerve	16 Psoas major
4 Greater omentum	17 Quadratus lumborum
5 Hypogastric plexus	18 Renal artery
6 Ilio-inguinal nerve	19 Renal vein
7 Iliohypogastric nerve	20 Spleen
8 Inferior mesenteric vein	21 Splenic vein
9 Inferior vena cava	22 Stomach
10 Left colic vein	23 Suprarenal gland
11 Liver	24 Suprarenal vein
12 Lower pole of kidney	25 Transversus abdominis
13 Lumbar part of thoracolumbar fascia	26 Ureter

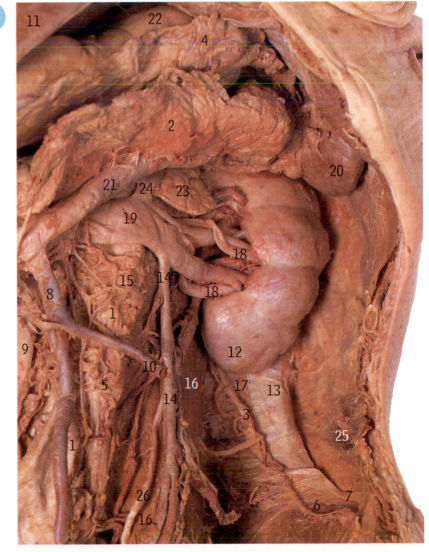

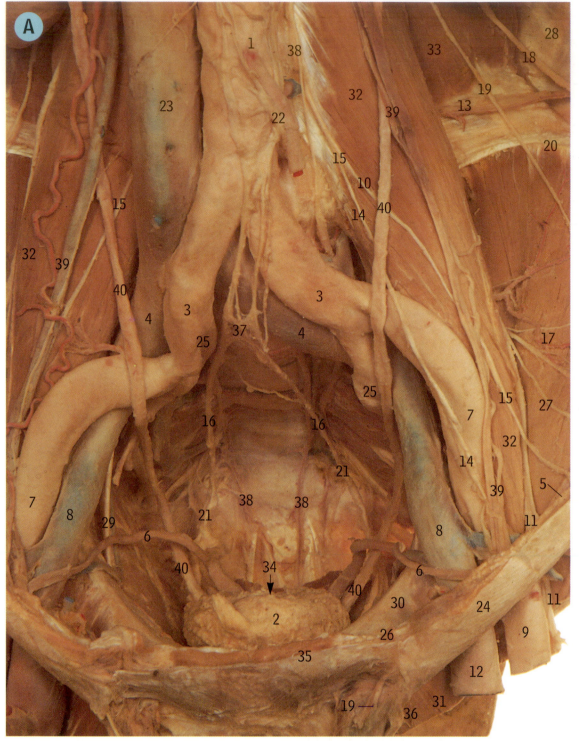

1 Aorta and aortic plexus
2 Bladder
3 Common iliac artery
4 Common iliac vein
5 Deep circumflex iliac artery
6 Ductus deferens
7 External iliac artery
8 External iliac vein
9 Femoral artery
10 Femoral branch of genitofemoral nerve
11 Femoral nerve
12 Femoral vein
13 Fourth lumbar artery
14 Genital branch of genitofemoral nerve
15 Genitofemoral nerve
16 Hypogastric nerve
17 Iliacus and branches from femoral nerve and
   iliolumbar artery
18 Iliohypogastric nerve
19 Ilio-inguinal nerve
20 Iliolumbar ligament
21 Inferior hypogastric (pelvic) plexus and pelvic
   splanchnic nerves
22 Inferior mesenteric artery and plexus
23 Inferior vena cava
24 Inguinal ligament
25 Internal iliac artery
26 Lacunar ligament
27 Lateral femoral cutaneous nerve arising from
   femoral nerve
28 Lumbar part of thoracolumbar fascia
29 Obturator nerve and vessels
30 Pectineal ligament
31 Position of femoral canal
32 Psoas major
33 Quadratus lumborum
34 Rectum (cut edge)
35 Rectus abdominis
36 Spermatic cord
37 Superior hypogastric plexus
38 Sympathetic trunk and ganglia
39 Testicular vessels
40 Ureter

A Posterior abdominal and pelvic walls
All peritoneum and viscera (except for the
bladder, 2, ureter, 40, and ductus deferens or vas
deferens, 6) have been removed, to display
vessels and nerves

- The aorta (A1) bifurcates into the common iliac
  arteries (A3) at the level of the fourth lumbar
  vertebra.
- The common iliac veins (A4) unite at the level of the
  fifth lumbar vertebra to form the inferior vena cava
  (A23), which lies on the right of the aorta (A1).
- In the pelvis the ureter (A40) is crossed superficially
  by the ductus deferens (A6).
- The single midline superior hypogastric plexus (A37)
  divides to form the right and left hypogastric nerves
  (A16) which enter the pelvis to contribute to the
  right and left inferior hypogastric plexuses (A21),
  collectively known as the pelvic plexus.

Psoas sign. The psoas major muscle passes
from the posterior abdominal wall to the lesser
trochanter of the femur. Infections or
haemorrhage of the posterior vertebral column
(i.e. tuberculosis) drain laterally into the psoas,
allowing pus to travel down the muscle and
present as a swelling in the groin below the
inguinal ligament.

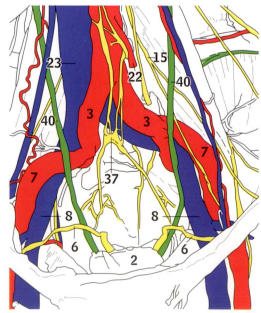

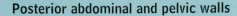

B Left lumbar plexus, from the front
Psoas major has been removed to show the constituent nerves of the plexus which are embedded within the muscle. Because of the removal of most of the anterolateral abdominal wall (except for the lowest parts of the external oblique, 1, internal oblique, 9, and transversus, 18), the iliohypogastric (6) and ilio-inguinal (7) nerves have fallen too far medially; they do not overlie iliacus (5)

1 External oblique
2 External oblique aponeurosis
3 Femoral nerve
4 Genitofemoral nerve
5 Iliacus
6 Iliohypogastric nerve
7 Ilio-inguinal nerve
8 Iliolumbar ligament
9 Internal oblique
10 Lateral femoral cutaneous nerve
11 Lumbosacral trunk
12 Obturator nerve
13 Quadratus lumborum
14 Rami communicantes
15 Superficial inguinal ring
16 Sympathetic trunk and ganglia
17 Third lumbar vertebra and anterior longitudinal ligament
18 Transversus abdominis
19 Upper surface of inguinal ligament
20 Ventral ramus of fifth lumbar nerve
21 Ventral ramus of first sacral nerve
22 Ventral ramus of fourth lumbar nerve

Lumbar sympathectomy is selective transection of the sympathetic trunk to reduce vasoconstriction in the lower limbs, for patients with poor circulation. Usually performed at the L2 level, it is easier to perform on the left than the right, where the inferior vena cava is immediately anterior.

A Muscles of the left half of the pelvis and upper thigh, from the front

All fasciae have been removed and the inguinal ligament (9), formed from part of the external oblique aponeurosis, has been preserved. Psoas major (14) and iliacus (8) are seen entering the thigh deep to the inguinal ligament. On the front of the thigh there is an unusually large gap between the adjacent borders of pectineus (11) and adductor longus (2), revealing part of the adductor brevis (1)

1 Adductor brevis
2 Adductor longus
3 Anterior superior iliac spine
4 Coccygeus
5 Fifth lumbar intervertebral disc
6 Gracilis
7 Iliac crest
8 Iliacus
9 Inguinal ligament
10 Obturator internus
11 Pectineus
12 Piriformis
13 Promontory of sacrum
14 Psoas major
15 Pubic tubercle
16 Rectus femoris
17 Sartorius
18 Tensor fasciae latae
19 Vastus lateralis

- The medial border of psoas major (A14) overlaps the side of the pelvic brim.
- Above the inguinal ligament (A9), iliacus (A8) forms the floor of the iliac fossa. On the right side, this is where the caecum and appendix lie (page 218, A4, A2, B4 and B3).
- Piriformis (A12) and obturator internus (A10) are muscles of the posterior and lateral walls of the pelvis; they are also classified as muscles of the lower limb.
- Coccygeus (A4, B5 and C5) and levator ani (C20 and 11) are the muscles of the pelvic floor, otherwise known as the pelvic diaphragm.
- Below the inguinal ligament (A9), iliacus (A8), psoas major (A14), pectineus (A11), adductor brevis (A1) and adductor longus (A2) form the floor of the femoral triangle (page 280), whose lateral boundary is the medial border of sartorius (A17) and medial boundary is the medial border of adductor longus (A2). Adductor longus is usually adjacent to pectineus (A11), so excluding adductor brevis (A1) from the floor of the triangle.
- Gracilis (A6, B9) is the most medial muscle of the thigh.

- The anterior superior iliac spine (A3) and the pubic tubercle (A15), which give attachment to the ends of the inguinal ligament (A9), are important palpable landmarks in the inguinal region (see page 192).
- The part of obturator internus (B15) *above* the attachment of levator ani (interrupted line in B) is part of the lateral wall of the pelvic cavity, while the part *below* the attachment is in the perineum and forms part of the lateral wall of the ischio-anal (ischiorectal) fossa (pages 245 and 246).
- Piriformis (B16) passes out of the pelvis into the gluteal region through the *greater* sciatic foramen *above* the ischial spine (B13), while obturator internus (B15) passes out through the *lesser* sciatic foramen *below* the ischial spine (B13).
- The posterior part of the iliococcygeus part of the levator ani (C11) arises from the *ischial* spine (B13, C13), not from any part of the ileum; the name is derived from animals in which the muscle has a higher origin.

B

F
R
O
N
T

Muscles of the left half of the pelvis, from the right, **B** with most of levator ani removed, **C** with levator ani intact in the female
Piriformis (16) is on the posterior pelvic wall, and obturator internus (15) on the lateral wall. Coccygeus (5) forms the posterior part of the pelvic floor (pelvic diaphragm), with levator ani (20 and 11) at the side and in front; in B most of levator ani has been removed (from the attachment indicated by the interrupted line) to show more of obturator internus (15), from whose overlying fascia (7) much of the levator ani arises. Here the iliococcygeus part of the levator ani (11) is more fibrous than usual. In C the lower ends of the urethra (24), vagina (25) and rectum (21) have been preserved.

1 Adductor longus
2 Adductor magnus
3 Anterior superior iliac spine
4 Branch of fourth sacral nerve
5 Coccygeus
6 Coccyx
7 Fascia over obturator internus
8 Gluteus maximus
9 Gracilis
10 Iliacus
11 Iliococcygeus part of levator ani
12 Inguinal ligament
13 Ischial spine
14 Lacunar ligament
15 Obturator internus
16 Piriformis
17 Promontory of sacrum
18 Psoas major
19 Pubic symphysis
20 Pubococcygeus part of levator ani
21 Rectum
22 Sacral canal
23 Sacrotuberous ligament
24 Urethra
25 Vagina

• The obturator artery appears abnormal, passing over the pelvic brim.

C

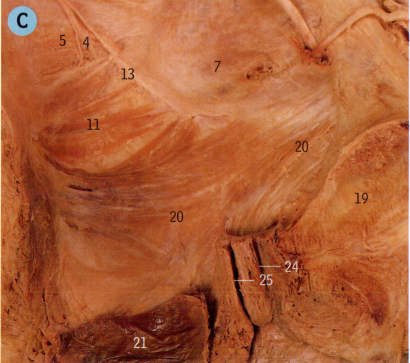

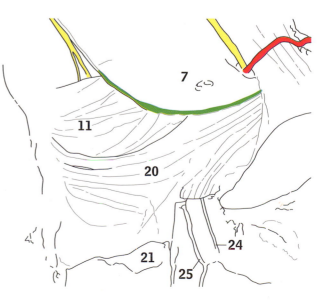

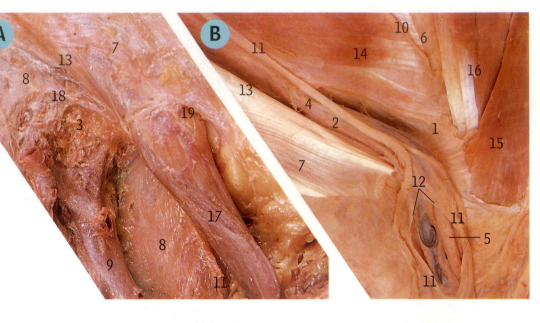

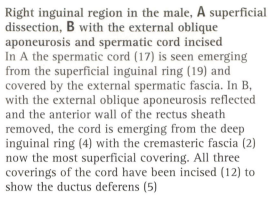

Right inguinal region in the male, A superficial dissection, B with the external oblique aponeurosis and spermatic cord incised

In A the spermatic cord (17) is seen emerging from the superficial inguinal ring (19) and covered by the external spermatic fascia. In B, with the external oblique aponeurosis reflected and the anterior wall of the rectus sheath removed, the cord is emerging from the deep inguinal ring (4) with the cremasteric fascia (2) now the most superficial covering. All three coverings of the cord have been incised (12) to show the ductus deferens (5)

1	Conjoint tendon
2	Cremasteric fascia and cremaster muscle over spermatic cord
3	Cribriform fascia
4	Deep inguinal ring
5	Ductus deferens
6	Edge of rectus sheath
7	External oblique aponeurosis
8	Fascia lata
9	Great saphenous vein
10	Iliohypogastric nerve
11	Ilio-inguinal nerve
12	Incised margin of coverings of cord
13	Inguinal ligament
14	Internal oblique
15	Pyramidalis
16	Rectus abdominis
17	Spermatic cord
18	Upper margin of saphenous opening
19	Upper margin of superficial inguinal ring

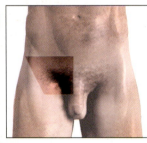

Varicoceles are enlarged varicose gonadal veins (pampiniform plexus of either the ovary or testis).

Hydrocele is an accumulation of fluid around the testis between the parietal and visceral layers of the tunica vaginalis.

Vasectomy is a surgical procedure that produces infertility in males by removal of a section of the vas deferens between sutured/clipped ends.

C Right testis and epididymis, and the penis, from the right

1	Appendix epididymis	9	Head of epididymis	17	Superficial dorsal vein
2	Body of epididymis	10	Lateral superficial vein	18	Superficial scrotal (dartos) fascia
3	Body of penis	11	Pampiniform venous plexus	19	Tail of epididymis
4	Corona of glans	12	Sac of tunica vaginalis	20	Testis
5	Ductus deferens	13	Scrotal sac	21	Tunica vaginalis, parietal
6	External urethral orifice	14	Spermatic cord	22	Tunica vaginalis, visceral
7	Foreskin	15	Superficial dorsal artery		
8	Glans penis	16	Superficial dorsal nerve		

D Right inguinal region, in the female

The external oblique aponeurosis (2) has been incised and reflected to show the position of the deep inguinal ring (7) which marks the lateral end of the inguinal canal. The round ligament of the uterus (9) emerges from the superficial inguinal ring (8), which marks the medial end of the canal, and becomes lost in the fat of the labium majus (3). The ilio-inguinal nerve (5) also passes through the canal and out of the superficial ring

- In the female the inguinal canal contains the round ligament of the uterus and the ilio-inguinal nerve.
- The processus vaginalis is normally obliterated, but if it remains patent within the female inguinal canal it is sometimes known as the canal of Nuck.

1 Conjoint tendon
2 External oblique aponeurosis
3 Fat of labium majus
4 Great saphenous vein
5 Ilio-inguinal nerve
6 Internal oblique
7 Position of deep inguinal ring
8 Position of superficial inguinal ring
9 Round ligament of uterus
10 Upper surface of inguinal ligament

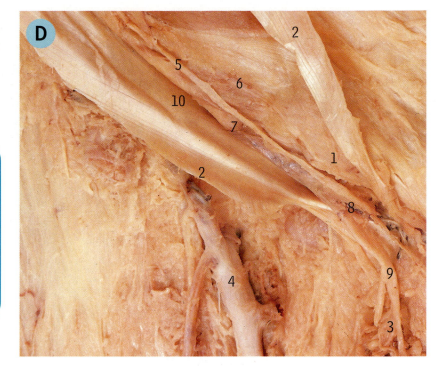

Cremasteric reflex tests nerve roots L1 and L2 in males and involves contraction of the cremaster muscle raising the testis after stroking the ipsilateral, medial thigh.

E Left inguinal and femoral regions, in the female

Part of the fascia lata of the thigh has been removed to show the femoral nerve (9), artery (8) and vein (10) beneath the inguinal ligament (13), and also the position of the femoral canal (20), medial to the vein (10). The femoral structures have been included here because of the importance of the femoral canal as a site for hernia in the female (see page 235)

1 Accessory saphenous vein
2 Adductor longus
3 Anterior superior iliac spine
4 External oblique aponeurosis
5 External oblique muscle
6 Fascia lata, cut edge
7 Fascia lata overlying tensor fasciae latae
8 Femoral artery
9 Femoral nerve
10 Femoral vein
11 Great saphenous vein
12 Iliotibial tract
13 Inguinal ligament
14 Intermediate femoral cutaneous nerve
15 Internal oblique muscle
16 Labum majus
17 Lymph vessels
18 Mons pubis
19 Pectineus
20 Position of femoral canal
21 Profunda femoris artery
22 Round ligament of uterus
23 Sartorius
24 Superficial circumflex iliac vessels
25 Superficial epigastric vein
26 Superficial external pudendal vessels
27 Superficial inguinal lymph nodes
28 Superficial inguinal ring

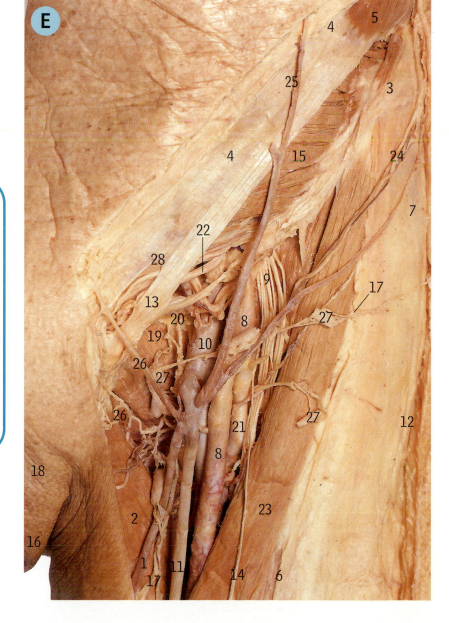

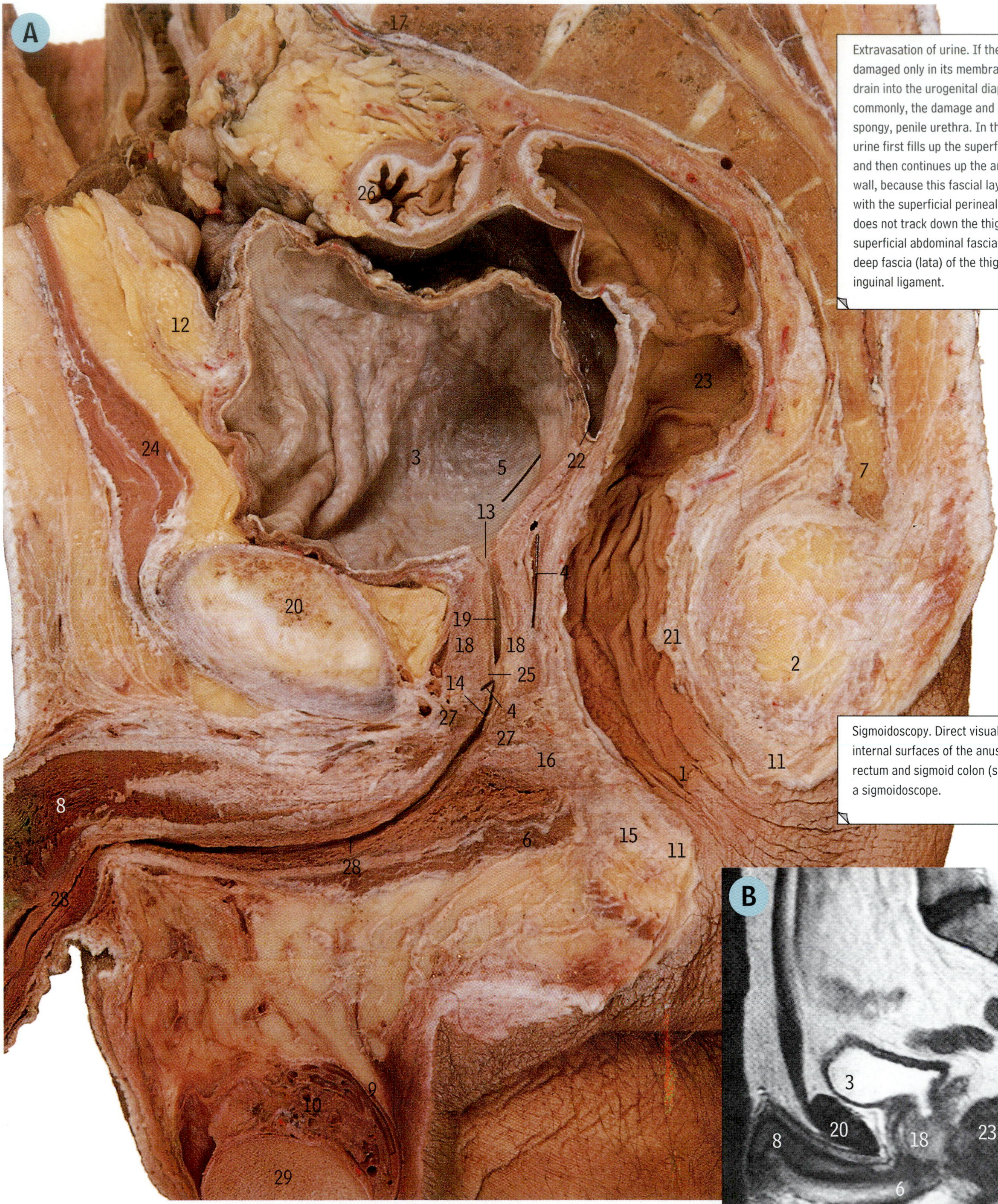

A

26

17

12

24

3

5

22

13

7

23

4

19

18   18

20

25

21

2

14

27

4

27

16

11

1

8

15

6

11

28

28

10

9

29

Extravasation of urine. If the urethra is damaged only in its membranous part, urine will drain into the urogenital diaphragm. Much more commonly, the damage and leakage is from the spongy, penile urethra. In this situation the urine first fills up the superficial perineal pouch and then continues up the anterior abdominal wall, because this fascial layer is continuous with the superficial perineal membrane. Urine does not track down the thighs because the superficial abdominal fascia is fused with the deep fascia (lata) of the thigh just below the inguinal ligament.

Sigmoidoscopy. Direct visualization of the internal surfaces of the anus (proctoscopy), rectum and sigmoid colon (sigmoidoscopy) using a sigmoidoscope.

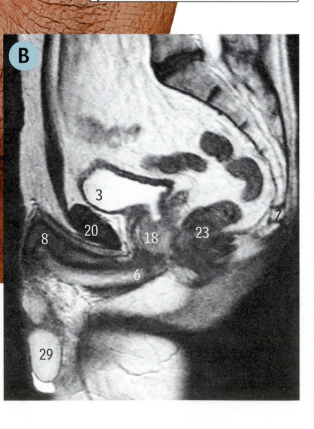

B

3

7

8

20

18

23

6

29

Male pelvis, A right half of a midline sagittal section, B sagittal MR image

The section (A) has passed exactly through the midline of the anal canal (1) and the prostatic, membranous and spongy parts of the urethra (19, 14 and 28) but has transected the left side of the scrotum and the left testis (29) and epididymis (10). The prostate (18) and bladder (3) are somewhat higher than usual; the empty bladder should not extend above the pubic symphysis (20). Compare the features seen in the MR image with the section

1 Anal canal with anal columns of mucous membrane
2 Anococcygeal body
3 Bladder
4 Bristle in ejaculatory duct
5 Bristle passing up into right ureteral orifice
6 Bulbospongiosus
7 Coccyx
8 Corpus cavernosum
9 Ductus deferens
10 Epididymis
11 External anal sphincter
12 Extraperitoneal fat
13 Internal urethral orifice
14 Membranous part of urethra
15 Perineal body
16 Perineal membrane
17 Promontory of sacrum
18 Prostate
19 Prostatic part of urethra
20 Pubic symphysis
21 Puborectalis fibres of levator ani
22 Rectovesical pouch
23 Rectum
24 Rectus abdominis
25 Seminal colliculus
26 Sigmoid colon
27 Sphincter urethrae
28 Spongy part of urethra and corpus spongiosum
29 Testis

- The lowest part of the peritoneal cavity is the rectovesical pouch (A22), between the front of the rectum (A23) and the posterior surface (base) of the bladder (A3).
- The lower end of the rectum (A23) and the anal canal (A1) are maintained at right angles to one another by a sling formed by the puborectalis fibres of both levator ani muscles (A21), which become continuous with the upper end of the external anal sphincter (A11).
- The various constituents of the spermatic cord come together at the deep inguinal ring (C3), which is in the transversalis fascia (C19) lateral to the inferior epigastric vessels (C8). The ductus deferens (C4) therefore appears to emerge from the ring by hooking round the lateral side of the vessels.
- The inguinal triangle (Hesselbach's triangle) is the area bounded laterally by the inferior epigastric vessels (C8), medially by the lateral border of rectus abdominis (C16) and below by the inguinal ligament (C9). A direct inguinal hernia passes forwards through this triangle, medial to the inferior epigastric vessels.
- An indirect inguinal hernia passes through the deep inguinal ring (C3) lateral to the inferior epigastric vessels (C8).
- A femoral hernia passes into the femoral canal through the femoral ring (C7), bounded medially by the lacunar ligament (C10) and laterally by the femoral vein (which becomes the external iliac vein, C6, as it passes beneath the inguinal ligament).

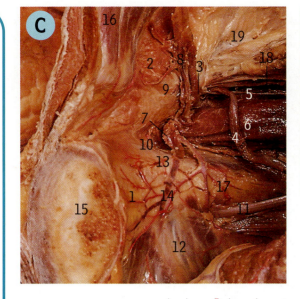

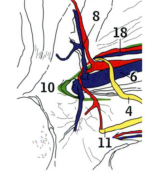

C Right deep inguinal ring and inguinal triangle, in the male

This is the view looking into the right half of the pelvis from the left, showing the posterior surface of the lower part of the anterior abdominal wall, above the pubic symphysis. The femoral ring (7), the entrance to the femoral canal, is below the medial end of the inguinal ligament (9). The inferior epigastric vessels (8) lie medial to the deep inguinal ring (3)

- The anastomosis between the pubic branches of the inferior epigastric and obturator arteries may be unusually large, forming the vessel known as the accessory or abnormal obturator artery (D1), in which case the normal obturator branch from the internal iliac may be absent.
- The accessory obturator artery usually lies at the lateral margin of the femoral ring (D7) but may lie at the medial edge of the ring, i.e. at the lateral margin of the lacunar ligament (D11), where it may

D Left accessory obturator artery in the male, from the right

This is a similar view to that in C but on the left side, showing an accessory obturator artery (1) passing from the inferior epigastric (9) over the superior pubic ramus (15) to enter the obturator foramen with the obturator nerve (12)

1 Accessory obturator artery
2 Bladder
3 Deep circumflex iliac vein
4 Ductus deferens
5 External iliac artery
6 External iliac vein (cut end)
7 Femoral ring
8 Iliacus
9 Inferior epigastric artery
10 Inguinal ligament
11 Lacunar ligament
12 Obturator nerve
13 Psoas major
14 Right common iliac artery and vein
15 Superior ramus of pubis and pectineal ligament
16 Testicular vessels

be at risk if the ligament has to be incised to enlarge the femoral ring in operations to reduce a femoral hernia.

1 Body of pubis
2 Conjoint tendon
3 Deep inguinal ring
4 Ductus deferens
5 External iliac artery
6 External iliac vein
7 Femoral ring
8 Inferior epigastric vessels
9 Inguinal ligament
10 Lacunar ligament
11 Obturator nerve
12 Origin of levator ani from fascia overlying obturator internus
13 Pectineal ligament
14 Pubic branches of inferior epigastric vessels
15 Pubic symphysis
16 Rectus abdominis
17 Superior ramus of pubis
18 Testicular vessels
19 Transversalis fascia overlying transversus abdominis

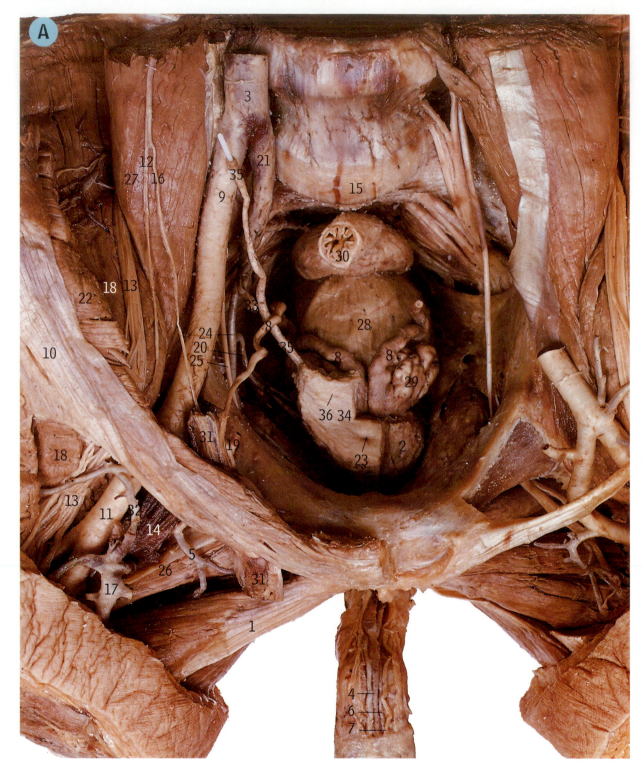

A Pelvis, right inguinal region and penis, from above

In the pelvis, most of the bladder (34) has been removed to show part of the base (upper surface) of the prostate (2), and the left seminal vesicle (29) lying lateral to the ductus deferens (8). The ductus in the pelvis crosses superficial to the ureter (35). The external iliac artery (9) passes under the inguinal ligament (10) to become the femoral artery (11). On the dorsum of the penis the fascia has been removed, showing the single midline deep dorsal vein (4) with a dorsal artery (6) and dorsal nerve (7) on each side

Carcinoma of the rectum. Carcinomata at the lower end of the hindgut tend to present relatively early, owing to a change in bowel habit, a sensation of incomplete evacuation or rectal bleeding. Surgical treatments (anterior resection and abdominoperineal excision) tend to have a good prognosis.

- The trigone of the bladder (A34, C36), at the lower part of the base or posterior surface, is the relatively fixed area with smooth mucous membrane between the internal urethral orifice (A23, C16) and the two ureteral openings (A36 on the right side, C38 on the left).
- In the male pelvis the ureter (A35, C37) is crossed superficially by the ductus deferens (A8, C7). (In the female pelvis it is crossed superficially by the uterine artery—page 243, A23 and 24.)
- The ureter (A35, C37) enters the pelvis at the bifurcation of the common iliac artery (A3), crossing the external iliac artery and vein (C9 and 10) and running down the side wall of the pelvis in front of the internal iliac artery (A21, C14).

1 Adductor longus	13 Femoral nerve	26 Pectineus
2 Base of prostate	14 Femoral vein	27 Psoas major
3 Common iliac artery	15 Fifth lumbar intervertebral disc	28 Rectum
4 Deep dorsal vein of penis	16 Genital branch of genitofemoral nerve	29 Seminal vesicle
5 Deep external pudendal artery	17 Great saphenous vein	30 Sigmoid colon (cut lower end)
6 Dorsal artery of penis	18 Iliacus	31 Spermatic cord
7 Dorsal nerve of penis	19 Inferior epigastric artery	32 Superficial circumflex iliac vein
8 Ductus deferens	20 Inferior vesical artery	33 Superior vesical artery
9 External iliac artery	21 Internal iliac artery	34 Trigone of bladder
10 External oblique aponeurosis and inguinal ligament	22 Internal oblique	35 Ureter
11 Femoral artery	23 Internal urethral orifice	36 Ureteral orifice
12 Femoral branch of genitofemoral nerve	24 Obturator artery	
	25 Obturator nerve	

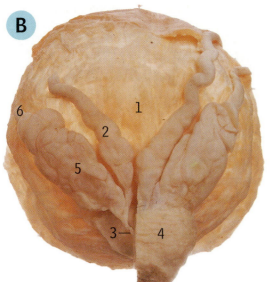

B Bladder and prostate, from behind

1 Base of bladder
2 Ductus deferens
3 Left ejaculatory duct
4 Posterior surface of prostate
5 Seminal vesicle
6 Ureter

Benign prostatic hyperplasia. In most males over the age of 60 this is a common occurrence and is normally diagnosed on a rectal examination where prostate enlargement can be felt easily. The patient usually complains of getting up at night to pass urine, passing too much urine or too often, and is unable to empty the bladder. It is the commonest cause of bladder outflow obstruction.

C Left side of the male pelvis, from the right
In this midline sagittal section, the prostate (24) is enlarged, lengthening the prostatic urethra (25) and accentuating the trabeculae of the bladder. The mucous membrane of the bladder (whose trigone is labelled at 36) has been removed to show muscular trabeculae in the wall. Variations in the branches of the internal iliac artery (14) are common, and here the obturator artery (22) gives origin to the superior vesical (34) and inferior vesical (13) as well as the middle rectal (20).

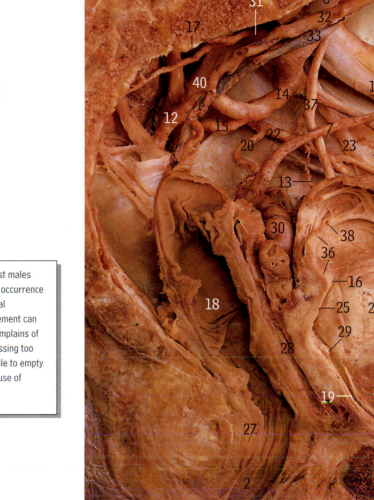

Carcinoma of the prostate. This common condition in men over the age of 70 is usually diagnosed on a rectal examination by the absence of a median prostatic sulcus and a hard palpable mass. Further spread from within the pelvis is often via the vertebral venous plexus and metastases to bone are common.

1 Accessory obturator vein	16 Internal urethral orifice	31 Superior gluteal artery
2 Anal canal	17 Lateral sacral artery	32 Superior rectal artery
3 Bulb of penis	18 Lower end of rectum	33 Superior rectal vein
4 Bulbar part of spongy urethra	19 Membranous part of urethra	34 Superior vesical artery
5 Bulbospongiosus	20 Middle rectal artery	35 Testicular vessels and deep inguinal ring
6 Common iliac artery	21 Obliterated umbilical artery	36 Trigone of bladder
7 Ductus deferens	22 Obturator artery	37 Ureter
8 External anal sphincter	23 Obturator nerve	38 Ureteral orifice
9 External iliac artery	24 Prostate (enlarged)	39 Urogenital diaphragm
10 External iliac vein	25 Prostatic part of urethra	40 Ventral ramus of first sacral nerve
11 Inferior epigastric vessels	26 Pubic symphysis	41 Vesicoprostatic venous plexus
12 Inferior gluteal artery	27 Puborectalis part of levator ani	
13 Inferior vesical artery	28 Rectovesical fascia	
14 Internal iliac artery	29 Seminal colliculus	
15 Internal pudendal artery	30 Seminal vesicle	

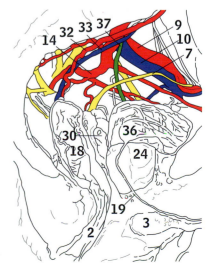

A Arteries and nerves of the pelvis, left side

In this left half section of the pelvis, all peritoneum, fascia, veins and visceral arteries have been removed together with the left levator ani, so displaying the whole of the internal surface of obturator internus (12). On the posterior pelvic wall, the vessels in general lie superficial to the nerves. In this specimen the external iliac artery (3) is unusually tortuous, and the anterior trunk of the internal iliac artery (1) has divided unusually high up into its terminal branches, the internal pudendal (8) and the inferior gluteal (5). The superior gluteal artery (19) has perforated the lumbosacral trunk

1 Anterior trunk of internal iliac artery	12 Obturator internus
2 Coccygeus and sacrospinous ligament	13 Obturator nerve and artery
3 External iliac artery	14 Piriformis
4 Inferior epigastric artery	15 Posterior trunk of internal iliac artery
5 Inferior gluteal artery	16 Pubic symphysis
6 Inguinal ligament	17 Sacral promontory
7 Internal iliac artery	18 Sacrococcygeal joint
8 Internal pudendal artery	19 Superior gluteal artery piercing lumbosacral trunk
9 Ischial tuberosity	20 Union of ventral rami of second and third sacral nerves
10 Lacunar ligament	21 Ventral ramus of first sacral nerve
11 Lateral sacral artery	

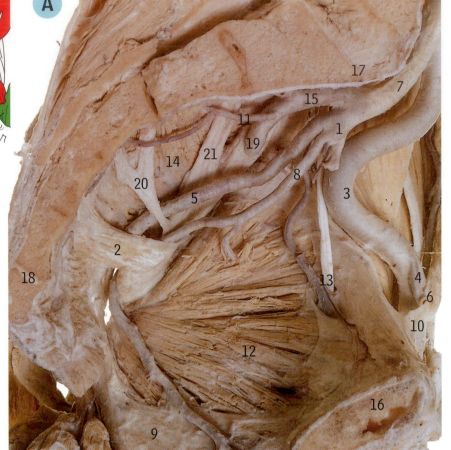

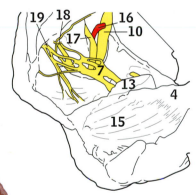

B Left inferior hypogastric plexus, from the right

In this view of the left side of the pelvis from the right, the right pelvic wall has been removed but the right levator ani (15) forming part of the pelvic floor (pelvic diaphragm) has been preserved and is seen from its right (perineal) side. Pelvic splanchnic nerves (12) arise from the ventral rami of the second and third sacral nerves (18 and 19) and contribute to the inferior hypogastric plexus (7)

1 Arcuate line of ilium
2 Fascia overlying obturator internus
3 Ischial spine
4 Lateral surface of fascia overlying right obturator internus
5 Left coccygeus and nerves to levator ani
6 Left ductus deferens
7 Left inferior hypogastric plexus
8 Left levator ani
9 Left seminal vesicle
10 Lumbosacral trunk
11 Part of left sympathetic trunk
12 Pelvic splanchnic nerves (nervi erigentes)
13 Rectum
14 Right ischiopubic ramus
15 Right levator ani and ischio-anal (ischiorectal) fossa
16 Superior gluteal artery
17 Ventral ramus of first sacral nerve
18 Ventral ramus of second sacral nerve
19 Ventral ramus of third sacral nerve

C Pelvic ligaments, left side, from the right
In this median sagittal section of the pelvis all soft tissues have been removed except the ligaments

Interrupted line = position of origin of levator ani

1 Anterior inferior iliac spine and origin of straight head of rectus femoris
2 Anterior superior iliac spine
3 Falciform process of sacrotuberous ligament
4 Greater sciatic foramen
5 Iliac fossa
6 Inguinal ligament
7 Ischial spine
8 Ischial tuberosity
9 Lacunar ligament
10 Lesser sciatic foramen
11 Obturator foramen
12 Obturator membrane
13 Pectineal ligament
14 Pubic symphysis
15 Sacral promontory
16 Sacrospinous ligament
17 Sacrotuberous ligament
18 Ventral sacro-iliac ligament

- The ligaments classified as 'the ligaments of the pelvis' (vertebropelvic ligaments) are the sacrotuberous (C17), sacrospinous (C16) and iliolumbar (seen in the posterior view on page 285, C7).
- The sacrotuberous and sacrospinous ligaments convert the greater and lesser sciatic notches of the hip bone (page 249, 7 and 15) into foramina (C4 and C10).
- The lacunar ligament (C9) passes backwards from the medial end of the inguinal ligament (C6) to the medial end of the pectineal line of the pubis, to which the pectineal ligament (C13) is attached.
- The lower attachment of the sacrotuberous ligament is to the medial side of the ischial tuberosity, but it gives off two slips. One is the falciform process (C3), which passes towards the ischial ramus to form the lower boundary of the pudendal canal (page 244, D25). The other runs into the ischial attachment of the long head of biceps (page 279, C9).

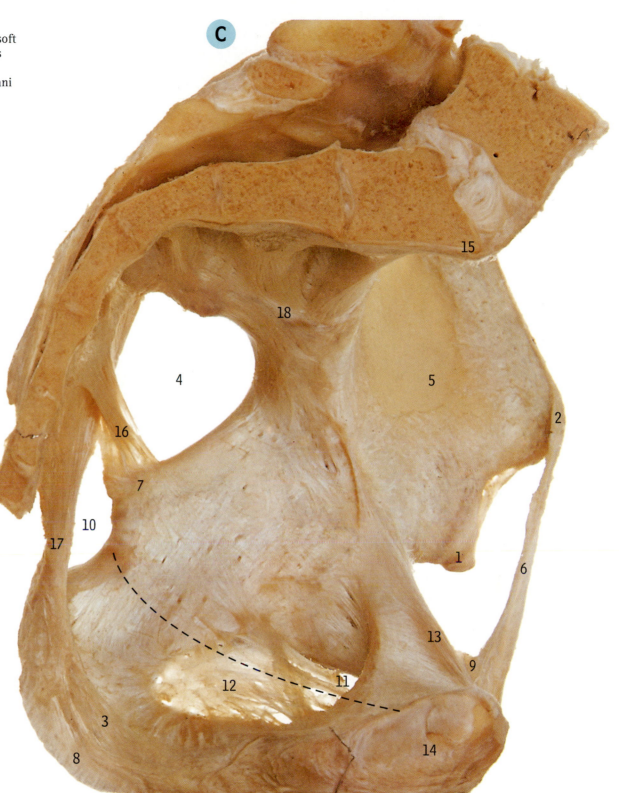

C

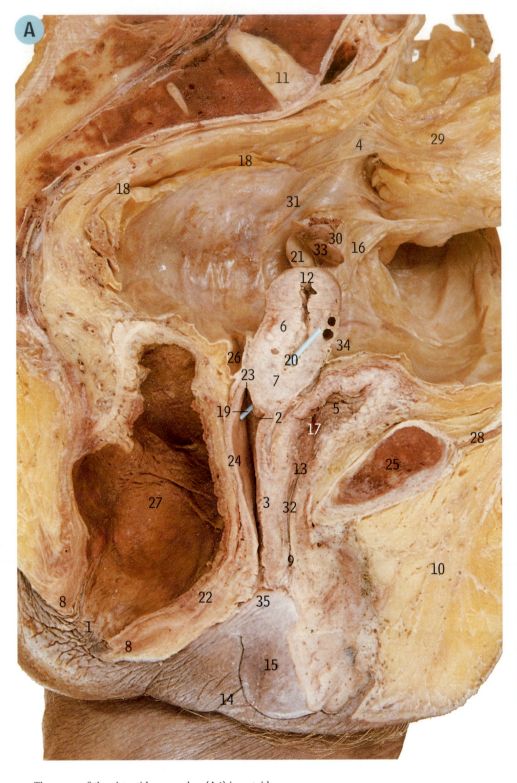

A

Female pelvis, A left half of a midline sagittal section

In A the lower end of the rectum (27) is dilated, and the bladder (5), uterus (6) and vagina (3 and 24) are contracted. The section has opened up the whole length of the urethra (32), but the cervix of the uterus (7) is rarely exactly in the midline and the line of the cervical canal is indicated by the marker in the internal and external os (20 and 19). Compare features in the MR image in B with the section

1	Anal canal
2	Anterior fornix of vagina
3	Anterior wall of vagina
4	Apex of sigmoid mesocolon
5	Bladder
6	Body of uterus
7	Cervix of uterus
8	External anal sphincter
9	External urethral orifice
10	Fat of mons pubis
11	Fifth lumbar intervertebral disc
12	Fundus of uterus
13	Internal urethral orifice
14	Labium majus
15	Labium minus
16	Left limb of sigmoid mesocolon overlying external iliac vessels
17	Left ureteral orifice
18	Line of attachment of right limb of sigmoid mesocolon
19	Marker in external os
20	Marker in internal os
21	Ovary
22	Perineal body
23	Posterior fornix of vagina
24	Posterior wall of vagina
25	Pubic symphysis
26	Recto-uterine pouch (of Douglas)
27	Rectum
28	Rectus abdominis (turned forwards)
29	Sigmoid colon (reflected to left and upwards)
30	Suspensory ligament of ovary containing ovarian vessels
31	Ureter underlying peritoneum
32	Urethra
33	Uterine tube
34	Vesico-uterine pouch
35	Vestibule of vagina

- The apex of the sigmoid mesocolon (A4) is a guide to the left ureter (A31) which enters the pelvis under the peritoneum at this point (in both sexes).
- The recto-uterine pouch (A26, pouch of Douglas) overlies the posterior fornix of the vagina (A23), but the vesico-uterine pouch (A34) does not reach the anterior fornix (A2).

Faecal continence is dependent on a complex mechanism involving both the internal and external anal sphincters and the anorectal angle, maintained by the puborectalis fibres of the levator ani muscle. This angle is normally about 90°; if it is altered to over 100°, incontinence ensues.

Intrauterine contraceptive devices (IUCDs) are small plastic or metal tubes inserted into the uterine cavity which prevent implantation of fertilized eggs.

Rectal examination (PR). Examination per rectum is obviously different in male and female patients. The major structures palpable in the male are the median sulcus of the prostate gland, the sacral concavity and coccyx, and in the female the sacrum, coccyx and cervix. This examination may be used during labour as a way of assessing the dilatation of the cervix, and in acute appendicitis a tender pelvic appendix may be reached by the examining finger.

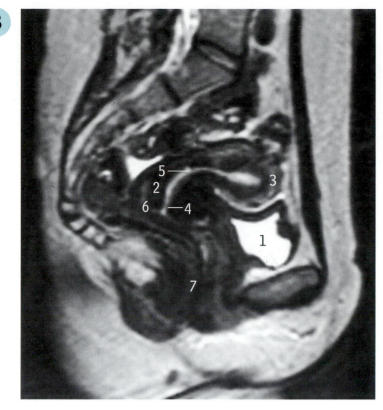

B

Female pelvis, **B** sagittal MR image

1 Bladder
2 Cervix of uterus
3 Fundus of uterus
4 Marker in external os
5 Marker in internal os
6 Posterior fornix of vagina
7 Rectum

Vaginal examination (PV). Often performed in the lithotomy position and preceded by a speculum examination, PV examination is used to assess the state of the cervix, uterus and ovaries. This examination combined with a hand on the abdominal wall (bimanual examination) reveals that the normal uterus is in an anteverted and anteflexed position; however, approximately one-fifth of normal women have a retroverted or retroflexed uterus. The non-pregnant cervix feels firm, like the end of the nose: the pregnant cervix feels warm and softer, rather like warm lips. When feeling the lateral fornices, occasionally one can detect an ovary, or (theoretically) a stone in the ureter.

Cystitis (inflammation of the bladder) often presents as pain on passing urine. Owing to the shortness of the female urethra it is much more common in young girls and women than in men, as organisms can enter the bladder more easily.

C

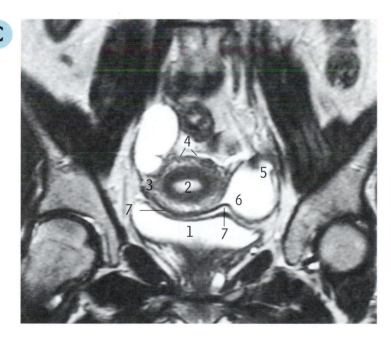

Female pelvis, **C** coronal MR image

Looking down into the pelvis from the front, the fundus of the uterus (2) overlies the bladder (1) with the peritoneum of the vesico-uterine pouch (7) intervening. These relationships are seen in the MR image.

1 Bladder
2 Fundus of uterus
3 Mesosalpinx
4 Recto-uterine pouch
5 Tubal extremity of ovary
6 Uterine extremity of ovary
7 Vesico-uterine pouch

Urinary continence has a complex, poorly understood mechanism involving both the bladder neck and the distal urethral mechanism, which includes the sphincter urethrae. The major control is by the pelvic splanchnic nerves, which are motor to the detrusor muscle and inhibitory to the internal sphincter. When these fibres are stimulated by a full bladder, the bladder contracts, the sphincter relaxes and urine flows into the urethra.

Cervical smear (Pap smear). Named after Dr George N. Papanicolaou, who developed it, this is a simple test to examine epithelial cells from the uterine cervix. Cells are removed by gently scraping, placing on a slide and examining microscopically for abnormalities of shape and size, to detect early uterine/cervical cancer.

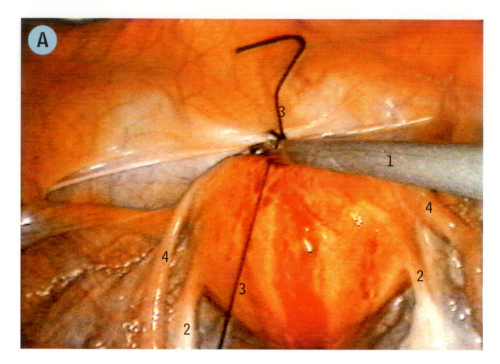

A Laparoscopic view into female pelvis

1 Forceps
2 Ligament of ovary
3 Silk sling into fundus
4 Uterine tube

Carcinoma of the ovary. The ovary drains to the para-aortic lymph nodes and although carcinoma of the ovary is rare in the young, it is a common cause of peritoneal metastases in older women.

Rupture of an ectopic pregnancy occurs most commonly about 6–8 weeks after conception and is an acute medical emergency normally presenting as vaginal bleeding and acute abdominal pain due to rupture of the uterine tube and bleeding into the peritoneal cavity. Occasionally, this may present as shoulder pain from diaphragmatic irritation, an excellent example of referred pain.

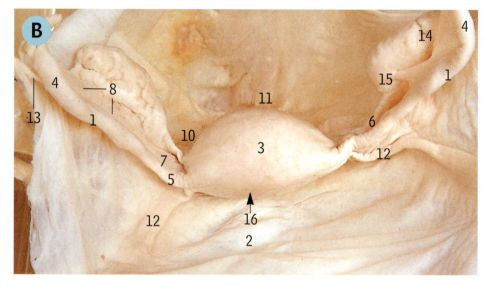

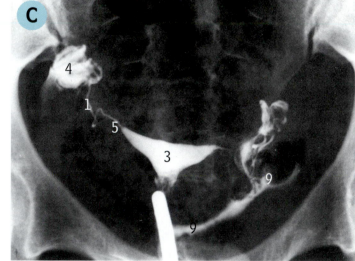

**Female pelvis, B uterus and ovaries, from above and in front,
C hysterosalpingogram**
Looking down into the pelvis from the front in B, the fundus of the uterus (3) overlies the bladder (2) with the peritoneum of the vesico-uterine pouch (16) intervening. In C, contrast medium has filled the uterus and tubes (3, 5, 1 and 4) and spilled out into the peritoneal cavity (9)

1 Ampulla of uterine tube	10 Posterior surface of broad
2 Bladder	ligament
3 Fundus of uterus	11 Recto-uterine pouch
4 Infundibulum of uterine tube	12 Round ligament of uterus
5 Isthmus of uterine tube	13 Suspensory ligament of ovary
6 Ligament of ovary	with ovarian vessels
7 Mesosalpinx	14 Tubal extremity of ovary
8 Mesovarium	15 Uterine extremity of ovary
9 Overspill of contrast into	16 Vesico-uterine pouch
recto-uterine pouch	

A

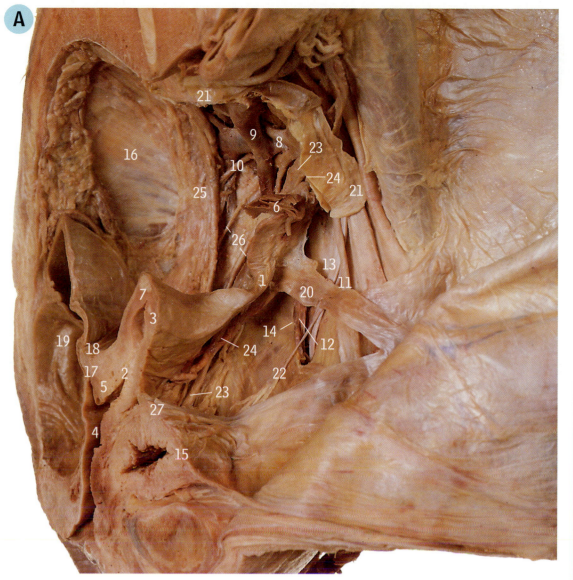

Female pelvis, A left half, obliquely from the front

Looking obliquely into the left half of the pelvis from the front, with the anterior abdominal wall turned forwards, the peritoneum of the vesico-uterine pouch (27) has been incised and the uterus (3) displaced backwards to show the ureter (23) running towards the bladder and being crossed by the uterine artery (24). The uterosacral ligament (25) passes backwards at the side of the rectum (19) towards the pelvic surface of the sacrum. The root of the sigmoid mesocolon (21) has been left in place to emphasize that the left ureter (23) passes from the abdomen into the pelvis beneath it

1	Ampulla of uterine tube
2	Anterior fornix of vagina
3	Body of uterus
4	Cavity of vagina
5	Cervix of uterus
6	Fimbriated end of uterine tube
7	Fundus of uterus
8	Internal iliac artery
9	Internal iliac vein
10	Middle rectal artery
11	Obliterated umbilical artery
12	Obturator artery
13	Obturator nerve
14	Obturator vein
15	Peritoneum overlying bladder
16	Peritoneum overlying piriformis
17	Posterior fornix of vagina
18	Recto-uterine pouch (of Douglas)
19	Rectum
20	Round ligament of uterus
21	Sigmoid mesocolon
22	Superior vesical artery
23	Ureter
24	Uterine artery
25	Uterosacral ligament
26	Vaginal artery (double)
27	Vesico-uterine pouch

Carcinoma of the uterus. Found normally in the elderly, often in those who have not had children; this condition may spread within the pelvis or, very occasionally, to the superficial inguinal lymph nodes along the round ligament of the uterus which is carrying its accompanying lymphatics.

Support of pelvic viscera. Pelvic structures are supported by various parts of the levator ani, sphincter urethrae, sphincter vaginae and puborectalis muscles and the ligamentous supports of the uterus and vagina.

Pudendal block produces anaesthesia of the perineum by injecting an anaesthetic agent around the ischial spine and thus affecting the pudendal nerve (S2, 3 and 4) as it travels over this structure.

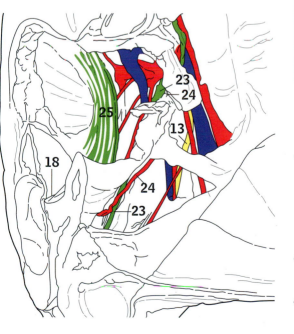

- Because the body and cervix of the uterus (A3 and 5) are rarely exactly in the midline, the section in A has not passed through the cervical canal, so the continuity between the cavity of the uterus and the vagina (A4) cannot be seen. The projection of the cervix into the vagina gives rise to the anterior and posterior fornices (A2 and 17).
- In the pelvis the ureter (A23) is crossed superficially by the uterine artery (A24). In the male pelvis it is crossed by the ductus deferens (pages 236 and 237).
- The uterosacral ligaments (A25), pass backwards on either side of the rectum to the sacrum. The lateral cervical ligaments (tissue underlying the ureter and uterine artery, A23 and 24, often called the cardinal or Mackenrodt's ligaments and passing to the lateral pelvic wall) are condensations of retroperitoneal tissue of great importance in supporting the uterine cervix (A5) in its normal position.

Female perineum

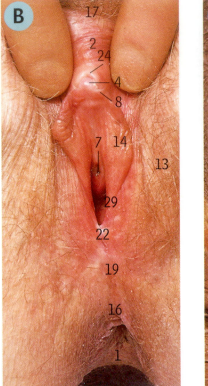

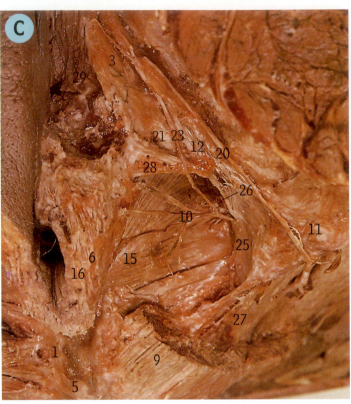

1. Anococcygeal body
2. Anterior commissure
3. Bulbospongiosus overlying bulb of vestibule
4. Clitoris
5. Coccyx
6. External anal sphincter
7. External urethral orifice
8. Frenulum of clitoris
9. Gluteus maximus
10. Inferior rectal nerve
11. Ischial tuberosity
12. Ischiocavernosus overlying crus of clitoris
13. Labium majus
14. Labium minus
15. Levator ani
16. Margin of anus
17. Mons pubis
18. Obturator internus and fascia
19. Perineal body
20. Perineal branch of posterior femoral cutaneous nerve
21. Perineal membrane
22. Posterior commissure
23. Posterior labial nerve
24. Prepuce of clitoris
25. Pudendal canal
26. Pudendal nerve
27. Sacrotuberous ligament
28. Superficial transverse perineal muscle overlying posterior border of perineal membrane
29. Vagina

Female perineum, B surface features, C left ischio-anal fossa, from below, D left ischio-anal fossa, from behind
In B the labia minora (14) have been separated to show the orifice of the vagina (29) with the urethra (7) opening into the vestibule anteriorly, 2.5 cm (1 in) behind the clitoris (4). In C and D fat and vessels have been removed from the ischio-anal fossa to show the pudendal canal (25) in the lateral wall, with levator ani (15) sloping downwards and medially to the external anal sphincter (6). The inferior rectal nerve (10) leaves the pudendal nerve (26) by piercing the wall of the pudendal canal (25) and crosses the fossa to reach the external anal sphincter (6)

Anal and rectal abscesses can drain centrally into the lumen or laterally into the ischio-anal fossae (below the level of the levator ani muscle) or pelvis (above this muscle). Successful treatment requires an understanding of these anatomical relationships.

Haemorrhoids are varicosities of the superior rectal veins which may protrude through the external anal sphincter or into the rectum, often causing bright red bleeding on defaecation.

Episiotomy is a small incision in the posterolateral vaginal wall, performed in the final stages of childbirth to enlarge the vaginal canal. It is usually performed after a pudendal nerve block has anaesthetized the S2, 3 and 4 dermatomes – the skin of that region. This procedure minimizes labour-induced lacerations in the midline which may damage the central tendon of the perineum or extend into the rectal mucosa.

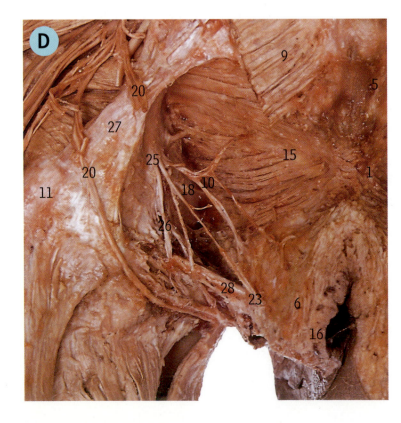

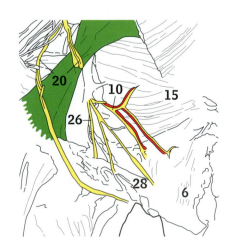

- The ischiorectal fossa is now properly and more correctly called the ischio-anal fossa; the anal canal (C and D16), not the rectum, is its lower medial boundary. The walls and contents are similar in both sexes.
- The vulva is the anterior part of the female perineum containing the external genitalia.
- The external genitalia consist of the mons pubis (B17), labia majus (B13), labia minus (B14), clitoris (B4), vestibule of the vagina, the bulb of the vestibule (C3), the greater vestibular (Bartholin's) glands (under the posterior end of the bulb of the vestibule), and the lesser vestibular glands (small mucous glands in the labia minus).
- The vestibule of the vagina is bounded by the labia minus (B14), and contains the external urethral orifice (B7), the vaginal orifice (B29, with the hymen at its margin in the virgin) and the ducts of the greater and lesser vestibular glands.
- The pudendal cleft is the region between the two labia majora (B13).

A Male perineum

The central area is shown, with the scrotum (5) pulled upwards and forwards

1 Anococcygeal body
2 Margin of anus, with skin tags
3 Perineal body
4 Raphe overlying bulb of penis
5 Scrotum overlying left testis

• Skin tags are often the remnants of previous haemorrhoids.

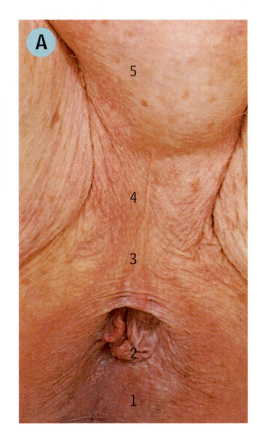

Chordee of the penis is an abnormal bend in the erect penis, often associated with hypospadias or priapism.

Haemorrhoids (dilatations of the veins in the lower rectum and upper anal regions) are a common cause of bleeding from the rectum.

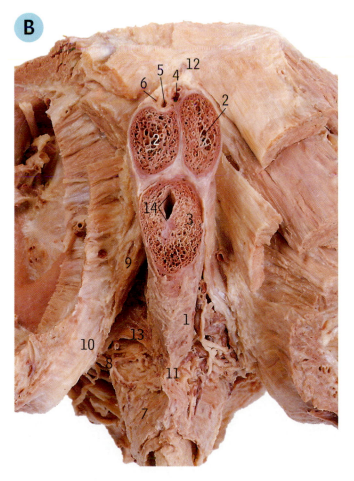

B Root of the penis, from below and in front

The front part of the penis has been removed to show the root, formed by the two corpora cavernosa dorsally (2) and the single corpus spongiosum ventrally (3) containing the urethra (14)

1 Bulbospongiosus
2 Corpus cavernosum
3 Corpus spongiosum
4 Deep dorsal vein of penis
5 Dorsal artery of penis
6 Dorsal nerve of penis
7 External anal sphincter
8 Inferior rectal vessels and nerve crossing ischio-anal fossa
9 Ischiocavernosus
10 Ischiopubic ramus
11 Perineal body
12 Pubic symphysis
13 Superficial transverse perineal muscle overlying perineal membrane
14 Urethra

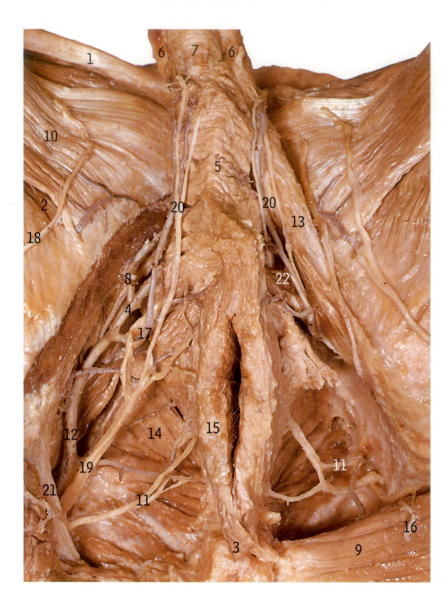

Male perineum and ischio-anal (ischiorectal) fossae

All the fat has been removed from the ischio-anal fossae so that a clear view is obtained of the perineal surface of levator ani (14) and of the vessels and nerves within the fossae. On the left side (right of the picture) the perineal membrane (22) is intact but on the right side it, and the underlying muscle (urogenital diaphragm), have been removed

1	Adductor longus	13	Ischiocavernosus overlying crus of penis
2	Adductor magnus	14	Levator ani
3	Anococcygeal body	15	Margin of anus
4	Artery to bulb	16	Perforating cutaneous nerve
5	Bulbospongiosus overlying bulb of penis	17	Perineal artery
6	Corpus cavernosum of penis	18	Perineal branch of posterior femoral cutaneous nerve
7	Corpus spongiosum of penis	19	Perineal nerve
8	Dorsal nerve and artery of penis	20	Posterior scrotal vessels and nerves
9	Gluteus maximus	21	Sacrotuberous ligament
10	Gracilis	22	Superficial transverse perineal muscle overlying posterior border of perineal membrane
11	Inferior rectal vessels and nerve in ischio-anal fossa		
12	Internal pudendal artery		

- In both sexes the ischio-anal (ischiorectal) fossa has the pudendal canal in its lateral wall. The canal has been opened up to display its contents: the internal pudendal artery (12) and the terminal branches of the pudendal nerve – the perineal nerve (19) and the dorsal nerve of the penis (8) or clitoris.

Erection. When stimulated, the pelvic parasympathetics S2–S4 cause relaxation of the coiled arteries of the penis and clitoris and engorgement of the cavernous spaces. The bulbospongiosus and ischiocavernosus muscles compress the venous caverns of these spaces and impede the venous return, causing erection.

Priapism is a permanent painful erection, often due to thrombosis within the cavernous tissue of the penis.

Hypospadias is a developmental abnormality of the penis in which the external urethral opening appears somewhere along its ventral surface.

Ejaculation. The expulsion of semen through the urethra is the result of at least three mechanisms: closure of the bladder neck; contraction of the urethral musculature (sympathetic control); and contraction of the bulbospongiosus muscle (pudendal nerve). A major factor is the sympathetic supply from L1 and L2.

Chapter

6

Lower limb

Left hip bone, lateral surface

1   Acetabular notch
2   Acetabulum
3   Anterior gluteal line
4   Anterior inferior iliac spine
5   Anterior superior iliac spine
6   Body of ilium
7   Body of ischium
8   Body of pubis
9   Greater sciatic notch
10 Iliac crest
11 Iliopubic eminence
12 Inferior gluteal line
13 Inferior ramus of pubis
14 Ischial spine
15 Ischial tuberosity
16 Junction of 25 and 13
17 Lesser sciatic notch
18 Obturator crest
19 Obturator foramen
20 Obturator groove
21 Posterior gluteal line
22 Posterior inferior iliac spine
23 Posterior superior iliac spine
24 Pubic tubercle
25 Ramus of ischium
26 Rim of acetabulum
27 Superior ramus of pubis
28 Tubercle of iliac crest

- The hip (innominate) bone is formed by the union of the ilium (6), ischium (7) and pubis (8).
- It bears on its lateral surface the cup-shaped acetabulum (2), to which the ilium, ischium and pubis each contribute a part (see page 274).
- The two hip bones articulate in the midline anteriorly at the pubic symphysis; posteriorly they are separated by the sacrum, forming the sacro-iliac joints. The two hip bones with the sacrum and coccyx constitute the pelvis.
- The ischiopubic ramus is formed by the union (16) of the ramus of the ischium (25) with the inferior ramus of the pubis (13).

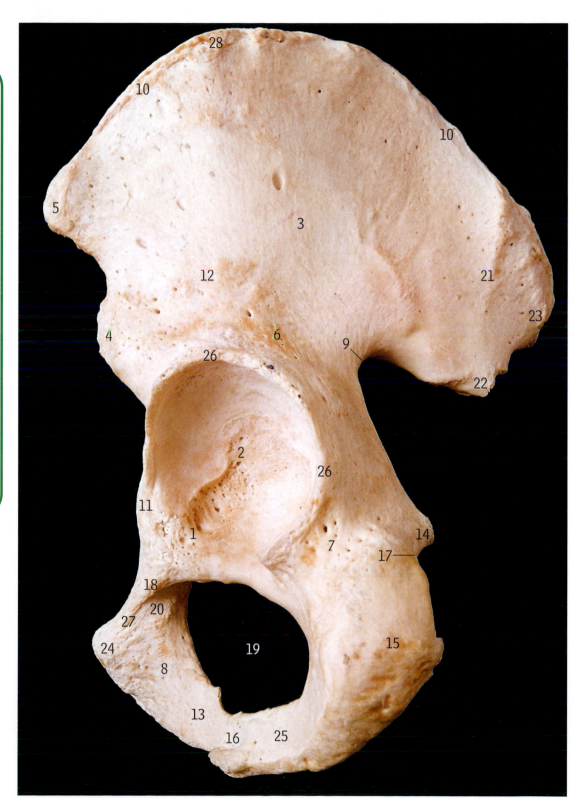

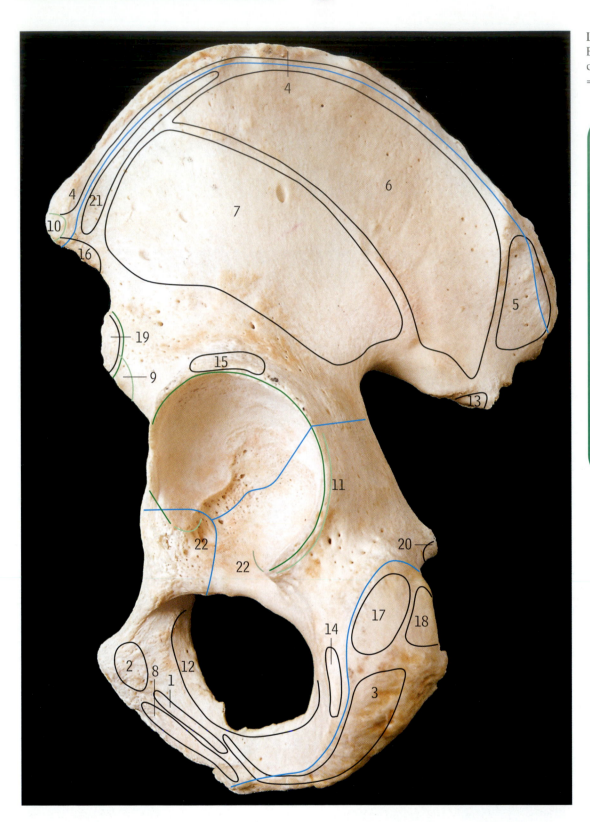

Left hip bone, lateral surface. Attachments
Blue lines = epiphysial lines; green lines =
capsular attachment of hip joint; pale green lines
= ligament attachments

1   Adductor brevis
2   Adductor longus
3   Adductor magnus
4   External oblique
5   Gluteus maximus
6   Gluteus medius
7   Gluteus minimus
8   Gracilis
9   Iliofemoral ligament
10   Inguinal ligament
11   Ischiofemoral ligament
12   Obturator externus
13   Piriformis
14   Quadratus femoris
15   Reflected head of rectus femoris
16   Sartorius
17   Semimembranosus
18   Semitendinosus and long head of biceps
19   Straight head of rectus femoris
20   Superior gemellus
21   Tensor fasciae latae
22   Transverse ligament

Left hip bone, medial surface

1 Anterior inferior iliac spine
2 Anterior superior iliac spine
3 Arcuate line
4 Auricular surface
5 Body of ischium
6 Body of pubis
7 Greater sciatic notch
8 Iliac crest
9 Iliac fossa
10 Iliac tuberosity
11 Iliopubic eminence
12 Ischial spine
13 Ischial tuberosity
14 Ischiopubic ramus
15 Lesser sciatic notch
16 Obturator foramen
17 Obturator groove
18 Pecten of pubis (pectineal line)
19 Posterior inferior iliac spine
20 Posterior superior iliac spine
21 Pubic crest
22 Pubic tubercle
23 Superior ramus of pubis

• The auricular surface of the ilium (4) is the articular surface for the sacro-iliac joint.

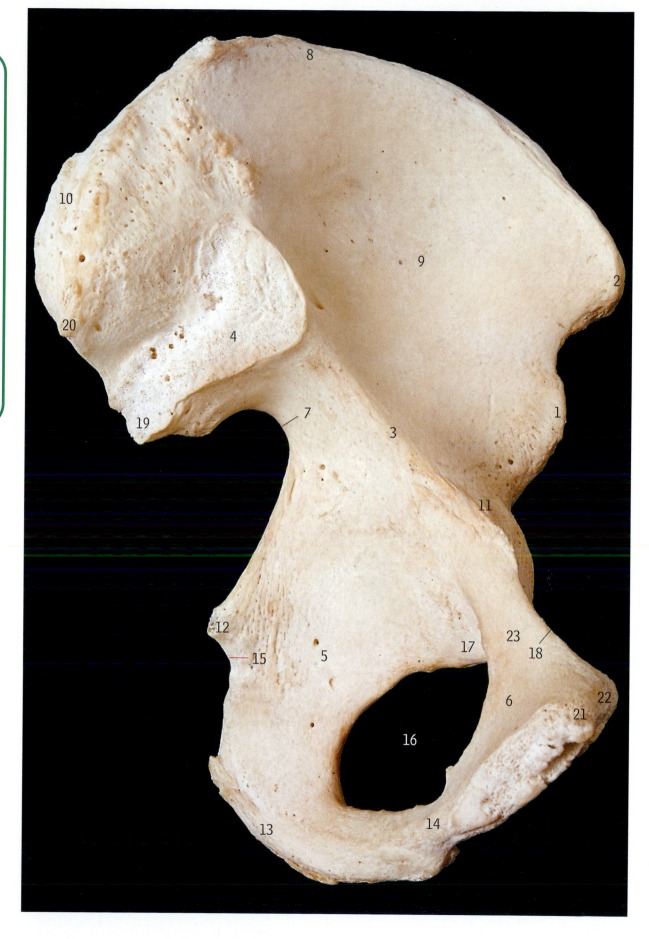

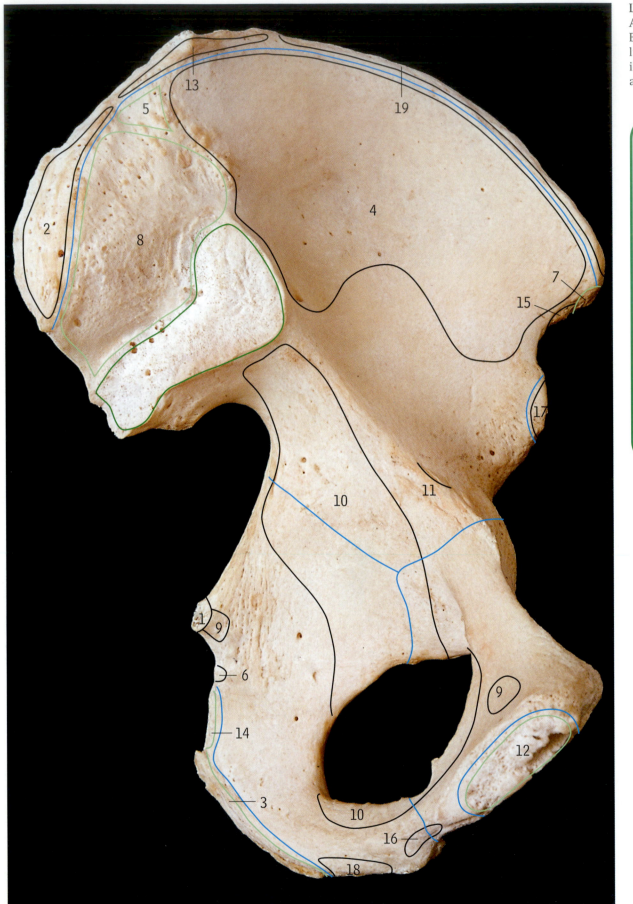

Left hip bone, medial surface.
Attachments
Blue lines = epiphysial lines; green
line = capsular attachment of sacro-
iliac joint; pale green lines = ligament
attachments

1　Coccygeus and sacrospinous ligament
2　Erector spinae
3　Falciform process of sacrotuberous
　　ligament
4　Iliacus
5　Iliolumbar ligament
6　Inferior gemellus
7　Inguinal ligament
8　Interosseous sacro-iliac ligament
9　Levator ani
10　Obturator internus
11　Psoas minor
12　Pubic symphysis
13　Quadratus lumborum
14　Sacrotuberous ligament
15　Sartorius
16　Sphincter urethrae
17　Straight head of rectus femoris
18　Superficial transverse perineal and
　　ischiocavernosus
19　Transversus abdominis

Left hip bone, from above

1 Anterior inferior iliac spine
2 Anterior superior iliac spine
3 Arcuate line
4 Auricular surface
5 Iliac crest
6 Iliac fossa
7 Iliopubic eminence
8 Ischial spine
9 Pecten of pubis (pectineal line)
10 Posterior inferior iliac spine
11 Posterior superior iliac spine
12 Pubic crest
13 Pubic tubercle
14 Tubercle of iliac crest

- The arcuate line on the ilium (3) and the pecten and crest of the pubis (9 and 12) form part of the brim of the pelvis (the rest of the brim being formed by the promontory and upper surface of the lateral part of the sacrum—see page 73).
- The pecten of the pubis (9) is more commonly called the pectineal line.

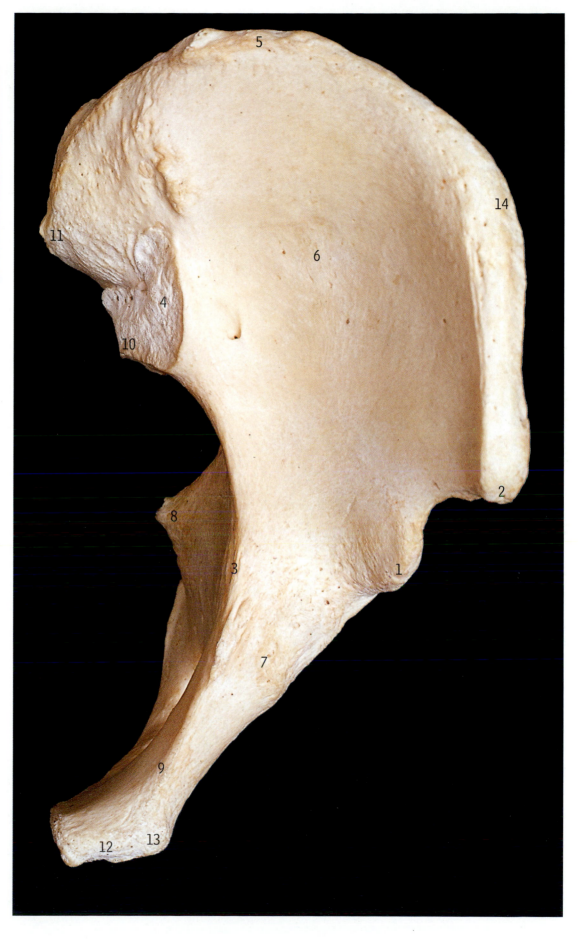

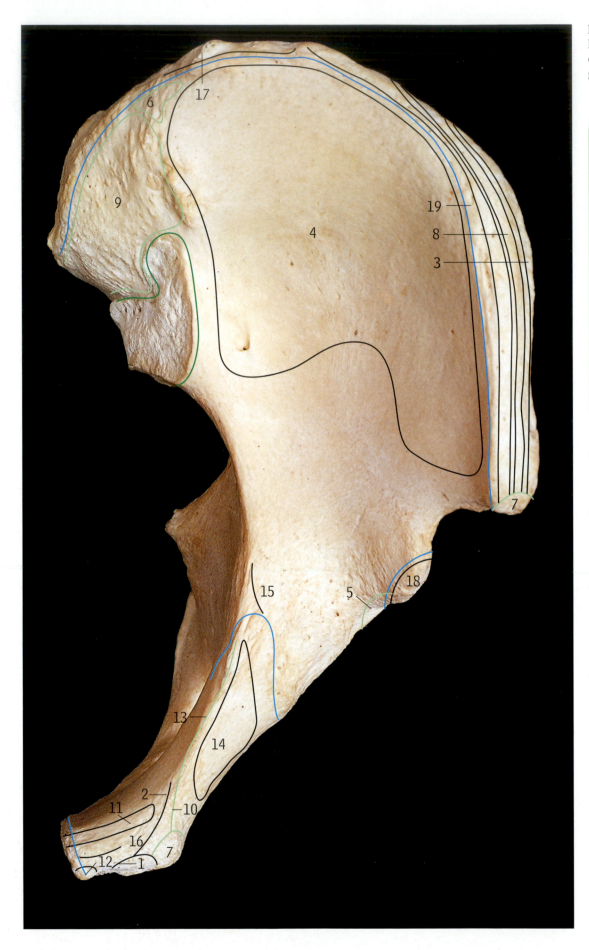

Left hip bone, from above. Attachments
Blue lines = epiphysial lines; green line = capsular attachment of sacro-iliac joint; pale green lines = ligament attachments

1　Anterior wall of rectus sheath
2　Conjoint tendon
3　External oblique
4　Iliacus
5　Iliofemoral ligament
6　Iliolumbar ligament
7　Inguinal ligament
8　Internal oblique
9　Interosseous sacro-iliac ligament
10　Lacunar ligament
11　Lateral head of rectus abdominis
12　Medial head of rectus abdominis
13　Pectineal ligament
14　Pectineus
15　Psoas minor
16　Pyramidalis
17　Quadratus lumborum
18　Straight head of rectus femoris
19　Transversus abdominis

- The inguinal ligament (7) is formed by the lower border of the aponeurosis of the external oblique muscle, and extends from the anterior superior iliac spine to the pubic tubercle.
- The lacunar ligament (10, sometimes called the pectineal part of the inguinal ligament) is the part of the inguinal ligament that extends backwards from the medial end of the inguinal ligament to the pecten of the pubis.
- The pectineal ligament (13) is the lateral extension of the lacunar ligament along the pecten. It is not classified as a part of the inguinal ligament, and must not be confused with the alternative name for the lacunar ligament, i.e. with the pectineal part of the inguinal ligament.
- The conjoint tendon (2) is formed by the aponeuroses of the internal oblique and transversus muscles, and is attached to the pubic crest and the adjoining part of the pecten, blending medially with the anterior wall of the rectus sheath.

A Left hip bone. Ischial tuberosity, from behind and below

1 Acetabular notch
2 Acetabulum
3 Ischial spine
4 Ischiopubic ramus
5 Lesser sciatic notch
6 Longitudinal ridge
7 Lower part of tuberosity
8 Obturator groove
9 Rim of acetabulum
10 Transverse ridge
11 Upper part of tuberosity

B Left hip bone, from the front

1 Acetabular notch
2 Anterior inferior iliac spine
3 Anterior superior iliac spine
4 Body of pubis
5 Iliac fossa
6 Iliopubic eminence
7 Ischial tuberosity
8 Ischiopubic ramus
9 Obturator crest
10 Obturator foramen
11 Obturator groove
12 Pecten of pubis (pectineal line)
13 Pubic crest
14 Pubic tubercle
15 Rim of acetabulum
16 Tubercle of iliac crest

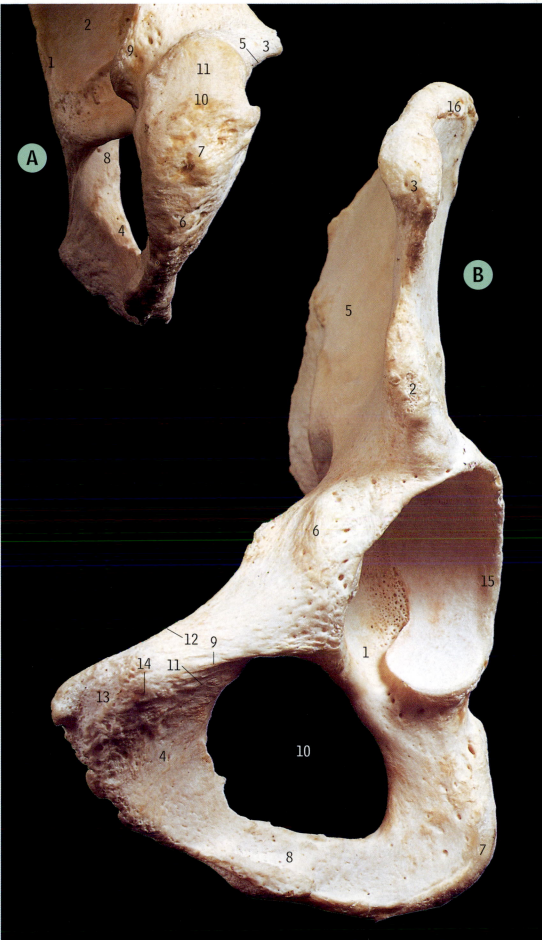

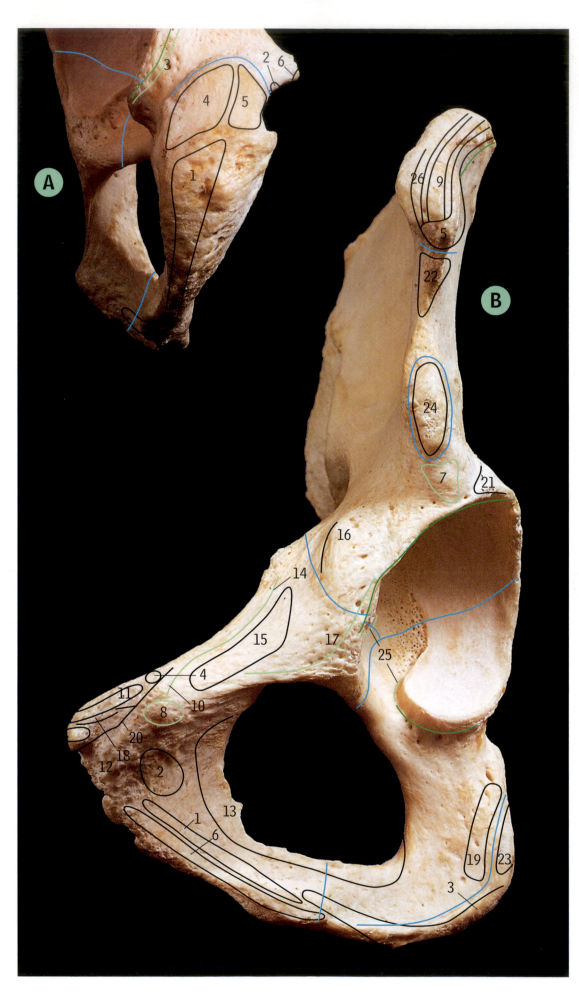

A Left hip bone. Ischial tuberosity, from behind and below. Attachments
Blue lines = epiphysial lines; green line = capsular attachment of hip joint; pale green lines = ligament attachments

1	Adductor magnus
2	Inferior gemellus
3	Ischiofemoral ligament
4	Semimembranosus
5	Semitendinosus and long head of biceps
6	Superior gemellus

- The area on the ischial tuberosity medial to the adductor magnus attachment (1) is covered by fibrofatty tissue and the ischial bursa underlying gluteus maximus.

B Left hip bone, from the front. Attachments
Blue lines = epiphysial lines; green line = capsular attachment of hip joint; pale green lines = ligament attachments

1	Adductor brevis
2	Adductor longus
3	Adductor magnus
4	Conjoint tendon
5	External oblique and inguinal ligament
6	Gracilis
7	Iliofemoral ligament
8	Inguinal ligament
9	Internal oblique
10	Lacunar ligament
11	Lateral head of rectus abdominis
12	Medial head of rectus abdominis
13	Obturator externus
14	Pectineal ligament
15	Pectineus
16	Psoas minor
17	Pubofemoral ligament
18	Pyramidalis
19	Quadratus femoris
20	Rectus sheath
21	Reflected head of rectus femoris
22	Sartorius
23	Semimembranosus
24	Straight head of rectus femoris
25	Transverse ligament
26	Transversus abdominis

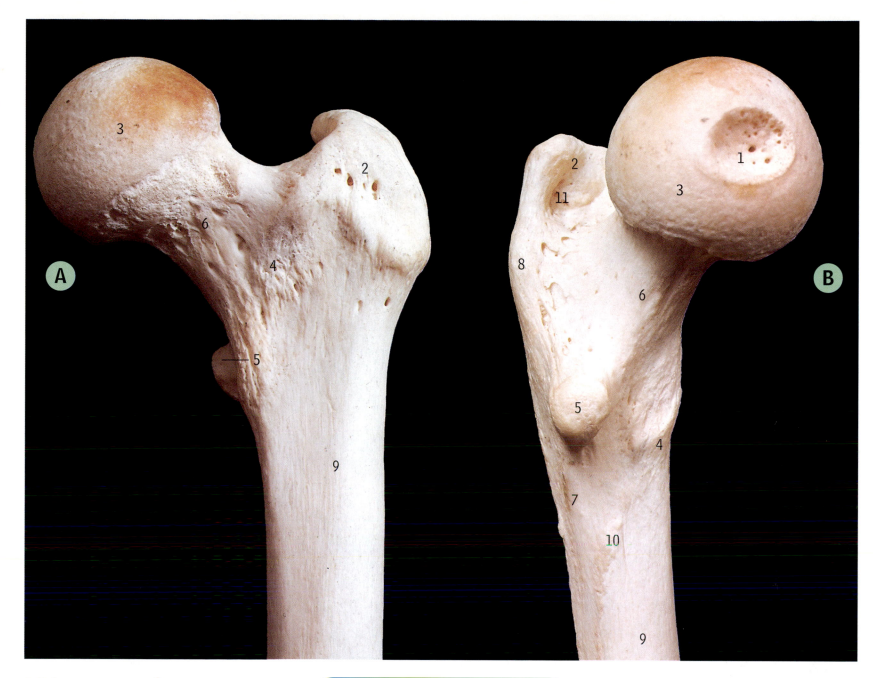

Left femur, upper end, A from the front, **B** from the medial side

1 Fovea of head
2 Greater trochanter
3 Head
4 Intertrochanteric line
5 Lesser trochanter
6 Neck
7 Pectineal line
8 Quadrate tubercle on intertrochanteric crest
9 Shaft
10 Spiral line
11 Trochanteric fossa

Hip fractures occur at the upper end of the femur and result from an indirect twisting force, most commonly in the frail and elderly, and the most common are considered under separate titles.

- The intertrochanteric *line* (4) is at the junction of the neck (6) and shaft (9) on the anterior surface; the intertrochanteric *crest* is in a similar position on the posterior surface (8, and page 256, A5).
- The neck makes an angle with the shaft of about 125°.
- The pectineal line of the femur (7) must not be confused with the pectineal line (pecten) of the pubis (page 251), nor with the spiral line of the femur (10) which is usually more prominent than the pectineal line.

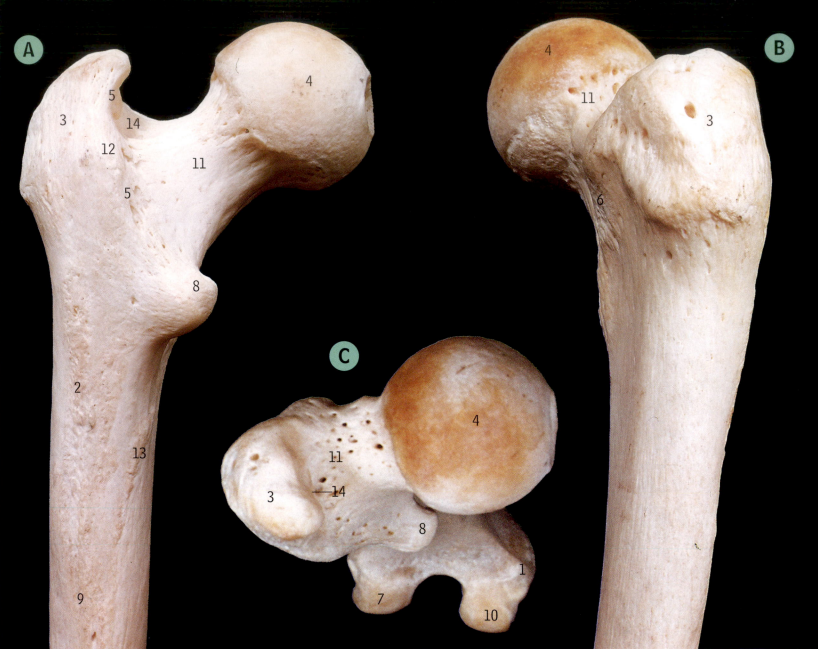

Left femur, upper end, A from behind, B from the lateral side, C from above

- The neck of the femur passes forwards as well as upwards and medially (C11), making an angle of about 15° with the transverse axis of the lower end (the angle of femoral torsion).
- The lesser trochanter (8) projects backwards and medially.

1	Adductor tubercle at lower end
2	Gluteal tuberosity
3	Greater trochanter
4	Head
5	Intertrochanteric crest
6	Intertrochanteric line
7	Lateral condyle at lower end
8	Lesser trochanter
9	Linea aspera
10	Medial condyle at lower end
11	Neck
12	Quadrate tubercle
13	Spiral line
14	Trochanteric fossa

Subcapital fracture of the femoral neck, commonly seen in frail osteoporotic women, can be impacted when the head of the femur is driven into the neck (sometimes with no pain or shortening of the leg) or unimpacted (severe pain with the affected leg shortened, and externally rotated and no active movement of the hip). If untreated, avascular necrosis of the femoral head may occur.

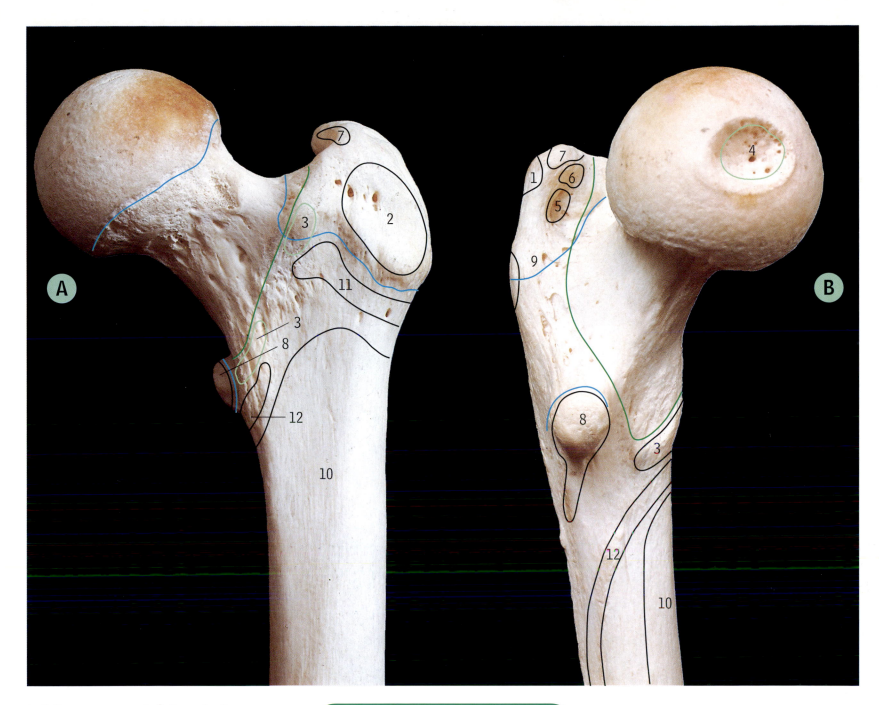

Left femur, upper end, A from the front, B from the medial side. Attachments
Blue lines = epiphysial lines; green line = capsular attachment of hip joint; pale green lines = ligament attachments

1 Gluteus medius
2 Gluteus minimus
3 Iliofemoral ligament
4 Ligament of head of femur
5 Obturator externus
6 Obturator internus and gemelli
7 Piriformis
8 Psoas major and iliacus
9 Quadratus femoris
10 Vastus intermedius
11 Vastus lateralis
12 Vastus medialis

- The iliofemoral ligament has the shape of an inverted V, with the stem attached to the anterior inferior iliac spine of the hip bone (page 254, B7), and the lateral and medial bands attached to the upper (lateral) and lower (medial) ends of the intertrochanteric line (3), blending with the capsule of the hip joint.
- The tendon of psoas major is attached to the lesser trochanter (8); many of the muscle fibres of iliacus are inserted into the psoas tendon but some reach the femur below the trochanter.

Intertrochanteric fracture of the femur is commonly seen in old people of both sexes and often is associated with an avulsion of the lesser trochanter due to pull of the iliopsoas muscle.

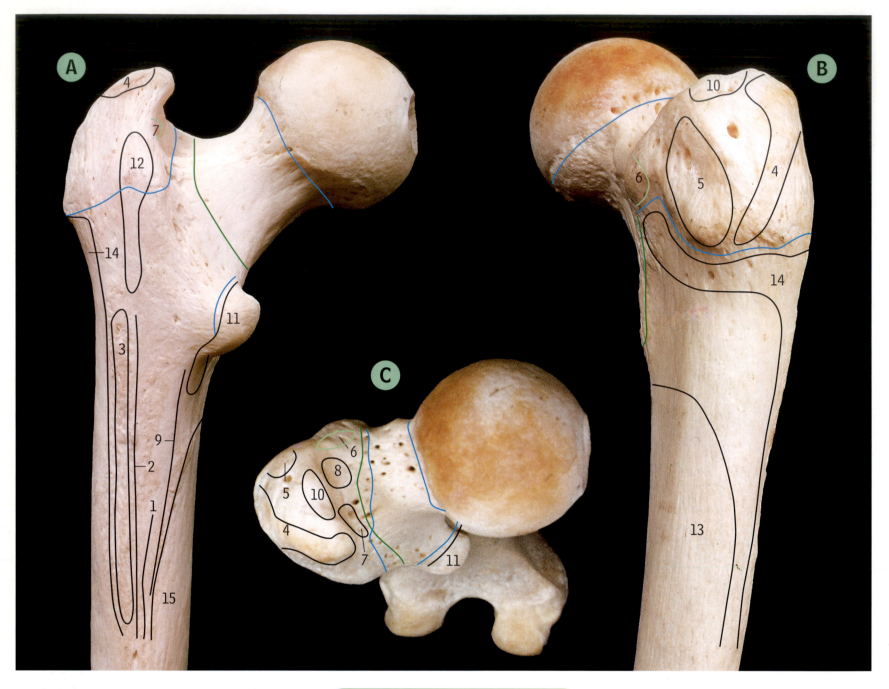

Left femur, upper end, A from behind, B from the lateral side, C from above. Attachments
Blue lines = epiphysial lines; green line = capsular attachment of hip joint; pale green lines = ligament attachments

- On the front of the femur (page 257) the capsule of the hip joint is attached to the intertrochanteric line, but at the back the capsule is attached to the neck of the femur and does not extend as far laterally as the intertrochanteric crest (page 256, A5).

1   Adductor brevis
2   Adductor magnus
3   Gluteus maximus
4   Gluteus medius
5   Gluteus minimus
6   Iliofemoral ligament (lateral band)
7   Obturator externus
8   Obturator internus and gemelli
9   Pectineus
10 Piriformis
11 Psoas major and iliacus
12 Quadratus femoris
13 Vastus intermedius
14 Vastus lateralis
15 Vastus medialis

D Left femur, upper end, from the front
This is the posterior half of a cleared and bisected specimen, to show the major groups of bone trabeculae

1   Calcar femorale
2   From lateral surface of shaft to greater trochanter
3   From lateral surface of shaft to head
4   From medial surface of shaft to greater trochanter
5   From medial surface of shaft to head
6   Triangular area of few trabeculae

- The calcar femorale (1) is a dense concentration of trabeculae passing from the region of the lesser trochanter to the under-surface of the neck.

E Left femur, shaft, from behind

1 Gluteal tuberosity
2 Lateral supracondylar line
3 Lesser trochanter
4 Linea aspera
5 Medial supracondylar line
6 Pectineal line

- The rough linea aspera (4) often shows distinct medial and lateral lips; the lateral lip continues upwards as the gluteal tuberosity (1).

F Left femur, shaft, from behind. Attachments

1 Adductor brevis
2 Adductor longus
3 Adductor magnus
4 Gluteus maximus
5 Pectineus
6 Psoas and iliacus
7 Quadratus femoris
8 Short head of biceps
9 Vastus intermedius
10 Vastus lateralis
11 Vastus medialis

- For diagrammatic clarity the muscle attachments to the linea aspera have been slightly separated.

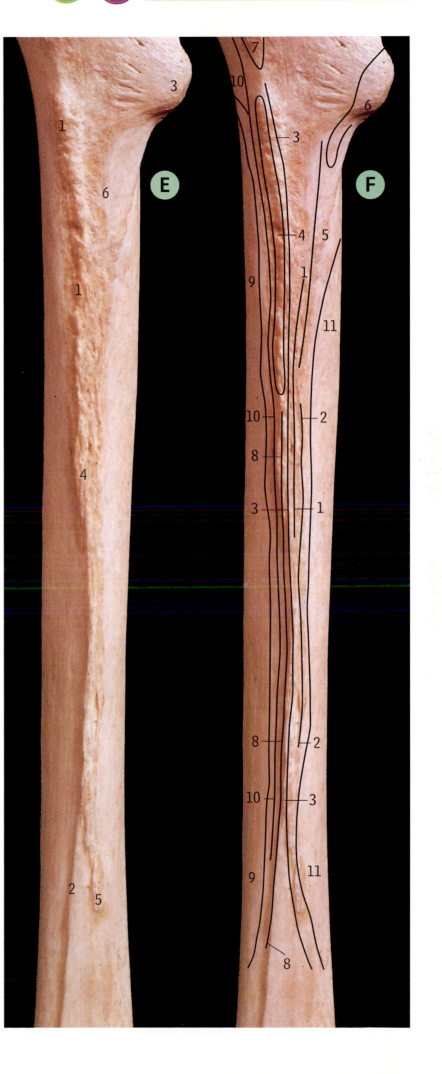

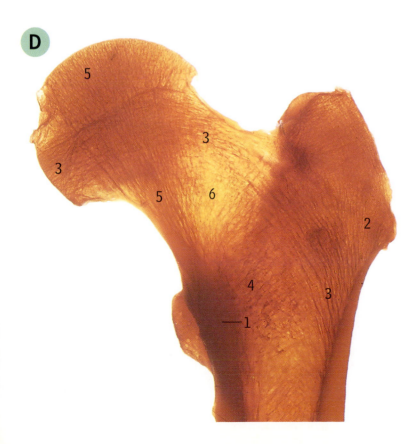

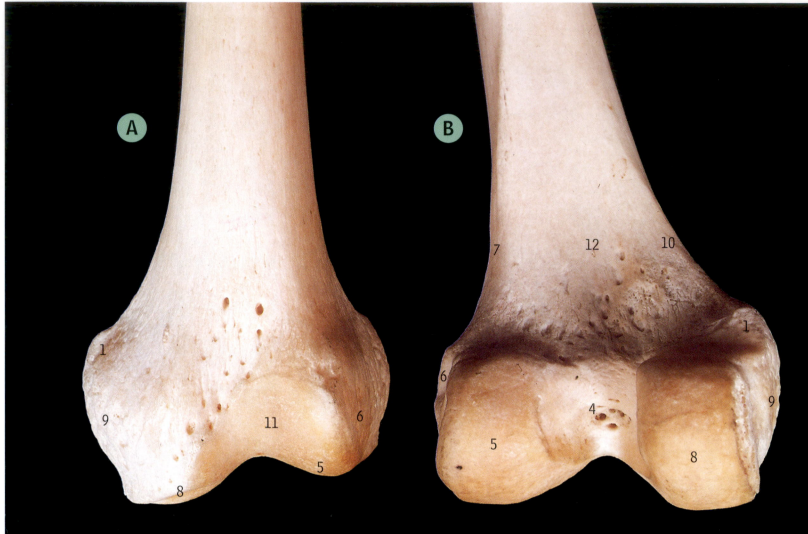

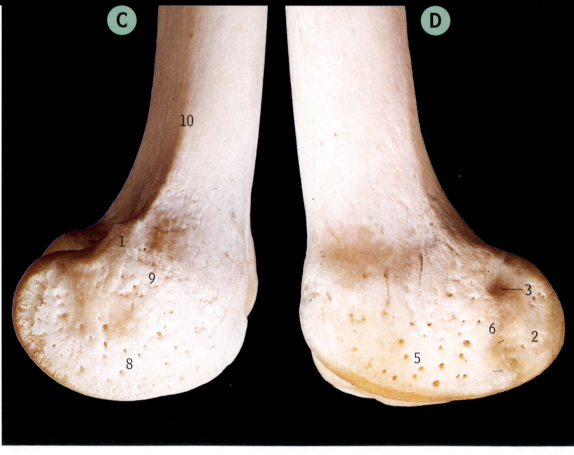

Left femur, lower end, **A** from the front, **B** from behind, **C** from the medial side, **D** from the lateral side

1 Adductor tubercle
2 Groove for popliteus tendon
3 Impression for lateral head of gastrocnemius
4 Intercondylar fossa
5 Lateral condyle
6 Lateral epicondyle
7 Lateral supracondylar line
8 Medial condyle
9 Medial epicondyle
10 Medial supracondylar line
11 Patellar surface
12 Popliteal surface

- The condyles (8 and 5) bear the articular surfaces for the tibia, and project backwards (B8 and 5); the epicondyles (9 and 6) are the most prominent points on the (non-articular) sides of the condyles.
- The lower ends of the condyles (A8 and 5) lie in the same horizontal plane in order to rest squarely on the condyles of the tibia at the knee joint. The shaft therefore passes obliquely outwards and upwards from the knee towards the hip.

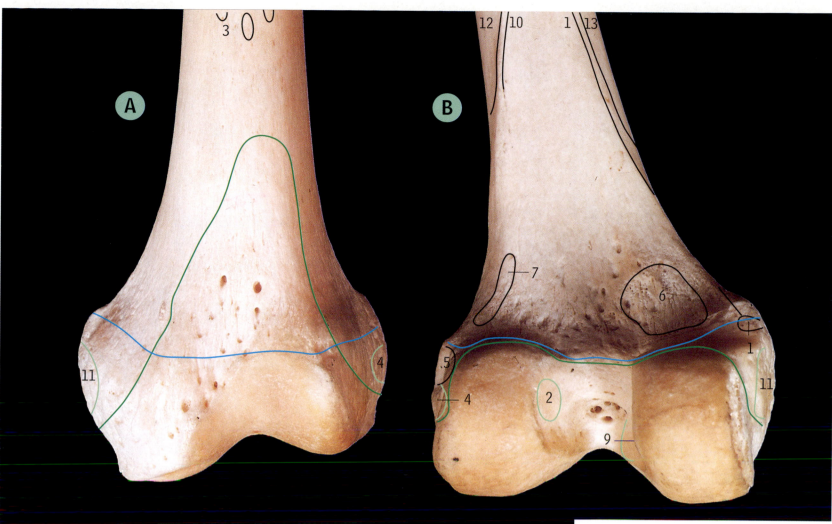

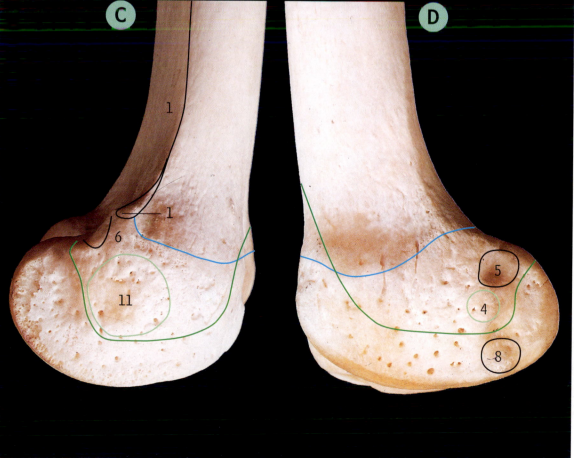

Left femur, lower end, **A** from the front,
B from behind, **C** from the medial side,
D from the lateral side. Attachments
Blue lines = epiphysial lines; green line =
capsular attachment of knee joint; pale green
lines = ligament attachments

1 Adductor magnus
2 Anterior cruciate ligament
3 Articularis genu
4 Fibular collateral ligament
5 Lateral head of gastrocnemius
6 Medial head of gastrocnemius
7 Plantaris
8 Popliteus
9 Posterior cruciate ligament
10 Short head of biceps
11 Tibial collateral ligament
12 Vastus intermedius
13 Vastus medialis

● The medial head of gastrocnemius (B6) arises
from the popliteal surface of the femur above
the medial condyle and from the adjacent part
of the capsule; the lateral head (D5) arises from
an impression on the lateral surface of the
lateral condyle above the lateral epicondyle
(not from the popliteal surface of the femur)
and from the adjacent part of the capsule.

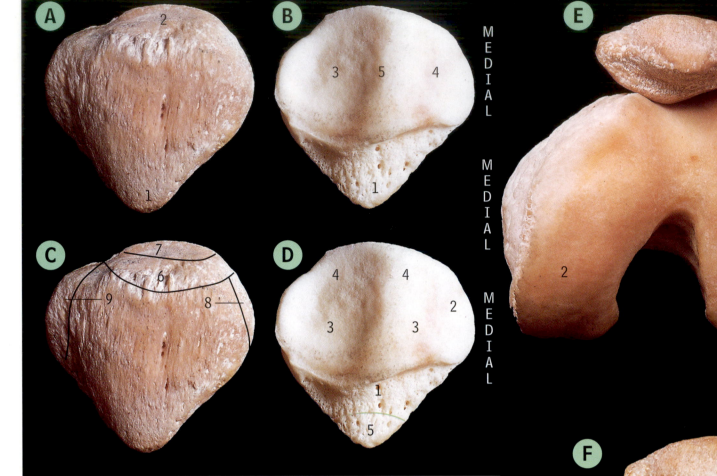

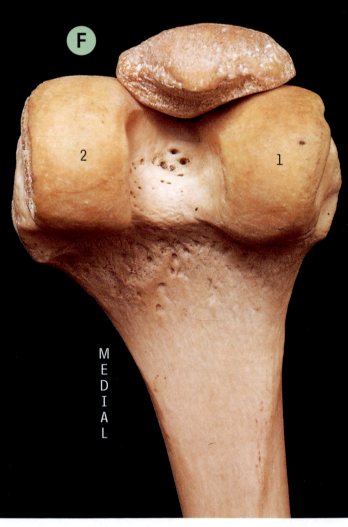

**Left patella, A anterior surface,
B articular (posterior) surface**

1 Apex
2 Base
3 Facet for lateral condyle of femur
4 Facet for medial condyle of femur
5 Vertical ridge

- The lateral part of the articular surface (B3) is larger than the medial (B4).
- The articular surface does not extend on to the apex (B1).

**Left patella, C anterior surface,
D articular (posterior) surface.
Attachments**
Pale green line = ligament attachment

1 Area for infrapatellar fat pad
2 Area for medial condyle in extreme flexion
3 Facets for femur in extension
4 Facets for femur in flexion
5 Patellar ligament
6 Rectus femoris of quadriceps tendon
7 Vastus intermedius of quadriceps tendon
8 Vastus lateralis of quadriceps tendon
9 Vastus medialis of quadriceps tendon

- The most medial facet of the patella (D2) only comes into contact with the medial condyle in extreme flexion as in F.

Dislocation of the patella is most commonly seen as a lateral dislocation in young women or overweight young boys who have a wide pelvis. Anatomically, it is often due to a flat lateral femoral condyle, an abnormality of the knee (genu valgum) or weakness in the vastus medialis (lower fibres).

**Left femur and patella articulated,
E from below with knee extended, F
from below and behind with knee flexed**
In flexion note the increased area of contact between the medial condyle of the femur (2) and the patella

1 Lateral condyle
2 Medial condyle

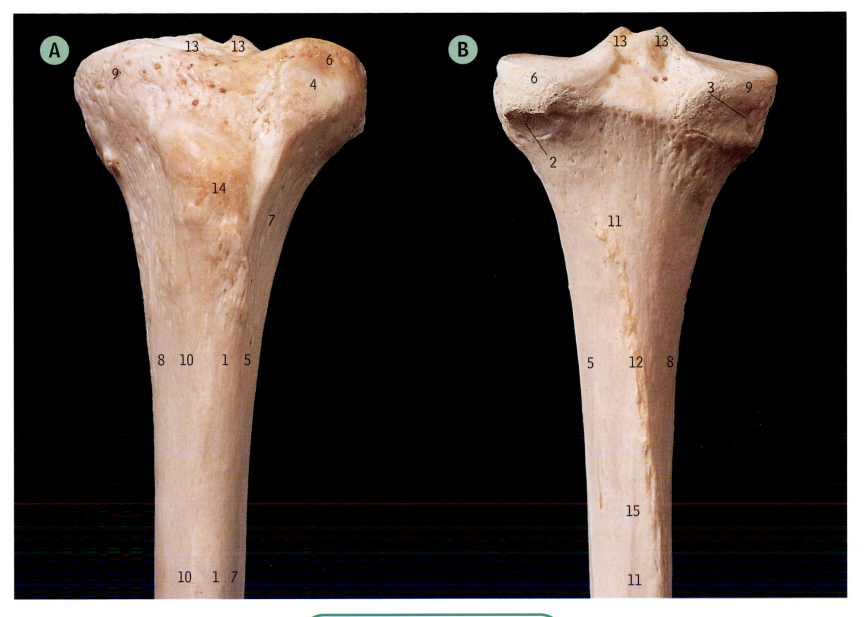

Left tibia, upper end, A from the front, B from behind

1 Anterior border
2 Articular facet for fibula
3 Groove for semimembranosus
4 Impression for iliotibial tract
5 Interosseous border
6 Lateral condyle
7 Lateral surface
8 Medial border
9 Medial condyle
10 Medial surface
11 Posterior surface
12 Soleal line
13 Tubercles of intercondylar eminence
14 Tuberosity
15 Vertical line

- The shaft of the tibia has three borders – anterior (1), medial (8) and interosseous (5) – and three surfaces – medial (10), lateral (7) and posterior (11).
- Much of the anterior border (1) forms a slightly curved crest commonly known as the shin. Most of the smooth medial surface (10) is subcutaneous. The posterior surface contains the soleal and vertical lines (12 and 15).
- The tuberosity (14) is at the upper end of the anterior border.

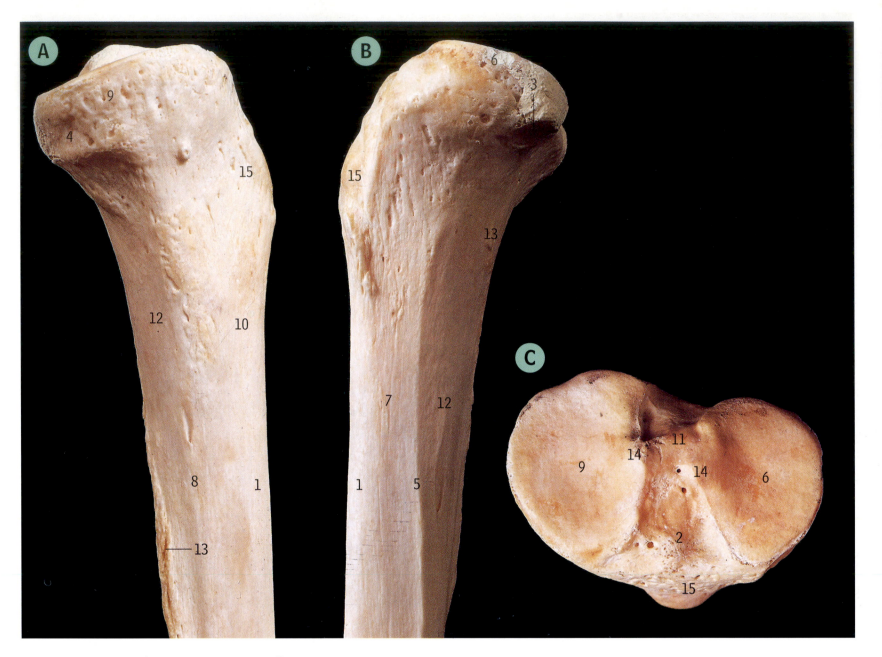

Left tibia, upper end, **A** from the medial side, **B** from the lateral side, **C** from above

- The medial condyle (C9) is larger than the lateral condyle (C6).
- The articular facet for the fibula is on the postero-inferior aspect of the lateral condyle (B3).

1	Anterior border	9	Medial condyle
2	Anterior intercondylar area	10	Medial surface
3	Articular facet for fibula	11	Posterior intercondylar area
4	Groove for semimembranosus	12	Posterior surface
5	Interosseous border	13	Soleal line
6	Lateral condyle	14	Tubercles of intercondylar eminence
7	Lateral surface	15	Tuberosity
8	Medial border		

Avulsion of the tibial tuberosity (Osgood–Schlatter disease) is seen in late childhood when a small part of the upper tibial epiphysis to which the ligamentum patellae is attached is pulled superiorly.

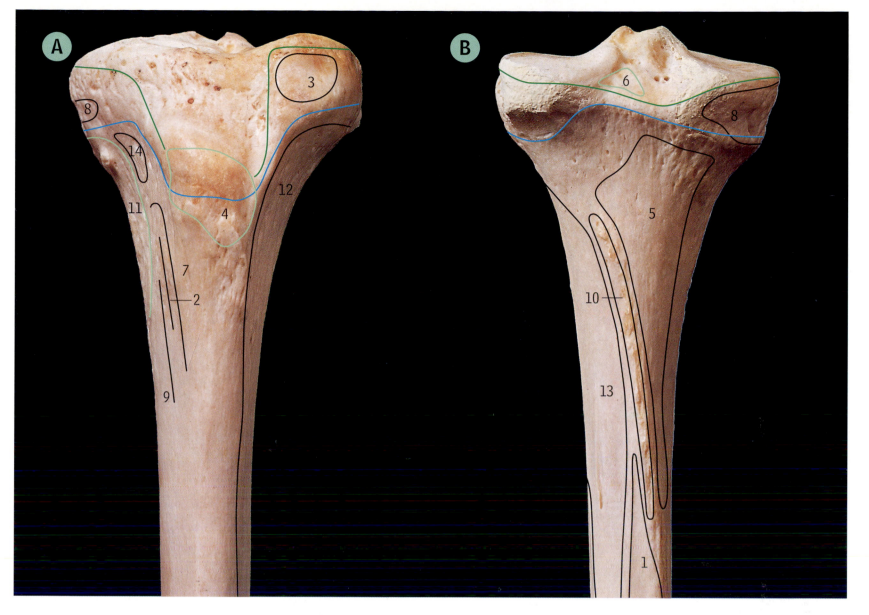

Left tibia, upper end, **A** from the front, **B** from behind. Attachments
Blue lines = epiphysial lines; green line = capsular attachment of knee joint; pale green lines = ligament attachments

1 Flexor digitorum longus	8 Semimembranosus
2 Gracilis	9 Semitendinosus
3 Iliotibial tract	10 Soleus
4 Patellar ligament	11 Tibial collateral ligament
5 Popliteus	12 Tibialis anterior
6 Posterior cruciate ligament	13 Tibialis posterior
7 Sartorius	14 Vastus medialis

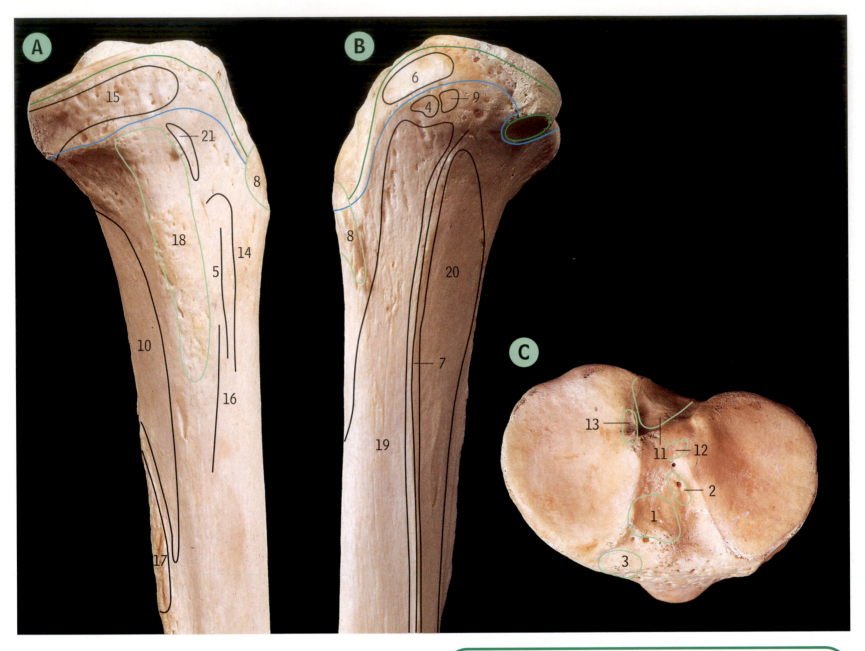

Left tibia, upper end, **A** from the medial side, **B** from the lateral side, **C** from above. Attachments

Blue lines = epiphysial lines; green lines = capsular attachments of knee joint and superior tibiofibular joint; pale green lines = ligament attachments

1	Anterior cruciate ligament	12	Posterior horn of lateral meniscus
2	Anterior horn of lateral meniscus	13	Posterior horn of medial meniscus
3	Anterior horn of medial meniscus	14	Sartorius
4	Extensor digitorum longus	15	Semimembranosus
5	Gracilis	16	Semitendinosus
6	Iliotibial tract	17	Soleus
7	Interosseous membrane	18	Tibial collateral ligament
8	Patellar ligament	19	Tibialis anterior
9	Peroneus longus	20	Tibialis posterior
10	Popliteus	21	Vastus medialis
11	Posterior cruciate ligament		

- Although arising mainly from the fibula (see page 269), extensor digitorum longus (B4) and peroneus longus (B9) have a small attachment to the tibia above tibialis anterior (B19).

- The horns of the lateral meniscus (C12 and 2) are attached close to one another on either side of the intercondylar eminence, but the horns of the medial meniscus (C13 and 3) are widely separated (see page 293).

- The tibial attachment of the anterior cruciate ligament (C1) is to the top of the intercondylar area, but the attachment of the posterior cruciate ligament (C11) extends 'over the top' on to the posterior surface.

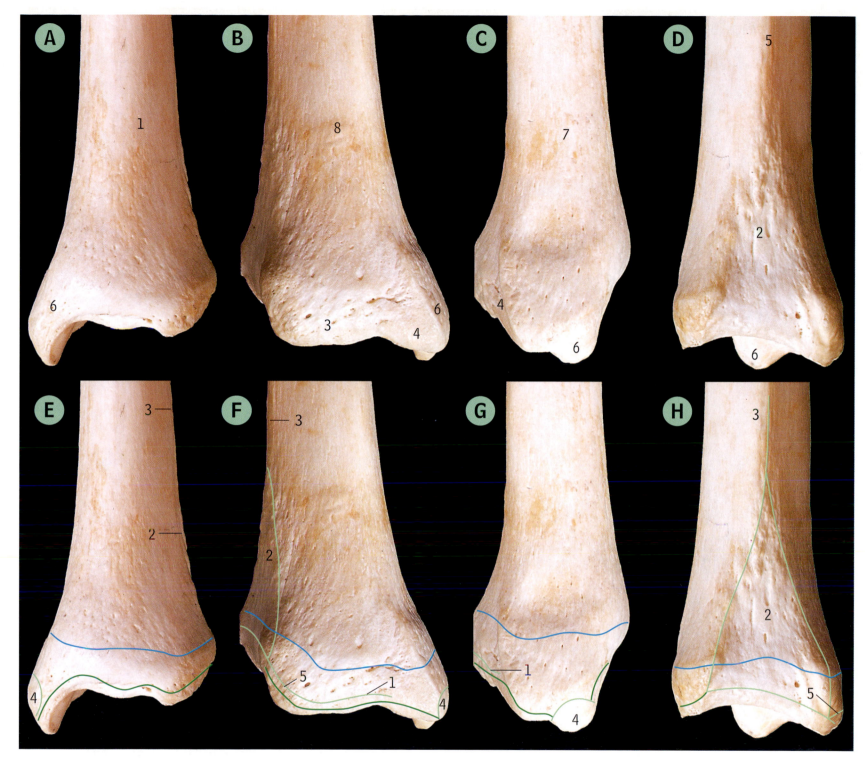

Left tibia, lower end, **A** from the front, **B** from behind, **C** from the medial side, **D** from the lateral side

- The lower end of the tibia has five surfaces – anterior, posterior, medial, lateral and inferior (for the inferior surface see page 270).
- The medial surface (C7) is continuous below with the medial surface of the medial malleolus (C6) (the lateral malleolus is the lower end of the fibula, see page 268).
- The fibular notch (D2) is triangular and constitutes the lateral surface of the lower end.

- The medial collateral ligament (G4) is commonly known as the deltoid ligament.
- The lowest fibres of the posterior tibiofibular ligament (attached most medially to the tibia) are known as the inferior transverse ligament (F5 and 1).

Left tibia, lower end, **E** from the front, **F** from behind, **G** from the medial side, **H** from the lateral side. Attachments
Blue line = epiphysial line; green line = capsular attachment of ankle joint; pale green lines = ligament attachments

1 Anterior surface
2 Fibular notch
3 Groove for flexor hallucis longus
4 Groove for tibialis posterior
5 Interosseous border
6 Medial malleolus
7 Medial surface
8 Posterior surface

1 Inferior transverse ligament
2 Interosseous ligament
3 Interosseous membrane
4 Medial collateral ligament
5 Posterior tibiofibular ligament

Lower limb bones

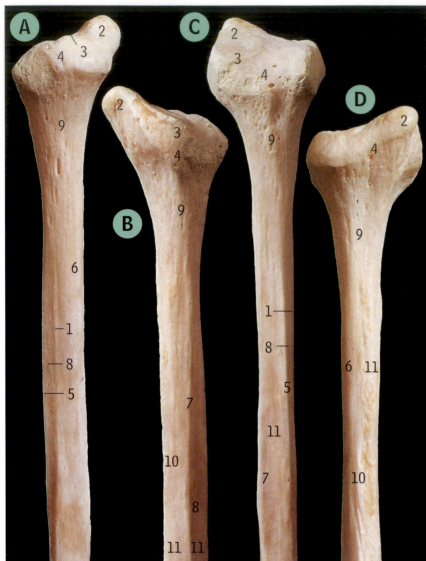

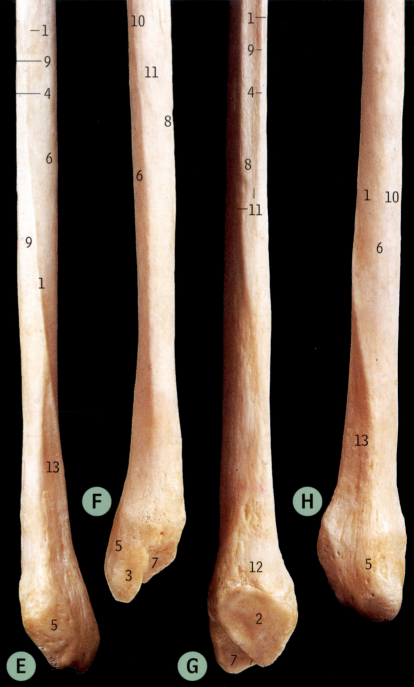

Left fibula, upper end, **A** from the front, **B** from behind, **C** from the medial side, **D** from the lateral side

1 Anterior border
2 Apex (styloid process)
3 Articular facet on upper surface
4 Head
5 Interosseous border
6 Lateral surface
7 Medial crest
8 Medial surface
9 Neck
10 Posterior border
11 Posterior surface

- The fibula has three borders—anterior (A1), interosseous (A5) and posterior (B10) – and three surfaces – medial (A8), lateral (A6) and posterior (B11).
- At first sight much of the shaft appears to have four borders and four surfaces, but this is because the posterior surface (B11) is divided into two parts (medial and lateral) by the medial crest (B7).

Left fibula, lower end, **E** from the front, **F** from behind, **G** from the medial side, **H** from the lateral side

1 Anterior border
2 Articular surface of lateral malleolus
3 Groove for peroneus brevis
4 Interosseous border
5 Lateral malleolus
6 Lateral surface
7 Malleolar fossa
8 Medial crest
9 Medial surface
10 Posterior border
11 Posterior surface
12 Surface for interosseous ligament
13 Triangular subcutaneous area

- At the lower end the lateral surface (H6) comes to face posteriorly, so leaving the triangular subcutaneous area (H13) above the lateral malleolus (H5).
- The anterior border (E1) is easily identified by following it upwards from the apex of the triangular subcutaneous area (E13); the interosseous border (E4) is usually 2–3 mm behind the anterior border (although in the upper part of the shaft these two borders may fuse into one).
- The malleolar fossa (G7) is posterior to the articular surface (G2).

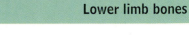

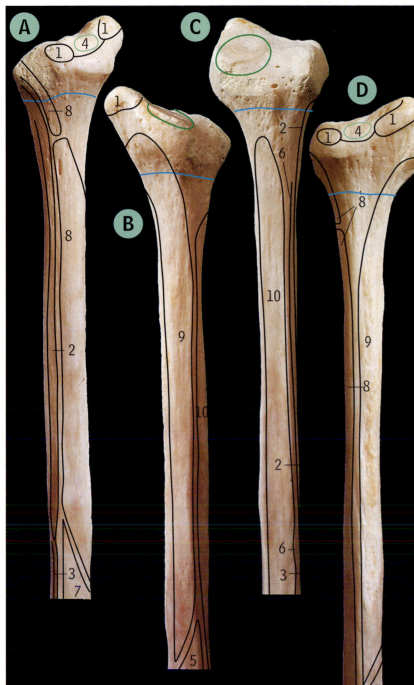

Left fibula, upper end, **A** from the front, **B** from behind, **C** from the medial side, **D** from the lateral side. Attachments

Blue line = epiphysial line; green line = capsular attachment of superior tibiofibular joint; pale green lines = ligament attachments

1	Biceps	6	Interosseous membrane
2	Extensor digitorum longus	7	Peroneus brevis
3	Extensor hallucis longus	8	Peroneus longus
4	Fibular collateral ligament	9	Soleus
5	Flexor hallucis longus	10	Tibialis posterior

- The posterior surface (between the interosseous and posterior borders) gives origin to flexor muscles – soleus (B9) and flexor hallucis longus (B5) lateral to the medial crest, and tibialis posterior (B10) medial to the medial crest.

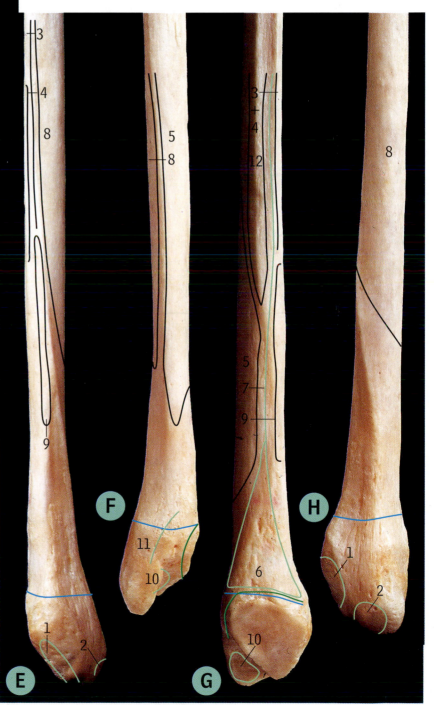

Left fibula, lower end, **E** from the front, **F** from behind, **G** from the medial side, **H** from the lateral side. Attachments

Blue line = epiphysial line; green line = capsular attachment of ankle joint; pale green lines = ligament attachments

1	Anterior talofibular ligament	7	Interosseous membrane
2	Calcaneofibular ligament	8	Peroneus brevis
3	Extensor digitorum longus	9	Peroneus tertius
4	Extensor hallucis longus	10	Posterior talofibular ligament
5	Flexor hallucis longus	11	Posterior tibiofibular ligament
6	Interosseous ligament	12	Tibialis posterior

- The medial surface (between the anterior and interosseous borders) gives origin to extensor muscles – extensor digitorum longus (A2), extensor hallucis longus (A3) and peroneus tertius (E9).
- The lateral surface (between the anterior and posterior borders) gives origin to peroneus longus (A8) and peroneus brevis (A7).

Left tibia and fibula articulated, **A** upper ends from behind, **B** upper ends from above, **C** upper end of fibula from above, **D** lower ends from behind, **E** lower ends from below

1 Apex of head (styloid process)
2 Articular facet (for superior tibiofibular joint)
3 Articular facet of lateral malleolus (for ankle joint)
4 Head of fibula
5 Inferior surface of tibia (for ankle joint)
6 Inferior tibiofibular joint
7 Lateral (articular) surface of medial malleolus (for ankle joint)
8 Lateral condyle of tibia
9 Lateral malleolus
10 Malleolar fossa
11 Medial malleolus
12 Superior tibiofibular joint

- The superior tibiofibular joint (A12) is synovial, with the tibial facet of the joint on the posterolateral and lower aspect of the lateral condyle. The facet on the fibula is towards the posterior and medial part of the upper surface of the head (C2).
- The inferior tibiofibular joint (D6) is fibrous.
- The lateral malleolus (D9) extends lower than the medial malleolus (D11). The articular surfaces of the malleoli (D3, E7) together with the inferior surface of the tibia (E5) embrace the talus to form the ankle (talocrural) joint.

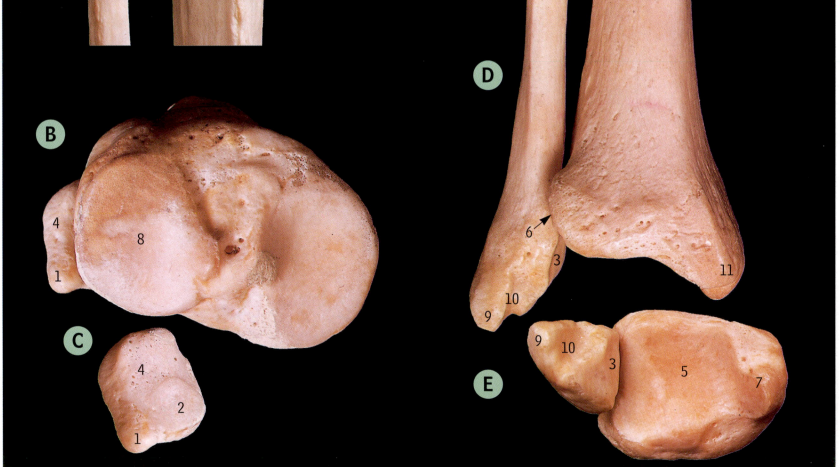

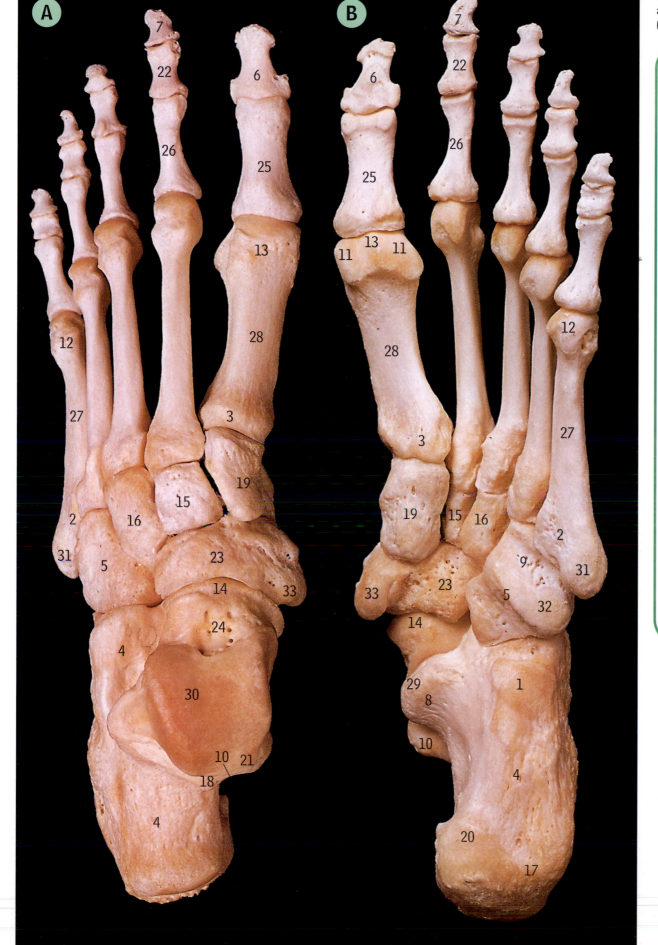

Bones of the left foot, **A** from above (dorsum), **B** from below (plantar surface)

1 Anterior tubercle of calcaneus
2 Base of fifth metatarsal
3 Base of first metatarsal
4 Calcaneus
5 Cuboid
6 Distal phalanx of great toe
7 Distal phalanx of second toe
8 Groove on calcaneus for flexor hallucis longus
9 Groove on cuboid for peroneus longus
10 Groove on talus for flexor hallucis longus
11 Grooves for sesamoid bones in flexor hallucis brevis
12 Head of fifth metatarsal
13 Head of first metatarsal
14 Head of talus
15 Intermediate cuneiform
16 Lateral cuneiform
17 Lateral process of calcaneus
18 Lateral tubercle of talus
19 Medial cuneiform
20 Medial process of calcaneus
21 Medial tubercle of talus
22 Middle phalanx of second toe
23 Navicular
24 Neck of talus
25 Proximal phalanx of great toe
26 Proximal phalanx of second toe
27 Shaft of fifth metatarsal
28 Shaft of first metatarsal
29 Sustentaculum tali of calcaneus
30 Trochlear surface of body of talus
31 Tuberosity of base of fifth metatarsal
32 Tuberosity of cuboid
33 Tuberosity of navicular

Hallux valgus, lateral displacement of the great toe, usually presents as pain over a prominent metatarsal head due to rubbing from shoes and it may be associated with deformity of the second toe which tends to override the great toe.

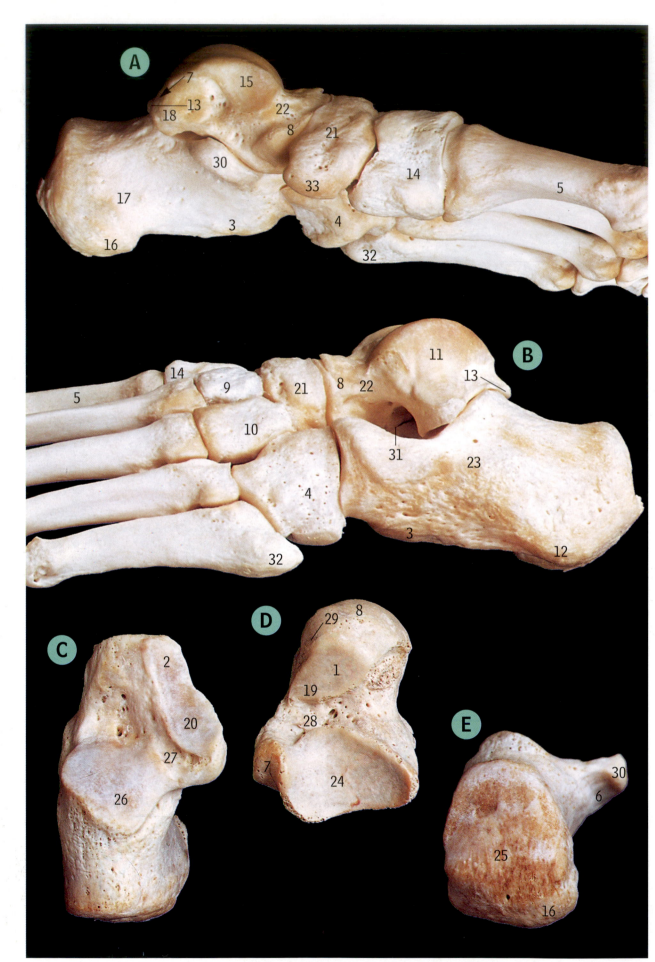

Bones of the left foot, **A** from the medial side, **B** from the lateral side, **C** calcaneus from above, **D** talus from below, **E** calcaneus from behind

1 Anterior calcanean articular surface of talus
2 Anterior talal articular surface of calcaneus
3 Anterior tubercle of calcaneus
4 Cuboid
5 First metatarsal
6 Groove for flexor hallucis longus of calcaneus
7 Groove for flexor hallucis longus of talus
8 Head of talus
9 Intermediate cuneiform
10 Lateral cuneiform
11 Lateral malleolar surface of talus
12 Lateral process of calcaneus
13 Lateral tubercle of talus
14 Medial cuneiform
15 Medial malleolar surface of talus
16 Medial process of calcaneus
17 Medial surface of calcaneus
18 Medial tubercle of talus
19 Middle calcanean articular surface of talus
20 Middle talal articular surface of calcaneus
21 Navicular
22 Neck of talus
23 Peroneal trochlea of calcaneus
24 Posterior calcanean articular surface of talus
25 Posterior surface of calcaneus
26 Posterior talal articular surface of calcaneus
27 Sulcus of calcaneus
28 Sulcus of talus
29 Surface for plantar calcaneonavicular (spring) ligament of talus
30 Sustentaculum tali of calcaneus
31 Tarsal sinus
32 Tuberosity of base of fifth metatarsal
33 Tuberosity of navicular

Fracture of the fifth metatarsal is a common foot fracture; the styloid process of the fifth metatarsal is often avulsed, owing to attachment of the peroneus brevis muscle at this site.

Bones of the left foot, **A** from above, **B** from below. Attachments Joint capsules and minor ligaments have been omitted. Pale green lines = ligament attachments

1 Abductor digiti minimi
2 Abductor hallucis
3 Adductor hallucis
4 Calcaneocuboid part of bifurcate ligament
5 Calcaneonavicular part of bifurcate ligament
6 Extensor digitorum brevis
7 Extensor digitorum longus
8 Extensor digitorum longus and brevis
9 Extensor hallucis brevis
10 Extensor hallucis longus
11 First dorsal interosseous
12 First plantar interosseous
13 Flexor accessorius
14 Flexor digiti minimi brevis
15 Flexor digitorum brevis
16 Flexor digitorum longus
17 Flexor hallucis brevis
18 Flexor hallucis longus
19 Fourth dorsal interosseous
20 Long plantar ligament
21 Opponens digiti minimi (part of 14)
22 Peroneus brevis
23 Peroneus longus
24 Peroneus tertius
25 Plantar calcaneocuboid (short plantar) ligament
26 Plantar calcaneonavicular (spring) ligament
27 Plantaris
28 Second dorsal interosseous
29 Second plantar interosseous
30 Tendo calcaneus (Achilles tendon)
31 Third dorsal interosseous
32 Third plantar interosseous
33 Tibialis anterior
34 Tibialis posterior

Metatarsal fractures. Following a 'sprained ankle', occasionally the peroneus brevis tendon pulls off the base of the fifth metatarsal. However, much more common are stress fractures of the middle metatarsals, often known as 'march fractures'.

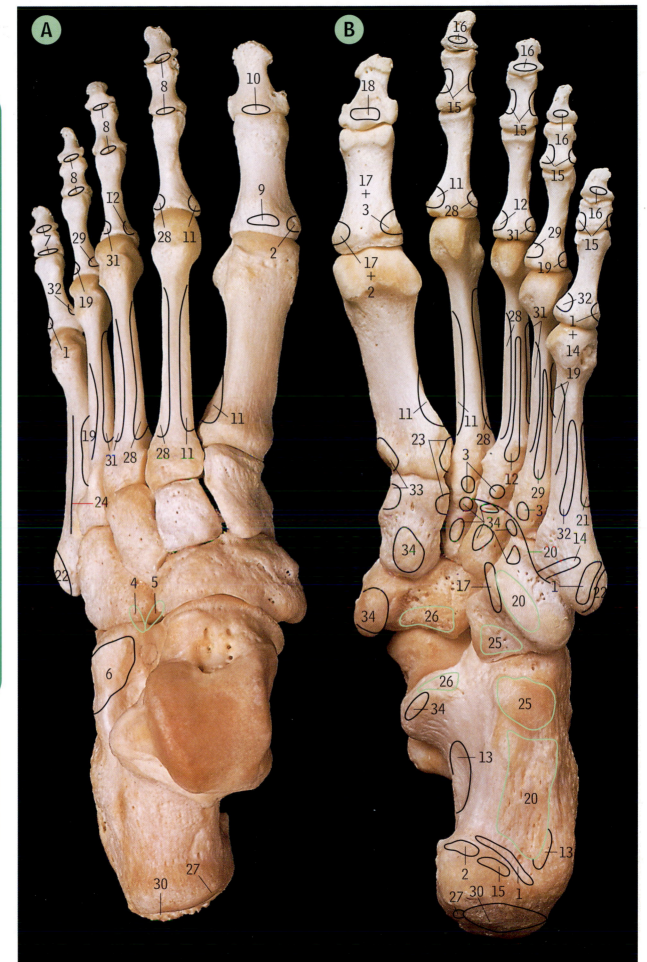

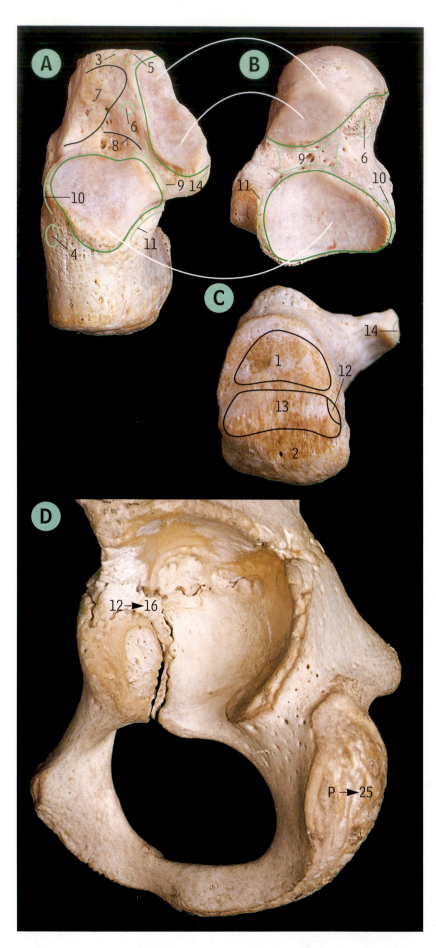

Left calcaneus, **A** from above, **C** from behind. **B** Left talus, from below
Curved lines indicate corresponding articular surfaces: green = capsular attachment of talocalcanean (subtalar) and talocalcaneonavicular joints; pale green lines = ligament attachments

1 Area for bursa
2 Area for fibrofatty tissue
3 Calcaneocuboid part of bifurcate ligament
4 Calcaneofibular ligament
5 Calcaneonavicular part of bifurcate ligament
6 Cervical ligament
7 Extensor digitorum brevis
8 Inferior extensor retinaculum
9 Interosseous talocalcanean ligament
10 Lateral talocalcanean ligament
11 Medial talocalcanean ligament
12 Plantaris
13 Tendo calcaneus (Achilles tendon)
14 Tibiocalcanean part of deltoid ligament

- The interosseous talocalcanean ligament (9) is formed by thickening of the adjacent capsules of the talocalcanean and talocalcaneonavicular joints.
- For different interpretations of the term 'subtalar joint' see the notes on page 308.

Secondary centres of ossification of the left lower limb bones
D Hip bone, lower lateral part
E and **F** Femur, upper and lower ends
G and **H** Tibia, upper and lower ends
J and **K** Fibula, upper and lower ends
L Calcaneus
M Metatarsal and phalanges of second toe
N Metatarsal and phalanges of great toe

Figures in years, commencement of ossification → fusion.
P = puberty, B = ninth intra-uterine month. See introduction on page 104
- In the hip bone (D) one or more secondary centres appear in the Y-shaped cartilage between ilium, ischium and pubis. Other centres (not illustrated) are usually present for the iliac crest, anterior inferior iliac spine, and (possibly) the pubic tubercle and pubic crest (all P → 25).
- The patella (not illustrated) begins to ossify from one or more centres between the third and sixth year.
- All the phalanges, and the first metatarsal, have a secondary centre at their proximal ends; the other metatarsals have one at their distal ends.
- Of the tarsal bones, the largest, the calcaneus, begins to ossify in the third intra-uterine month and the talus about three months later. The cuboid may begin to ossify either just before or just after birth, with the lateral cuneiform in the first year, medial cuneiform at two years and the intermediate cuneiform and navicular at three years.
- The calcaneus (L) is the only tarsal bone to have a secondary centre.

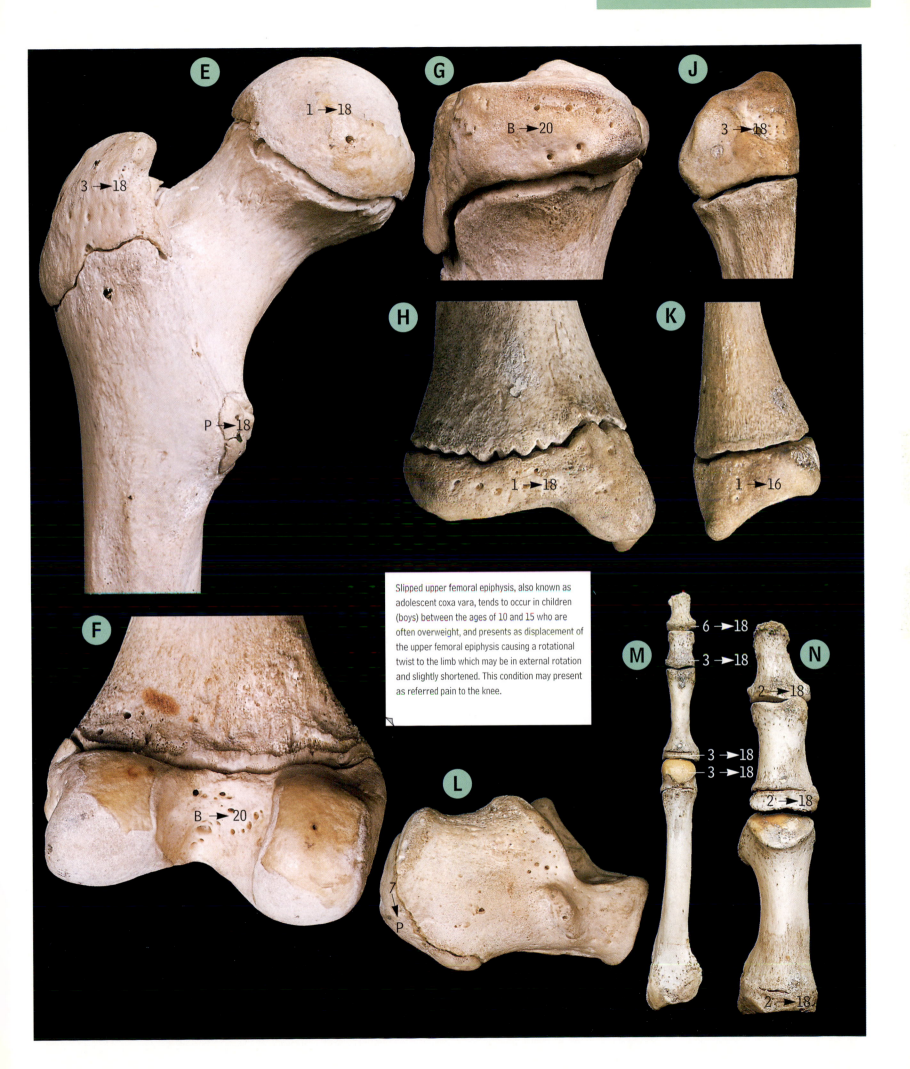

E

1 → 18

3 → 18

P → 18

G

B → 20

J

3 → 18

H

1 → 18

K

1 → 16

F

B → 20

Slipped upper femoral epiphysis, also known as adolescent coxa vara, tends to occur in children (boys) between the ages of 10 and 15 who are often overweight, and presents as displacement of the upper femoral epiphysis causing a rotational twist to the limb which may be in external rotation and slightly shortened. This condition may present as referred pain to the knee.

L

7
P

M

6 → 18

3 → 18

3 → 18
3 → 18

N

2 → 18

2 → 18

2 → 18

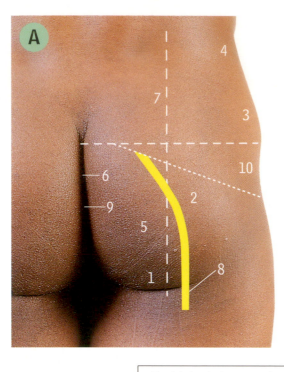

A Right gluteal region. Surface features
The interrupted lines divide the gluteal region into quadrants (see the note below). The iliac crest (4) with the posterior superior iliac spine (7), the tip of the coccyx (9), the ischial tuberosity (5) and the tip of the greater trochanter of the femur (10) are palpable landmarks. A line drawn from a point midway between the posterior superior iliac spine (7) and the tip of the coccyx (9) to the tip of the greater trochanter (10) marks the lower border of piriformis (dotted white line) which is a key feature of the gluteal region, where the most important structure is the sciatic nerve (indicated here in yellow, 8; see dissections and notes opposite).

1 Fold of buttock
2 Gluteus maximus
3 Gluteus medius
4 Iliac crest
5 Ischial tuberosity
6 Natal cleft
7 Posterior superior iliac spine
8 Sciatic nerve
9 Tip of coccyx
10 Tip of greater trochanter of femur

Sciatica is a clinical condition presenting as pain down the leg anywhere from the buttock to the heel and may be from any cause irritating the sciatic trunk or its nerve roots. Commonly it is due to intervertebral disc pathology in the L4/L5 or L5/S1 level. In the younger patient, back problems are the most common cause of this presenting symptom; however, in the older patient it is important to exclude neoplastic pathology within the pelvis, such as spread from rectal, uterine or prostate malignancies. A useful test for its diagnosis is the straight leg raising test (see page 319) or Achilles tendon reflex (see page 304), which tests the S1/2 nerve root.

Trendelenburg's sign occurs when a patient stands on one leg, and the contralateral hip droops. The causes are numerous but any damage to the abductors (tensor fasciae latae, gluteus medius and gluteus minimus), the superior gluteal nerve, or an abnormality of the hip joint (congenital dislocation) will give this positive sign. It is often associated with a 'waddling' gait as the person is unable to raise the pelvis on the swing-through leg during normal walking.

B Right gluteal region. Superficial nerves
Skin and subcutaneous tissue have been removed, preserving cutaneous branches from the first three lumbar (3) and first three sacral (4) nerves, the cutaneous branches of the posterior femoral cutaneous nerve (5) and the perforating cutaneous nerve (11). The curved line near the bottom of the picture indicates the position of the gluteal fold (fold of the buttock). The muscle fibres of gluteus maximus (7) run downwards and laterally, and its lower border does not correspond to the gluteal fold

1 Adductor magnus
2 Coccyx
3 Cutaneous branches of dorsal rami of first three lumbar nerves
4 Gluteal branches of dorsal rami of first three sacral nerves
5 Gluteal branches of the posterior femoral cutaneous nerve
6 Gluteal fascia overlying gluteus medius
7 Gluteus maximus
8 Gracilis
9 Iliac crest
10 Ischio-anal fossa and levator ani
11 Perforating cutaneous nerve
12 Posterior layer of lumbar fascia overlying erector spinae
13 Semitendinosus

- The first three lumbar nerves and the first three sacral nerves supply skin over the gluteal region (by the lateral branches of their dorsal rami, 3 and 4) but the intervening fourth and fifth lumbar nerves do not have a cutaneous distribution in this region.
- The gluteal region or buttock is sometimes used as a site for intramuscular injections. The correct site is in the upper outer quadrant of the buttock, and for delimiting this quadrant it is essential to remember that the upper boundary of the buttock is the uppermost part of the iliac crest. The lower boundary is the fold of the buttock. Dividing the area between these two boundaries by a vertical line midway between the midline and the lateral side of the body indicates that the upper outer quadrant is well above and to the right of the label 7 in B, and this is the safe site for injection—well above and to the right of the sciatic nerve which is displayed in the dissections opposite.

Right gluteal region, **C** with most of gluteus maximus removed, **D** with the sciatic trunk displaced

1	Common peroneal part of sciatic nerve	13	Obturator externus
2	Gluteus maximus	14	Obturator internus
3	Gluteus medius	15	Piriformis
4	Gluteus minimus	16	Posterior femoral cutaneous nerve
5	Greater trochanter of femur	17	Pudendal nerve
6	Inferior gemellus	18	Quadratus femoris
7	Inferior gluteal artery	19	Sacrotuberous ligament
8	Inferior gluteal nerve	20	Superior gemellus
9	Internal pudendal artery	21	Superior gluteal artery
10	Ischial tuberosity	22	Superior gluteal nerve
11	Nerve to obturator internus	23	Tibial part of sciatic nerve
12	Nerve to quadratus femoris		

Gluteal nerve paralysis. Superior gluteal nerve damage affects the three abductor muscles (tensor fasciae latae, gluteus medius and gluteus minimus), resulting in a waddling gait (sometimes called a Trendelenburg gait) seen in patients with congenital dislocation of the hip. Inferior gluteal nerve damage, which affects only gluteus maximus, is a disabling condition because gluteus maximus is the strongest muscle used in extending the leg on the trunk, such as in running or in climbing stairs. These nerves may also be damaged during intramuscular injections into the buttock.

- The two parts of the sciatic trunk (common peroneal and tibial, 1 and 23) usually divide from one another at the top of the popliteal fossa (page 290) but are sometimes separate as they emerge beneath piriformis, and the common peroneal may even perforate piriformis.

Sciatic trunk paralysis may be the unfortunate result of a badly placed intramuscular injection. When injecting into the buttock it is important to remember the surface markings of the sciatic trunk which passes through the middle of the buttock swelling. Injections should be placed in the upper outer quadrant of the buttock above a line joining the greater trochanter and the posterior superior iliac spine. Injections are given in the lateral thigh instead of the buttock because of this potential complication. A complete sciatic trunk lesion will paralyse the hamstrings and all the muscles below the knee, as well as causing sensory loss below the knee except for the area supplied by the saphenous nerve.

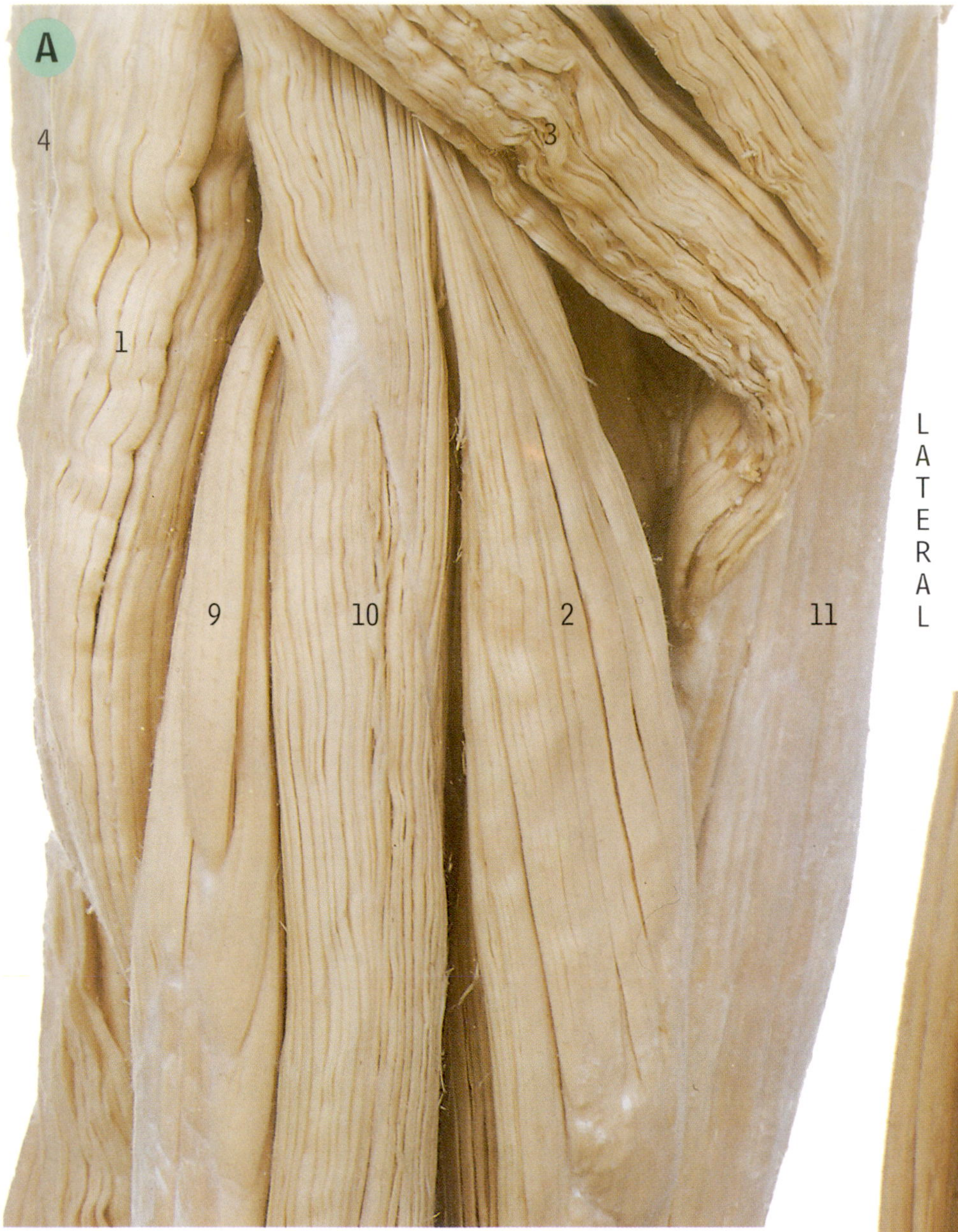

A

Back of the right thigh. Muscles, **A** in the upper part, **B** in the lower part bordering the popliteal fossa

1 Adductor magnus
2 Biceps
3 Gluteus maximus
4 Gracilis
5 Lateral head of gastrocnemius
6 Medial head of gastrocnemius
7 Plantaris
8 Sartorius
9 Semimembranosus
10 Semitendinosus
11 Vastus lateralis

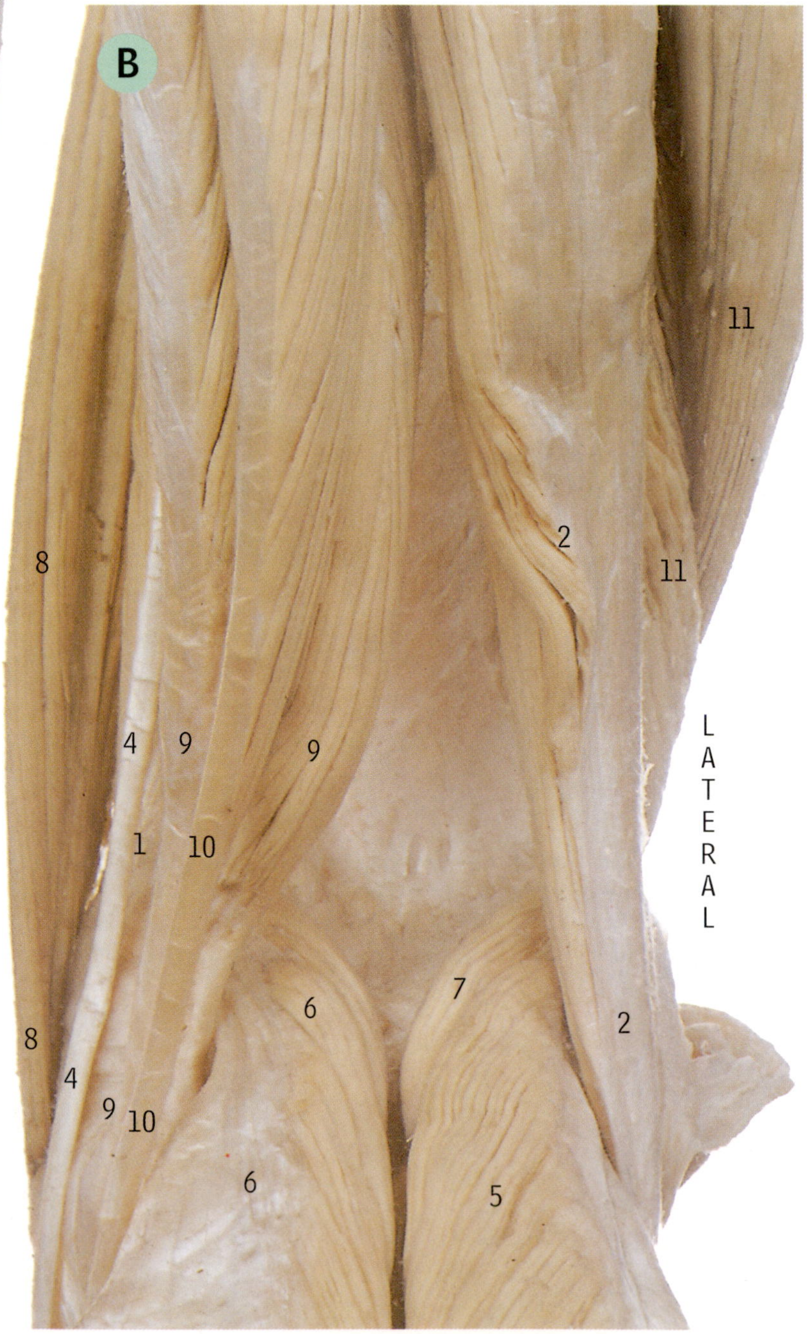

LATERAL

B

LATERAL

• The long head of biceps (the part seen in A, 2), semimembranosus (9) and semitendinosus (10) are commonly called the hamstrings. The short head of biceps, which is under cover of the long head and arises from the back of the shaft of the femur and not from the ischial tuberosity (as the other muscles do), is not classified as a hamstring. The true hamstrings span both the hip and knee joint; they extend the hip and flex the knee.

Torn hamstrings is a common injury seen when the unwarmed muscles are suddenly put into violent contraction, tearing fibres within one of the four major hamstrings (adductor magnus, biceps femoris, semitendinosus or semimembranosus). Stretching exercises are often the best treatment, though occasionally surgery is indicated.

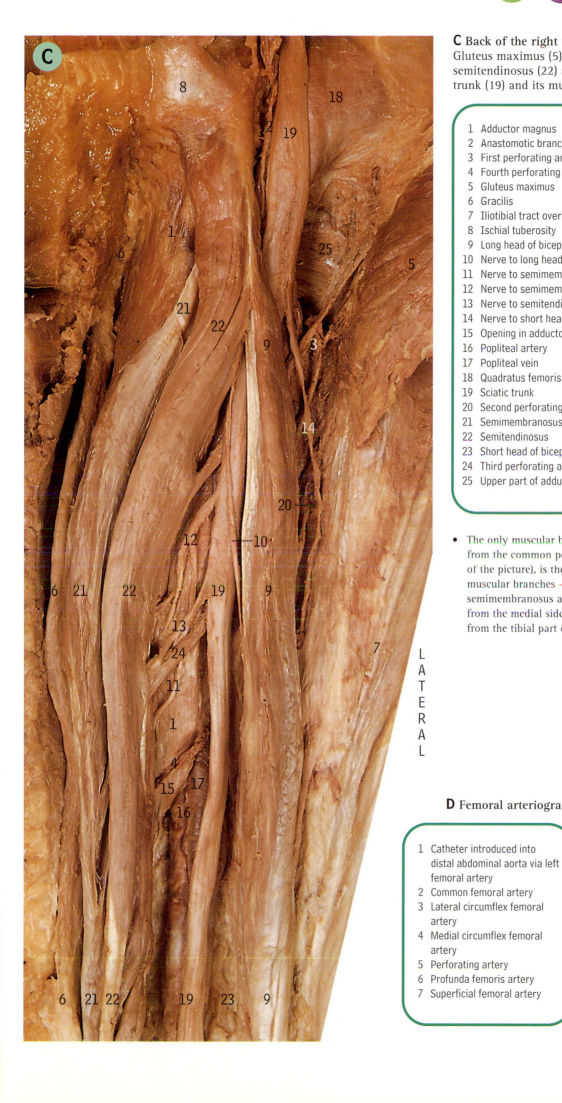

C Back of the right upper thigh

Gluteus maximus (5) has been reflected laterally and the gap between semitendinosus (22) and biceps (9) has been opened up to show the sciatic trunk (19) and its muscular branches

1 Adductor magnus
2 Anastomotic branch of inferior gluteal artery
3 First perforating artery
4 Fourth perforating artery
5 Gluteus maximus
6 Gracilis
7 Iliotibial tract overlying vastus lateralis
8 Ischial tuberosity
9 Long head of biceps
10 Nerve to long head of biceps
11 Nerve to semimembranosus
12 Nerve to semimembranosus and adductor magnus
13 Nerve to semitendinosus
14 Nerve to short head of biceps
15 Opening in adductor magnus
16 Popliteal artery
17 Popliteal vein
18 Quadratus femoris
19 Sciatic trunk
20 Second perforating artery
21 Semimembranosus
22 Semitendinosus
23 Short head of biceps
24 Third perforating artery
25 Upper part of adductor magnus ('adductor minimus')

- The only muscular branch to arise from the lateral side of the sciatic trunk (i.e. from the common peroneal part of the nerve – 19, uppermost label near the top of the picture), is the nerve to the short head of biceps (14). All the other muscular branches – to the long head of biceps (10), semimembranosus (11), semimembranosus and adductor magnus (12) and semitendinosus (13) – arise from the medial side of the sciatic trunk (19, near the centre of the picture) (i.e. from the tibial part of the nerve).

D Femoral arteriogram

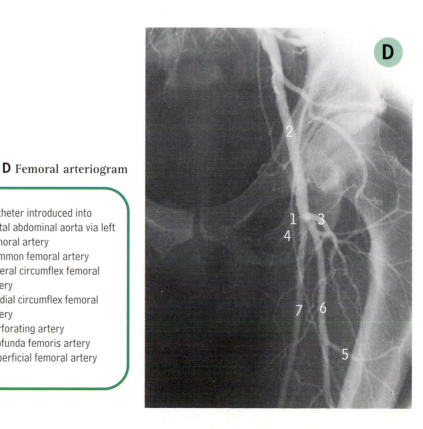

1 Catheter introduced into distal abdominal aorta via left femoral artery
2 Common femoral artery
3 Lateral circumflex femoral artery
4 Medial circumflex femoral artery
5 Perforating artery
6 Profunda femoris artery
7 Superficial femoral artery

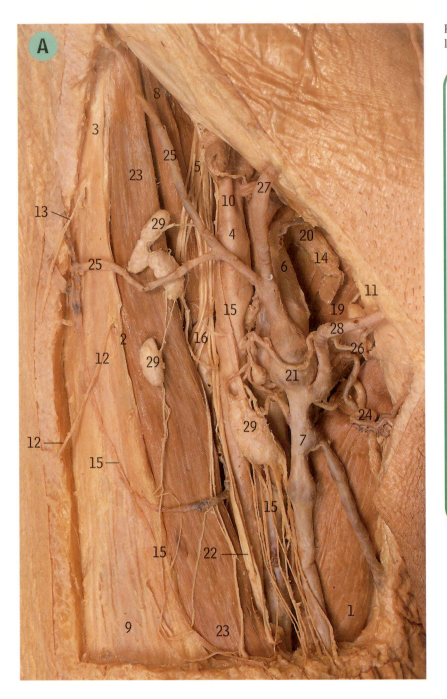

A

Right femoral region, A femoral vessels and lymphatics, B branches of femoral nerve

1 Adductor longus
2 Fascia lata, cut edge
3 Fascia lata overlying tensor fasciae latae
4 Femoral artery
5 Femoral nerve
6 Femoral vein
7 Great saphenous vein
8 Iliacus
9 Iliotibial tract overlying vastus lateralis
10 Inferior epigastric vessels
11 Inguinal ligament
12 Intermediate cutaneous nerve of the thigh
13 Lateral cutaneous nerve of the thigh
14 Lymph node (Cloquet)
15 Lymph vessels
16 Muscular branches of femoral nerve overlying lateral circumflex femoral vessels
17 Nerve to sartorius
18 Nerve to vastus lateralis
19 Pectineus
20 Position of femoral canal
21 Saphena varix
22 Saphenous nerve
23 Sartorius
24 Scrotal veins
25 Superficial circumflex iliac vein
26 Superficial external pudendal artery
27 Superficial epigastric vein
28 Superficial external pudendal vein
29 Vertical chain of superficial inguinal lymph nodes

- The boundaries of the femoral triangle are the inguinal ligament (11), the medial border of sartorius (23) and the medial border of adductor longus (1).
- The femoral canal (20) is the medial compartment of the femoral sheath (removed) which contains in its middle compartment the femoral vein (6), and in the lateral compartment the femoral artery (4). The femoral nerve (5) is lateral to the sheath, not within it.

Femoral hernia is a protrusion of peritoneum and/or abdominal contents through the femoral ring into the thigh. The swelling usually lies inferior and lateral to the pubic tubercle.

The Trendelenburg test for varicose veins checks the competence of the valves in the superficial venous system, especially at the saphenofemoral junction just before the great saphenous vein pierces the cribriform fascia of the leg to enter the femoral vein. Lack of competence of this valve may lead to severe varicosities throughout the great saphenous system.

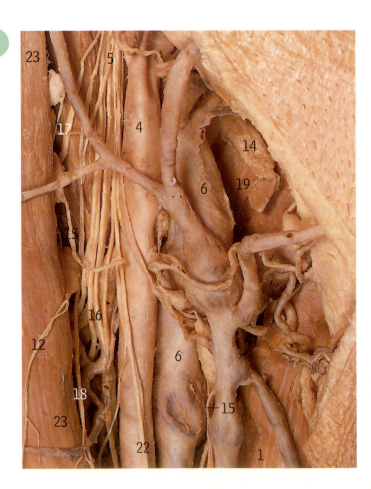

B

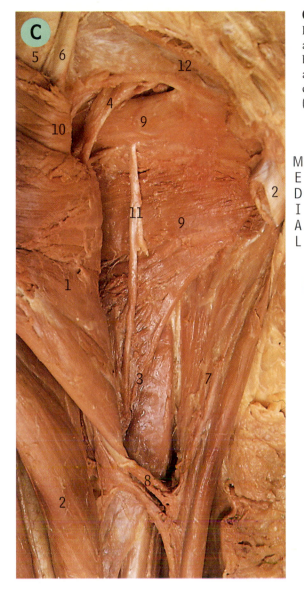

C Right obturator nerve

In this right femoral region, pectineus (10), adductor longus (2, lower label) and adductor brevis (1) have been detached from their origins and reflected laterally to display obturator externus (9) and the anterior (4) and posterior (11) branches of the obturator nerve

1	Adductor brevis	8	Nerve and vessels to gracilis
2	Adductor longus	9	Obturator externus
3	Adductor magnus	10	Pectineus
4	Anterior branch of obturator nerve	11	Posterior branch of obturator nerve
5	Femoral artery	12	Superior ramus of pubis
6	Femoral vein		
7	Gracilis		

> Obturator nerve paralysis is a rare condition causing pain down the medial side of the upper thigh and weakness of the adductor group of muscles. It is seen in lateral pelvic wall pathology, often related to malignancy.

> Femoral nerve paralysis, the result of pressure on the roots of the femoral nerve in high lumbar disc lesions or compression beneath the inguinal ligament by a tumour or swelling within the psoas muscle, presents clinically as quadriceps weakness or atrophy and makes walking downstairs extremely difficult. Weakness in these muscles may lead to instability of the knee joint. Quadriceps exercises are the most important physiotherapy for any knee pathology or femoral nerve damage.

D Right femoral nerve

Sartorius (16) and rectus femoris (14) have been displaced laterally to open up the upper part of the adductor canal and show the lateral circumflex femoral vessels (3, 18 and 4) between branches of the femoral nerve (6)

1	Adductor brevis and nerve	11	Nerve to vastus lateralis
2	Adductor longus	12	Pectineus
3	Ascending branch of lateral circumflex femoral artery	13	Profunda femoris artery
4	Descending branch of lateral circumflex femoral artery	14	Rectus femoris
5	Femoral artery	15	Saphenous nerve
6	Femoral nerve	16	Sartorius
7	Femoral vein	17	Tensor fasciae latae
8	Iliacus	18	Transverse branch of lateral circumflex femoral artery
9	Nerve to rectus femoris	19	Vastus intermedius and nerve
10	Nerve to sartorius	20	Vastus medialis and nerves

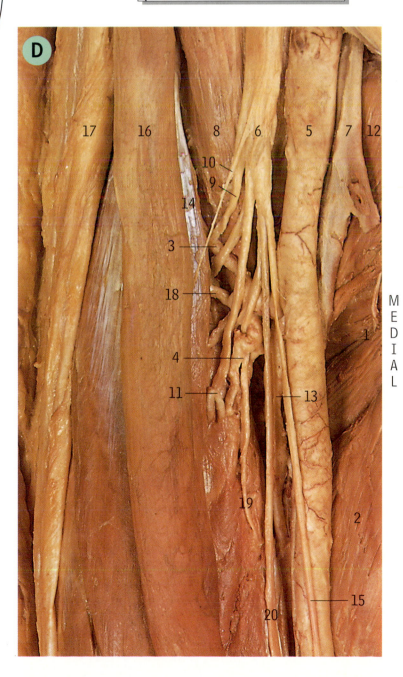

A Right femoral artery

1 Adductor brevis	16 Nerve to vastus medialis
2 Adductor longus	17 Pectineus
3 Anterior branch of obturator nerve	18 Profunda femoris artery
4 Ascending branch of lateral circumflex femoral artery	19 Rectus femoris
	20 Saphenous nerve
5 Descending branch of lateral circumflex femoral artery	21 Sartorius
	22 Spermatic cord
	23 Superficial circumflex iliac artery (double)
6 Femoral artery	24 Superficial epigastric artery
7 Femoral nerve	
8 Femoral vein	25 Superficial external pudendal artery (low origin)
9 Gracilis	
10 Iliacus	
11 Inguinal ligament	26 Tensor fasciae latae
12 Lateral circumflex femoral artery	27 Transverse branch of lateral circumflex femoral artery
13 Lateral femoral cutaneous nerve	
14 Medial circumflex femoral artery	28 Vastus intermedius
	29 Vastus medialis
15 Nerve to rectus femoris	

Femoral hernia, a protrusion through the femoral canal, has the lacunar ligament medially, the inguinal ligament anteriorly, the pectineal ligament posteriorly and the femoral vein laterally. It is a common cause of a strangulated hernia and tends to be seen more commonly in women, who have a wider pelvis and a slightly larger femoral canal. The swelling usually lies lateral and inferior to the pubic tubercle.

Meralgia paraesthetica, sensory loss confined to the lateral, proximal thigh, occurs when the lateral femoral cutaneous nerve is trapped, usually just deep to the inguinal ligament near the anterior superior iliac spine. It may be due to increased size of the psoas muscle, such as in professional cyclists, bleeding into that muscle or a complication of laparoscopic hernia repair.

Femoral vein catheterization. Lying just medial to the femoral artery is the femoral vein which is easily catheterized. Medial to this point of puncture lies the femoral canal, a site of herniation.

Femoral artery catheterization. The femoral artery lies in the midinguinal region just below the inguinal ligament. It is here that catheters are passed into the femoral artery for catheterization of abdominopelvic and thoracic structures and it is a site where arterial blood can be obtained for gas analysis.

B Right lower thigh, from the front and medial side
The lower part of sartorius (13) has been displaced medially to open up the lower part of the adductor canal and expose the femoral artery (2) passing through the opening in adductor magnus (7) to enter the popliteal fossa behind the knee and become the popliteal artery (page 290)

1 Adductor magnus	8 Patella
2 Femoral artery	9 Quadriceps tendon
3 Gracilis	10 Rectus femoris
4 Iliotibial tract	11 Saphenous branch of descending
5 Lowest (horizontal) fibres of vastus	genicular artery
medialis	12 Saphenous nerve
6 Medial patellar retinaculum	13 Sartorius
7 Opening in adductor magnus	14 Vastus medialis and nerve

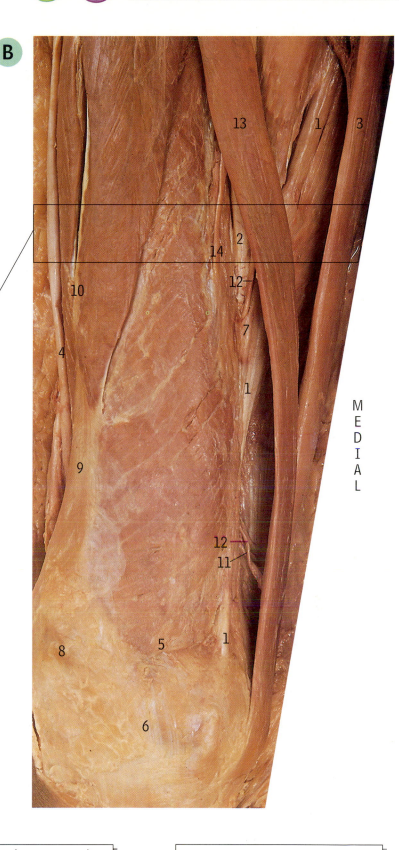

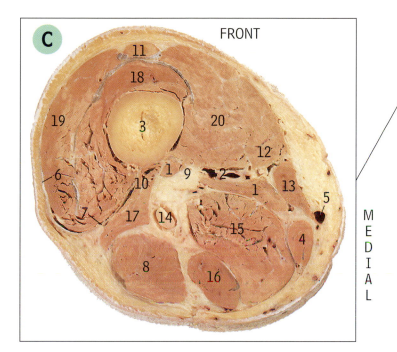

Right lower thigh, **C** cross section

1 Adductor magnus
2 Femoral vessels
3 Femur
4 Gracilis
5 Great saphenous vein
6 Iliotibial tract of fascia lata
7 Lateral intermuscular septum
8 Long head of biceps
9 Opening in adductor magnus
10 Profunda femoris vessels
11 Rectus femoris
12 Saphenous nerve
13 Sartorius
14 Sciatic nerve
15 Semimembranosus
16 Semitendinosus
17 Short head of biceps
18 Vastus intermedius
19 Vastus lateralis
20 Vastus medialis

Femoropopliteal bypass is a common vascular operation in the lower limb. The blockage nearly always occurs where the femoral artery passes through the adductor hiatus and following occlusion it is common to find that a superior genicular arterial branch, part of the knee anastomosis, is enlarged. Nowadays, instead of a complete bypass, various balloon or stent techniques are being used to enlarge stenotic arteries.

Intermittent claudication is limping due to pain, usually experienced in the calf muscles, due to a generalized ischaemia of the lower limb. The patient may be able to walk a fixed distance before this pain occurs. Commonly it is due to atherosclerosis in the region of the adductor hiatus where the femoral artery becomes the popliteal artery.

A

Right hip joint, **A** from front and below, **B** from front and above

1 Anterior inferior iliac spine	18 Ischial tuberosity
2 Anterior superior iliac spine	19 Lesser trochanter
3 Bursa for psoas tendon	20 Lumbosacral trunk
4 First sacral nerve root	21 Median sacral artery
5 Fourth lumbar nerve root	22 Obturator externus
6 Gluteus minimus muscle	23 Obturator internus tendon
7 Greater trochanter	24 Obturator nerve, anterior
8 Hamstring origin	branch
9 Iliac crest	25 Obturator nerve, posterior
10 Iliacus muscle	branch
11 Iliofemoral ligament	26 Obturator vessels
12 Iliolumbar ligament	27 Piriformis muscle
13 Iliopsoas tendon	28 Pubofemoral ligament
14 Iliopubic eminence	29 Pudendal nerve
15 Inferior gemellus muscle	30 Rectus femoris muscle
16 Inguinal ligament	31 Sacrospinous ligament
17 Intertrochanteric line and	32 Second sacral nerve root
capsule attachment	33 Superior gluteal artery

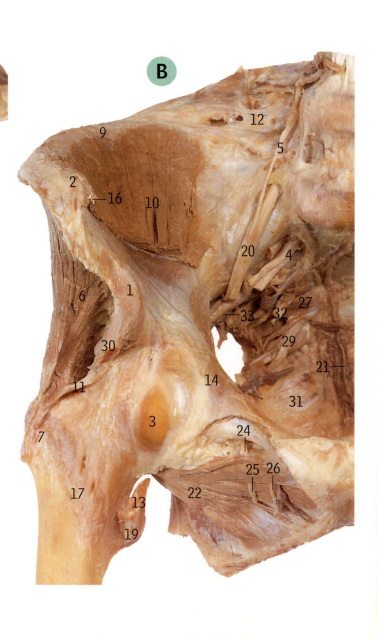

B

- The iliofemoral ligament (11) has the shape of an inverted V. It and the interosseous sacro-iliac ligament are the two strongest ligaments in the body.
- Some of the fibres of the ischiofemoral ligament help to form the zona orbicularis – circular fibres of the capsule that form a collar round the neck of the femur.
- Posteriorly the capsule is attached to the neck of the femur, not to the intertrochanteric crest. (Anteriorly it is attached to the intertrochanteric line.) See pages 257 and 258.

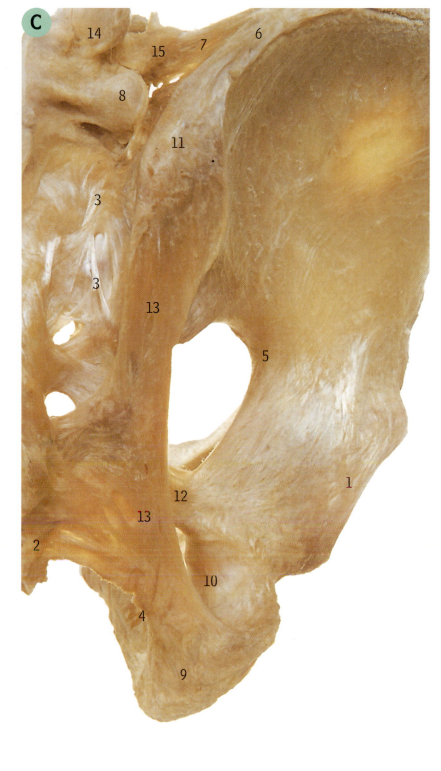

C Right vertebropelvic and sacro-iliac ligaments, from behind

1 Acetabular labrum
2 Coccyx
3 Dorsal sacro-iliac ligaments
4 Falciform process of
   sacrotuberous ligament
5 Greater sciatic notch
6 Iliac crest
7 Iliolumbar ligament
8 Inferior articular process of
   fifth lumbar vertebra
9 Ischial tuberosity
10 Lesser sciatic notch
11 Posterior superior iliac spine
12 Sacrospinous ligament and
   ischial spine
13 Sacrotuberous ligament
14 Superior articular process of
   fifth lumbar vertebra
15 Transverse process of fifth
   lumbar vertebra

Avascular necrosis of the head of the femur can
occur following subcapital fractures of the
femoral neck or less commonly with
transcervical fractures. In the young the
femoral head receives an abundant blood
supply from the ligamentum teres and the
obturator artery. In old age this supply is
insufficient to maintain the head of the femur,
most of the blood coming from anastomoses
along the retinacular fibres of the capsule of the
neck which are interrupted in these fractures.

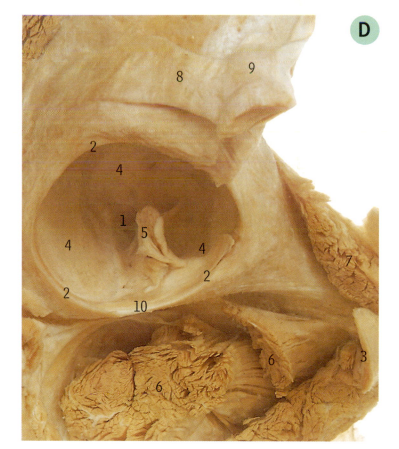

D Right hip joint with femur removed, from the right
The femur has been disarticulated from the acetabulum and removed,
leaving the acetabular labrum (2), transverse ligament (10) and the ligament
of the head of the femur (5)

1 Acetabular fossa (non-articular)
2 Acetabular labrum
3 Adductor longus
4 Articular surface
5 Ligament of head of femur
6 Obturator externus
7 Pectineus
8 Reflected head of rectus femoris
9 Straight head of rectus femoris
10 Transverse ligament

• The acetabular labrum (2) is attached to the margin of the acetabulum and is
  composed of fibrocartilage.
• The transverse ligament (10) fills in the acetabular notch and the gap between the
  two ends of the labrum (2), and is composed of fibrous tissue, not fibrocartilage.
• The ligament of the head of the femur (5) extends from the transverse ligament
  (10) and the margins of the acetabular notch to the fovea or pit on the medial
  side of the head of the femur. Like the transverse ligament it is composed of
  fibrous tissue.

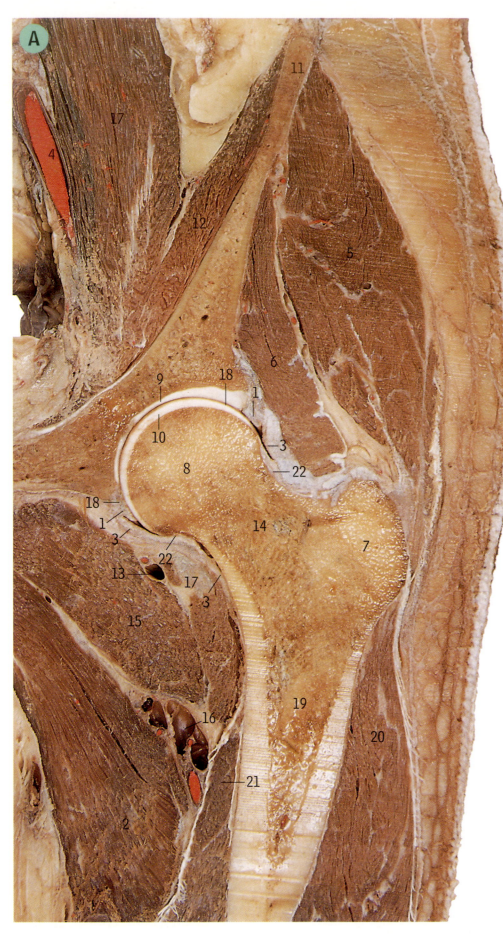

A

Left hip joint, **A** coronal section, from the front, **B** coronal MR image

The section has almost passed through the centre of the head (8) of the femur and the centre of the greater trochanter (7). Above the neck of the femur (14), gluteus minimus (6) with gluteus medius (5) above it run down to their attachments to the greater trochanter (7), while below the neck the tendon of psoas major (17) and muscle fibres of iliacus (12) pass backwards towards the lesser trochanter. The circular fibres of the zona orbicularis (22) constrict the capsule (3) around the intracapsular part of the neck of the femur

1	Acetabular labrum
2	Adductor longus
3	Capsule of hip joint
4	External iliac artery
5	Gluteus medius
6	Gluteus minimus
7	Greater trochanter
8	Head of femur
9	Hyaline cartilage of acetabulum
10	Hyaline cartilage of head
11	Iliac crest
12	Iliacus
13	Medial circumflex femoral vessels
14	Neck of femur
15	Pectineus
16	Profunda femoris vessels
17	Psoas major
18	Rim of acetabulum
19	Shaft of femur
20	Vastus lateralis
21	Vastus medialis
22	Zona orbicularis of capsule

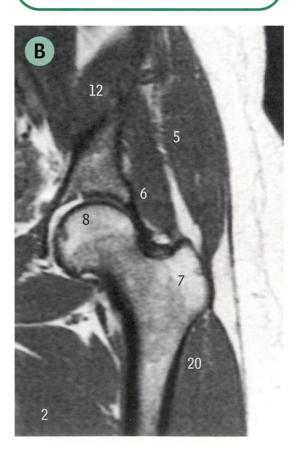

B

- The convergence of gluteus medius and minimus (5 and 6) on to the greater trochanter is well displayed in this section. These muscles are classified as abductors of the femur at the hip joint, but their more important action is in walking, where they act to prevent adduction – preventing the pelvis from tilting to the opposite side when the opposite limb is off the ground.

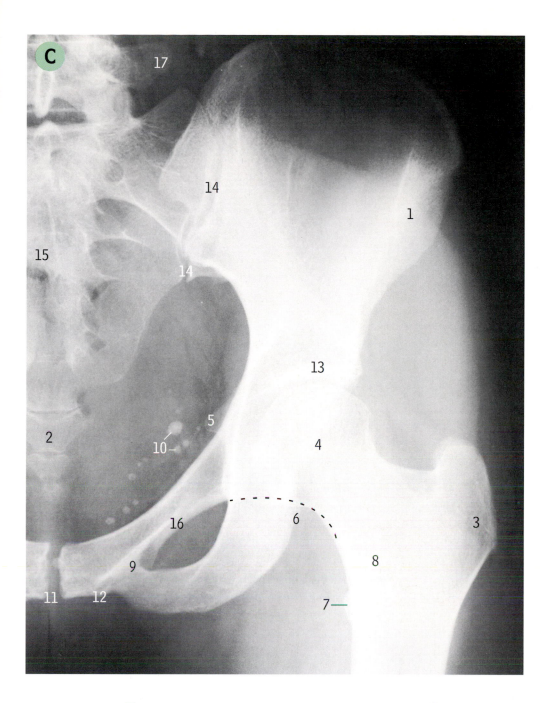

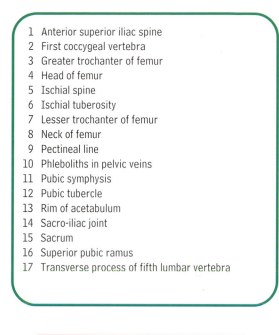

C Left hip and sacro-iliac joint radiograph
In this standard anteroposterior view of the hip joint (13 and 4), much of the joint line of the sacro-iliac joint can also be seen (14). The dashed line on C is Shenton's line

1   Anterior superior iliac spine
2   First coccygeal vertebra
3   Greater trochanter of femur
4   Head of femur
5   Ischial spine
6   Ischial tuberosity
7   Lesser trochanter of femur
8   Neck of femur
9   Pectineal line
10 Phleboliths in pelvic veins
11 Pubic symphysis
12 Pubic tubercle
13 Rim of acetabulum
14 Sacro-iliac joint
15 Sacrum
16 Superior pubic ramus
17 Transverse process of fifth lumbar vertebra

Dislocation of the hip. A congenital problem, an abnormally shallow acetabulum, may present later in life as subluxation of the femoral head when the child becomes more active. Radiologically, Shenton's line is usually a useful guide. Traumatic dislocation of a normal hip is seen most commonly following automobile accidents. This posterior dislocation may be associated with fractures of the acetabulum or damage to the sciatic trunk, with resultant foot drop. Avulsion of the ligamentum teres may lead to avascular necrosis of the femoral head.

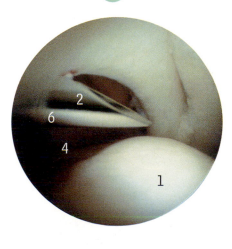

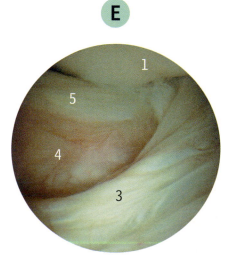

D and **E** Endoscopic views of the hip joint
Reproduced with kind permission of Richard N. Villar from *Hip Arthroscopy* (Butterworth Heinemann).

1   Femoral head
2   Irrigation needle
3   Ligamentum teres
4   Synovium
5   Transverse ligament
6   Zona orbicularis

A

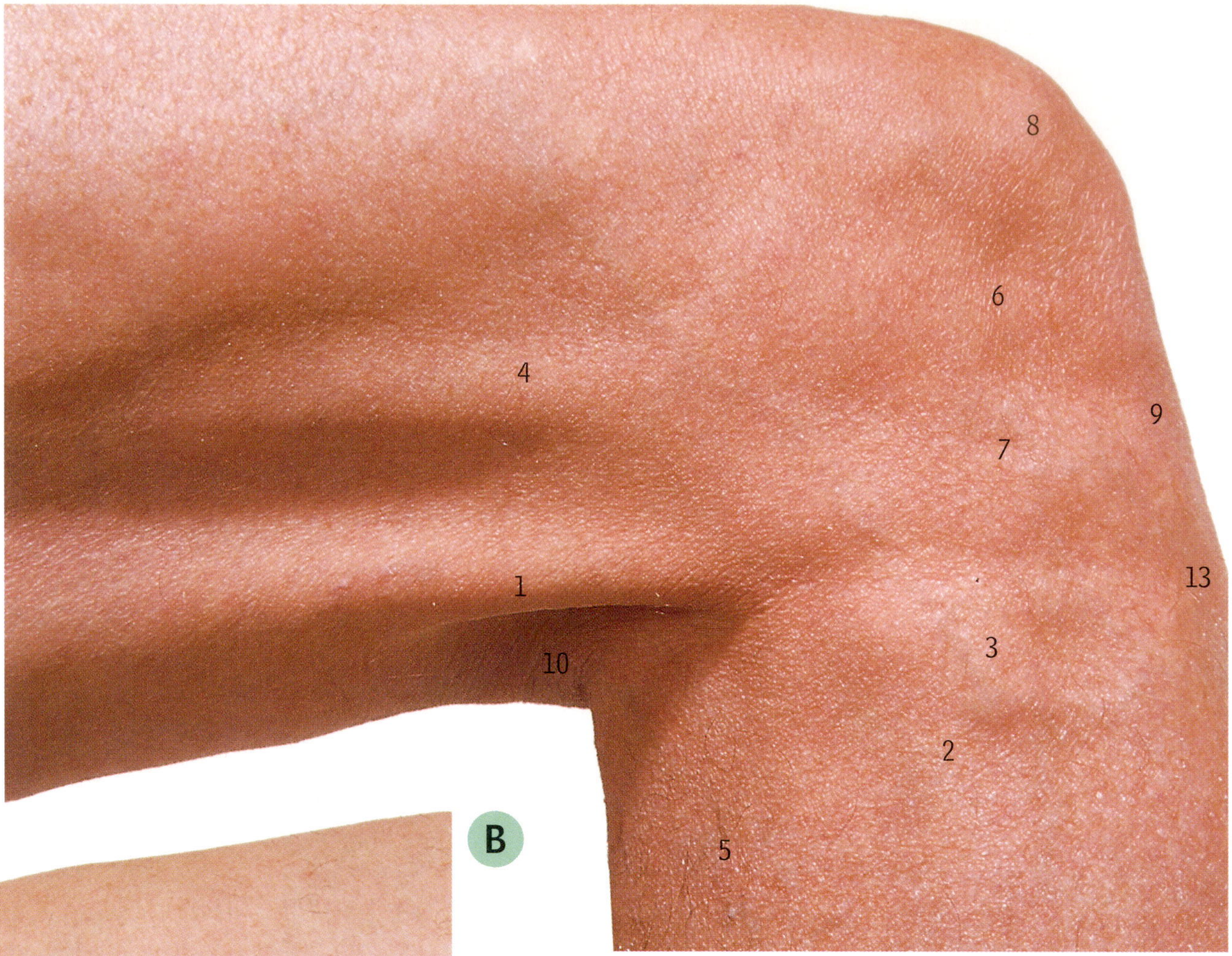

B

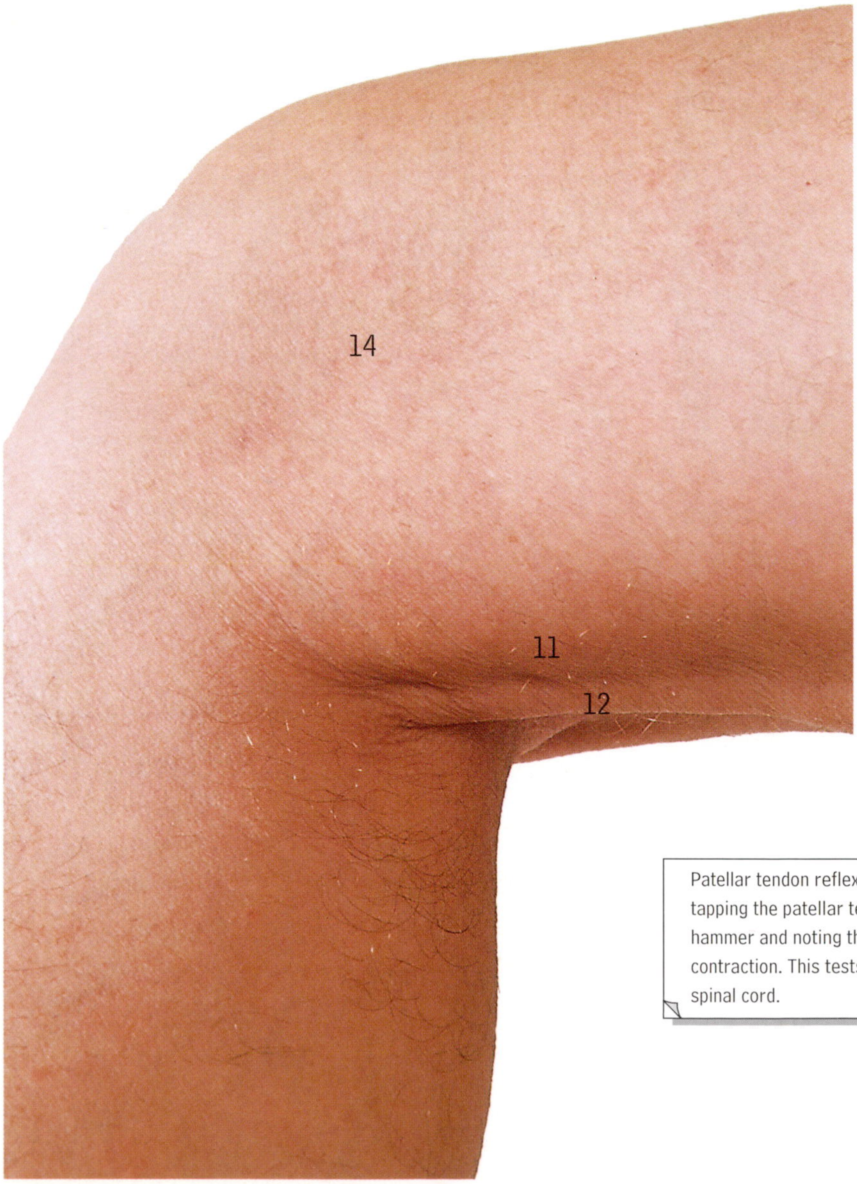

Right knee, partially flexed, A from the lateral side, B from the medial side

Behind the knee on the lateral side the rounded tendon of biceps (1) can be felt easily, with the broad strap-like iliotibial tract (4) in front of it, with a furrow between them. On the medial side two tendons can be felt – the narrow rounded semitendinosus (12) just behind the broader semimembranosus (11). At the front the patellar ligament (9) keeps the patella (8) at a constant distance from the tibial tuberosity (13), while at the side the adjacent margins of the femoral and tibial condyles (6 and 7) can be palpated

1	Biceps femoris
2	Common peroneal (fibular) nerve
3	Head of fibula
4	Iliotibial tract
5	Lateral head of gastrocnemius
6	Margin of condyle of femur
7	Margin of condyle of tibia
8	Patella
9	Patellar ligament
10	Popliteal fossa
11	Semimembranosus
12	Semitendinosus
13	Tuberosity of tibia
14	Vastus medialis

Patellar tendon reflex (knee jerk) is elicited by tapping the patellar tendon with a patellar hammer and noting the quadriceps muscle contraction. This tests the L3 segment of the spinal cord.

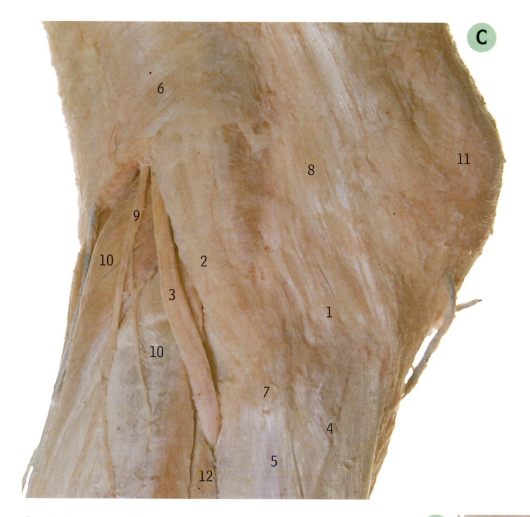

C

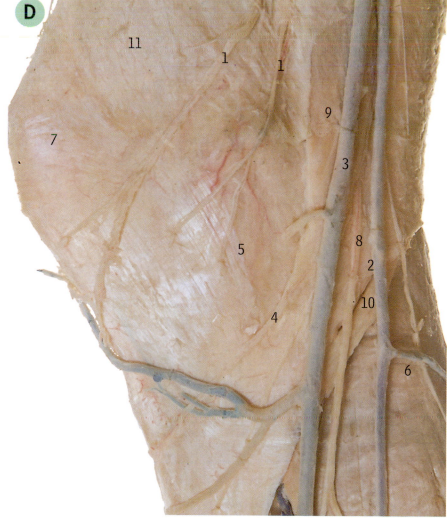

D

D Right knee, superficial dissection, from the medial side
The great saphenous vein (3) runs upwards about a hand's breadth behind the medial border of the patella (7). The saphenous nerve (8) becomes superficial between the tendons of sartorius (9) and gracilis (2), and its infrapatellar branch (4) curls forwards a little below the upper margin of the tibial condyle

1 Branches of medial femoral cutaneous nerve
2 Gracilis
3 Great saphenous vein
4 Infrapatellar branch of saphenous nerve
5 Level of margin of medial condyle of tibia
6 Medial head of gastrocnemius
7 Patella
8 Saphenous nerve
9 Sartorius
10 Semitendinosus
11 Vastus medialis

C Right knee, superficial dissection, from the lateral side
The fascia behind biceps (2) has been removed to show the common peroneal (fibular) nerve (3) passing downwards immediately behind the tendon, and then running between the adjacent borders of soleus (12) and peroneus longus (5), under cover of which it lies against the neck of the fibula. Minor superficial vessels and nerves have been removed

1 Attachment of iliotibial tract to tibia
2 Biceps
3 Common peroneal (fibular) nerve
4 Deep fascia overlying extensor muscles
5 Deep fascia overlying peroneus longus
6 Fascia lata
7 Head of fibula
8 Iliotibial tract
9 Lateral cutaneous nerve of calf
10 Lateral head of gastrocnemius
11 Patella
12 Soleus

- The iliotibial tract (8) is the thickened lateral part of the fascia lata (6). At its upper part the tensor fasciae latae and most of gluteus maximus are inserted into it; its lower end is attached to the lateral condyle of the tibia (1).
- Its subcutaneous position and contact with the neck of the fibula make the common peroneal (fibular) nerve (3) the most commonly injured nerve in the lower limb.

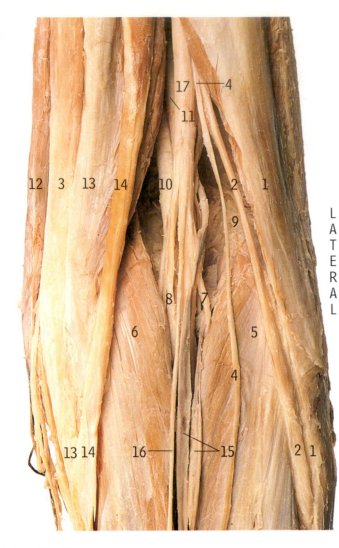

A Right popliteal fossa
The fascia that forms the roof of the diamond-shaped fossa and the fat within it have been removed but the small saphenous vein which pierces the fascia is preserved (15, double in this specimen).

1 Biceps	9 Plantaris
2 Common peroneal (fibular) nerve	10 Popliteal artery
3 Gracilis	11 Popliteal vein
4 Lateral cutaneous nerve of calf	12 Sartorius
5 Lateral head of gastrocnemius	13 Semimembranosus
6 Medial head of gastrocnemius	14 Semitendinosus
7 Nerve to lateral head of	15 Small saphenous vein (double)
gastrocnemius	16 Sural nerve
8 Nerve to medial head of	17 Tibial nerve
gastrocnemius	

A Baker's cyst is a large swelling in the popliteal fossa due to herniation of synovial membrane from the knee joint. It can be mistaken for a popliteal aneurysm and on occasions may rupture, allowing fluid to track down the calf muscles where it may be mistaken for deep venous thrombosis of the calf.

B Right popliteal fossa. Vessels and nerves
Here the margins of the fossa have been displaced and markers hold various structures apart. The upper red marker passes between the tibial nerve (15) and the underlying popliteal vein (seen lower down, 9). The uppermost blue marker is behind an unlabelled muscular branch of the popliteal artery (seen lower down, 8). The middle blue marker is behind the superior lateral genicular artery (13), and the lateral head of gastrocnemius (5). The lowest blue marker displaces the popliteal vein (9) medially to show the underlying popliteal artery (8). The lower red marker holds the two heads of gastrocnemius apart (5 and 6). The lower white marker (outside the fossa) is between the tendons of gracilis (4) and semitendinosus (12)

1 Biceps	8 Popliteal artery
2 Branches of sural nerve	9 Popliteal vein
3 Common peroneal (fibular) nerve	10 Sartorius
4 Gracilis	11 Semimembranosus
5 Lateral head of gastrocnemius	12 Semitendinosus
6 Medial head of gastrocnemius and	13 Superior lateral genicular artery
nerve	14 Sural arteries
7 Plantaris	15 Tibial nerve

• The most lateral branch of the sural nerve (B2) may here take the place of the lateral cutaneous nerve of the calf which normally arises from the common peroneal (fibular) nerve.

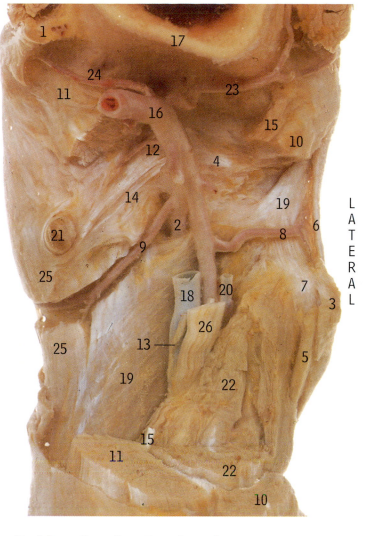

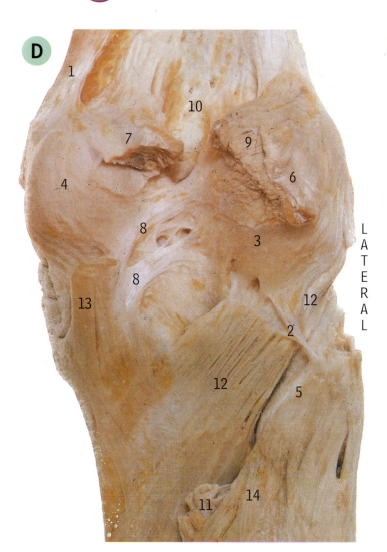

C Right popliteal fossa. Deep dissection of vessels

Most of the superficial muscles have been removed to show the branches of the popliteal artery (16). The knee is flexed so that the artery and the cut end of the femur (17) are seen 'end on'. The artery divides into anterior tibial (2) and posterior tibial (20) branches; usually the anterior tibial passes superficial to popliteus (19) but here the popliteal artery has divided at a high level and the anterior tibial has passed deep to the muscle. The upper pair of superior genicular arteries (24 and 23) run above the heads of gastrocnemius (11 and 10); the inferior pair (9 and 8) pass deep to the tibial and fibular collateral ligaments (25 and 6)

D Right popliteal fossa. Joint capsule and popliteal ligaments

1	Adductor magnus	8	Oblique popliteal ligament
2	Arcuate popliteal ligament	9	Plantaris
3	Capsule of knee joint	10	Popliteal surface of femur
4	Capsule overlying medial condyle of femur	11	Popliteal vessels and tibial nerve
		12	Popliteus
5	Head of fibula	13	Semimembranosus
6	Lateral head of gastrocnemius	14	Soleus
7	Medial head of gastrocnemius		

1	Adductor magnus	14	Oblique popliteal ligament
2	Anterior tibial artery	15	Plantaris
3	Biceps	16	Popliteal artery
4	Capsule of knee joint	17	Popliteal surface of femur in section
5	Common peroneal nerve	18	Popliteal vein
6	Fibular collateral ligament	19	Popliteus
7	Head of fibula	20	Posterior tibial artery and vein
8	Inferior lateral genicular artery	21	Semimembranosus
9	Inferior medial genicular artery	22	Soleus
10	Lateral head of gastrocnemius	23	Superior lateral genicular artery
11	Medial head of gastrocnemius	24	Superior medial genicular artery
12	Middle genicular artery	25	Tibial collateral ligament
13	Nerve to popliteus	26	Tibial nerve

- The oblique popliteal ligament (8) is derived from the semimembranosus tendon (13) and reinforces the central posterior part of the joint capsule (3); it is pierced by the middle genicular artery which passes through the capsule to supply the cruciate ligaments.

Popliteal aneurysm. The popliteal pulse is difficult to feel because the popliteal artery is the deepest structure in the popliteal fossa, lying against the oblique popliteal ligament – the reinforcement of the posterior capsule of the knee. To feel the pulse it is often necessary to flex the knee and press all fingers into the popliteal fossa hard against the bone. Owing to the constant bending and stretching of this artery with flexion of the knee, it is a site where aneurysms develop. A popliteal aneurysm may not be palpated until fairly large and may be mistaken for a Baker's cyst.

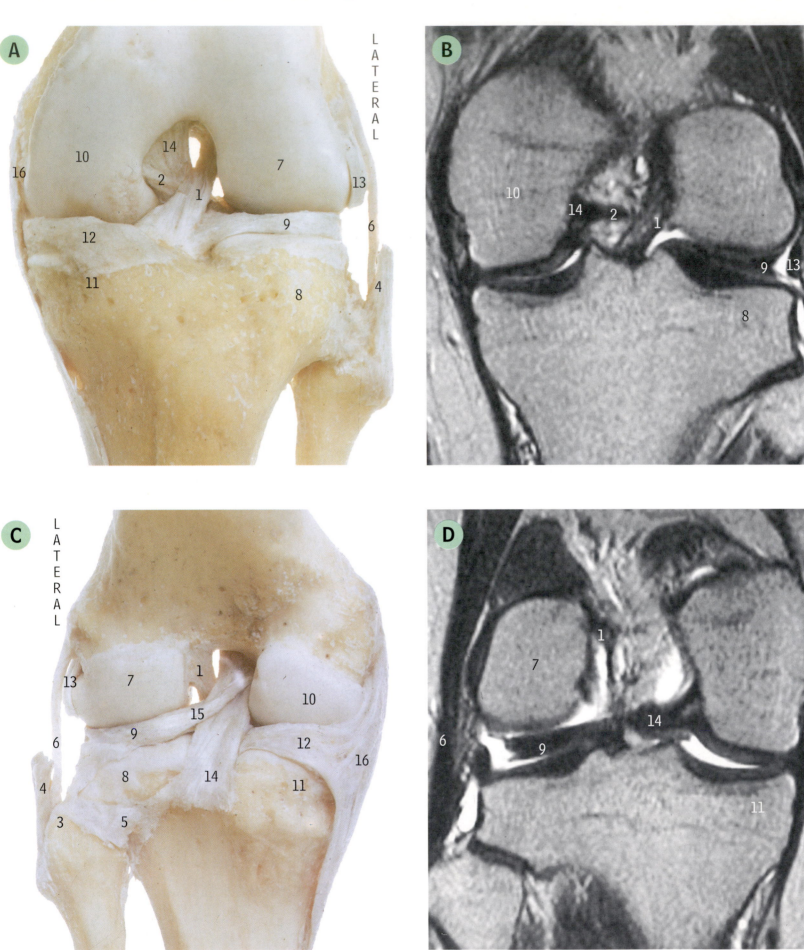

Left knee joint. Ligaments, **A** from the front, **B** coronal MR image, **C** from behind, **D** coronal MR image
The capsule of the knee joint and all surrounding tissues have been removed, leaving only the ligaments of the joint, which is partially flexed

1 Anterior cruciate ligament
2 Anterior meniscofemoral ligament
3 Apex of head of fibula
4 Biceps tendon
5 Capsule of superior tibiofemoral joint
6 Fibular collateral ligament
7 Lateral condyle of femur
8 Lateral condyle of tibia
9 Lateral meniscus
10 Medial condyle of femur
11 Medial condyle of tibia
12 Medial meniscus
13 Popliteus
14 Posterior cruciate ligament
15 Posterior meniscofemoral ligament
16 Tibial collateral ligament

- The fibular collateral (lateral) ligament (A6) is a rounded cord about 5 cm long, passing from the lateral epicondyle of the femur to the head of the fibula just in front of its apex (C3), largely under cover of the tendon of biceps (C4).
- The medial meniscus (E12 and F12) is attached to the deep part of the tibial collateral ligament (E19 and F20). This helps to anchor the meniscus but makes it liable to become trapped and torn by rotatory movements between the tibia and femur.
- The lateral meniscus (A9) is not attached to the fibular collateral ligament (A6), but is attached posteriorly to the popliteus muscle (F5, and C2 on page 294).
- The tibial collateral (medial) ligament (E19) is a broad flat band about 12 cm long, passing from the medial epicondyle of the femur (E11) to the medial condyle of the tibia (E10) and an extensive area of the medial surface of the tibia below the condyle (as in the lower part of E).
- The cruciate ligaments are named from their attachments to the tibia.
- The anterior cruciate ligament (A1 and F1) passes upwards, backwards and laterally to be attached to the medial side of the lateral condyle of the femur (C7).

- The posterior cruciate ligament (C14 and F13) passes upwards, forwards and medially to be attached to the lateral surface of the medial condyle of the femur (A10).

Meniscal tears. Both the medial and lateral menisci are subject to rotational injuries and may be torn. The medial is much more liable to injury as it is attached to the medial collateral ligament, whereas the lateral meniscus or semilunar cartilage is separated from the fibular collateral ligament. Commonly seen in footballers' knees, these injuries are nowadays diagnosed by MR imaging or on direct arthroscopy. Presenting symptoms may be pain and swelling of the knee or locking of the knee when a partly detached cartilage wedges between the tibia and femur. Sometimes a momentary click may be heard in flexion/extension movements of the knee. Meniscectomy is a successful operation but nowadays there is greater emphasis on repairing small tears.

Left knee joint. Ligaments, E from the medial side, F from above
The same specimen as in A and C is seen from the medial side in E, to show the broad tibial collateral ligament (19). F is the view looking down on the upper surface of the tibia after removing the femur by cutting through the capsule, the collateral ligaments, and the cruciate ligaments. The medial and lateral menisci (12 and 8) remain at the periphery of the articular surfaces of the tibial condyles. The horns of the menisci (3 and 15; 2 and 14) and the cruciate ligaments (1 and 13) are attached to the non-articular intercondylar area of the tibia. Compare with C on page 266

1 Anterior cruciate ligament
2 Anterior horn of lateral meniscus
3 Anterior horn of medial meniscus
4 Anterior meniscofemoral ligament
5 Attachment of lateral meniscus to popliteus (with underlying marker)
6 Fibular collateral ligament
7 Lateral condyle of tibia
8 Lateral meniscus
9 Medial condyle of femur
10 Medial condyle of tibia
11 Medial epicondyle of femur
12 Medial meniscus
13 Posterior cruciate ligament
14 Posterior horn of lateral meniscus
15 Posterior horn of medial meniscus
16 Posterior meniscofemoral ligament
17 Tendon of biceps
18 Tendon of popliteus
19 Tibial collateral ligament
20 Tibial collateral ligament attached to medial meniscus
21 Transverse ligament

Anterior cruciate ligament rupture may occur after violent abduction and twisting of the knee (sliding tackle in soccer) which tears the medial meniscus and the anterior cruciate ligament. This is tested clinically by the 'drawer test', moving the tibia with respect to the femur – excessive anterior movement is indicative of an anterior cruciate ligament tear.

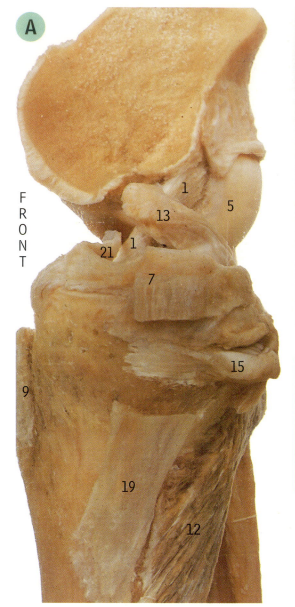

A

F R O N T

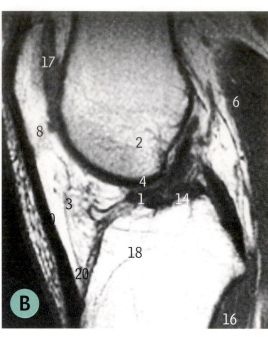

B

Right knee joint, **A** from the medial side with the medial femoral condyle removed, **B** sagittal MR image

Removal of the medial half of the lower end of the femur enables the X-shaped crossover of the cruciate ligaments to be seen; the anterior cruciate (1) is passing backwards and laterally, while the posterior cruciate (13) passes forwards and medially. The MR image in B shows the backward projection of the infrapatellar fat pad (3)

Suprapatellar bursitis. The suprapatellar bursa lies a hand's breath superior to the upper border of the patella and communicates with the knee joint. A knee joint effusion can therefore be 'milked' into the suprapatellar region and a patellar tap may be an indication of increased fluid within the whole knee joint complex.

1	Anterior cruciate ligament	8	Patella	16	Soleus
2	Femur	9	Patellar ligament	17	Tendon of quadriceps
3	Infrapatellar fat pad	10	Patellar tendon	18	Tibia
4	Intercondylar notch	11	Popliteal artery and vein	19	Tibial collateral ligament
5	Lateral condyle of femur	12	Popliteus	20	Tibial tubercle
6	Lateral head of gastrocnemius muscle	13	Posterior cruciate ligament	21	Transverse ligament (displaced backwards)
7	Medial meniscus and attachment of tibial collateral ligament	14	Posterior meniscofemoral ligament		
		15	Semimembranosus		

C Left knee joint, from behind with the femur removed
This view demonstrates the attachment of the lateral meniscus (2) to popliteus (9). There are markers underneath the attachment and behind the popliteus tendon

1	Anterior cruciate ligament
2	Attachment of lateral meniscus to popliteus
3	Biceps
4	Fibular collateral ligament
5	Head of fibula
6	Interosseous membrane
7	Medial meniscus attached to tibial collateral ligament
8	Popliteus
9	Popliteus tendon
10	Posterior cruciate ligament
11	Posterior meniscofemoral ligament
12	Semimembranosus
13	Soleus

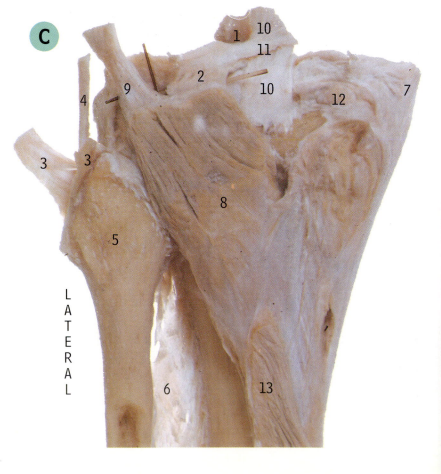

C

L A T E R A L

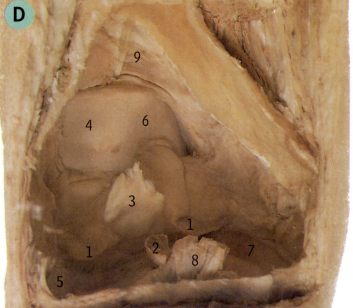

D Left knee joint, opened from behind with the femur removed
By looking into the joint from behind after removal of the femur, the articular surfaces of the patella (4 and 6) are seen, while below them are the alar and infrapatellar folds (1 and 3)

E Left knee joint, from the medial side, with synovial and bursal cavities injected
The resin injection has distended the synovial cavity of the joint (3) and extends into the suprapatellar bursa (10), the bursa round the popliteus tendon (2) and the semimembranosus bursa (9)

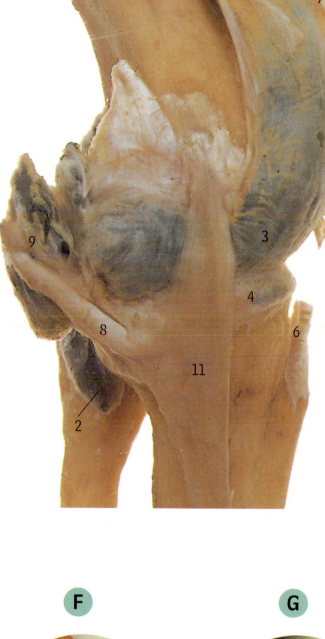

1 Alar fold
2 Anterior cruciate ligament
3 Infrapatellar fold (ligamentum mucosum)
4 Lateral articular surface of patella
5 Lateral meniscus
6 Medial articular surface of patella
7 Medial meniscus
8 Posterior cruciate ligament
9 Suprapatellar bursa (supported by glass rod)

1 Articularis genu
2 Bursa of popliteus tendon
3 Capsule
4 Medial meniscus
5 Patella
6 Patellar ligament
7 Quadriceps tendon
8 Semimembranosus
9 Semimembranosus bursa
10 Suprapatellar bursa
11 Tibial collateral ligament

• The normal knee joint (the largest of all synovial joints) contains less than 1 ml of synovial fluid; the joint illustrated in E contains about 80 ml of injected resin which has distended the synovial cavity.
• The suprapatellar bursa (E10) always communicates with the joint cavity. The bursa around the popliteus tendon (E2) usually does so. The semimembranosus bursa (E9) may do so.

Arthroscopic views of the left knee,
F anterolateral approach,
G posteromedial approach
Reproduced with kind permission of David J. Dandy from *Current Problems in Orthopaedics: Arthroscopic Management of the Knee,* 2nd Edition, Churchill Livingstone

1 Lateral condyle of femur
2 Lateral condyle of tibia
3 Lateral meniscus
4 Medial condyle of femur
5 Medial meniscus
6 Posterior cruciate ligament
7 Posterior part of capsule

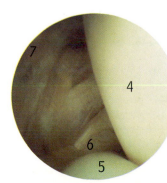

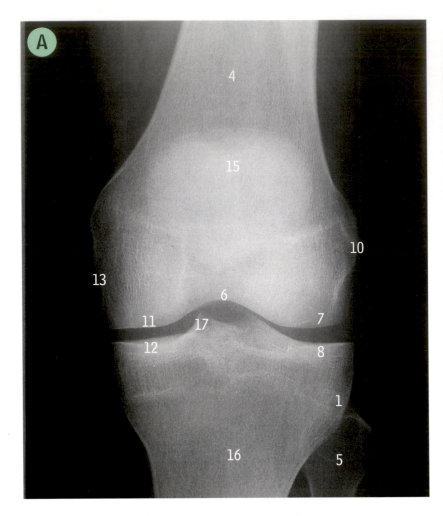

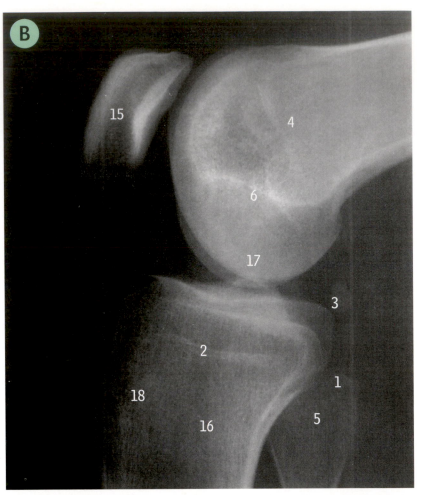

Radiographs and arthroscopic views of the knee, **A** from the front, **B** from the lateral side in partial flexion, **C** skyline view projection, **D** anterolateral approach, **E** lateral view of patella

In A the shadow of the patella (15) is superimposed on that of the femur. The regular space between the condyles of the femur and tibia (7 and 8, 11 and 12) is due to the thickness of the hyaline cartilage on the articulating surface, with the menisci at the periphery. In C with the knee flexed, the view should be compared with the bones seen on page 262, E, and the lateral edge of the patella (9) is seen in the arthroscopic view in E. (D and E reproduced with kind permission of David J Dandy from *Current Problems in Orthopaedics: Arthroscopic Management of the Knee*, 2nd Edition, Churchill Livingstone.)

1	Apex (styloid process) of fibula	10	Lateral epicondyle of femur
2	Epiphysial line	11	Medial condyle of femur
3	Fabella	12	Medial condyle of tibia
4	Femur	13	Medial epicondyle of femur
5	Head of fibula	14	Medial meniscus
6	Intercondylar fossa	15	Patella
7	Lateral condyle of femur	16	Tibia
8	Lateral condyle of tibia	17	Tubercles of intercondylar eminence
9	Lateral edge of patella	18	Tuberosity of tibia

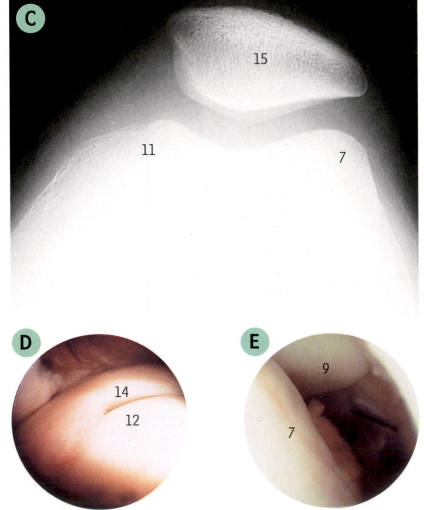

A Left leg, from the front and lateral side
Most of the deep fascia has been removed, and segments of extensor digitorum longus (4) and peroneus longus (9) have been cut out to display the deep (3) and superficial (11) branches of the common peroneal (fibular) nerve just below the head of the fibula (6). The gap between tibialis anterior (12) and extensor digitorum longus (4) has been opened up to show the anterior tibial artery (1)

1 Anterior tibial artery overlying interosseous membrane
2 Branch of deep peroneal (fibular) nerve to tibialis anterior
3 Deep peroneal (fibular) nerve
4 Extensor digitorum longus
5 Extensor hallucis longus
6 Head of fibula
7 Lateral branch of superficial peroneal (fibular) nerve
8 Medial branch of superficial peroneal (fibular) nerve
9 Peroneus longus
10 Recurrent branch of common peroneal (fibular) nerve
11 Superficial peroneal (fibular) nerve
12 Tibialis anterior and overlying fascia
13 Tuberosity of tibia and patellar ligament

- The common peroneal (fibular) nerve (page 298, B2) divides into its superficial and deep branches (A11 and A3) below the lateral side of the head of the fibula (A6), where it lies in contact with the neck of the bone under cover of peroneus longus (A9). Just before dividing into its two main branches it gives off a small recurrent branch (A10) (to the knee and superior tibiofibular joints).
- The deep peroneal (fibular) nerve (A3) supplies the muscles of the anterior compartment of the leg – tibialis anterior (A12 and B8, to which two branches are shown here, A2), extensor digitorum longus (A4 and B3), extensor hallucis longus (A5 and B4) and peroneus tertius (page 307, C12).
- The superficial peroneal (fibular) nerve (A11) supplies the muscles of the lateral compartment – peroneus longus (A9) and peroneus brevis (under cover of peroneus longus in A, but its lower end is seen at D7 on page 301). After supplying the peroneal muscles, the nerve pierces the deep fascia between extensor digitorum longus and peroneus longus (A4 and 9), and divides into medial and lateral (cutaneous) branches (A8 and 7).

Deep peroneal nerve paralysis. The deep peroneal nerve supplies the anterior tibial compartment and damage is often due to trauma or compression in this compartment. The result is numbness over the dorsum of the foot in the region of the first web space and an inability to dorsiflex the foot with extensor digitorum longus.

B Left lower leg and ankle, from the front and lateral side

1 Anterior tibial vessels
2 Deep peroneal (fibular) nerve
3 Extensor digitorum longus
4 Extensor hallucis longus
5 Lateral malleolus
6 Medial branch of superficial peroneal (fibular) nerve
7 Medial malleolus
8 Tibialis anterior

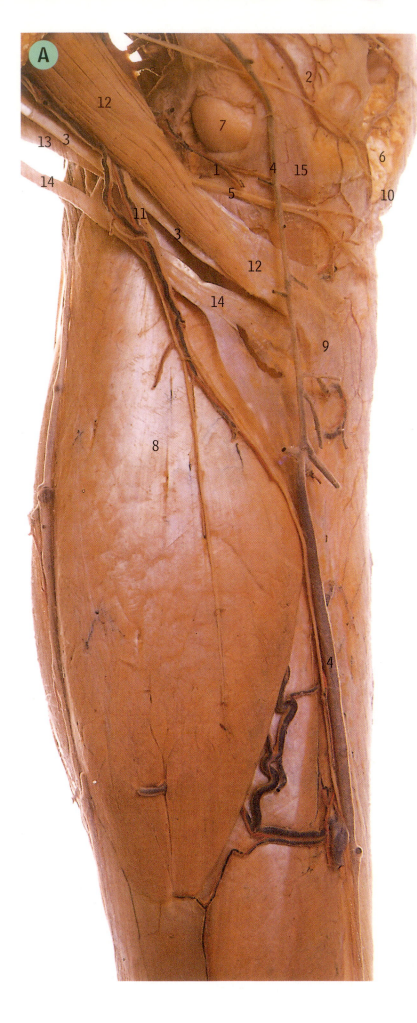

A Left knee and leg, from the medial side and behind

A small window has been cut in the capsule of the knee joint to show part of the medial condyle of the femur (7) and the medial meniscus (1).

1 Branch of saphenous artery overlying medial meniscus
2 Branches of superior medial genicular artery
3 Gracilis
4 Great saphenous vein
5 Infrapatellar branch of saphenous nerve
6 Infrapatellar fat pad
7 Medial condyle of femur (part of capsule removed)
8 Medial head of gastrocnemius
9 Medial surface of tibia
10 Patellar ligament
11 Saphenous nerve and artery
12 Sartorius
13 Semimembranosus
14 Semitendinosus
15 Tibial collateral ligament

B Left knee and leg, from the lateral side

A small window has been cut in the capsule of the knee joint to show the tendon of popliteus (14) passing deep to the fibular collateral ligament (5). The common peroneal (fibular) nerve (2) runs down behind biceps (1) to pass through the gap between peroneus longus (13) and soleus (15). The superficial peroneal (fibular) nerve becomes superficial between peroneus longus (13) and extensor digitorum longus (3)

1 Biceps
2 Common peroneal (fibular) nerve
3 Extensor digitorum longus
4 Fascia overlying tibialis anterior
5 Fibular collateral ligament
6 Head of fibula
7 Iliotibial tract
8 Infrapatellar fat pad
9 Lateral cutaneous nerve of calf
10 Lateral head of gastrocnemius
11 Lateral meniscus
12 Patellar ligament
13 Peroneus longus
14 Popliteus
15 Soleus
16 Superficial peroneal (fibular) nerve

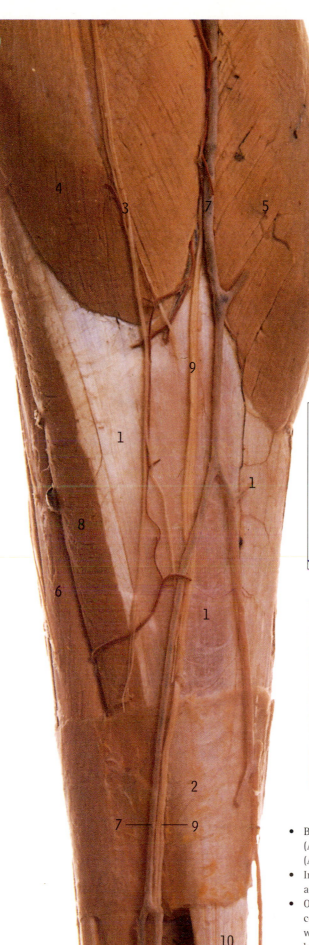

C Left calf, superficial dissection

1 Aponeurosis of gastrocnemius
2 Deep fascia
3 Lateral cutaneous nerve of calf
4 Lateral head of gastrocnemius
5 Medial head of gastrocnemius
6 Peroneus longus
7 Small saphenous vein
8 Soleus
9 Sural nerve
10 Tendo calcaneus (Achilles tendon)

Vein harvest for coronary artery bypass grafting (CABG) . This procedure has become a routine practice in cardiovascular units that are carrying out CABG – known as a 'cabbage' operation. During removal of the long saphenous vein for use by the cardiac surgeon, the saphenous nerve (the terminal branch, L4, of the femoral nerve) which lies entwined with the branches of the long saphenous vein may be damaged and result in loss of sensation on the medial side of the foot and ankle.

Common peroneal nerve paralysis is normally the result of trauma or pressure on the neck of the fibula, often following a fracture when the lower leg is put in a cast. Pressure on the common peroneal nerve will denervate the anterior and lateral compartments of the lower leg, causing sensory loss on the lateral shin and dorsum of the foot and motor loss which makes it impossible to evert and dorsiflex the foot (foot drop). Patients with foot drop scuff their toes on the ground, so this nerve lesion can often be diagnosed by looking at the patient's shoes or gait.

• Below knee level the great saphenous vein (A4) is accompanied by the saphenous nerve (A11).
• In the calf the small saphenous vein (C7) is accompanied by the sural nerve (C9).
• On the lateral side of the upper leg, biceps (B1) converges on to the head of the fibula (B6), with the common peroneal (fibular) nerve (B2) behind them. The nerve divides under cover of peroneus longus (B13) and in contact with the neck of the fibula into superficial and deep branches (page 297, A11 and 3).

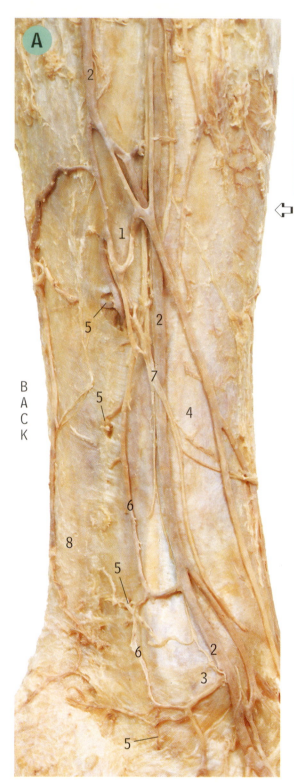

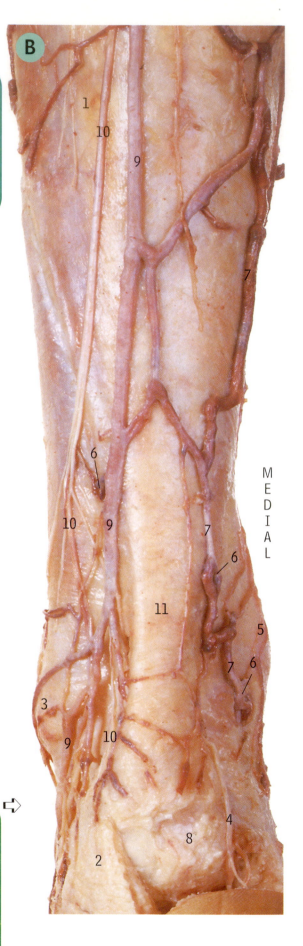

A Left leg and ankle. Superficial veins and nerves, from the medial side

1 Deep fascia over soleus
2 Great saphenous vein
3 Medial malleolus
4 Medial (subcutaneous) surface of tibia
5 Perforating veins
6 Posterior arch vein
7 Saphenous nerve
8 Tendo calcaneus (Achilles tendon)

• The perforating veins are communications between the superficial veins (outside the deep fascia) and the deep veins (below the fascia). The commonest sites for them are just behind the tibia, behind the fibula and in the adductor canal. These communicating vessels possess valves which direct the blood flow from superficial to deep; venous return from the limb is then brought about by the pumping action of the deep muscles (which are all below the deep fascia). If the valves become incompetent or the deep veins blocked, pressure in the superficial veins increases and they become varicose (dilated and tortuous).

Varicose veins. Normally, the superficial venous supply below the knee drains via numerous perforators into the deep system. If the valves controlling this flow are damaged, which is often a congenital weakness, then the blood will flow from the deep system to the superficial system and many dilated, tortuous veins will be seen in the calf region. These unsightly veins may be cosmetically unacceptable; physiologically, the leg may begin to swell and eventually ulceration may occur in the region of the medial malleolus. Removal (stripping) or injection of these veins will improve the vascular physiology of the lower extremity.

B Left leg and ankle. Superficial veins and nerves, from behind
In this specimen (different from that in A), the posterior arch vein (7) on the medial side is large and becoming varicose.

1 Deep fascia
2 Fibrofatty tissue of heel
3 Lateral malleolus
4 Medial calcanean nerve
5 Medial malleolus
6 Perforating vein
7 Posterior arch vein
8 Posterior surface of calcaneus
9 Small saphenous vein
10 Sural nerve
11 Tendo calcaneus (under fascia)

C Right popliteal fossa and upper calf

1 Attachment of popliteus to lateral
   meniscus
2 Biceps
3 Capsule of knee joint
4 Fibular collateral ligament
5 Flexor digitorum longus
6 Flexor hallucis longus
7 Gracilis
8 Lateral head of gastrocnemius
9 Medial condyle of femur
10 Medial head of gastrocnemius
11 Peroneus longus
12 Plantaris
13 Popliteus
14 Posterior surface of fibula (soleus
   removed)
15 Sartorius
16 Semimembranosus
17 Semitendinosus
18 Soleus
19 Tibial collateral ligament
20 Tibialis posterior

Thrombophlebitis, inflammation of the veins
with resultant thrombosis, may occur in either
the deep or the superficial veins. In the
superficial veins, serious complications are
rare. However, if the deep veins of the leg are
blocked there is risk of a thrombus breaking off
and causing a pulmonary embolism. Diagnosis
is made using venography or colour Doppler
ultrasound and the treatment usually involves
pharmacological thinning of the blood.

D Right lower calf and ankle

1 Fascia overlying tibialis posterior
2 Flexor digitorum longus
3 Flexor hallucis longus
4 Lateral malleolus
5 Medial malleolus
6 Part of flexor retinaculum
7 Peroneus brevis
8 Peroneus longus
9 Position of posterior tibial vessels and tibial nerve
10 Posterior talofibular ligament
11 Superior peroneal retinaculum
12 Tendo calcaneus (Achilles tendon)
13 Tibialis posterior

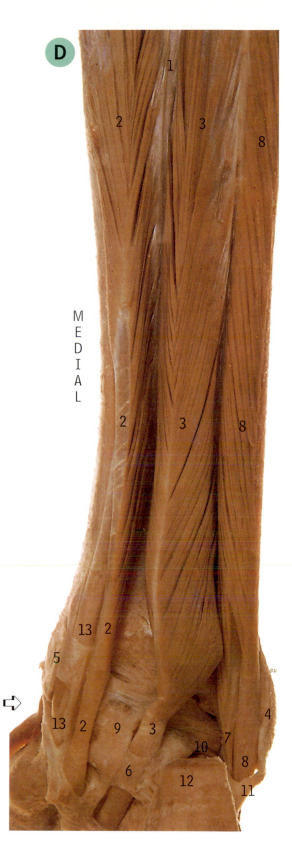

- Tibialis posterior (C20, D13) is the deepest muscle of
  the calf.
- Flexor hallucis longus (C6, D3), although passing to
  the great toe on the *medial* side of the foot, arises
  from the fibula on the *lateral* side of the leg.

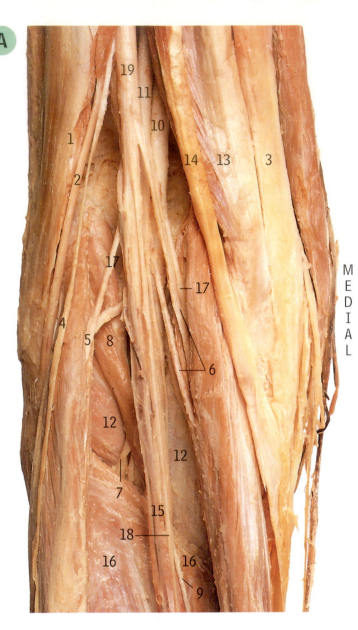

A Left popliteal fossa and upper calf
Gastrocnemius has been incised longitudinally and the two heads (6 and 5) split apart to reveal plantaris (8) and its thin tendon (9), popliteus (12) and the upper part of soleus (16).

1 Biceps	10 Popliteal artery
2 Common peroneal (fibular) nerve	11 Popliteal vein
3 Gracilis	12 Popliteus
4 Lateral cutaneous nerve of calf	13 Semimembranosus
5 Lateral head of gastrocnemius and nerve	14 Semitendinosus
6 Medial head of gastrocnemius and nerves	15 Small saphenous vein (double)
7 Nerve to soleus	16 Soleus
8 Plantaris	17 Sural artery
9 Plantaris tendon	18 Sural nerve
	19 Tibial nerve

B Left calf. Deep dissection of muscles and arteries

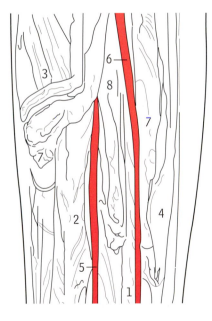

1 Flexor digitorum longus
2 Flexor hallucis longus
3 Lateral head of gastrocnemius
4 Medial head of gastrocnemius
5 Peroneal (fibular) artery
6 Posterior tibial artery
7 Soleus
8 Tibialis posterior

• Flexor hallucis longus, going to the great toe on the medial side of the foot, arises from the fibula on the lateral side of the leg; the tendon crosses to the medial side in the sole.

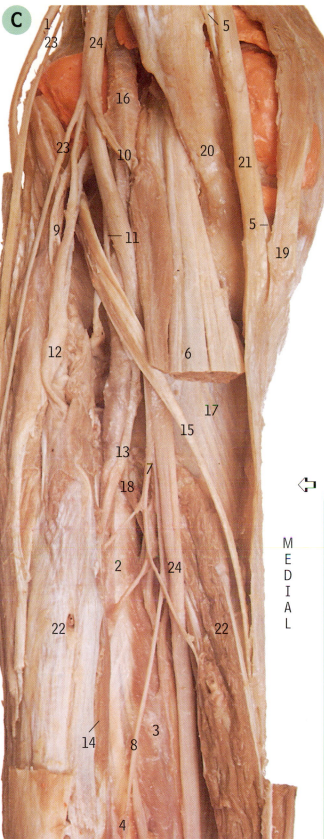

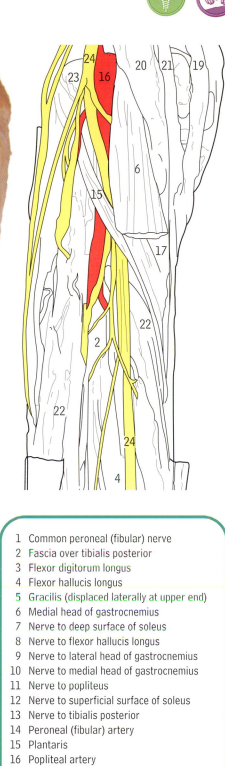

C Left popliteal fossa and calf. Deep dissection

M
E
D
I
A
L

1 Common peroneal (fibular) nerve
2 Fascia over tibialis posterior
3 Flexor digitorum longus
4 Flexor hallucis longus
5 Gracilis (displaced laterally at upper end)
6 Medial head of gastrocnemius
7 Nerve to deep surface of soleus
8 Nerve to flexor hallucis longus
9 Nerve to lateral head of gastrocnemius
10 Nerve to medial head of gastrocnemius
11 Nerve to popliteus
12 Nerve to superficial surface of soleus
13 Nerve to tibialis posterior
14 Peroneal (fibular) artery
15 Plantaris
16 Popliteal artery
17 Popliteus
18 Posterior tibial artery
19 Sartorius
20 Semimembranosus
21 Semitendinosus
22 Soleus
23 Sural nerve (double origin)
24 Tibial nerve

• The deep veins of the calf, deep to and within soleus, are sites for potentially dangerous venous thrombosis.

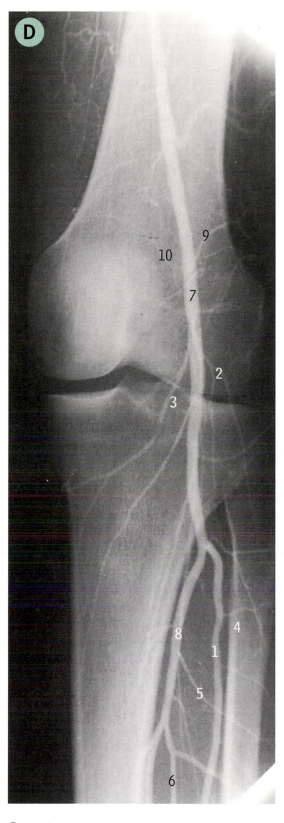

D Popliteal arteriogram

1 Anterior tibial artery
2 Inferior lateral genicular artery
3 Inferior medial genicular artery
4 Muscular branches of anterior tibial artery
5 Muscular branches of posterior tibial artery
6 Peroneal artery
7 Popliteal artery
8 Posterior tibial artery
9 Superior lateral genicular artery
10 Superior medial genicular artery

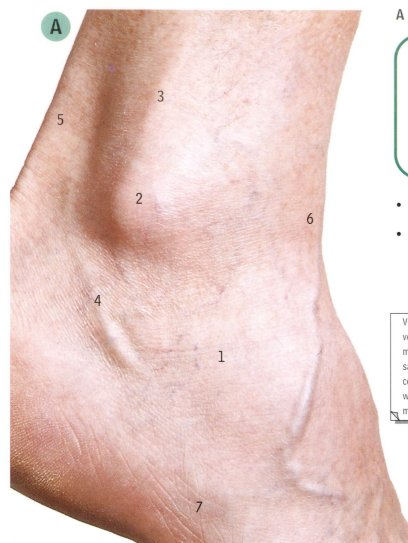

A Right ankle and foot, from the lateral side

1 Extensor digitorum brevis
2 Lateral malleolus
3 Peroneus longus and brevis
4 Small saphenous vein
5 Tendo calcaneus (Achilles tendon)
6 Tibialis anterior
7 Tuberosity of base of fifth metatarsal

- The great saphenous vein (B7) runs upwards in front of the medial malleolus (B9).
- The small saphenous vein (A4) runs upwards behind the lateral malleolus (A2).

Venous cutdowns. The most reliable site for a venous cutdown is just anterior to the medial malleolus of the ankle where the great saphenous vein is found. Prior to the advent of central venous catheterization in the 1960s this was, and in many parts of the world still is, the most common access to a vein in an emergency.

Achilles tendon rupture. The Achilles tendon (the weak point in Achilles' anatomy, as he was held by the heel when dipped in the River Styx) is the combined calcaneal insertion of the gastrocnemius and soleus muscles and occasionally the plantaris tendon. Rupture is usually due to an acute contraction in an unexercised muscle. Rupture may be partial or complete; conservative treatment is adequate for the partial tears but surgery is normally required for complete rupture.

B Right ankle and foot, from the front and medial side

The most prominent surface features are the medial malleolus (9), the tendo calcaneus (11) at the back and the tendons of tibialis anterior (12) and extensor hallucis longus (6) at the front. The dorsalis pedis artery (3) can be palpated where labelled, as may the long tendons

1	Calcaneus	8	Head of first metatarsal
2	Dorsal venous arch	9	Medial malleolus
3	Dorsalis pedis artery	10	Posterior tibial artery
4	Extensor digitorum brevis	11	Tendo calcaneus (Achilles tendon)
5	Extensor digitorum longus	12	Tibialis anterior
6	Extensor hallucis longus	13	Tibialis posterior
7	Great saphenous vein	14	Tuberosity of navicular

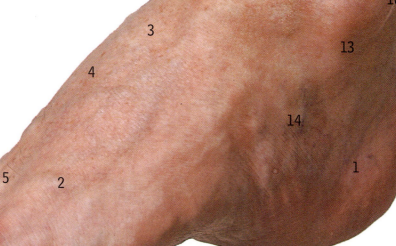

Achilles tendon reflex (ankle jerk) is elicited by tapping the Achilles tendon with a patellar hammer and noting plantar flexion, a good test of the S1 segment of the spinal cord.

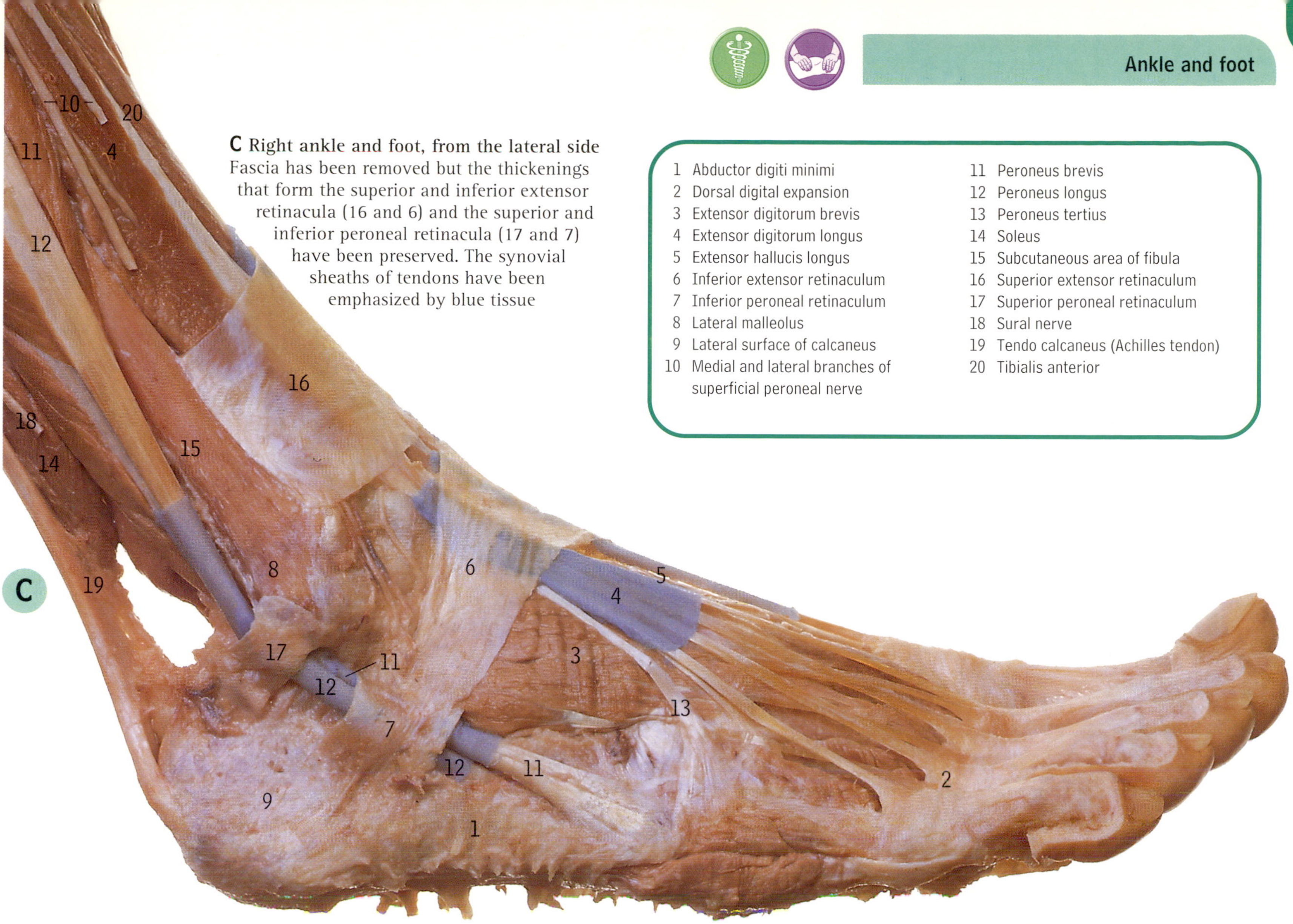

C Right ankle and foot, from the lateral side
Fascia has been removed but the thickenings that form the superior and inferior extensor retinacula (16 and 6) and the superior and inferior peroneal retinacula (17 and 7) have been preserved. The synovial sheaths of tendons have been emphasized by blue tissue

1 Abductor digiti minimi
2 Dorsal digital expansion
3 Extensor digitorum brevis
4 Extensor digitorum longus
5 Extensor hallucis longus
6 Inferior extensor retinaculum
7 Inferior peroneal retinaculum
8 Lateral malleolus
9 Lateral surface of calcaneus
10 Medial and lateral branches of superficial peroneal nerve
11 Peroneus brevis
12 Peroneus longus
13 Peroneus tertius
14 Soleus
15 Subcutaneous area of fibula
16 Superior extensor retinaculum
17 Superior peroneal retinaculum
18 Sural nerve
19 Tendo calcaneus (Achilles tendon)
20 Tibialis anterior

D Right ankle and foot, from the medial side

1 Abductor hallucis
2 Extensor hallucis longus
3 Flexor digitorum longus
4 Flexor hallucis longus
5 Flexor retinaculum
6 Inferior extensor retinaculum (lower band)
7 Inferior extensor retinaculum (upper band)
8 Medial calcanean nerve
9 Medial malleolus
10 Medial surface of tibia
11 Plantaris tendon
12 Posterior surface of calcaneus
13 Posterior tibial artery and venae comitantes
14 Soleus
15 Tendo calcaneus (Achilles tendon)
16 Tibial nerve
17 Tibialis anterior
18 Tibialis posterior

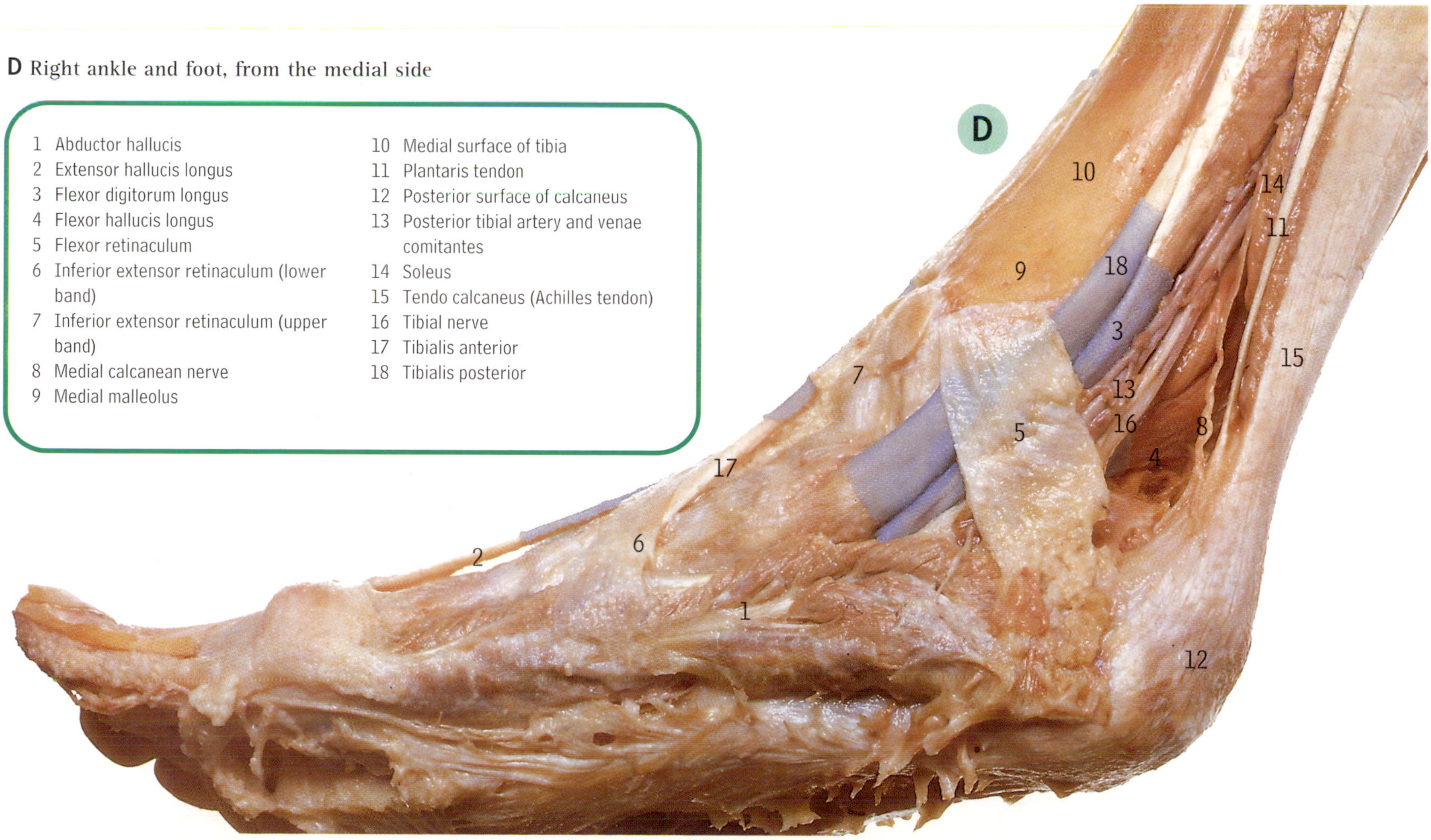

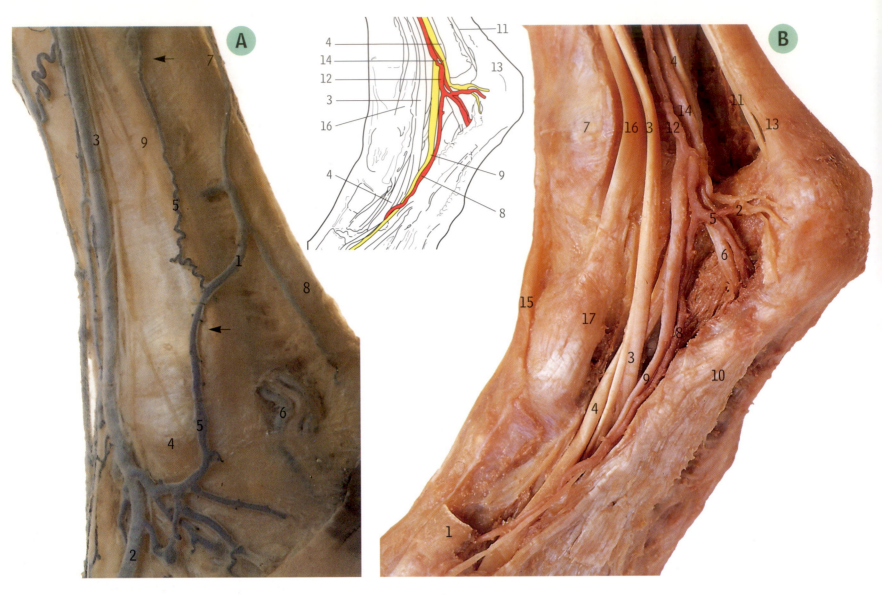

A Right lower leg and ankle, from the medial side and behind
The deep fascia remains intact apart from a small window cut to show the position of the posterior tibial vessels and tibial nerve (6). The great saphenous vein (3) runs upwards in front of the medial malleolus (4) with the posterior arch vein (5) behind it. The arrows indicate common levels for perforating veins (page 300, A5 and B6)

B Right ankle and sole, from the medial side and below
The foot is in plantar flexion, and the flexor retinaculum and most of abductor hallucis (1) have been removed to show how the tendon of flexor hallucis longus (4) passes deep to flexor digitorum longus (3) in the sole to run towards the great toe

1 Communication with small saphenous vein
2 Dorsal venous arch
3 Great saphenous vein and saphenous nerve
4 Medial malleolus
5 Posterior arch vein
6 Posterior tibial vessels and tibial nerve
7 Small saphenous vein
8 Tendo calcaneus (Achilles' tendon)
9 Tibialis posterior and flexor digitorum longus underlying deep fascia

Varicose veins. Normally, the superficial venous supply below the knee drains via numerous perforators into the deep system. If the valves controlling this flow are damaged, which is often a congenital weakness, then the blood will flow from the deep system to the superficial system and many dilated, tortuous veins will be seen in the calf region. These unsightly veins may be cosmetically unacceptable; physiologically, the leg may begin to swell and eventually ulceration may occur in the region of the medial malleolus. Removal (stripping) or injection of these veins will improve the vascular physiology of the lower extremity.

1 Abductor hallucis
2 Calcanean nerves and vessels
3 Flexor digitorum longus
4 Flexor hallucis longus
5 Lateral plantar artery
6 Lateral plantar nerve
7 Medial malleolus
8 Medial plantar artery
9 Medial plantar nerve
10 Plantar aponeurosis overlying flexor digitorum brevis
11 Plantaris tendon
12 Posterior tibial artery
13 Tendo calcaneus (Achilles tendon)
14 Tibial nerve
15 Tibialis anterior
16 Tibialis posterior
17 Tuberosity of navicular

C

C Left ankle and foot, from the front and lateral side
The foot is plantar flexed and part of the capsule of the ankle joint has been removed to show the talus (1). The tendons of peroneus tertius (12) and extensor digitorum longus (5) lie superficial to extensor digitorum brevis (4). The sural nerve and small saphenous vein (13) pass behind the lateral malleolus (8)

1	Anterior lateral malleolar artery overlying talus (ankle joint capsule removed)	8	Lateral malleolus
		9	Perforating branch of peroneal artery
2	Anterior tibial vessels and deep peroneal (fibular) nerve	10	Peroneus brevis
		11	Peroneus longus
3	Deep fascia forming superior extensor retinaculum	12	Peroneus tertius
		13	Small saphenous vein and sural nerve
4	Extensor digitorum brevis	14	Superficial peroneal (fibular) nerve
5	Extensor digitorum longus	15	Tarsal sinus
6	Extensor hallucis longus	16	Tendo calcaneus (Achilles tendon)
7	Inferior extensor retinaculum (partly removed)	17	Tibialis anterior

D

E

Left ankle, **D** cross section, **E** axial MR image
This section, looking down from above, emphasizes the positions of tendons, vessels and nerves in the ankle region. The talus (18) is in the centre, with the medial malleolus (9) on the left of the picture and the lateral malleolus (8) on the right. The great saphenous vein (7) and saphenous nerve (15) are in front of the medial malleolus, with the tendon of tibialis posterior (22) immediately behind it. The small saphenous vein (16) and the sural nerve (17) are behind the lateral malleolus, with the tendons of peroneus longus (11) and peroneus brevis (10) intervening. At the front of the ankle the dorsalis pedis vessels (2) and deep peroneal (fibular) nerve (1) are between the tendons of extensor hallucis longus (4) and extensor digitorum longus (3). Behind the medial malleolus (9) and tibialis posterior (22), the posterior tibial vessels (14) and tibial nerve (20) are between the tendons of flexor digitorum longus (5) and flexor hallucis longus (6)

1	Deep peroneal (fibular) nerve	12	Peroneus tertius
2	Dorsalis pedis artery and venae comitantes	13	Posterior talofibular ligament
3	Extensor digitorum longus	14	Posterior tibial artery and venae comitantes
4	Extensor hallucis longus	15	Saphenous nerve
5	Flexor digitorum longus	16	Small saphenous vein
6	Flexor hallucis longus	17	Sural nerve
7	Great saphenous vein	18	Talus
8	Lateral malleolus of fibula	19	Tendo calcaneus (Achilles tendon)
9	Medial malleolus of tibia	20	Tibial nerve
10	Peroneus brevis	21	Tibialis anterior
11	Peroneus longus	22	Tibialis posterior

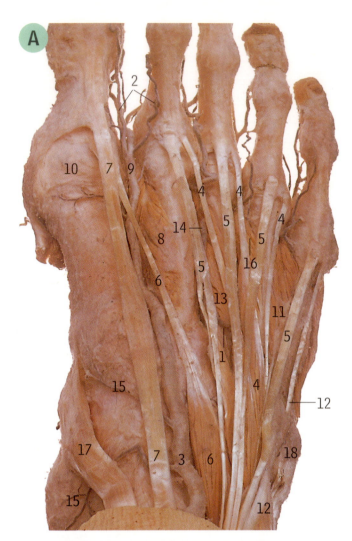

A Dorsum of the right foot

1	Arcuate artery	11	Fourth dorsal interosseous
2	Digital arteries	12	Peroneus tertius
3	Dorsalis pedis artery	13	Second dorsal interosseous
4	Extensor digitorum brevis	14	Second dorsal metatarsal artery
5	Extensor digitorum longus	15	Tarsal arteries
6	Extensor hallucis brevis	16	Third dorsal interosseous
7	Extensor hallucis longus	17	Tibialis anterior
8	First dorsal interosseous	18	Tuberosity of base of fifth metatarsal and
9	First dorsal metatarsal artery		peroneus brevis
10	First metatarsophalangeal joint		

- Extensor digitorum longus (5) sends its tendons to the four lateral toes, while extensor digitorum brevis (4) sends its tendons to the four medial toes; the part of extensor digitorum brevis that goes to the great toe is often known as extensor hallucis brevis (6).

Sprained ankle. Because the normal range of inversion is greater than that of eversion, the lateral ligaments of the ankle tend to suffer most in inversion rotational injuries. The lateral ligament has three components – the anterior talofibular ligament, the calcaneofibular ligament and the posterior talofibular ligament; the most commonly torn is the anterior talofibular ligament. Clinically, there is tenderness and swelling over the lateral side of the ankle and a particularly tender spot just anterior to the tip of the lateral malleolus.

B Right talocalcanean and talocalcaneonavicular joints
The talus has been removed to show the articular surfaces of the calcaneus (21, 17 and 2), navicular (3) and plantar calcaneonavicular (spring) ligament (20)

1	Abductor hallucis	15	Inferior extensor retinaculum
2	Anterior articular surface on calcaneus for talus	16	Interosseous talocalcanean ligament
		17	Middle articular surface on calcaneus for talus
3	Articular surface on navicular for talus	18	Peroneus brevis
4	Calcaneonavicular part of bifurcate ligament	19	Peroneus longus
5	Cervical ligament	20	Plantar calcaneonavicular (spring) ligament
6	Deep peroneal (fibular) nerve	21	Posterior articular surface on calcaneus for talus
7	Deltoid ligament		
8	Dorsal venous arch	22	Posterior tibial vessels and medial and lateral plantar nerves
9	Dorsalis pedis artery and vena comitans		
10	Extensor digitorum brevis	23	Small saphenous vein
11	Extensor digitorum longus	24	Sural nerve
12	Extensor hallucis longus	25	Tendo calcaneus (Achilles tendon)
13	Flexor digitorum longus	26	Tibialis anterior
14	Flexor hallucis longus	27	Tibialis posterior

- There are two joints beneath the talus (page 274, A and B). The more posterior is the talocalcanean joint, between the posterior articular surfaces of the calcaneus and talus; this joint is sometimes known anatomically as the subtalar joint.
- The more anterior joint, the talocalcaneonavicular, is between (a) the middle and anterior articular surfaces of the talus and calcaneus and the upper surface of the plantar calcaneonavicular (spring) ligament, all of which constitute the talocalcanean part of the joint, and (b) the head of the talus and the posterior articular surface of the navicular, which constitute the talonavicular part of the joint. The two parts of the joint share one synovial cavity.
- Do not confuse the talocalcanean joint with the talocalcanean part of the talocalcaneonavicular joint.
- Clinicians sometimes use the term subtalar joint as a combined name for both the talocalcanean joint and the talocalcanean part of the talocalcaneonavicular joint, because it is at both these joints beneath the talus that most of the movements of inversion and eversion of the foot occur.

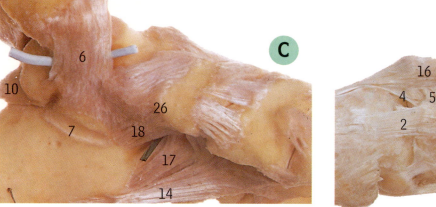

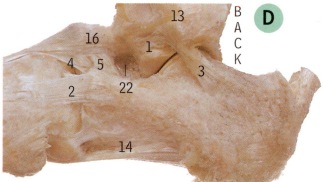

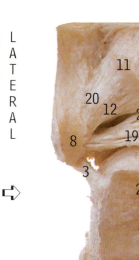

Ligaments of the left ankle and foot, **C** from the medial side, **D** from the lateral side, **E** from behind

In C the marker below the medial malleolus (15) passes between the superficial and deep parts of the deltoid ligament (6). The marker below the tuberosity of the navicular (26) passes between the plantar calcaneonavicular (spring) and calcaneocuboid (short plantar) ligaments (18 and 17)

1 Anterior talofibular ligament
2 Calcaneocuboid part of bifurcate ligament
3 Calcaneofibular ligament
4 Calcaneonavicular part of bifurcate ligament
5 Cervical ligament
6 Deltoid ligament
7 Groove below sustentaculum tali for flexor hallucis longus
8 Groove on lateral malleolus for peroneus brevis
9 Groove on medial malleolus for tibialis posterior
10 Groove on talus for flexor hallucis longus
11 Groove on tibia for flexor hallucis longus
12 Inferior transverse ligament
13 Lateral malleolus

14 Long plantar ligament
15 Medial malleolus
16 Neck of talus
17 Plantar calcaneocuboid (short plantar) ligament
18 Plantar calcaneonavicular (spring) ligament
19 Posterior talofibular ligament
20 Posterior tibiofibular ligament
21 Posterior tibiotalal part of deltoid ligament
22 Tarsal sinus
23 Tendo calcaneus (Achilles tendon)
24 Tibial slip of posterior talofibular ligament
25 Tibiocalcanean part of deltoid ligament
26 Tuberosity of navicular

F Sagittal section of the left foot, from the right

Deltoid ligament rupture. The strong deltoid ligament attaches the medial malleolus to the talus, navicular and calcaneus and also connects to the spring ligament to maintain the medial longitudinal arch. It is occasionally torn in severe twisting injuries, such as when a ballerina or ski jumper falls from a normally controlled position.

1 Abductor digiti minimi
2 Abductor hallucis
3 Calcaneus
4 Cuneonavicular joint
5 Distal phalanx
6 Extensor hallucis longus
7 Fat pad
8 First metatarsal
9 First tarsometatarsal (cuneometatarsal) joint
10 Flexor accessorius
11 Flexor digitorum brevis
12 Flexor hallucis brevis
13 Flexor hallucis longus
14 Great saphenous vein
15 Interosseous talocalcanean ligament
16 Interphalangeal joint
17 Lateral plantar nerve and vessels

18 Medial cuneiform
19 Medial plantar artery
20 Metatarsophalangeal joint of great toe
21 Navicular
22 Plantar aponeurosis
23 Plantar calcaneonavicular (spring) ligament
24 Proximal phalanx
25 Soleus muscle
26 Talocalcanean (subtalar) joint
27 Talonavicular part of talocalcaneonavicular joint
28 Talus
29 Tendo calcaneus (Achilles tendon)
30 Tendon of flexor hallucis
31 Tibia
32 Tibialis posterior muscle
33 Tibiotalal part of ankle joint

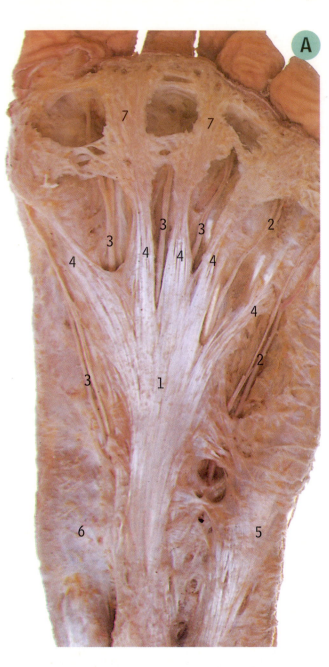

A Sole of the left foot. Plantar aponeurosis
Skin and connective tissue of the sole have been removed to show the tough central part of the plantar aponeurosis (1) which divides into slips (4) for each toe, and thinner medial and lateral parts overlying abductor hallucis (6) and abductor digiti minimi (5) respectively

1　Central part of aponeurosis overlying flexor digitorum brevis
2　Digital branches of lateral plantar nerve and artery
3　Digital branches of medial plantar nerve and artery
4　Digital slip of central part of aponeurosis
5　Lateral part of aponeurosis overlying abductor digiti minimi
6　Medial part of aponeurosis overlying abductor hallucis
7　Superficial stratum of digital slip of aponeurosis

Plantar fasciitis is inflammation of the plantar aponeurosis, an important structure in maintaining the longitudinal arches of the foot. Inflammation of these strong fibrous bands may be extremely painful at the calcaneal end and is particularly common in people who have been walking in thin-soled shoes over rough ground.

B Sole of the left foot, with the plantar aponeurosis removed
The central muscle is flexor digitorum brevis (8), and passing forwards from beneath its borders are various digital vessels and nerves (such as 5 and 11) which arise from the medial and lateral plantar vessels and nerves still largely under cover of the muscle

Flat foot (pes planus) is due to flattening of the longitudinal arch. Often congenital, it may be associated with minor structural anomalies of the tarsal bones. This condition can be seen in wet footprints where the medial arch (normally raised) is visible. Treatment may be assisted by intensive foot exercises or by arch supports worn in the shoes. Occasionally, surgery is needed in the form of arthrodesis (fusion of the tarsal bones).

1	Abductor digiti minimi	13	Fourth lumbrical
2	Abductor hallucis	14	Lateral plantar artery
3	Deep branch of lateral plantar nerve	15	Plantar aponeurosis
4	Fibrous flexor sheath	16	Plantar digital nerve of fifth toe
5	First common plantar digital nerve	17	Plantar digital nerve of great toe
6	First lumbrical	18	Plantar digital nerves of first cleft
7	Flexor digiti minimi brevis	19	Second lumbrical
8	Flexor digitorum brevis	20	Superficial digital branch of medial plantar artery
9	Flexor hallucis brevis	21	Superficial transverse metatarsal ligament
10	Flexor hallucis longus	22	Third lumbrical
11	Fourth common plantar digital nerve	23	Third plantar interosseous
12	Fourth dorsal interosseous	24	Third plantar metatarsal artery

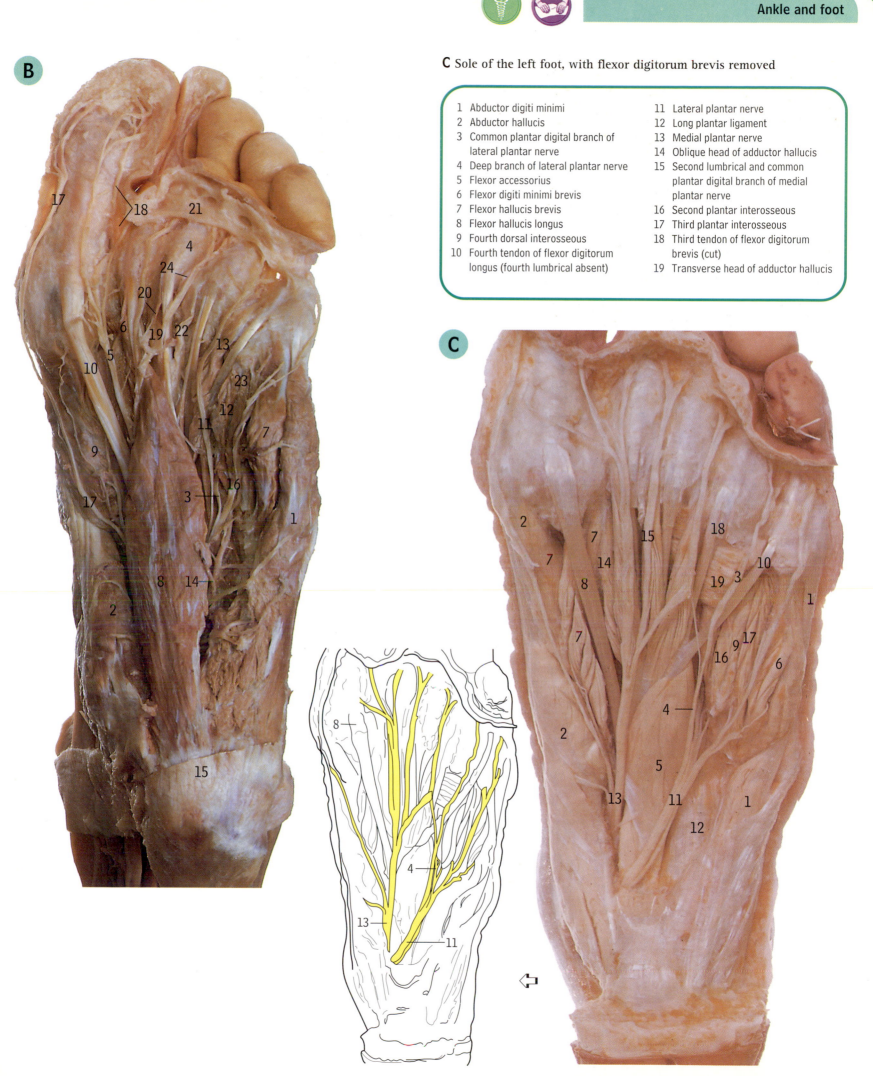

B

C Sole of the left foot, with flexor digitorum brevis removed

1 Abductor digiti minimi
2 Abductor hallucis
3 Common plantar digital branch of lateral plantar nerve
4 Deep branch of lateral plantar nerve
5 Flexor accessorius
6 Flexor digiti minimi brevis
7 Flexor hallucis brevis
8 Flexor hallucis longus
9 Fourth dorsal interosseous
10 Fourth tendon of flexor digitorum longus (fourth lumbrical absent)

11 Lateral plantar nerve
12 Long plantar ligament
13 Medial plantar nerve
14 Oblique head of adductor hallucis
15 Second lumbrical and common plantar digital branch of medial plantar nerve
16 Second plantar interosseous
17 Third plantar interosseous
18 Third tendon of flexor digitorum brevis (cut)
19 Transverse head of adductor hallucis

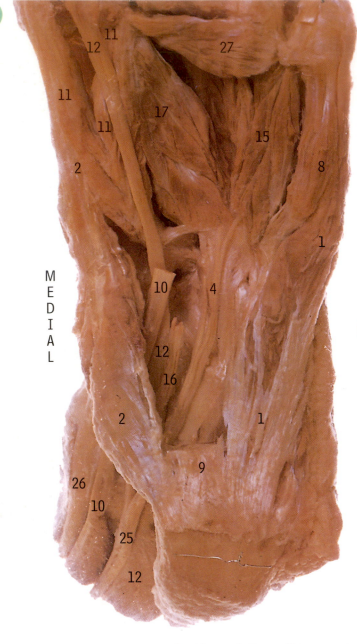

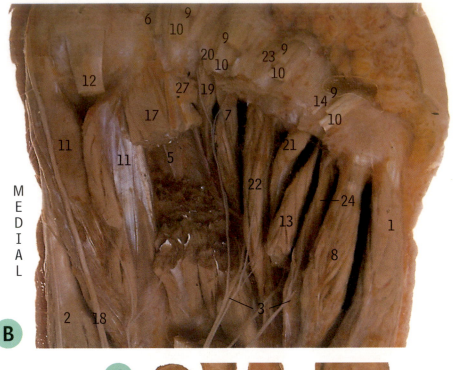

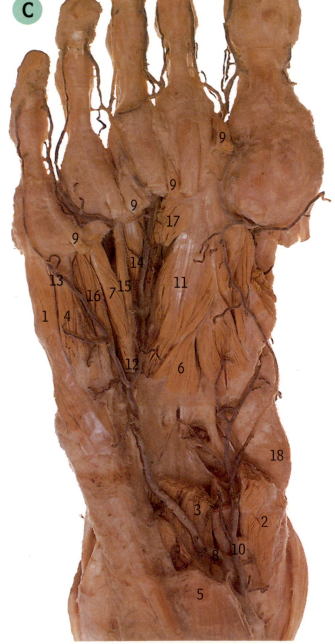

Sole of the left foot. Deep muscles, A adductor hallucis, B interossei

Most of the flexor muscles and tendons have been removed to show in A the two heads of adductor hallucis (17 and 27), and in B (which corresponds to the front part of A) the dorsal interossei (5, 19, 22 and 13) and plantar interossei (7, 21 and 24), with the ends of the lumbricals (6, 20, 23 and 14).

1	Abductor digiti minimi	14	Fourth lumbrical
2	Abductor hallucis	15	Interossei
3	Branches of deep branch of lateral plantar nerve	16	Medial plantar nerve
4	Deep branch of lateral plantar nerve	17	Oblique head of adductor hallucis
5	First dorsal interosseous	18	Plantar digital nerve of great toe
6	First lumbrical	19	Second dorsal interosseous
7	First plantar interosseous	20	Second lumbrical
8	Flexor digiti minimi brevis	21	Second plantar interosseous
9	Flexor digitorum brevis	22	Third dorsal interosseous
10	Flexor digitorum longus	23	Third lumbrical
11	Flexor hallucis brevis	24	Third plantar interosseous
12	Flexor hallucis longus	25	Tibial nerve
13	Fourth dorsal interosseous	26	Tibialis posterior
		27	Transverse head of adductor hallucis

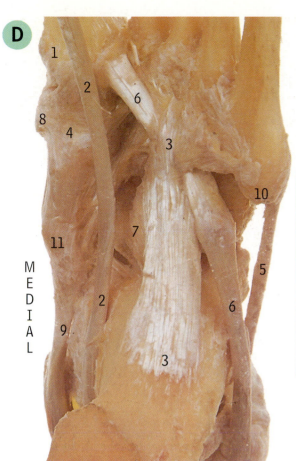

D Sole of the left foot. Ligaments and tendons
The anterior end of the long plantar ligament (3) forms with the groove of the cuboid (E6) a tunnel for the peroneus longus tendon (6) which runs to the medial cuneiform (4) and the base of the first metatarsal (1).

1 Base of first metatarsal
2 Flexor hallucis longus
3 Long plantar ligament
4 Medial cuneiform
5 Peroneus brevis
6 Peroneus longus
7 Plantar calcaneocuboid (short plantar) ligament
8 Tibialis anterior
9 Tibialis posterior
10 Tuberosity of base of fifth metatarsal
11 Tuberosity of navicular

• The plantar calcaneonavicular ligament (E9), commonly called the spring ligament, is one of the most important in the foot. It stretches between the sustentaculum tali (E7) and the tuberosity of the navicular (E16), blending on its medial side with the deltoid ligament of the ankle joint and supporting on the upper surface (page 308, B20) part of the head of the talus.

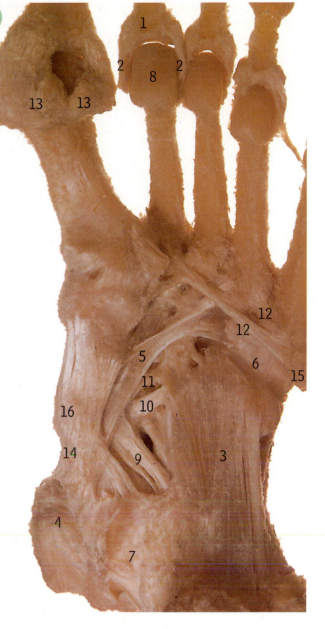

E Sole of the left foot. Ligaments
The anterior end of the long plantar ligament (3) has been removed to show the groove for peroneus longus on the cuboid (6).

1 Base of proximal phalanx
2 Collateral ligament of metatarsophalangeal joint
3 Deep fibres of long plantar ligament
4 Deltoid ligament
5 Fibrous slip from tibialis posterior
6 Groove on cuboid for peroneus longus
7 Groove on sustentaculum tali for flexor hallucis longus
8 Head of second metatarsal
9 Plantar calcaneonavicular (spring) ligament
10 Plantar cuboideonavicular ligament
11 Plantar cuneonavicular ligament
12 Plantar metatarsal ligament
13 Sesamoid bone
14 Tibialis posterior
15 Tuberosity of base of fifth metatarsal
16 Tuberosity of navicular

C Sole of the right foot. Plantar arch
Most of the flexor muscles and tendons have been removed to show the lateral plantar artery (8) crossing flexor accessorius (3) to become the plantar arch (12) which would lie deep to the flexor tendons

1 Abductor digiti minimi
2 Abductor hallucis
3 Flexor accessorius
4 Flexor digiti minimi brevis
5 Flexor digitorum brevis
6 Flexor hallucis brevis
7 Fourth dorsal interosseous
8 Lateral plantar artery
9 Lumbrical
10 Medial plantar artery and nerve
11 Oblique head of adductor hallucis
12 Plantar arch
13 Plantar digital artery
14 Plantar metatarsal artery
15 Second plantar interosseous
16 Third plantar interosseous
17 Transverse head of adductor hallucis
18 Tuberosity of navicular

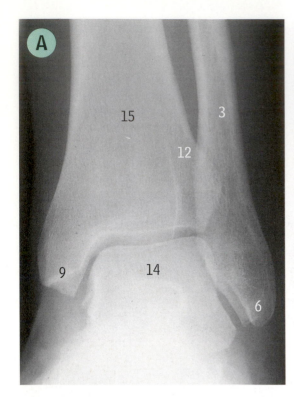

Ankle, A anteroposterior projection; B calcaneus, axial projection
The side view in B shows a small calcanean spur (about 2 cm below the label 1)

1 Calcaneus	10 Medial tubercle of talus
2 Cuboid	
3 Fibula	11 Navicular
4 Head of talus	12 Region of inferior tibiofibular joint
5 Lateral cuneiform	
6 Lateral malleolus of fibula	13 Sustentaculum tali of calcaneus
7 Lateral tubercle of talus	14 Talus
	15 Tibia
8 Medial malleolus	16 Tuberosity of base of fifth metatarsal
9 Medial malleolus of tibia	

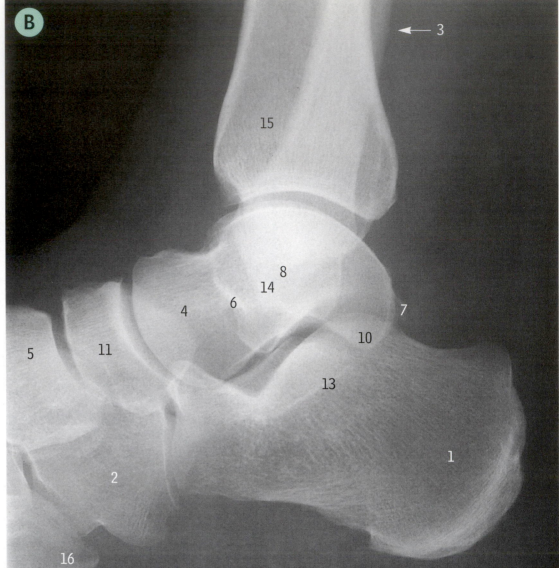

Foot, C axial MR image

1 Abductor digiti minimi muscle	9 Lateral cuneiform
	10 Medial cuneiform
2 Base of metatarsal	11 Medial malleolus
3 Base of proximal phalanx	12 Neck of talus
	13 Shaft of metatarsal
4 Calcaneus	14 Talus
5 Cuboid	15 Tarsal sinus
6 Dorsal interossei muscle	16 Tendon of peroneus brevis muscle
7 Head of talus	17 Tendon of tibialis anterior muscle
8 Intermediate cuneiform	18 Tibia

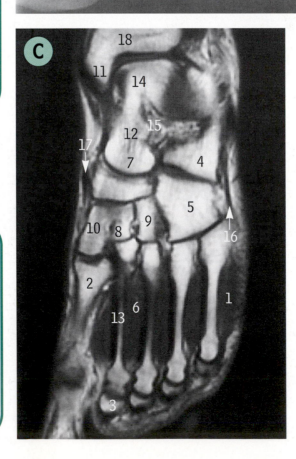

Pott's fracture (ankle). External rotation injuries to the ankle are often associated with fractures. Identified in 1769 by Sir Percival Pott, they are more commonly described as first-degree, second-degree and third-degree ankle injuries. In a first-degree fracture a single structure – the lateral malleolus – is damaged. In a second-degree fracture the two structures damaged are the lateral malleolus (obliquely fractured and displaced) and the medial malleolus (fractured transversely and usually moved laterally with the talus). In a third-degree fracture the talus is displaced both outwards and backwards, in addition to being externally rotated, and three structures are damaged (lateral malleolus, the medial malleolus with displacement of the talus and a vertical fracture through the tibial articular surface).

Chapter 7

Systemic review

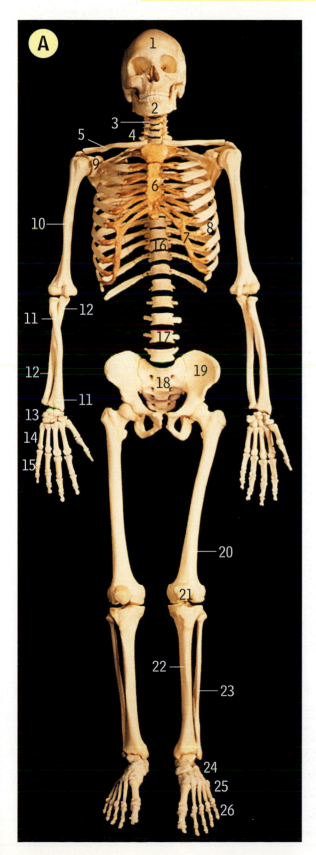

Skeleton

A from the front, **B** from behind. The left forearm is in the position of supination, the right in pronation

1 Skull
2 Mandible
3 Hyoid bone
4 Cervical vertebrae
5 Clavicle
6 Sternum
7 Costal cartilages
8 Ribs
9 Scapula
10 Humerus
11 Radius
12 Ulna
13 Carpal bones
14 Metacarpal bones
15 Phalanges of thumb and fingers
16 Thoracic vertebrae
17 Lumbar vertebrae
18 Sacrum
19 Hip bone
20 Femur
21 Patella
22 Tibia
23 Fibula
24 Tarsal bones
25 Metatarsal bones
26 Phalanges of toes
27 Coccyx

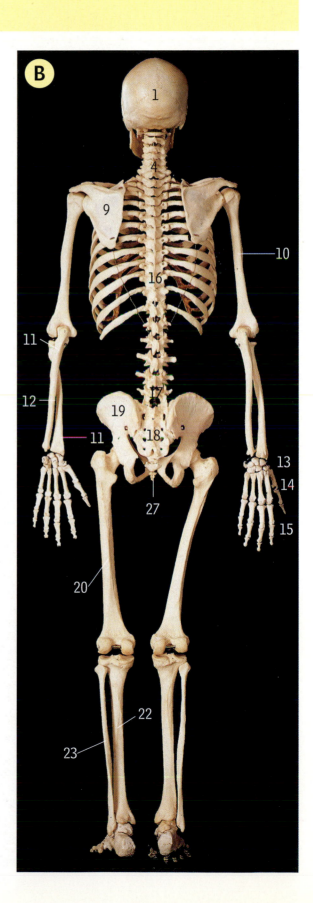

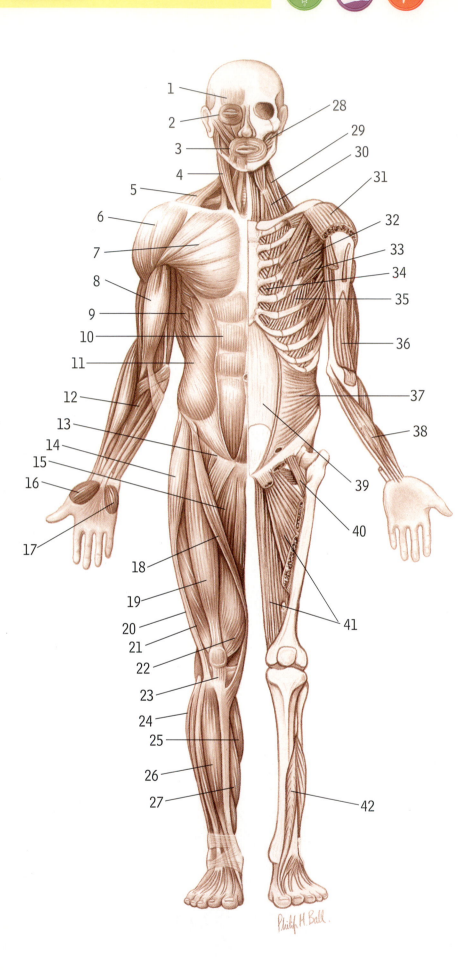

Muscles

From the front. Superficial muscles on the right side of the body, deep muscles on the left side

1 Frontalis part of occipitofrontalis
2 Orbicularis oculi
3 Orbicularis oris
4 Sternocleidomastoid
5 Trapezius
6 Deltoid
7 Pectoralis major
8 Biceps brachii
9 Serratus anterior
10 Rectus abdominis
11 External oblique
12 Superficial flexor muscles of forearm
13 Inguinal ligament
14 Tensor fasciae latae
15 Adductor muscles
16 Thenar muscles
17 Hypothenar muscles
18 Sartorius
19 Rectus femoris
20 Iliotibial tract
21 Vastus lateralis
22 Vastus medialis
23 Patellar ligament
24 Peroneal muscles
25 Gastrocnemius
26 Extensor muscles of leg
27 Soleus
28 Buccinator
29 Levator scapulae
30 Scalenus anterior
31 Deltoid
32 Pectoralis minor
33 Serratus anterior
34 Internal intercostal
35 External intercostal
36 Brachialis
37 Internal oblique
38 Deep flexor muscles of forearm
39 Rectus sheath (posterior wall)
40 Psoas major and iliacus
41 Adductor magnus
42 Extensor hallucis longus

Muscles

From the back. Superficial muscles on the left side of the body, deep muscles on the right side

1   Sternocleidomastoid
2   Trapezius
3   Spine of scapula
4   Deltoid
5   Infraspinatus
6   Latissimus dorsi
7   Triceps
8   External oblique
9   Iliac crest
10 Gluteus medius
11 Superficial extensor muscles of forearm
12 Gluteus maximus
13 Iliotibial tract
14 Biceps femoris
15 Semimembranosus
16 Semitendinosus
17 Gastrocnemius
18 Soleus
19 Tendo calcaneus (Achilles tendon)
20 Semispinalis capitis
21 Splenius
22 Levator scapulae
23 Supraspinatus
24 Rhomboid minor
25 Infraspinatus
26 Teres minor
27 Rhomboid major
28 Teres major
29 Erector spinae
30 Triceps
31 Deep extensor muscles of forearm
32 Gluteus medius
33 Piriformis
34 Obturator internus
35 Quadratus femoris
36 Adductor magnus
37 Semimembranosus
38 Biceps femoris
39 Popliteus
40 Soleus
41 Deep flexor muscles of leg
42 Flexor hallucis longus

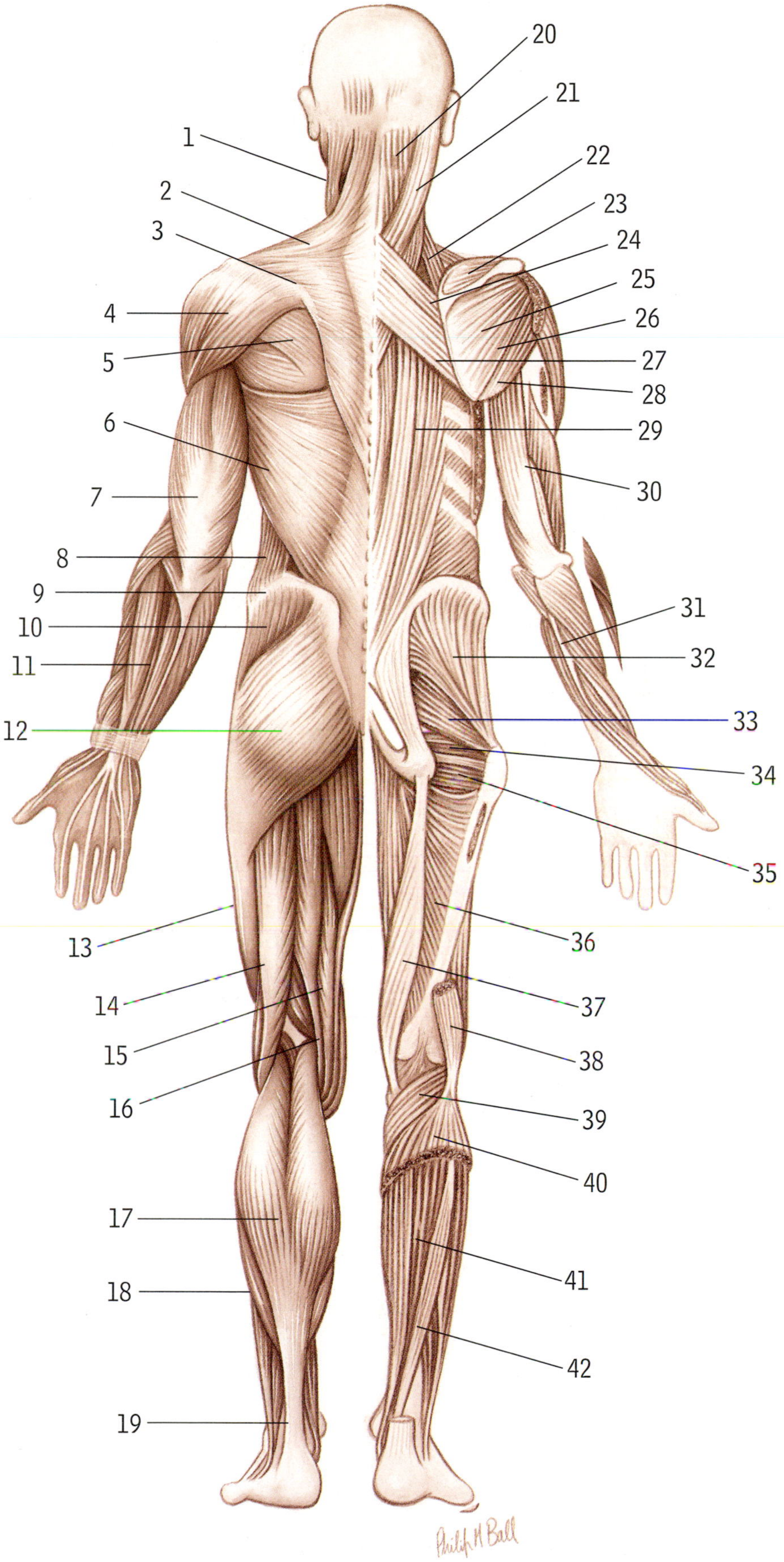

Arteries

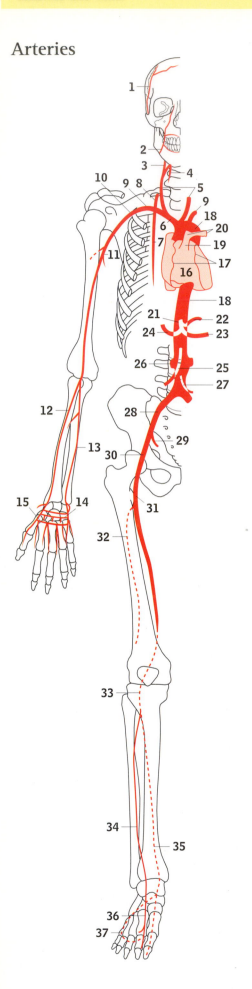

Veins

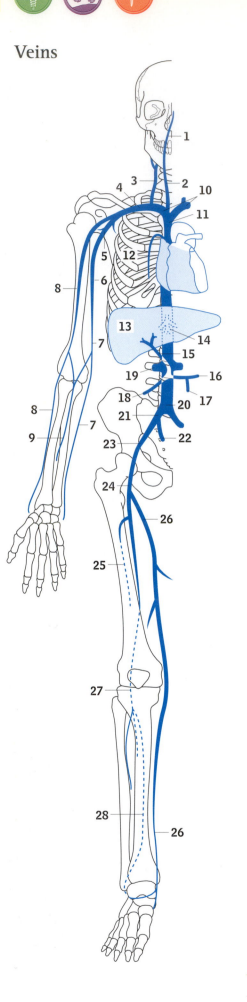

Arteries

Some major arteries, from the front

1	Superficial temporal a.	20	Pulmonary a.
2	Facial a.	21	Coeliac trunk
3	Internal carotid a.	22	Left gastric a.
4	External carotid a.	23	Splenic a.
5	Common carotid a.	24	Common hepatic a.
6	Brachiocephalic trunk	25	Superior mesenteric a.
7	Internal thoracic a.	26	Renal a.
8	Vertebral a.	27	Inferior mesenteric a.
9	Subclavian a.	28	Common iliac a.
10	Axillary a.	29	Internal iliac a.
11	Brachial a.	30	External iliac a.
12	Radial a.	31	Femoral a.
13	Ulnar a.	32	Profunda femoris a.
14	Deep palmar arch	33	Popliteal a.
15	Superficial palmar arch	34	Anterior tibial a.
16	Heart	35	Posterior tibial a.
17	Coronary a.	36	Dorsalis pedis a.
18	Aorta	37	Plantar arch
19	Pulmonary trunk		

Veins

Some major veins, from the front
(The pulmonary veins enter the left atrium at the back of the heart and are not shown)

1	Facial v.	15	Portal v.
2	Internal jugular v.	16	Splenic v.
3	External jugular v.	17	Inferior mesenteric v.
4	Subclavian v.	18	Superior mesenteric v.
5	Axillary v.	19	Renal v.
6	Brachial v.	20	Inferior vena cava
7	Basilic v.	21	Common iliac v.
8	Cephalic v.	22	Internal iliac v.
9	Median forearm v.	23	External iliac v.
10	Brachiocephalic v.	24	Femoral v.
11	Superior vena cava	25	Profunda femoris v.
12	Azygos v.	26	Great saphenous v.
13	Liver	27	Popliteal v.
14	Hepatic v.	28	Small saphenous v.

Nerves

The facial nerve and some major branches of
the brachial, lumbar and sacral plexuses,
A from the front, **B** from the back

1   Facial n.
2   Brachial plexus
3   Musculocutaneous n.
4   Median n.
5   Ulnar n.
6   Lumbar plexus
7   Obturator n.
8   Femoral n.
9   Saphenous n.
10 Common peroneal (fibular) n.
11 Superficial peroneal (fibular) n.
12 Deep peroneal (fibular) n.
13 Axillary n.
14 Radial n.
15 Sacral plexus
16 Superior gluteal n.
17 Inferior gluteal n.
18 Pudendal n.
19 Posterior femoral cutaneous n.
20 Sciatic n.
21 Tibial n.
22 Sural n.

Straight leg raising test (SLR) (Lasègue's test) is
a standard orthopaedic manoeuvre used for the
diagnosis of pain associated with lower back
pathology. The test consists of raising the lower
leg in a supine patient with the knee extended.
If this is painful with the leg straight but
painless when the knee is flexed, it is
considered a positive test. Measurements are
made of the degrees of angle before the pain is
perceived and this can be used to assess the
prognosis of 'sciatica'.

Sciatica is a painful condition along the course
of the sciatic trunk, commonly from disc
prolapse and nerve impingement in the lower
lumbar region, particularly L4/5 or L5/S1. The
nerve affected can be diagnosed from the
distribution of sensory loss and muscle
weakness.

Nerves

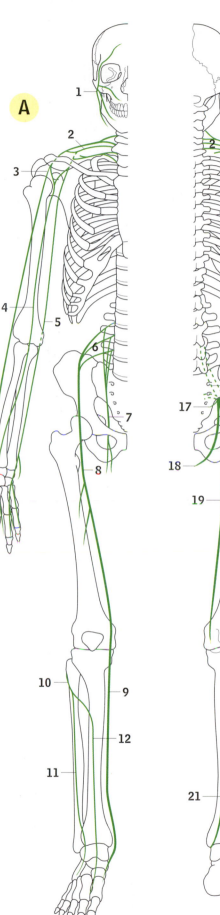

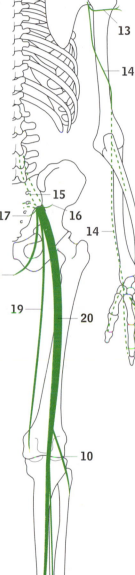

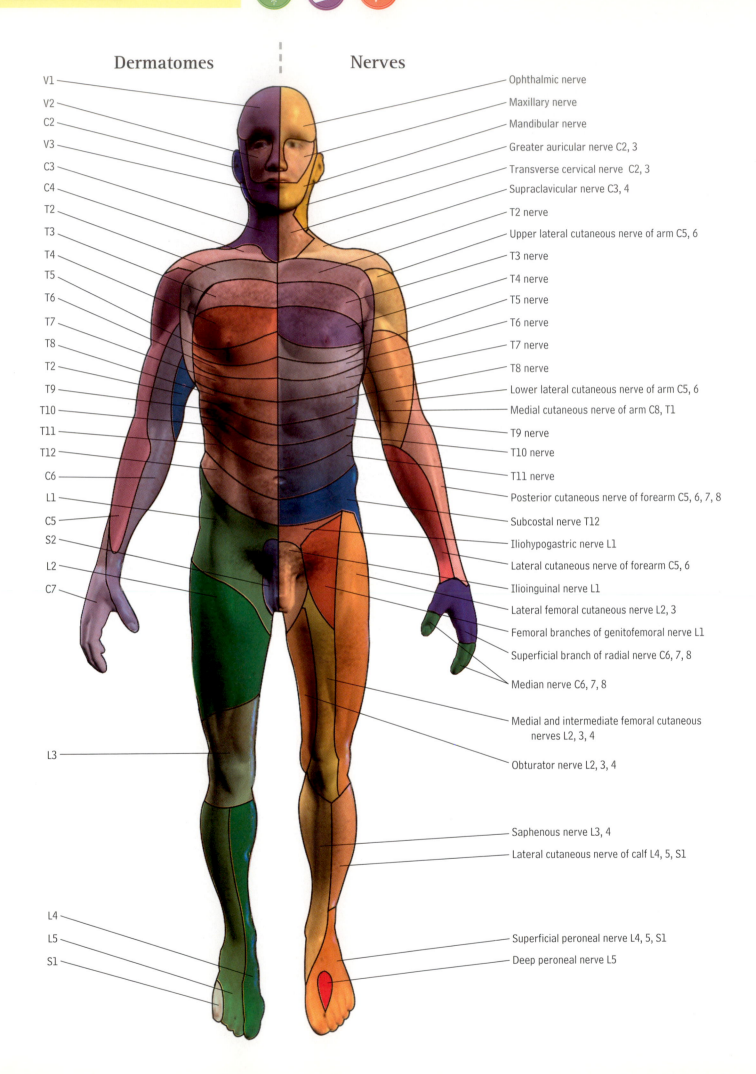

Dermatomes

Nerves

V1 — Ophthalmic nerve
V2 — Maxillary nerve
C2 — Mandibular nerve
V3 — Greater auricular nerve C2, 3
C3 — Transverse cervical nerve C2, 3
C4 — Supraclavicular nerve C3, 4
T2 — T2 nerve
T3 — Upper lateral cutaneous nerve of arm C5, 6
T4 — T3 nerve
T5 — T4 nerve
T6 — T5 nerve
T7 — T6 nerve
T8 — T7 nerve
T2 — T8 nerve
T9 — Lower lateral cutaneous nerve of arm C5, 6
T10 — Medial cutaneous nerve of arm C8, T1
T11 — T9 nerve
T12 — T10 nerve
C6 — T11 nerve
L1 — Posterior cutaneous nerve of forearm C5, 6, 7, 8
C5 — Subcostal nerve T12
S2 — Iliohypogastric nerve L1
L2 — Lateral cutaneous nerve of forearm C5, 6
C7 — Ilioinguinal nerve L1
— Lateral femoral cutaneous nerve L2, 3
— Femoral branches of genitofemoral nerve L1
— Superficial branch of radial nerve C6, 7, 8
— Median nerve C6, 7, 8
— Medial and intermediate femoral cutaneous nerves L2, 3, 4
L3 — Obturator nerve L2, 3, 4
— Saphenous nerve L3, 4
— Lateral cutaneous nerve of calf L4, 5, S1
L4 — Superficial peroneal nerve L4, 5, S1
L5 — Deep peroneal nerve L5
S1

Greater occipital nerve C2
Greater auricular nerve C2, 3
Lesser occipital nerve C2
Supraclavicular nerve C3, 4
Dorsal rami C3, 4, 5
T2 nerve
T3 nerve
T4 nerve
Upper lateral cutaneous nerve of arm C5, 6
T5 nerve
T6 nerve
T7 nerve
Medial cutaneous nerve of arm C8, T1
Posterior cutaneous nerve of arm C5, 6, 7, 8
T8 nerve
T9 nerve
Posterior cutaneous nerve of forearm C5, 6, 7, 8
T10 nerve
T11 nerve
T12 nerve
Medial cutaneous nerve of forearm C8, T1
Subcostal nerve T12
Iliohypogastric nerve L1
Dorsal rami L1, 2, 3
Lateral cutaneous nerve of forearm C5, 6
Ulnar nerve C8, T1
Median nerve C6, 7, 8
Dorsal rami S1, 2, 3
Lateral cutaneous nerve of thigh L2, 3
Posterior cutaneous nerve of thigh S1, 2, 3
Lateral cutaneous nerve of calf L4, 5, S1
Saphenous nerve L3, 4
Sural nerve S1
Medial calcaneal branches of tibial nerve S1
Superficial peroneal nerve L4, 5, S1

Nerves

Dermatomes

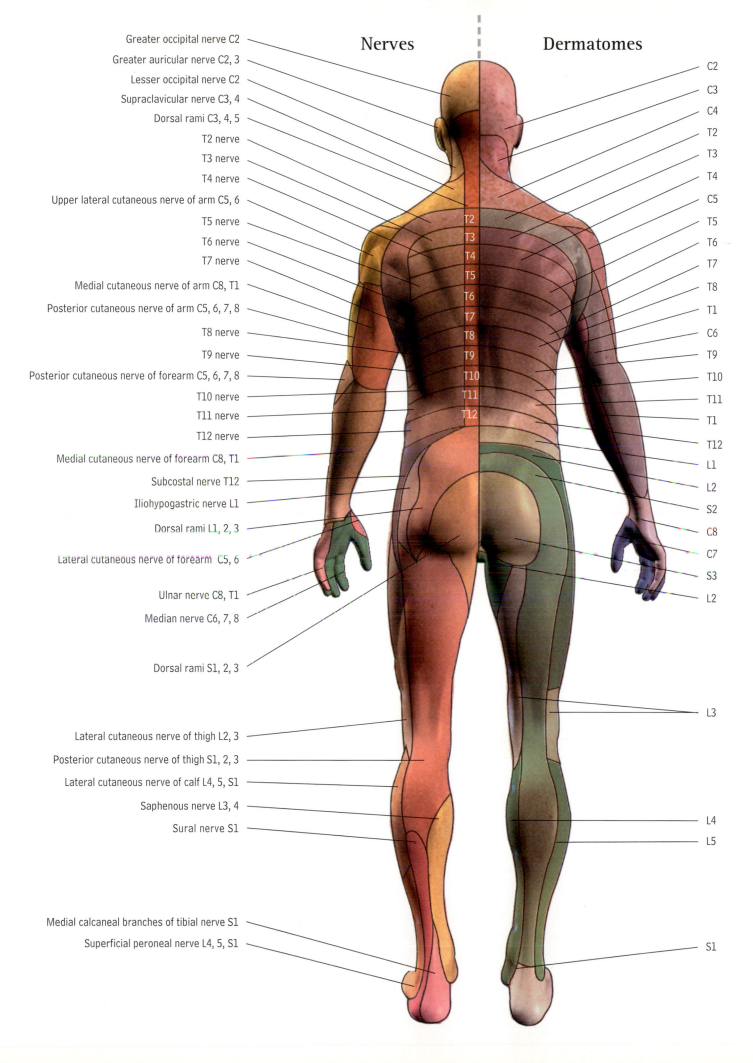

C2
C3
C4
T2
T3
T4
C5
T5
T6
T7
T8
T1
C6
T9
T10
T11
T1
T12
L1
L2
S2
C8
C7
S3
L2
L3
L4
L5
S1

Cranial Nerve I – Olfactory

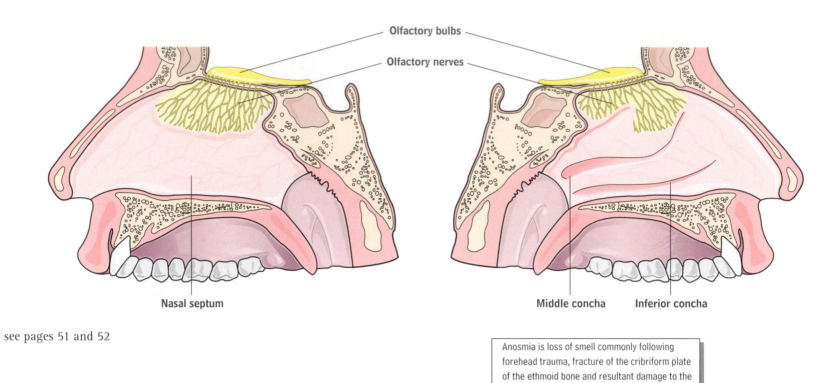

Olfactory bulbs

Olfactory nerves

Nasal septum

Middle concha Inferior concha

see pages 51 and 52

Anosmia is loss of smell commonly following forehead trauma, fracture of the cribriform plate of the ethmoid bone and resultant damage to the olfactory nerves.

Cranial Nerve II – Optic

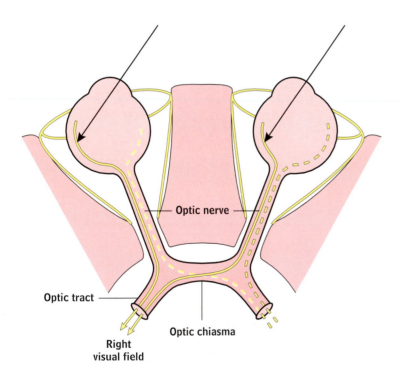

Optic nerve

Optic tract

Optic chiasma

Right visual field

see pages 51, 52 and 54

Cranial Nerves III – Oculomotor, IV – Trochlear, VI – Abducens

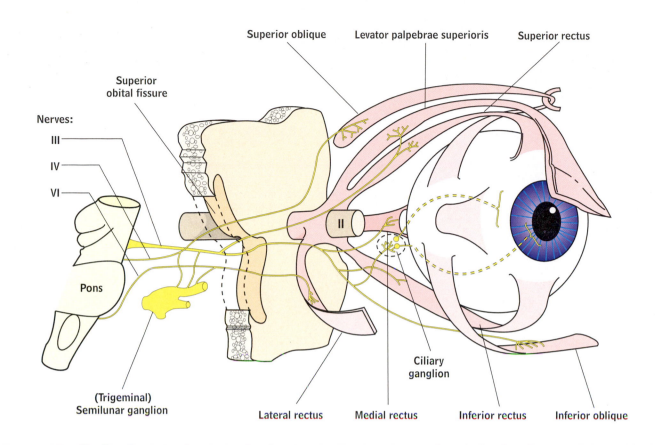

see pages 51–54 for III
see pages 51–54 for IV
see pages 50, 51 and 54 for VI

Abducent nerve paralysis. The abducent nerve innervates only the lateral rectus muscle of the eye, and damage results in an inability to move the eye laterally in the horizontal plane. This sign may be an indication of increased intracranial pressure, owing to the long intracranial course of the sixth nerve.

Trochlear nerve paralysis. This rare condition affects only the superior oblique muscle of the eye and commonly presents as a squint or with the patient complaining of diplopia on looking downwards or at the end of the nose.

Oculomotor nerve paralysis. If complete, this condition affects most eye muscles, and especially the levator palpebrae superioris and the sphincter pupillae. Consequently, the upper eyelid droops (ptosis), there is a fully dilated non-reactive pupil, and the eyeball tends to be looking downwards and outwards owing to the unopposed action of the lateral rectus and superior oblique muscles.

Accommodation reflex is contraction of the pupil when trying to focus on a near object and is controlled by the parasympathetic nerve fibres carried in the third cranial nerve from the Edinger–Westphal nucleus of the midbrain (synapse in ciliary ganglion) which act on the sphincter pupillae muscle to cause reduction in pupil diameter and on the ciliary muscle to cause relaxation of the suspensory ligament, allowing the lens to adopt a more spherical shape for near focusing.

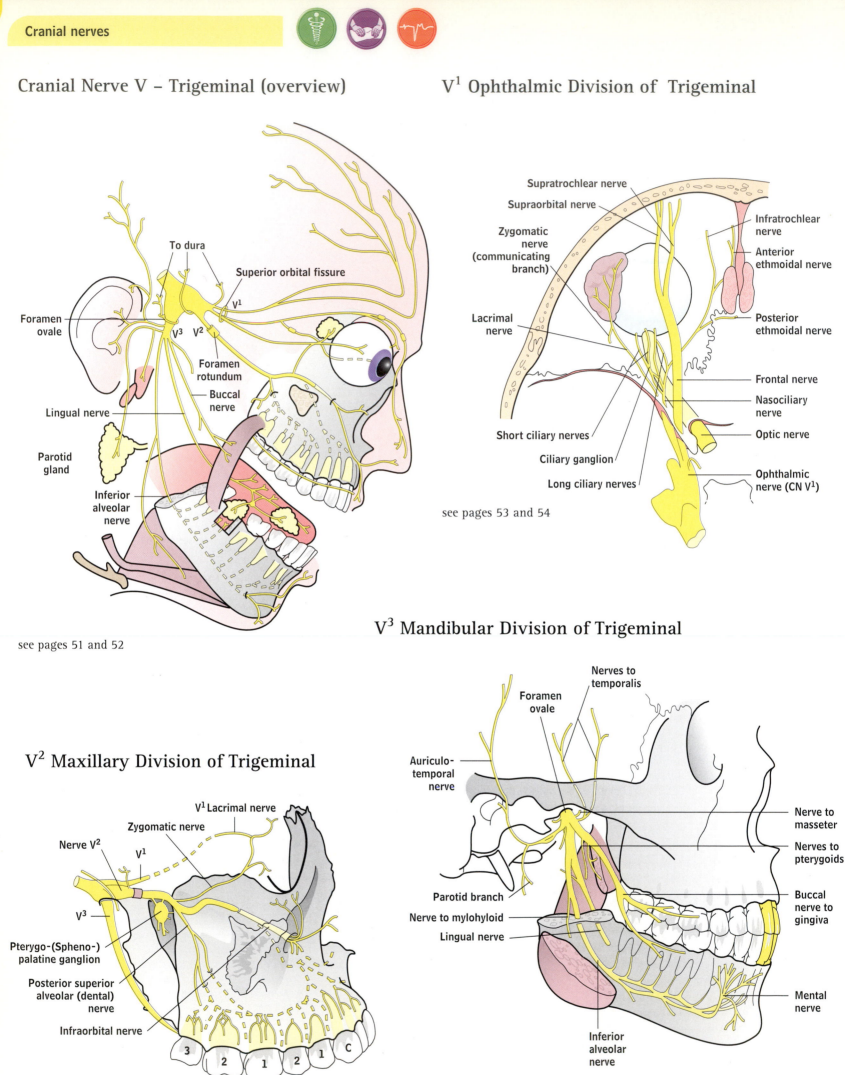

Cranial Nerve V – Trigeminal (overview)

To dura

Superior orbital fissure

V¹

Foramen ovale

V³ V²

Foramen rotundum

Buccal nerve

Lingual nerve

Parotid gland

Inferior alveolar nerve

see pages 51 and 52

V¹ Ophthalmic Division of Trigeminal

Supratrochlear nerve

Supraorbital nerve

Infratrochlear nerve

Zygomatic nerve (communicating branch)

Anterior ethmoidal nerve

Lacrimal nerve

Posterior ethmoidal nerve

Frontal nerve

Nasociliary nerve

Short ciliary nerves

Optic nerve

Ciliary ganglion

Long ciliary nerves

Ophthalmic nerve (CN V¹)

see pages 53 and 54

V³ Mandibular Division of Trigeminal

Nerves to temporalis

Foramen ovale

Auriculo-temporal nerve

Nerve to masseter

Nerves to pterygoids

Parotid branch

Buccal nerve to gingiva

Nerve to mylohyoid

Lingual nerve

Mental nerve

Inferior alveolar nerve

V² Maxillary Division of Trigeminal

V¹ Lacrimal nerve

Zygomatic nerve

Nerve V²

V¹

V³

Pterygo-(Spheno-) palatine ganglion

Posterior superior alveolar (dental) nerve

Infraorbital nerve

3 2 1 2 1 C

see pages 43 and 44

see pages 31, 37 and 40

Cranial Nerve VII – Facial

Bell's palsy is a facial nerve palsy of unknown aetiology first describedby Sir Charles Bell. The site of this lower motor neurone lesion may be diagnosed precisely by careful evaluation as to whether the stapedius, petrosal nerves and chorda tympani are involved.

Hyperacusis is increased sensitivity to sound (lowered threshold), most commonly caused by damage to the stapedius muscle (seventh cranial nerve) or the tensor tympani muscle (fifth cranial nerve). Occasionally this is a symptom associated with Bell's palsy.

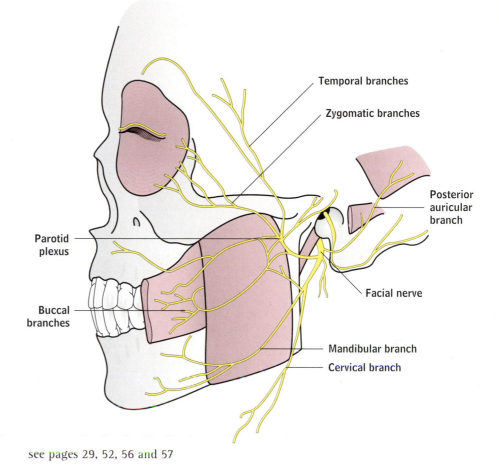

Temporal branches

Zygomatic branches

Posterior auricular branch

Parotid plexus

Facial nerve

Buccal branches

Mandibular branch

Cervical branch

see pages 29, 52, 56 and 57

Cranial Nerve VIII– Vestibulocochlear

Otalgia (pain in the ear). Pain from the ear itself is a common complaint but referred otalgia can be a diagnostic nightmare. Any structure that has the same nerve supply as the pinna or middle ear may have its pain referred to the ear. These include numerous branches of C2, C3 of the cervical plexus, and the fifth, seventh, ninth and tenth cranial nerves. Conditions that can cause referred otalgia include myocardial infarction, oesophagitis, tonsilitis, arthritis of the cervical spine, malocclusion, dental caries, sinusitis and carcinoma of the larynx or pharynx. A painful ear with a normal ear examination is therefore an anatomical challenge.

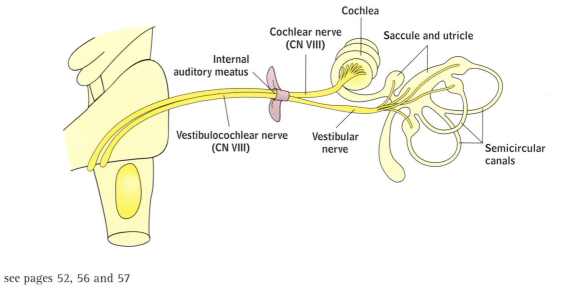

Cochlea

Cochlear nerve (CN VIII)

Saccule and utricle

Internal auditory meatus

Vestibulocochlear nerve (CN VIII)

Vestibular nerve

Semicircular canals

see pages 52, 56 and 57

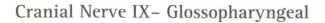

Cranial Nerve IX– Glossopharyngeal

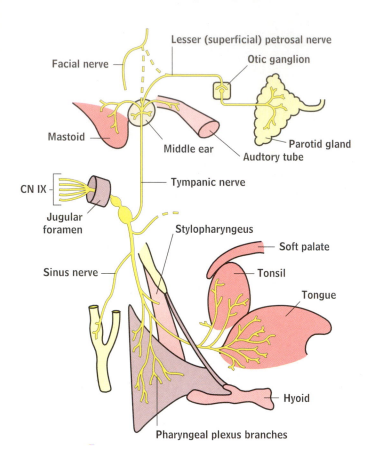

Facial nerve

Lesser (superficial) petrosal nerve

Otic ganglion

Mastoid

Middle ear

Parotid gland

Audtory tube

CN IX

Tympanic nerve

Jugular foramen

Stylopharyngeus

Soft palate

Sinus nerve

Tonsil

Tongue

Hyoid

Pharyngeal plexus branches

see pages 45, 46, 47 and 52

Recurrent laryngeal nerve damage. A complication of thyroid surgery, this condition causes paralysis of the vocal cords. When bilateral the voice is almost absent as the two vocal folds cannot be adducted. A unilateral recurrent laryngeal nerve injury may not be detected in normal speech.

Parotid tumours may involve the retromandibular vein or superficial temporal artery which lie within its substance, but the most common effect is involvement of the facial nerve as its numerous branches pass from deep to superficial through this gland. A facial paralysis associated with a swelling in the parotid gland is a condition to be taken seriously. The pain of a tumour here may often be referred to the temporomadibular joint via the auriculotemporal nerve which carries the parotid's secretomotor fibres.

Cranial Nerve X– Vagus

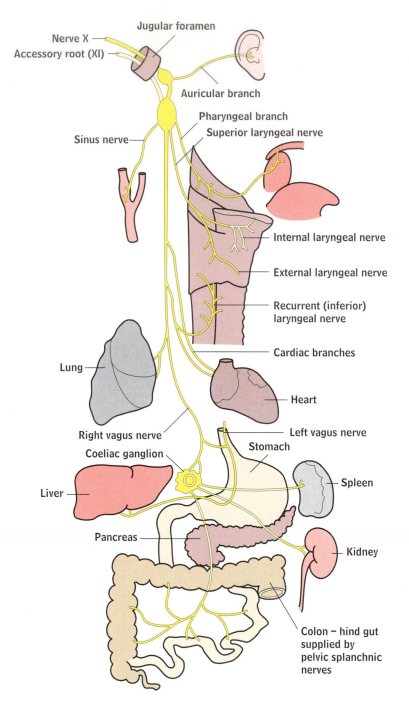

Nerve X

Jugular foramen

Accessory root (XI)

Auricular branch

Sinus nerve

Pharyngeal branch

Superior laryngeal nerve

Internal laryngeal nerve

External laryngeal nerve

Recurrent (inferior) laryngeal nerve

Cardiac branches

Lung

Heart

Right vagus nerve

Left vagus nerve

Stomach

Coeliac ganglion

Spleen

Liver

Pancreas

Kidney

Colon – hind gut supplied by pelvic splanchnic nerves

see pages 45, 51, 52, 165, 167, 179, 181 and 208

Vagus nerve injuries are most commonly seen associated with surgery in the region of the carotid sheath. The resultant injury may cause a speech problem (recurrent laryngeal branch) or gastrointestinal disturbances.

Cranial Nerve XI – Accessory

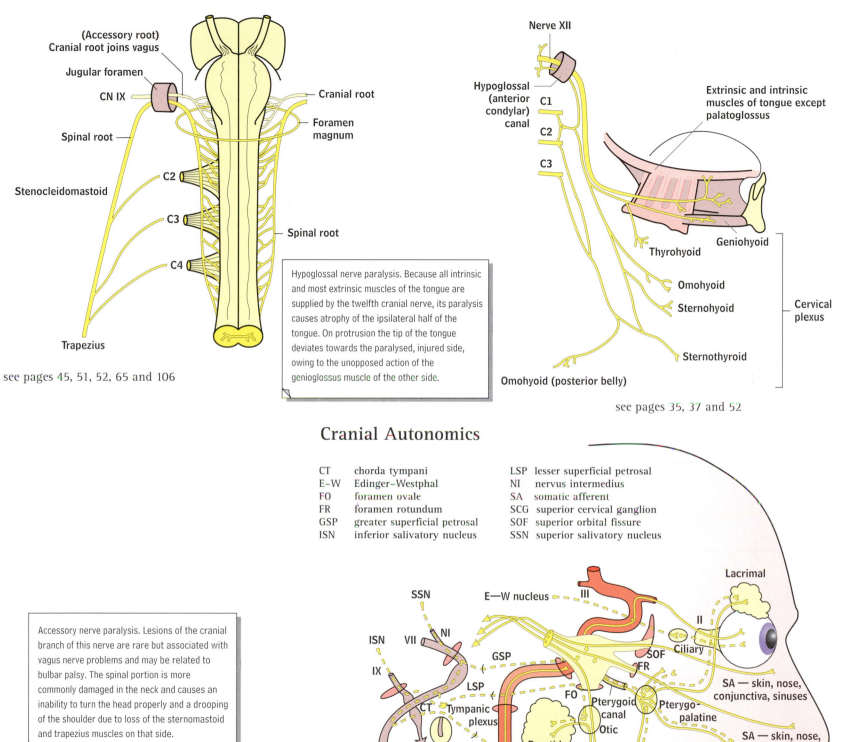

(Accessory root)
Cranial root joins vagus

Jugular foramen

CN IX

Spinal root

Stenocleidomastoid

C2

C3

C4

Trapezius

Cranial root

Foramen magnum

Spinal root

see pages 45, 51, 52, 65 and 106

Accessory nerve paralysis. Lesions of the cranial branch of this nerve are rare but associated with vagus nerve problems and may be related to bulbar palsy. The spinal portion is more commonly damaged in the neck and causes an inability to turn the head properly and a drooping of the shoulder due to loss of the sternomastoid and trapezius muscles on that side.

Gag reflex. Stimulation of the back of the tongue or posterior oropharynx sends afferent stimuli via the glossopharyngeal nerve. The motor pathway involves the vagus and accessory (tenth and eleventh) cranial nerves and elevates the soft palate. Putting a spatula against the back of the mouth will normally elicit this reflex and thus test three different cranial nerves.

Cranial Nerve XII – Hypoglossal

Nerve XII

Hypoglossal (anterior condylar) canal

C1

C2

C3

Extrinsic and intrinsic muscles of tongue except palatoglossus

Geniohyoid

Thyrohyoid

Omohyoid

Sternohyoid

Cervical plexus

Sternothyroid

Omohyoid (posterior belly)

Hypoglossal nerve paralysis. Because all intrinsic and most extrinsic muscles of the tongue are supplied by the twelfth cranial nerve, its paralysis causes atrophy of the ipsilateral half of the tongue. On protrusion the tip of the tongue deviates towards the paralysed, injured side, owing to the unopposed action of the genioglossus muscle of the other side.

see pages 35, 37 and 52

Cranial Autonomics

CT	chorda tympani	LSP	lesser superficial petrosal
E–W	Edinger–Westphal	NI	nervus intermedius
FO	foramen ovale	SA	somatic afferent
FR	foramen rotundum	SCG	superior cervical ganglion
GSP	greater superficial petrosal	SOF	superior orbital fissure
ISN	inferior salivatory nucleus	SSN	superior salivatory nucleus

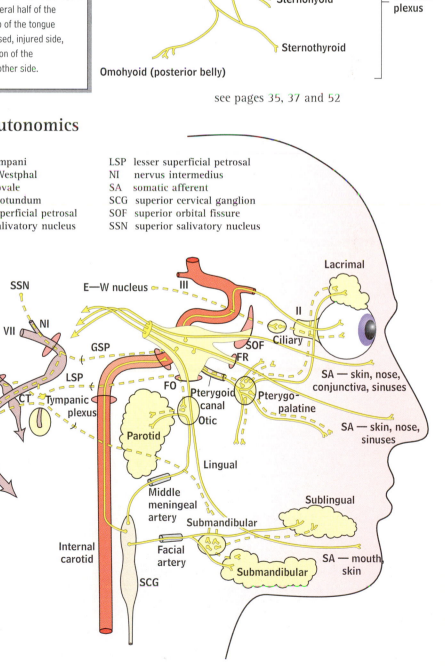

The reference lists of vessels and nerves have been arranged for quick identification of parent trunks and branches. Thus, the left common carotid artery is one of the three branches of the arch of the aorta, while the right common carotid is one of the branches of the brachiocephalic trunk.

The arrows indicate a continuity (instead of branching) with a change of name.

The generally accepted classical pattern has been given. For common variations, which are particularly frequent among veins, reference should be made to standard texts. (The articular and vascular branches of nerves have been omitted.)

The inclusion of items in these lists does not necessarily imply that they are illustrated in the atlas; many of the smaller and less important items are not shown but have been included in the lists for reference purposes.

For skull foramina, two lists are given: one of the principal foramina and their contents, which most students would be expected to know, and another giving more specialized details for those who require them.

Arteries

AORTA AND BRANCHES
Ascending aorta → arch of aorta → thoracic aorta
    → abdominal aorta

Ascending aorta
Right coronary
    Conus
    Sinuatrial nodal (55%)
    Atrial
    Right marginal
    Posterior interventricular
      Septal
    Atrioventricular nodal (90%)
Left coronary
    Anterior interventricular
      Conus
      Diagonal
      Septal
    Circumflex
      Sinuatrial nodal (45%)
      Atrioventricular nodal (10%)
      Atrial
      Left marginal

Arch of aorta
Brachiocephalic trunk
    Right common carotid
      Right internal carotid
      Right external carotid
    Right subclavian → axillary → brachial
    Thyroidea ima (occasional)
Left common carotid
    Left internal carotid
    Left external carotid
Left subclavian → axillary → brachial

Thoracic aorta
Pericardial
Left bronchial
Oesophageal
Mediastinal
Phrenic
Posterior intercostal (3 to 11)
    Right bronchial (from third)
Subcostal

Abdominal aorta
Coeliac trunk
Superior mesenteric
Inferior mesenteric
Inferior phrenic
    Superior suprarenal
Middle suprarenal
Renal
    Inferior suprarenal
Testicular (ovarian)
Lumbar
Median sacral
Common iliac
    Internal iliac
    External iliac → femoral

CAROTID ARTERIES AND BRANCHES

Internal carotid
Caroticotympanic
Pterygoid
Cavernous
Hypophysial
Meningeal
Ophthalmic
    Central of retina
    Lacrimal
      Lateral palpebral
      Zygomatic
      Recurrent meningeal
    Muscular
      Anterior ciliary
    Long posterior ciliary
    Short posterior ciliary
    Supra-orbital
    Posterior ethmoidal
    Anterior ethmoidal
    Anterior meningeal
    Medial palpebral
    Supratrochlear
    Dorsal nasal
Anterior cerebral
    Striate (and others)
Anterior communicating
Middle cerebral
    Striate (and others)
Posterior communicating
Anterior choroidal

External carotid
Ascending pharyngeal
Superior thyroid
    Infrahyoid
    Sternocleidomastoid
    Superior laryngeal
    Cricothyroid
Lingual
Facial
    Ascending palatine
    Tonsillar
    Glandular
    Submental
    Inferior labial
    Superior labial
      Septal
    Lateral nasal
Occipital
Posterior auricular
Superficial temporal
    Transverse facial
Maxillary → sphenopalatine
    Deep auricular
    Anterior tympanic
    Middle meningeal
    Accessory meningeal
    Inferior alveolar
      Dental
      Mylohyoid
      Mental
    Deep temporal
    Pterygoid
    Masseteric
    Buccal
    Posterior superior alveolar
      Dental
    Infra-orbital
      Anterior superior alveolar
        Dental
    Greater palatine
      Lesser palatine
    Pharyngeal
    Artery of pterygoid canal

SUBCLAVIAN ARTERY AND BRANCHES

Subclavian → axillary → brachial
Vertebral
    Spinal
    Meningeal
    Anterior spinal
    Posterior spinal
    Posterior inferior cerebellar
Internal thoracic
    Pericardiacophrenic
    Mediastinal
    Thymic
    Pericardial
    Sternal
    Perforating
        Mammary
    Anterior intercostal
    Musculophrenic
    Superior epigastric
Thyrocervical trunk
    Inferior thyroid
        Ascending cervical
        Inferior laryngeal
        Glandular
        Pharyngeal
        Oesophageal
        Tracheal
    Suprascapular
    Superficial cervical
Costocervical trunk
    Superior intercostal
    Deep cervical
Dorsal scapular (sometimes from superficial
    cervical)

Basilar (union of vertebrals)
Pontine
Labyrinthine
Anterior inferior cerebellar
Superior cerebellar
Posterior cerebral

AXILLARY ARTERY AND BRANCHES

Axillary → brachial
Superior thoracic
Thoraco-acromial
    Acromial
    Clavicular
    Deltoid
    Pectoral
Lateral thoracic
    Lateral mammary
Subscapular
    Circumflex scapular
    Thoracodorsal
Anterior circumflex humeral
Posterior circumflex humeral

Brachial
Profunda brachii
    Posterior descending
    Radial collateral
Nutrient
Superior ulnar collateral
Inferior ulnar collateral
Radial → deep palmar branch
    Radial recurrent
    Palmar carpal
    Superficial palmar
    Dorsal carpal
    Dorsal metacarpal
        Dorsal digital
    First dorsal metacarpal
    Princeps pollicis
    Radialis indicis
    Deep palmar arch
        Palmar metacarpal
        Perforating
Ulnar → superficial palmar branch
    Anterior ulnar recurrent
    Posterior ulnar recurrent
    Common interosseous
        Anterior interosseous
            Median
        Posterior interosseous
            Interosseous recurrent
    Palmar carpal
    Dorsal carpal
    Deep carpal
    Superficial palmar arch
        Common palmar arch
            Palmar digital

SOME BRANCHES OF THE ABDOMINAL AORTA

Coeliac trunk
Left gastric
    Oesophageal
Common hepatic
    Hepatic
        Right hepatic
            Cystic
        Left hepatic
    Gastroduodenal
        Right gastro-epiploic
        Superior pancreaticoduodenal
        Supraduodenal
    Right gastric
Splenic
    Pancreatic
    Short gastric
    Left gastro-epiploic

Superior mesenteric
Inferior pancreaticoduodenal
Jejunal and ileal
Ileocolic
    Ascending
    Anterior caecal
    Posterior caecal
        Appendicular
    Ileal
Right colic
    Ascending
    Descending
Middle colic
    Right
    Left

Inferior mesenteric → superior rectal
Left colic
    Ascending
    Descending
Sigmoid

Internal iliac
Anterior trunk
    Superior vesical → obliterated umbilical
    Artery to ductus deferens (sometimes from
    inferior vesical)
    Inferior vesical
    Middle rectal
    Uterine
    Vaginal
    Obturator
    Internal pudendal
        Inferior rectal
        Perineal
        Artery of the bulb
        Urethral
        Deep artery of the penis (clitoris)
        Dorsal artery of the penis (clitoris)
    Inferior gluteal
Posterior trunk
    Iliolumbar
    Lateral sacral
    Superior gluteal

External iliac → femoral
Inferior epigastric
    Cremasteric
    Pubic (accessory obturator)
Deep circumflex iliac

FEMORAL ARTERY AND BRANCHES

Femoral → popliteal
Superficial epigastric
Superficial circumflex iliac
Superficial external pudendal
Deep external pudendal
Profunda femoris
    Lateral circumflex femoral
    Medial circumflex femoral
    Perforating
Descending genicular

Popliteal
Sural
Superior genicular
Middle genicular
Inferior genicular
Anterior tibial → dorsalis pedis
    Posterior tibial recurrent
    Anterior tibial recurrent
    Anterior medial malleolar
    Anterior lateral malleolar
    Dorsalis pedis → plantar arch
        Tarsal
        First dorsal metatarsal
            Dorsal digital
        Arcuate
            Dorsal metatarsal (2–4)
                Dorsal digital
Posterior tibial
    Circumflex fibular
    Peroneal
        Nutrient
        Perforating
        Communicating
        Lateral malleolar
        Calcanean
    Nutrient
    Communicating
    Medial malleolar
    Calcanean
    Medial plantar
        Superficial digital
    Lateral plantar → plantar arch
        Superficial digital
        Plantar metatarsal
            Common plantar digital
                Plantar digital
        Perforating

Veins

TRIBUTARIES OF MAJOR VEINS

SUPERIOR VENA CAVA

Superior vena cava
Left brachiocephalic
    Left internal jugular
    Left subclavian
    Left vertebral
    Left supreme (first posterior) intercostal
    Left superior intercostal (2–4)
    Inferior thyroid
    Thymic
    Pericardial
Right brachiocephalic
    Right internal jugular
    Right subclavian
    Right vertebral
    Right supreme (first posterior) intercostal
Azygos
    Right superior intercostal (2–4)
    Right posterior intercostal (5–11)
    Right subcostal
    Right ascending lumbar and/or lumbar azygos
    Right bronchial
    Oesophageal
    Pericardial
    Mediastinal
    Vertebral venous plexuses
    Hemi-azygos
        Left ascending lumbar and/or lumbar azygos
        Left subcostal
        Left posterior intercostal (9–11)
        Oesophageal
        Pericardial
        Mediastinal
        Vertebral venous plexuses
    Accessory hemi-azygos
        Left posterior intercostal (5–8)
        Left bronchial
        Oesophageal
        Pericardial
        Mediastinal
        Vertebral venous plexuses

INFERIOR VENA CAVA

Inferior vena cava
Common iliac (right and left)
Fourth lumbar (right and left)
Third lumbar (right and left)
Testicular (ovarian) (right)
Renal (right and left)
Suprarenal (right)
Inferior phrenic (right and left)
Hepatic (right, middle and left)
(Upper lumbar veins join ascending lumbar. Left
    testicular or ovarian and suprarenal veins join
    left renal)

INTERNAL JUGULAR VEIN

Internal jugular
Inferior petrosal sinus
Pharyngeal
Lingual
Facial
Superior thyroid
Middle thyroid

EXTERNAL JUGULAR VEIN

External jugular
Posterior auricular
Posterior branch of retromandibular
Occipital
Posterior external jugular
Suprascapular
Superficial cervical
Anterior jugular

RETROMANDIBULAR VEIN

Retromandibular
Superficial temporal
Maxillary
Transverse facial
Pterygoid plexus
    Middle meningeal
    Greater palatine
    Sphenopalatine
    Buccal
    Dental
    Deep facial
    Inferior ophthalmic
Anterior branch to join facial
Posterior branch to external jugular

FACIAL VEIN

Facial
Supratrochlear
Supra-orbital
Superior ophthalmic
Palpebral
External nasal
Labial
Deep facial
Submental
Submandibular
Tonsillar
External palatine (paratonsillar)

GREAT SAPHENOUS VEIN

Great saphenous
Dorsal venous arch
Posterior arch
    Perforating
Perforating
Accessory saphenous
Anterior femoral cutaneous
Superficial epigastric
Superficial circumflex iliac
Superficial external pudendal
Deep external pudendal
(Small saphenous vein communicates with great
    saphenous but usually drains to popliteal vein)

CARDIAC VEINS

Coronary sinus
Great cardiac
Middle cardiac
Small cardiac
Posterior of left ventricle
Oblique of left atrium

Anterior cardiac

Vena cordis minimae

DURAL VENOUS SINUSES

Posterosuperior group
Superior sagittal
Inferior sagittal
Straight
Transverse
Sigmoid
Petrosquamous
Occipital

Antero-inferior group
Cavernous
Intercavernous
Inferior petrosal
Superior petrosal
Sphenoparietal
Basilar
Middle meningeal veins

HEPATIC PORTAL SYSTEM

Portal vein
Superior mesenteric
    Jejunal and ileal
    Right gastro-epiploic
    Pancreatic
    Pancreaticoduodenal
    Ileocolic
      Caecal
        Appendicular
    Right colic
    Middle colic
Splenic
    Pancreatic
    Short gastric
    Left gastro-epiploic
    Inferior mesenteric
      Left colic
      Sigmoid
      Superior rectal
Left gastric
Right gastric
    Prepyloric
Para-umbilical (to left branch)
Cystic (to right branch)

PORTAL-SYSTEMIC ANASTOMOSES

Oesophageal branches of the left gastric vein with
    the hemi-azygos vein
Superior rectal branch of the inferior mesenteric
    vein with the middle and inferior rectal veins
    (internal iliac)
Para-umbilical veins of the falciform ligament with
    anterior abdominal wall veins
Retroperitoneal colonic veins with posterior
    abdominal wall veins
Bare area of the liver with diaphragmatic veins

Lymphatic system

THORACIC DUCT AND CISTERNA CHYLI TRIBUTARIES

Thoracic duct
Left jugular trunk
Left subclavian trunk
Left bronchomediastinal trunk

Right lymphatic duct
Right jugular trunk
Right subclavian trunk
Right bronchomediastinal trunk

Cisterna chyli
Left lumbar trunk
Right lumbar trunk
Intestinal trunk

LYMPH NODES OF THE HEAD AND NECK

Deep cervical
Superior (including jugulodigastric)
Inferior (including jugulo-omohyoid)

Draining superficial tissues in the head
Occipital
Retro-auricular (mastoid)
Parotid
Buccal (facial)

Draining superficial tissues in the neck
Submandibular
Submental
Anterior cervical
Superficial cervical

Draining deep tissues in the neck
Retropharyngeal
Paratracheal
Lingual
Infrahyoid
Prelaryngeal
Pretracheal

LYMPH NODES OF THE UPPER LIMB AND MAMMARY GLAND

Draining the upper limb
Axillary
    Apical
    Central
    Lateral
    Pectoral (anterior)
    Subscapular (posterior)
Infraclavicular
Supratrochlear
Cubital

Draining the mammary gland
Pectoral
Subscapular
Apical
Parasternal
Intercostal

LYMPH NODES OF THE THORAX

Draining thoracic walls
Superficial
    Pectoral
    Subscapular
    Parasternal
    Inferior deep cervical
Deep
    Parasternal
    Intercostal
    Phrenic
    Diaphragmatic

Draining thoracic contents
Brachiocephalic
Posterior mediastinal
Tracheobronchial
    Paratracheal
    Superior tracheobronchial
    Inferior tracheobronchial
    Bronchopulmonary
    Pulmonary

LYMPH NODES OF THE ABDOMEN AND PELVIS

Lumbar
    Pre-aortic
      Coeliac
        Gastric
          Left gastric
          Right gastro-epiploic
          Pyloric
        Hepatic
        Pancreaticosplenic
      Superior mesenteric
      Inferior mesenteric
    Lateral aortic
      Common iliac
      External iliac
      Internal iliac
      Inferior epigastric
      Circumflex iliac
      Sacral
    Retro-aortic

LYMPH NODES OF THE LOWER LIMB

Superficial inguinal
    Upper
    Lower
Deep inguinal
Popliteal

Nerves

CRANIAL NERVES AND BRANCHES

I Olfactory (from olfactory mucous membrane)

II Optic (from retina)

III Oculomotor
Superior ramus (to superior rectus and levator palpebrae superioris)
Inferior ramus (to medial rectus, inferior rectus, inferior oblique and ciliary ganglion)

IV Trochlear (to superior oblique)

V Trigeminal
Ophthalmic
    Lacrimal
    Frontal
        Supratrochlear
        Supra-orbital
    Nasociliary → anterior ethmoidal → external nasal
        Internal nasal (from anterior ethmoidal)
        Ciliary ganglion
        Long ciliary
        Infratrochlear
        Posterior ethmoidal
Maxillary → infra-orbital
    Meningeal
    Pterygopalatine
        Orbital
        Palatine
        Nasal
        Pharyngeal
    Zygomatic
        Zygomaticotemporal
        Zygomaticofacial
    Posterior superior alveolar
    Middle superior alveolar
    Anterior superior alveolar
    Palpebral ⎫
    Nasal ⎬ (from infra-orbital)
    Superior labial ⎭
Mandibular
    Meningeal
    Nerve to medial pterygoid (and tensor veli palatini and tensor tympani)
    Anterior trunk
        Buccal
        Masseteric
        Deep temporal
        Nerve to lateral pterygoid
    Posterior trunk
        Auriculotemporal
        Lingual
        Inferior alveolar
            Nerve to mylohyoid
            Mental

VI Abducent (to lateral rectus)

VII Facial
Greater petrosal
Nerve to stapedius
Chorda tympani
Posterior auricular (to occipitalis and auricular muscles)
Nerve to posterior belly of digastric
Nerve to stylohyoid
Temporal ⎫
Zygomatic ⎬ to frontalis and
Buccal ⎬ muscles of facial
Mandibular ⎬ expression
Cervical ⎭

VIII Vestibulocochlear
Cochlear (from coils of cochlea)
Vestibular (from utricle, saccule and ampullae of semicircular ducts)

IX Glossopharyngeal
Tympanic
    Lesser petrosal
Carotid sinus
Pharyngeal
Muscular (to stylopharyngeus)
Tonsillar
Lingual

X Vagus
Meningeal
Auricular
Pharyngeal (to muscles of pharynx and soft palate except stylopharyngeus and tensor veli palatini)
Carotid body
Superior laryngeal
    Internal laryngeal
    External laryngeal (to cricothyroid)
Right recurrent laryngeal (to muscles of larynx except cricothyroid)
Cardiac (cervical)
Cardiac (thoracic)
Left recurrent laryngeal (to muscles of larynx except cricothyroid)
Pulmonary
Oesophageal
Anterior trunk
    Gastric
    Hepatic
Posterior trunk
    Coeliac
    Gastric

XI Accessory
Cranial root (to muscles of palate and possibly larynx via vagus)
Spinal root (to sternocleidomastoid and trapezius)

XII Hypoglossal
Meningeal
Descending (upper root of ansa cervicalis, from C1 to superior belly of omohyoid, then forming ansa cervicalis – see cervical plexus)
Nerve to thyrohyoid (from C1)
Muscular (to geniohyoid from C1) and to muscles of tongue except palatoglossus

SOME HEAD AND NECK NERVE SUPPLIES

All the muscles of	Supplied by	Except	Supplied by
Pharynx	Pharyngeal plexus	Stylo-pharyngeus	Glosso-pharyngeal nerve
Palate	Pharyngeal plexus	Tensor veli palatini	Nerve to medial pterygoid
Larynx	Recurrent laryngeal nerve	Cricothyroid	External laryngeal nerve
Tongue	Hypoglossal nerve	Palatoglossus	Pharyngeal plexus
Facial expression (including buccinator)	Facial nerve		
Mastication	Mandibular branch of trigeminal nerve		

Skull foramina

PRINCIPAL FORAMINA AND CONTENTS

Supra-orbital foramen
Supra-orbital nerve and vessels

Infra-orbital foramen
Infra-orbital nerve and vessels

Mental foramen
Mental nerve and vessels

Mandibular foramen
Inferior alveolar nerve and vessels

Optic canal
Optic nerve
Ophthalmic artery

Superior orbital fissure
Ophthalmic nerve and veins
Oculomotor, trochlear and abducent nerves

Inferior orbital fissure
Maxillary nerve

Sphenopalatine foramen
Sphenopalatine artery
Nasal branches of pterygopalatine ganglion and
   maxillary nerve

Foramen rotundum
Maxillary nerve

Foramen ovale
Mandibular and lesser petrosal nerve

Foramen spinosum
Middle meningeal vessels

Carotid canal
Internal carotid artery and sympathetic plexus

Jugular foramen
Inferior petrosal sinus
Glossopharyngeal, vagus and accessory nerves
Internal jugular vein (emerging below) as
   continuation of sigmoid sinus

Internal acoustic meatus
Facial and vestibulocochlear nerves
Labyrinthine artery

Hypoglossal canal
Hypoglossal nerve

Stylomastoid foramen
Facial nerve

Foramen magnum
Medulla oblongata and meninges
Vertebral and anterior and posterior spinal arteries
Accessory nerves (spinal parts)

INSIDE THE SKULL

MIDDLE CRANIAL FOSSA

Optic canal: in the sphenoid between the body and
   the two roots of the lesser wing
Optic nerve
Ophthalmic artery

Superior orbital fissure: in the sphenoid between
   the body and greater and lesser wings, with a
   fragment of the frontal bone at the lateral
   extremity
Oculomotor, trochlear and abducent nerves
Lacrimal, frontal and nasociliary nerves
Filaments from the internal carotid (sympathetic)
   plexus
Orbital branch of the middle meningeal artery
Recurrent branch of the lacrimal artery
Superior ophthalmic vein

Foramen rotundum: in the greater wing of the
   sphenoid
Maxillary nerve

Foramen ovale: in the greater wing of the
   sphenoid
Mandibular nerve
Lesser petrosal nerve (usually)
Accessory meningeal artery
Emissary veins (from cavernous sinus to pterygoid
   plexus)

Foramen spinosum: in the greater wing of the
   sphenoid
Middle meningeal vessels
Meningeal branch of the mandibular nerve

Venous (emissary sphenoidal) foramen: in 40% of
   skulls, in the greater wing of the sphenoid
   medial to the foramen ovale
Emissary vein (from the cavernous sinus to the
   pterygoid plexus)

Petrosal (innominate) foramen: occasional, in the
   greater wing of the sphenoid, medial to the
   foramen spinosum
Lesser petrosal nerve (if not through foramen ovale)

Foramen lacerum: between the sphenoid, apex of
   the petrous temporal and the basilar part of the
   occipital
A meningeal branch of the ascending pharyngeal
   artery
Emissary veins (from the cavernous sinus to the
   pterygoid plexus)

Hiatus for the greater petrosal nerve: in the
   tegmen tympani of the petrous temporal, in
   front of the arcuate eminence
Greater petrosal nerve
Petrosal branch of the middle meningeal artery

Hiatus for the lesser petrosal nerve: in the tegmen
   tympani of the petrous temporal, about 3 mm in
   front of the hiatus for the greater petrosal nerve
Lesser petrosal nerve

ANTERIOR CRANIAL FOSSA

Foramina in the cribriform plate of the ethmoid
Olfactory nerve filaments
Anterior ethmoidal nerve and vessels

Foramen caecum: between the frontal crest of the
   frontal bone and the ethmoid in front of the
   crista galli
Emissary vein (between nose and superior sagittal
   sinus)

POSTERIOR CRANIAL FOSSA

Internal acoustic meatus: in the posterior surface
   of the petrous temporal
Facial nerve
Vestibulocochlear nerve
Labyrinthine artery

Aqueduct of the vestibule: in the petrous temporal
   about 1 cm behind the internal acoustic meatus
Endolymphatic duct and sac
A branch from the meningeal branch of the
   occipital artery
A vein (from the labyrinth and vestibule to the
   sigmoid sinus)

Jugular foramen: between the jugular fossa of the
   petrous temporal and the occipital bone
Glossopharyngeal, vagus and accessory nerves
Meningeal branches of the vagus nerve
Inferior petrosal sinus
Internal jugular vein
A meningeal branch of the occipital artery

Hypoglossal canal: in the occipital bone above the
   anterior part of the condyle
Hypoglossal nerve and its (recurrent) meningeal
   branch
A meningeal branch of the ascending pharyngeal
   artery
Emissary vein (from the basilar plexus to the
   internal jugular vein)

Condylar canal: occasional, from the lower part of
   the sigmoid groove in the lateral part of the
   occipital bone to the condylar fossa on the
   external surface of the occipital bone behind the
   condyle
Emissary vein (from the sigmoid sinus to occipital
   veins)
A meningeal branch of the occipital artery

Mastoid foramen: in the petrous temporal near the
   posterior margin of the lower part of the
   sigmoid groove, passing backwards to open
   behind the mastoid process
Emissary vein (from the sigmoid sinus to occipital
   veins)
A meningeal branch of the occipital artery

Foramen magnum: in the occipital bone
Apical ligament of the dens of the axis
Tectorial membrane
Medulla oblongata and meninges (including first
   digitations of denticulate ligaments)
Spinal parts of the accessory nerves
Meningeal branches of the upper cervical nerves
Vertebral arteries
Anterior spinal artery
Posterior spinal arteries

IN THE BASE OF THE SKULL EXTERNALLY

Foramen lacerum
Foramen ovale
Foramen spinosum
Jugular foramen — see INSIDE THE SKULL
Hypoglossal canal
Condylar canal
Mastoid foramen
Foramen magnum

Inferior orbital fissure—see IN THE ORBIT

Lateral incisive foramen: opens into the incisive fossa, in the midline at the front of the hard palate
Nasopalatine nerve
Greater palatine vessels

Greater palatine foramen: between the maxilla and the palatine bone at the lateral border of the hard palate behind the palatomaxillary fissure
Greater palatine nerve and vessels

Lesser palatine foramina: two or three, in the inferior and medial aspects of the pyramidal process of the palatine bone
Lesser palatine nerves and vessels

Palatovaginal canal: between lower surface of the vaginal process of the root of the medial pterygoid plate and the upper surface of the sphenoidal process of the palatine bone
Pharyngeal branch of the pterygopalatine ganglion
Pharyngeal branch of the maxillary artery

Vomerovaginal canal: occasional, medial to the palatovaginal canal, between the upper surface of the vaginal process of the root of the medial pterygoid plate and the lower surface of the ala of the vomer
Pharyngeal branch of the sphenopalatine artery

Petrosquamous fissure: between the squamous temporal and the tegmen tympani
Petrosquamous vein

Petrotympanic fissure: between the tympanic part of the temporal bone and the tegmen tympani
Chorda tympani
Anterior ligament of the malleus
Anterior tympanic branch of the maxillary artery

Cochlear canaliculus: in the petrous temporal, at the apex of a notch in front of the medial part of the jugular fossa
Perilymphatic duct
Emissary vein (from the cochlea to the internal jugular vein or inferior petrosal sinus)

Carotid canal: in the inferior surface of the petrous temporal
Internal carotid artery
Internal carotid (sympathetic) plexus
Internal carotid venous plexus (from the cavernous sinus to the internal jugular vein)

Tympanic canaliculus: in the inferior surface of the petrous temporal, on the ridge of bone between the carotid canal and the jugular fossa
Tympanic branch of the glossopharyngeal nerve
Inferior tympanic branch of the ascending pharyngeal artery

Mastoid canaliculus: in the inferior surface of the petrous temporal, on the lateral wall of the jugular fossa
Auricular branch of the vagus nerve

Stylomastoid foramen: between the styloid and mastoid processes of the temporal bone
Facial nerve
Stylomastoid branch of the posterior auricular artery

IN THE ORBIT

Superior orbital fissure
Optic canal } see INSIDE THE SKULL

Frontal notch or foramen: in the supra-orbital margin of the frontal bone one finger's breadth from the midline
Supratrochlear nerve and vessels

Supra-orbital notch or foramen: in the supra-orbital margin of the frontal bone two fingers' breadths from the midline
Supra-orbital nerve and vessels

Anterior ethmoidal foramen: in the medial wall of the orbit between the orbital part of the frontal bone and the ethmoid labyrinth
Anterior ethmoidal nerve and vessels

Posterior ethmoidal foramen: occasional, 1 to 2 cm behind the anterior ethmoidal foramen
Posterior ethmoidal nerve and vessels

Zygomatico-orbital foramen: in the orbital surface of the zygomatic bone
Zygomatic branch of the maxillary nerve

Nasolacrimal canal: at the front, lower, medial corner of the orbit formed by the lacrimal bone and maxilla
Nasolacrimal duct

Inferior orbital fissure: towards the back of the orbit, between the maxilla and the greater wing of the sphenoid
Maxillary nerve
Zygomatic nerve
Orbital branches of the pterygopalatine ganglion
Infra-orbital vessels
Inferior ophthalmic veins

Infra-orbital canal: in the orbital surface of the maxilla
Infra-orbital nerve and vessels

MISCELLANEOUS

Infra-orbital foramen: the anterior opening of the infra-orbital canal, in the maxilla below the infra-orbital margin
Infra-orbital nerve and vessels

Mental foramen: on the outer surface of the body of the mandible below the second premolar tooth or slightly more anteriorly
Mental nerve and vessels

Mandibular foramen: on the inner surface of the ramus of the mandible, overlapped anteriorly and medially by the lingula
Inferior alveolar nerve and vessels

Foramina in the infratemporal (posterior) surface of the maxilla
Posterior superior alveolar nerves and vessels

Pterygomaxillary fissure: between the lateral pterygoid plate and the infratemporal (posterior) surface of the maxilla, and continuous above with the posterior end of the inferior orbital fissure
Maxillary artery (entering pterygopalatine fossa)
Maxillary nerve (entering inferior orbital fissure)
Sphenopalatine veins

Sphenopalatine foramen: at the upper end of the perpendicular plate of the palatine between its orbital and sphenoidal processes and (above) the body of the sphenoid; in the medial wall of the pterygopalatine fossa (viewed laterally through the pterygomaxillary fissure) and lateral wall of the nasal cavity (viewed medially)
Nasopalatine and posterior superior nasal nerves
Sphenopalatine vessels

Foramina in the perpendicular plate of the palatine
Posterior inferior nasal nerves

Pterygoid canal: at the root of the pterygoid process of the sphenoid in line with the medial pterygoid plate, leading from the anterior wall of the foramen lacerum to the posterior wall of the pterygopalatine fossa (and only clearly seen in a disarticulated sphenoid)
Nerve of the pterygoid canal
Artery of the pterygoid canal

Musculotubular canal: at the lateral side of the apex of the petrous temporal, at the junction of the petrous and squamous parts, and divided by a bony septum into upper and lower semicanals
Tensor tympani (upper semicanal)
Auditory tube (lower semicanal)

Parietal foramen: in the parietal bone near the posterosuperior (occipital) angle
Emissary vein (from the superior sagittal sinus to the scalp)

Index

A page number in parentheses indicates that the information is repeated because it is relevant in a different context. The clinical note for anosmia, for example, appears on page 11, where the the skull is featured, and on page 322, in connection with the olfactory cranial nerve.